KT PROJECT
SUCCESS ACHIEVED + DELIVERED

Glossary of Common Industry and Project Terminology

Guidelines for Successful Projects

Mr. Roy O. Christensen, RET

The first quality terminology resource written specifically for capital projects in the energy, industrial, mining, petrochemical, pipeline, power, and related sectors

ISBN: 978-1-0689766-1-2 (Hardcover)

978-1-0689766-0-5 (eBook)

Preface

Scope of Services

This Glossary of Common Industry and Project Terminology (this document) was prepared by the Knowledge Transfer (KT) Project (hereinafter referred to as the KT Project), which is owned and operated by Christensen Qualityworks Inc. (CQI). The information, ideas, concepts, and opinions contained in this document are presented in light of the scope, schedule, and other limitations stated in this document. This document is designed to provide accurate and reliable resources but is not guaranteed in that regard for the subject matter herein. The information, ideas, concepts, and opinions expressed in this document are based on conditions and information existing at the time this document was published and does not take into account any subsequent changes. In preparing this document, the KT Project did not verify information supplied to it by others and has assumed the accuracy, genuineness and the authenticity of all documents submitted to it as originals and the conformity with the originals of all documents submitted to it as copies thereof. As to any questions of fact which are or may be material to KT Project's opinion as hereinafter set forth, the KT Project has not verified independently any such information unless expressly stated herein.

Standard of Care

The work addressed herein has been performed according to the contributors and authors' knowledge and experience in accordance with commonly accepted standards of practice. No other warranty, expressed or implied, is made or intended by the KT Project. No opinion is provided relating to legal or other regulatory advice and, if such assistance is required, the reader is hereby advised to seek the services of a lawyer or other licensed professional. All sample forms, examples, or illustrations in this document are intended only as a guide for informational and educational purposes. A relevant third party, including government, regulatory or monitoring authorities, may set, change, amend, or remove any requirements or procedures from time to time.

Limitation and Disclaimer of Liability

It is expressly understood that all information, ideas, concepts, and expressions of opinion in this document are based on information and assumptions supplied to the KT Project by others or are derived by the KT Project and are not intended, nor can they be construed as, representations or warranties. The KT Project specifically disclaims any and all liability or risk whatsoever in the event any loss, injury or damage is incurred as a result of the application of any information or content offered herein, whether directly or indirectly used and howsoever arising. The KT Project, its directors, officers, shareholders, contractors, and employees or its subsidiaries' directors, officers, shareholders, contractors or employees (all of whom are KT Project for the purpose of this Preface Statement) does not assume responsibility for the engineering, fabrication, manufacture, testing, inspection, repair, or use of materials, equipment, supplies, processes, methods or decisions as a result of the information, ideas, concepts, and expressions of opinion provided in this document.

The analysis and conclusions provided in this document are for the sole use and benefit of KT Project clients and may not be transferred, sold, or assigned without the prior consent of the KT Project, in its sole discretion. Any use of, or reliance on this document, by any party other than KT Project clients is strictly prohibited. Any use that any third party makes of this document shall be at the sole risk of that third party and in no event will the KT Project be liable to any third party regarding any of the information, ideas, concepts, and expressions of opinion in this document, or for any use of, reliance on, accuracy, or adequacy of this document.

Executive Summary

This Glossary of *Common Industry and Project Terminology* document (this document) was prepared by the Knowledge Transfer (KT) Project (the KT Project), which is owned and operated by Christensen Qualityworks Inc. (CQI). This is the first comprehensive terminology resource written specifically for capital projects in energy, industrial, mining, petrochemical, pipeline, power, and other sectors.

This document describes and defines common project terminology (thousands of phrases and terms), and provides related resources. Use this document as a complete go-to resource for project terminology.

See KTP-GEN-001, *Introduction to the Knowledge Transfer Project* for a comprehensive list of guidelines that support the supply and delivery of successful projects.

The KT Project provides key documents (guidelines and best practices), like this one, to support the successful supply and service of energy projects and projects in other sectors (e.g., industrial, infrastructure, mining, and power). Knowledge transfer is the effective and efficient transfer of knowledge from subject matter experts to other professionals. KT Project documents capture decades of valuable, relevant subject matter expert knowledge and experience, and help organizations mitigate, minimize, and eliminate project challenges caused by knowledge gaps. KT Project documents save organizations significant time and money.

Table of Contents

Tables

Key Knowledge

Appendices

1 Introduction

This *Glossary of Common Industry and Project Terminology* document (this document) was prepared by the Knowledge Transfer (KT) Project (the KT Project), which is owned and operated by Knowledge Transfer (KT) Project (KT Project).

This document describes and defines common capital project terminology (thousands of phrases and terms), and provides related resources, including technical terms related to:

- Construction;
- Engineering;
- Fabrication;
- Manufacturing;
- Non-destructive examination;
- Other industries;
- Procurement;
- Supplier quality;
- Testing; and,
- Welding.

While anyone can search the Internet for a definition, results are often conflicting, confusing, or not related to projects; thus, much of the information gleaned from the Internet is not useful in the context of projects. Use this document as a complete go-to resource for project terminology. See:

- Section 3 for a list of common project abbreviations and their meanings;
- Section 4 for a list of common project terms with definitions; and,
- KTP-GEN-001, *Introduction to the Knowledge Transfer Project* for a comprehensive list of guidelines that support the supply and delivery of successful projects.

Some engineering and procurement (EP) companies, suppliers and clients use the same terms to describe different activities and functions. For clarity, this document describes some terms as *not preferred* (i.e., less descriptive or slang). The use of a *preferred* term may be more descriptive or technically correct than another; however, this does not preclude an organization from using any terminology that suits their needs.

This document and related KT Project documents may be used as-is and are also customizable. These documents can be rebranded with your organization's logo and formatting. In addition, document contents may be customized to suit the specialized needs of your project or organization to include project-specific best practices, work processes, and appropriate terminology.

This glossary is the first comprehensive terminology resource written specifically for capital projects in energy, industrial, mining, petrochemical, pipeline, power, and other sectors. It is a compendium of selected business, career, engineering, industry, procurement, and scientific terminology relevant to projects. It is a brilliant and industry-leading document developed over years of research, writing, and editing.

1.1 Summary of Document Contents

See:

- Section 2 to read about the importance of clear communication;
- Section 3 for a comprehensive list of common industry and project abbreviations;
- Section 4 for the KT Project glossary (this section comprises the bulk of this document);
- Appendix A for meanings of abbreviations and definitions of glossary terms used in this document;
- Appendix B for a list of references used in this document;
- Appendix C for a list of books referenced in this document;
- Appendix D for a list of resources that provide meanings of abbreviations and definitions of glossary terms for related industries;
- Appendix E for a description of the sequence of work and definitions of typical welding crews that construct a pipeline;
- Appendix F for a description of document review codes used to summarize the results of a supplier document review process;
- Appendix G for a general procedure to prepare, implement, and maintain an effective glossary for your organization; and,
- Appendix H for contact information to obtain assistance with this document, or for implementing, editing, and preparing an effective glossary.

1.2 Purpose

The purpose of this document is to define common language used in industry and on projects, and to provide a measure of assurance that project requirements are effectively communicated and clearly understood. Defining and using precise terminology is a key component of effective communication that supports the successful delivery of supplier services and goods, regardless of where the requirements originate or how they are fulfilled.

1.3 Scope

The scope of this document includes energy projects, and projects in other sectors (e.g., industrial, infrastructure, mining, and power). This document is for use by an EP company acting as an agent, or a consultant, for an owner-user or resource company. The guidelines contained herein may also be applied by other organizations (e.g., consultants, constructors, distributors, fabricators, and manufacturers), with modifications where required to suit specific needs and work processes.

1.4 About the Knowledge Transfer Project

The *KT Project* provides key documents (guidelines and best practices), like this one, to support the successful supply and service of energy projects, and projects in other sectors (e.g., industrial, infrastructure, mining and power). Knowledge transfer is the effective and efficient transfer of knowledge from subject matter experts to other professionals. KT Project documents capture decades of valuable, relevant subject matter expert knowledge and experience, and help organizations mitigate, minimize, and eliminate project challenges caused by knowledge gaps. KT Project documents save organizations significant time and money.

1.5 Terminology Related to Quality

Some companies, projects, suppliers, and clients use the same terms interchangeably to describe various activities and functions. This is not recommended; different terms have different meanings. Lack of precise communication can be costly and even mission critical. For example, in 1998, the NASA Mars Climate Orbiter disintegrated during orbit entry due to a communication error. A simple unit conversion mistake caused catastrophic mission failure and the loss of USD$ 327 million[1]. Do not let your project success fall victim to inferior communications.

Risk due to miscommunication is particularly impactful within the Quality discipline, including:

- Quality assurance (QA);
- Quality control (QC); and,
- Quality surveillance (QS).

Sections 1.5.1 to 1.5.2 discuss Quality and industry-specific terms that are commonly misunderstood and prone to ambiguity.

1.5.1 Quality Assurance, Quality Control and Quality Surveillance

QA, QC, and QS are methods used to facilitate successful project execution:

- QA is the planned and systematic activities used to provide confidence that a product or service will fulfill quality requirements.

[1] In 1998, NASA launched the Mars Climate Orbiter (the Orbiter) to study the Martian climate. After travelling nearly 300 days, the spacecraft was supposed to enter Mars' orbit at an altitude 226,000 m (m or metres being the SI unit for distance). The minimum altitude the spacecraft needed was just 80,000 m, so this gave the Orbiter a large margin of error. From what NASA could piece together after losing contact with the Orbiter, it entered orbit at 57,000 m and instantly disintegrated. The problem was that the output of a Lockheed Martin space software program was in pounds-seconds, but the SI unit needed was Newton-seconds. Had there been clearer communication between the software provider (supplier) and NASA (owner), a simple unit conversion mistake could have been avoided, and the USD$ 327 million mission might have succeeded.
Source: https://blog.beekeeper.io/4-historical-communication-failures-prevented-beekeeper/

For suppliers, QA refers to the fulfillment of a supplier's quality management system (QMS) requirements (e.g., ISO 9001, Quality Management System).

- QC is the operational techniques and activities used to fulfill requirements for quality.

  For suppliers, QC refers to the fulfillment of a supplier's:

  - Integrity Management System (IMS). For example, AB-512, *Owner-User Pressure Equipment Integrity Management Requirements*; or,
  - Code requirements. For example, electrical and pressure equipment must be certified to the code of construction.

- QS is the selective audit, evaluation, inspection, observation, and review, of engineering, procurement, fabrication, and manufacturing operations, and quality management systems, to determine a supplier's compliance with purchase order (PO) requirements.

  See Section 1.5.1.1 for more about QS.

1.5.1.1 Quality Surveillance Levels

An EP company performs QS for all procured equipment and materials, which may include inspection at the supplier's facility, depending on the QS level specified in the PO (i.e., QS levels 0 to 4). Table 1 defines QS levels.

Table 1: Quality Surveillance Levels

Level	Scope	Frequency	Description
0	None	N/A	Receiving inspection only at the final destination or site.
1	Final	Final inspection(s) only	Low dollar value. Multiple shipments may require multiple visits.
2	Limited	Random inspection(s)	Quality verification points (QVPs) with periodic inspections for lower complexity and dollar value. Some in-progress visits may be required (e.g., at the start of production or near production completion).
3	Progressive	Frequent inspection(s)	QVPs with progressive inspection for higher complexity and dollar value. In-progress visits are typically required (e.g., weekly).
4	Continuous	Continuous inspection(s)	Continual monitoring of production or construction.

QS levels may be alphabetic or numeric characters and may be used in ascending or descending order. Table 1 uses an numeric-descending order.

Each QS level is a shorthand means of describing QS activities. The activities that may be performed are described in detail in PO Section 4, *Inspection Requirements Document* as QVPs.

Inspection activities are performed by a supplier quality representative (SQR). The SQR is identified when the inspection assignment is issued at the start of fabrication or manufacturing. To fulfill inspection requirements, multiple SQRs may be required at multiple locations.

Some POs are assigned a QS level 0, which are only subject to a receiving inspection at the final destination or site as no inspection is planned at the supplier's facility.

Most POs are assigned QS level 1 to 3, which provides progressive levels of inspection from final only, to two types of in-process inspections. QS level 4, with continuous inspection, is more typical of inspection for pipe manufacturing and coating, as well as at construction sites.

Although a PO has a specified QS level, an inspection assignment at a sub-supplier's facility may have a different QS level applied. It may be higher or lower, depending upon the criticality ranking or criticality rating (CR) and other factors (e.g., cost and complexity).

1.5.2 Hold, Observe, and Witness Points

Hold, observe, and witness points are QVPs that are identified in the PO Section 4 *Inspection Requirements Document*. QVPs are then incorporated into the supplier's inspection and test plan (ITP). For the supplier, these are quality control activities. For the EP company, these are quality surveillance activities.

It is a mistake to assume that the terms *hold, observe, and witness points* have the same meaning to everyone. These terms may be defined differently and vary by project, especially in these project documents:

- Code(s) of Construction (e.g., API, ASME, CSA, and IEEE);
- Owner specifications;
- PO Section 1. *Scope of Supply*;
- PO Section 2. *Technical Documents List*, including technical notes;
- PO Section 4. *Inspection Requirements Document*;
- Project documents, EP company documents, or both (e.g., instructions and plans); and,
- Supplier:
    - Quality Management System (QMS); and,
    - ITP or Quality plans.

See Section 4 for definitions of the above terms and phrases.

Each code of construction typically includes a glossary that lists abbreviations and definitions for terminology used. For consistency, the equipment data sheet should always use the code of construction's terms for QVPs.

When a data sheet is attached to the PO Section 2 *Technical Document List*, it may conflict with and have precedence over the terms used in the PO Section 4 *Inspection Requirements Document* or the technical notes.

The PO Section 1 *Scope of Supply* may include line items with QVP descriptions. The scope of supply may also include equipment from different disciplines that use different terms (e.g., electrical and mechanical). In both cases, there may be a conflict with the precedence of documents and terms that are used.

The supplier's QMS and ITP or quality plans may have terms for QVPs that are different from, or not used in, the code of construction, PO, or both. The QVPs may be properly identified, and the activities can be completed according to project requirements.

The intended result is that project requirements are translated from the PO into supplier documents using the supplier's terms.

1.6 Industry- and Project-Specific Terminology

Projects may be complex and multi-disciplined, with many stakeholders, EP companies and EP company clients, and suppliers and sub-suppliers from many industry sectors. The meanings of terminology (abbreviations and terms) used by each may vary, sometimes to the point of confusion and misunderstanding. Different terminology may also be used by other stakeholders (e.g., jurisdictions) and in other areas or regions.

This is the nature of industry- and project-specific terminology, and this challenge must be addressed. If it is not, a project faces unknown risk. Nevertheless, a project can address this dilemma.

Precise terminology is the master key to effective communication. Effective communication is the master key to the successful delivery of products and services. These master keys can be used to unlock project success.

See Sections 1.6.1 and 1.6.2 for examples that offer insight into industry- and project-specific terminology.

1.6.1 Example of Pipeline Construction Terminology

What is a *poor-boy crew* working on a *spread*? If you have never heard this before, it may be difficult to picture. The terms *poor-boy crew* and *spread* are specific to pipeline construction and their meanings are typically not known outside of the pipeline construction industry. Section 4 defines these and other common project terms. See Appendix E for a typical sequence of work performed by welding crews when constructing a pipeline.

1.6.2 Example of Engineering and Procurement Company Terminology

What is a *document review code*? If you have never heard of this before, it may be difficult to understand. The phrase *document review code* is specific to an EP company's review of supplier documents for a project.

The PO Section 3 *Supplier Document Requirements* includes a list of supplier documents that are required to be submitted for review and record. An EP company will have a procedure for receiving and returning these documents through a Document Control department. An EP company will also have a procedure for the responsible engineer to review these documents. The purpose is to ensure that all PO and project requirements are incorporated, and that there are no conflicting or confusing requirements.

Some of the documents require engineering input or verification – e.g., drawings, ITPs, and welding procedure specifications (WPSs) – while others are only required for information – (e.g., catalogs and manuals). Supplier documents that have been submitted are assigned a code with a status to provide direction for the supplier and project team. These codes may be alphabetic or numeric characters, and may be used in ascending or descending order.

See Appendix F for a list of the document review codes, a description of each, and more information about these processes.

2 The Importance of Clear Communication

The importance of clear communication on a project cannot be understated. The point of clear communication is not to simply avoid confusion or misunderstandings; clear communication is the engine that drives project success. Without clear communication, expect direct negative impacts including schedule delays, increased costs, or worse.

An effective glossary, such as the KT Project Glossary (see Section 4), is key to clear communication because it serves as the authority for the meanings of industry and project terminology.

2.1 Responsibilities, Advantages, and Incentives

The owner is responsible to ensure that owner requirements are correctly stated in project documents and specifications. The EP company is responsible to ensure that project requirements are correctly stated in request for quotations (RFQs) and POs.

All authors, reviewers, and approvers of documents are responsible to ensure that the technical terms and requirements are properly described in project and procurement documents, and then properly translated into appropriate supplier documents.

2.2 Knowledge Transfer Project Key Knowledge: Supplier-Project Advantages and Incentives

Key Knowledge 01 describes how to anticipate and leverage supplier-project advantages and incentives.

KEY KNOWLEDGE 01

A supplier does not have the same incentive for success as a project; a supplier has a disincentive

A supplier can make additional profit by submitting a low bid to win the PO and escalating costs for change orders (COs).

Before the PO is awarded, the project has the advantage because suppliers must compete over scheduling and pricing. After the PO is awarded, the supplier has the advantage because the supplier can increase costs if COs are needed for any reason. For all changes with a reduced scope of work after PO award, there is no incentive for the supplier to provide any cost savings – even if the work is only partially completed and the minimum amount of materials is purchased.

The project can mitigate these concerns by using one or more of these strategies:

1. Ensure that all requirements for the scope of supply, including services, documentation, materials, and equipment are carefully reviewed and clearly stated, so COs become unnecessary or are minimized;
2. Negotiate appropriate pricing in advance if additional scope is expected to be added or if scope may change during future project stages;
3. Build long-term supplier relationships to exploit the benefits and efficiencies of repeated business. This may be established in a master service and supply agreement, or other commercial agreement; and,
4. Hold formal meetings with bidders and suppliers using a prepared agenda and recorded minutes to ensure detailed requirements are properly documented, communicated and understood. See KTP-GEN-003, *Project Supplier Meetings* for more details.

www.ktproject.ca KT Project: KK #002

2.3 Delivering Benefits to Your Project

A project benefits when all requirements are clearly stated in procurement documents – this eliminates the need for costly COs. COs are generated because of lack of clarity in project documents that increase or reduce the scope of work and put a project at a competitive disadvantage. By increasing comprehension of project terminology, COs can be avoided. Use this document to inform and educate project personnel, including:

- Personnel who are new to project work;
- English as a second language (ESL) professionals; and,
- Experienced project personnel (who can verify and update meanings of project terminology).

In cases of ambiguity or misunderstanding, the best way to understand the meaning of a term or phrase is to ask the author or originator. Document the clarification provided so others can also receive the same clarification.

An organization's personnel will benefit from the compilation and use of a glossary to define and standardize the use of common terms such as:

- Abbreviations and acronyms;
- Brand names, proprietary names, and trademarks;
- Company-, discipline-, industry-, and project-specific terminology;

- Expressions, historical use of terms, etymology, and social influencers of jargon and slang;
- Phrases and terms used by third parties (e.g., clients, stakeholders, and suppliers);
- Preferred and non-preferred terminology; and,
- Translation of phrases and terms from one language into another.

See Section 3 for a comprehensive list of common industry and project abbreviations.

See Section 4 for a comprehensive glossary of common industry phrases and terms.

2.4 Document Conventions

Sections 2.4.1 to 2.4.3 describe document conventions used in this document and all KT Project documents.

2.4.1 Spelling

This document uses Canadian English spelling. Canadian English is a mixture of American English and British English, with a Canadian twist. The differences between American, British, and Canadian English are usually slight and not immediately apparent to the uninitiated. Table 2 contains examples of spelling differences between these English dialects.

Table 2: Examples of Differences: American, British, and Canadian Spellings

No.	American	British	Canadian
1.	Ax	Axe	Axe
2.	Behavior	Behaviour	Behaviour
3.	Caliber	Calibre	Calibre
4.	Candor	Candour	Candour
5.	Centimeter	Centimetre	Centimetre
6.	Center	Centre	Centre
7.	Color	Colour	Colour
8.	Defense	Defence	Defence, defense
9.	Hemorrhage	Haemorrhage	Hemorrhage
10.	Judgment	Judgment, judgement	Judgment, judgement
11.	Labeled	Labelled	Labelled
12.	Meter	Metre	Metre
13.	Odor	Odour	Odour
14	Paneled	Panelled	Panelled
15.	Realization	Realisation, realization	Realization
16.	Smolder	Smoulder	Smoulder
17.	Theater	Theatre	Theatre

No.	American	British	Canadian
18.	Valor	Valour	Valour
19.	Vise	Vice	Vice
20.	Yogurt	Yoghurt	Yogurt

2.4.2 English is the Language of Business

English is the language of business, regardless of dialect. In fact, American specifications and standards are widely used by Canadian industry and projects. The American industry is also known to draw upon Canadian specifications and standards. Consequently, Canadian-American English spelling and terminology is often used interchangeably. British-English spelling and terminology may sometimes be used, especially by personnel with off-shore project experience based in non-North American countries.

In cases of ambiguity or misunderstanding, the best way to understand the meaning of a term or phrase is to ask the author or originator.

2.4.3 Finding Abbreviations, Phrases, and Terminology in this Document

Some industries and projects use the same terms and words interchangeably to describe various activities and functions. Also, some industries and projects use different terms to mean the same (or a similar) thing.

In Section 3 and Section 4, abbreviations, terms, and phrases may be defined as alternative, preferred or not preferred, and similar.

In some cases, similar terms are grouped together with the most common or preferred term listed first. For example, the terms *consulting company* and *consultancy firm* are similar and therefore grouped together using the same definition.

In other cases, the terminology is grouped with the definition. For example, there are several types of random length pipe, but these are all defined with the term *random length*. In the API SPEC 5L 44th Edition, six plain-end random lengths are listed in metres (with feet in parentheses). The traditional terms *single*, *double*, *triple*, and *quad random length pipe* have been deleted, but remain in popular use in references and specifications.

When searching for specific terminology in an open electronic document or website, press **Ctrl+F** to reveal the search dialog that allows you to search for a specific abbreviation, phrase, or term.

3 List of Common Industry and Project Abbreviations

Table 3 lists and provides the meanings for common industry and project abbreviations used in this and other KT Project documents.

Table 3: Common Industry and Project Abbreviations

Abbreviation	Meaning
+/-	Plus or minus
©	Copyright
®	Registered
1D	One dimensional
20/20	Twenty-twenty
24/7, 24-7, 24x7	24 hours a day and 7 days a week
2D	Two dimensional
300 SS	300 series stainless steel (all types of austenitic)
3BL	Triple bottom line
3D	Three dimensional
3PL	Third-party logistics
3Ps	Profit people planet
4D	Four dimensional
50/50	Fifty-fifty
5BX	Five basic exercises (for men)
5-D Cycle	Definition, discovery, dream, design, and destiny/deploy/delivery
5Ps	Proper preparation prevents poor performance
5S	Five S methodology
5W1H	What when where who why and how
6S	Five S methodology plus safety
6Ws	What when where who why and how
6σ	Six Sigma
7Ps	Proper planning and preparation prevent piss poor performance
A/V	Audio video
A2A	Alberta to Alaska
AB	Alberta
ABC	*Alberta Building Code*
ABCGA	*Alberta Common Ground Alliance*
ABSA	*Alberta Boilers Safety Association*
AC	Actual cost

Abbreviation	Meaning
ACA	Asset condition assessment
ACHX	Air cooled heat exchanger
AD	Anti-dumping
ADE	Approved deviations and exceptions
ADR	Alternative dispute resolution
AEC	Architectural, engineering, construction
AEPW	Alliance to End Plastic Waste
AER	*Alberta Energy Regulator* (formerly ERCB)
AESS	Architecturally exposed structural steel
AFE	Authorization for expenditure
AFFF	Aqueous firefighting foam
AG, A/G	Aboveground
AGI	Artificial general intelligence
AHJ	Authority having jurisdiction
AI	Appreciative inquiry, assignment instruction, authorized inspector, artificial intelligence, augmented intelligence
AIA	Authorized inspection agency
AIDA	Attention interest desire and action.
AIL	Action item list or log
AISI	*American Iron and Steel Institute*
aka	Also known as
AM	Asset integrity management
AM	Additive manufacturing
AMA	Ask me anything, American Motor Association
AML	Approved manufacturer's list
AMPP	*Association for Materials Protection and Performance* (formed by a merger of NACE and SSPC)
ANSI	*American National Standards Institute*
Anti-SLAPP	Anti-strategic lawsuit against public participation
AOGV	Add on gate valve
AP	Anodic protection, accounts payable
AP/AR	Accounts payable/accounts receivable
aPaaS	Application platform as a service
APD	Alternative project delivery
APEGA	*Association of Professional Engineers and Geoscientists of Alberta*
APF	Adaptive project framework
API	*American Petroleum Institute*, application programming interface

Abbreviation	Meaning
APM	Asset performance management
AQ	Adaptability quotient
AR	Accounts receivable
ARAD	After receipt of approved drawings
ARO	Abrasion resistant overcoat (dual-layer FBE coating), after receipt of order, asset retirement obligations
ASA	Alternative safety approach
ASAP	As soon as possible
AScT	Applied Science Technologist
ASET	*Association of Science and Engineering Technology Professionals of Alberta*
ASL	Approved supplier's list
ASM	*ASM International* (formerly *American Society for Metals*)
ASME	*American Society of Mechanical Engineers,* always sometimes maybe except
ASNT	*American Society for Nondestructive Testing*
ASQ	*American Society for Quality*
Assignment	Inspection assignment
ASTM	*ASTM International* (formerly *American Society for Testing and Materials*)
ATOMIC	Asset trust ownership money identity and control
ATS	Applicant tracking system
AutoCAD®	Computer-aided design software
AV	Audio video
AWOC	Alberta Welding Optimization Committee
AWP	Advanced work packaging
AWS	*American Welding Society*
AWWA	*American Water Works Association*
AZETEC	Alberta Zero Emissions Truck Electrification Collaboration
B2B, B to B	Business to business
B2C, B to C	Business to customer
B2G, B to G	Business to government
BANANA	Build absolutely nothing anywhere near anyone, build absolutely nothing anywhere near anything
BATNA	Best alternative to a negotiated agreement
BAU	Business as usual
BBB	*Better Business Bureau*, build back better, bar beach bed
BBC	Bitumen beyond combustion
bcc	Blind carbon copy

Abbreviation	Meaning
BCM	Bid clarification meeting
BCSM	Behavioral change stairway model
BCP	Business continuity planning
BCSA	British Columbia Safety Authority (now TSBC)
BD	Business Development
BECCS	Biomass energy with carbon capture and storage
BESS	Battery energy storage system
BHP	Brake horsepower
BI	Business intelligence
BIB	Business is business
BIG	Business is a game
BIM	Building information modelling
Bio	Biological, biography
BIPOC	Black Indigenous and People of Colour
BIW	Business is war
BLT	Biomass-to-liquids, build lease transfer
BLUF	Bottom line up front
BOE	Barrel of oil equivalent
BOK	Body of knowledge
BOM	Bill of material, bill of materials
BOO	Build own operate
BOOT	Build own operate transfer
BOP	Blowout preventer
BoP	Base of pyramid
BOPSAAT	Bunch of people sitting around a table
BOSCARD	Background objectives scope constraints assumptions risks deliverables
Bot	Robot
BOT	Build own transfer, build operate transfer
BPM	Business process modeling
BPO	Blanket purchase order
BPS	Bonding procedure specification
BPVC	*Boiler and Pressure Vessel Code*
BRICS	Brazil Russia India China and South Africa
BS	British Standards, bullshit
BUTS	Baseline ultrasonic thickness survey
BYOL	Bring your own lunch (pay for your own lunch)

Abbreviation	Meaning
C	Carbon
C&P	Contract and procurement
C_1	A molecule made of one carbon atom
C_2	A molecule made of two carbon atoms, diatomic carbon
C2B, C to B	Customer to business
C2C, C to C	Customer to customer
C_3	A molecule made of three carbon atoms, tricarbon
C_4	A molecule made of four carbon atoms
CA	Corrosion allowance, confidentiality agreement, California
CaaS	Content as a service, communications as a service
CAD	Canadian, Canadian dollar, computer-aided design
CADD	Computer-aided design and drafting
CAE	Computer-aided engineering
CAES	Compressed air energy storage
Cal sil	Calcium silicate (insulation)
CAP	*Certificate of Authorization Permit*
Capex	Capital expenditure
CAR	Corrective action recommendation or corrective action request
CAutoD	Computer-automated design
CAVE	Citizens against virtually everything
CBI	Condition-based inspection
CBM	Condition-based monitoring
cc	Carbon copy
CC	Contract change
CCC	Calgary Corporate Challenge
CCO	Chief commercial officer
CCRN	Canadian Central Registration Number
CCS	Carbon capture and storage, carbon capture and sequestration
CCUS	Carbon capture utilization and storage, carbon capture and utilization or storage
CDA	Confidential disclosure agreement
CE	Commercial bid evaluation
CE	Carbon equivalent
CEAR	Capital expenditure authorization request
CEDO	Certified exposure devise operator
CEN	Carbon equivalent number, European Committee for Standardization (French: Comité Européen de Normalisation)

Abbreviation	Meaning
CEMS	Continuous emission monitoring system
CEO	Chief executive officer, Chief everything officer
CET	Certified Engineering Technologist
CETL	*Canadian Electrical Testing Labs*
CEV	Clean energy vehicle
CF	Carbon fibre
CFD	Computational fluid dynamics
CFO	Chief financial officer
C-free	Carbon-free
CFS	*Clean Fuel Standard*
CGSB	*Canadian General Standards Board*
CH_4	Methane
CHP	Combined heat and power
CI	Continual improvement, continuous improvement, carbon intensity
CICIC	*Canadian Information Centre for International Credentials*
CINDE	*Canadian Institute for Non-destructive Evaluation*
CIO	Chief information officer
CIP	Capital improvement plan, continual improvement process, continuous improvement process
CJP	Complete joint penetration
CLB	Canadian language benchmark
CLTV	Customer lifetime value
CMA	*Chemical Manufacturers Association*
CML	Corrosion monitoring location
CMMIS	Computerized maintenance management information system
CMMS	Computerized maintenance management system
CMT	Construction management team
CMTR	Certified material test report
CN	Change notice
CNC	Computer numerically controlled
CNG	Compressed natural gas
CO	Change order, carbon monoxide
CO_2	Carbon dioxide
CO_2e	Carbon dioxide equivalent
Cobot	Collaborative robot
COC	Certificate of conformance or certificate of compliance
COF	Consequence of failure

Abbreviation	Meaning
Cogen	Cogeneration
COGS	Cost of goods sold
Comms	Communications
COO	Chief operating officer, conduct of operations
COO/OD	Conduct of operations/operational discipline
COOP	Cooperative
CO-OP	Cooperative
COP	Conference of the Parties
CORN	Change order request notice
COTC	Crude oil-to-chemicals
COVID-19	Coronavirus disease 2019
CP	Cathodic protection
CPD	Continuing professional development
CPI	Corruption perceptions index
CPM	Critical path method, corporate performance management
CPU	Central processing unit (a computer)
CR	Criticality ranking, criticality rating
CRA	Corrosion resistant alloy
CRM	Customer relationship management
CRMO	Chief risk management officer
CRN	Canadian Registration Number
CRO	Chief risk officer, conversion rate optimization
CRP	Capacity requirements planning
CS	Carbon steel
CSA	*Canadian Standards Association*
CSR	Corporate social responsibility
CSS	Cascading Style Sheets
CTA	Call to action
CTech, C.Tech.	Certified Technologist
CTL	Coal-to-liquid
CUB	Central utility block
CUI	Corrosion under insulation
CUPS	Corrosion under pipe supports
CUSMA	Canada-United States-Mexico Agreement
CV	Curriculum vitae (i.e., a resume or résumé)
CVD	Countervailing duties

Abbreviation	Meaning
CVM	Customer value model
CVN	Charpy V-notch
CWB	*Canadian Welding Bureau*
CWM	Collaborative workforce management
CX	Customer experience
CxO	Chief (title) officer (e.g., chief executive officer)
Daas	Desktop as a service
DBB	Double block and bleed
DBFO	Design build finance operate
DBM	Design basis memorandum
DBOT	Design build operate transfer
DC	Document control
DCMF	Design construct manage finance
DCN	Design change notice
DCS	Distributed control system
DD	Detailed design
DDOS	Distributed denial of service
DEI	Diversity equity and inclusion
Demo	Demonstrate, demonstration
DF	Design factor
DFMA	Designed for manufacturing and assembly
DFRAM	Design for reliability and maintenance
DFT	Dry film thickness
DI	Desk instruction
Dia	Diameter
DIB	Double isolation and bleed
DIE	Diversity inclusivity and equity
DIN	Deutsches Institut für Normung (German Institute for Standardization)
DIRFT	Do it right the first time
DIRTT	Do it right this time
Dis	Disrespect
Diss	Disrespect
DIY	Do it yourself
DJAI	Dow Jones Industrial Average
DM	Decision-making
DMADV	Define measure analyze design and verify

Abbreviation	Meaning
DMAIC	Define measure analyze improve and control
DMG	Document management group
DMR	Damage mechanism review
DMW	Dissimilar metal weld
DN	Dimensionless designator or diamètre nominal
DOA	Delegation of authority
Doc	Document
DoDoDab	Who does what who does not and avoid dabbling
DOTWIMP	Defects overproduction transportation waiting inventory motion and processing (i.e., the seven wastes)
DPD	Digital project delivery
DPI	Dye penetrant inspection, see also PT
DPM	Digital project management
DQR	Document quality review
DR	Digital radiography, digital reality
DRA	Drag-reducing agents
DRaaS	Disaster recovery as a service
DRIFT	Do it right the first time
DRIP	Data rich, information poor
DRIVE	Define review identify verify and execute
DRL	Double random length
DRP	Drag-reducing polymers
DTI	Direct tension indicating
DTR	Data transfer rate
Dwg	Drawing
e.g.	For example
E2E	End-to-end
EA	Environmental assessment
EAC	Estimate at complete
EAM	Enterprise asset management
EBDM	Evidence-based decision-making
EBITDA	Earnings before interest taxes depreciation and amortization
EBITDAC	Earnings before interest taxes depreciation amortization and COVID-19
EBM	Evidence-based management
EBP	Evidence-based planning
E-building	Electrical building

Abbreviation	Meaning
ECA	Educational credential assessment
ECI	Engineering construction and infrastructure, early contractor involvement
ECN	Engineering change notice
Eco	Ecological, ecology, economical
E-commerce	Electronic commerce
ECREE	Extraordinary claims require extraordinary evidence
EDMS	Electronic data-document management system
EDO	Exposure device operator
EDS	Engineering design specification
EFL	Energy Futures Lab
EGAP	Everything goes according to plan
E-house	Electrical house
EHT	Electric heat tracing
EHV	Extra high voltage
EI	Emotional intelligence, engineering instruction
EIA	Environmental impact assessment
EIQ	Emotional intelligence quotient
EL	Emotional leadership
ELV	Extra low voltage
EN	European Standards
ENGO	Environmental non-government organization
EOB	End of business
EOD	End of day
EOI	Expression of interest
EOR	End of record, engineer of record, enhanced oil recovery
EP	Engineering and procurement, exploration and production
EPC	Engineering procurement and construction, engineer procure construct
EPCM	Engineering procurement and construction management
EPE	Engineered pressure enclosure
EPM	Enterprise performance management
EQ	Emotional quotient
EQFM	European Quality Foundation Model
EQMS	Electronic quality management system
ERCB	*Energy Resources Conservation Board*, see AER
ERM	Enterprise risk management
ERMP	Electrical regulatory management plan

Abbreviation	Meaning
EROEE	Energy return on the amount of energy expended
EROEI	Energy return on the amount of energy invested
EROI	Energy return on investment
ERP	Enterprise resource planning
ERW	Electric resistance welding
ESC	Engineered structural composite
ESG	Environmental social governance
ESGI	Environmental social governance with Indigenous partnerships and participation
ESL	English as a second language, estimated service life, expected service life
ESO	Embedded service organization
ET	Eddy current test, engineering technology
ETC	Estimate to complete
EU	European Union
EULA	End user license agreement
eVTOL	Electric vertical take-off and landing aircraft
Expat	Expatriate
F3	Form fit and function
FAAP	Family accepted accounting principles
FAI	First article inspection
FAT	Factory acceptance test
FBE	Fusion-bonded epoxy (coating)
FCA	Free carrier
FCAW	Flux-cored arc welding
FCEV	Fuel cell electric vehicle
FCN	Field change notice
FCV	Fuel cell vehicle
FDPT	First day production testing
FEA	Finite Element Analysis
FEED	Front end engineering and design
FEL	Front end loading
FEMI	Fixed equipment mechanical integrity
FFH	Future fit hydrocarbons
FFS	Fitness for service
FFW	Field fit weld, field fit-up weld
FID	Final investment decision
Five Ws	What when where who and, why

Abbreviation	Meaning
FMA	First-mover advantage
FN	First Nation, First Nations
FOBO	Fear of better options
FOE	Finance one's existence
FOIP	Freedom of information and protection of privacy
FOMO	Fear of missing out
FOS	Factor of safety
FP&A	Financial planning and analysis
FPSO	Floating production storage and offloading
FPT	Fluorescent penetrant test
FPTP	First-past-the-post
FQR	Foreign qualification recognition
FRP	Fibre-reinforced plastic
FSM	Field service management
FST	Functional services team
ft	Feet, foot
FTE	Full-time equivalent
FTP	File Transfer Protocol
FV	Flow valve
FVR	Field verification report
FW	Field weld
G&A	General and administrative
G6	Group of Six
G7	Group of Seven
G8	Group of Eight
G8+5	Group of Eight + Five
G20	Group of Twenty
G77	Group of 77
GA	General arrangement
GAAP	Generally accepted accounting principles
GBA	Gender-based analysis
GBA+	Gender-based analysis-plus
GDP	Gross domestic product
Genset	Generator set
GES	Global Energy Show
GH_2	Gaseous hydrogen

Abbreviation	Meaning
GHG	Greenhouse gas
GHP	Gross happiness product
GHT	Glycol heat tracing
GIGO	Garbage in garbage out *or* garbage in gospel out
GIS	Geographical information system
GJP	*Good Judgment Project*
Glycol	Ethylene glycol
GMAT	Graduate management admission test
GMAW	Gas metal arc welding
GND	Green new deal
GNH	Gross national happiness
GOA	Grant of authority
GOM	Gulf of Mexico
GPS	Global positioning system, Global Petroleum Show (now the Global Energy Show)
GQP	Good quality practice
GRE	Glass reinforced epoxy
GTAW	Gas tungsten arc welding
GTL	Gas-to-liquids
GUI	Graphical user interface
H, H_2	Hydrogen
H2CNG	Hydrogen to compressed natural gas
H_2O	Water
H_2S	Hydrogen sulphide
HAC	Hydrogen-assisted cracking
HARP	Hydrogen assisted renewable power
HASD	Horizontal alternating steam drive
HAT	Hardware acceptance test
HAZ	Heat affected zone
Hazmat	Hazardous materials (i.e., dangerous goods)
Hazop	Hazard and operability
HCNG	Hydrogen-enriched compressed natural gas
HD	Heavy duty, high definition, hard drive
HDD	Horizontal directional drilling
HDG	Hot dip galvanized
HDPE	High density polyethylene
HE	Hydrogen embrittlement

Abbreviation	Meaning
HENG	Hydrogen enriched natural gas
HES	Hybrid energy system
HEX	Heat exchanger
HHC	Highly hazardous chemicals
HIC	Hydrogen-induced cracking
HICEV	Hydrogen internal combustion engine vehicle
HMI	Human-machine interface (monitor or touch screen)
HNS	Hazardous and noxious substance
HP	High pressure
HPI	Hydrocarbon processing industry
HR	Human resources
HRIS	Human resource information system
HRM	Human resources management
HRS	Hydrogen recharging station
HSE	Health, safety, and environment
HSFO	High-sulphur fuel oil
HSLA	High strength low alloy
HSS	Hollow structural section
HSSE	Health, safety, security, and environment
HTML	Hypertext markup language
HTTP	Hypertext transfer protocol
HUD	Heads up display
HV	Hand valve, high voltage
HVAC	Heating ventilating and air conditioning
I&C	Instrumentation and controls
i.e.	For instance
I2PD	Industrial integrated project delivery
I4.0, I4	Industry 4.0
IA	Interface agreement
IaaS	Infrastructure as a service
IAMS	Integrated asset management system
ICA	Independent contractor agreement
ICC	International Chamber of Commerce
ICE	Instrumentation controls and electrical
ICF	Initial contact form, insulating concrete form, insulated concrete from
ICMS	*International Construction Measurement Standards*

Abbreviation	Meaning
ICMSC	*International Construction Measurement Standards Coalition*
ICSS	Integrated control and safety system
ID	Inside diameter, intelligent design, internal diameter, identification
IEA	*International Energy Agency*
IEEE	*Institute of Electrical and Electronics Engineers*
IFC	Issued for construction
IFD	Issued for delivery, issued for design
IFE	International fisher effect theory
IFR	Issued for review
IGI	Incoming goods inspection
IHES	Integrated hybrid energy system
II	Incredibly important
IIOT	Industrial Internet of things
IIW	*International Institute of Welding*
ILI	In-line inspection
IM	Interface management
IMS	Integrated management system, integrity management system
in.	Inch
Info	Information
Intel	Intelligence
IO&M	Installation operation and maintenance
IOC	International oil company
IOT	Internet of things
IP	Intellectual property, interface point
IPD	Integrated project delivery
IPEIA	*International Pressure Equipment Integrity Association*
IPM	In production mode
IPP	Independent power producers
IPPC	*International Plant Protection Convention*
IPR	Independent project review, intellectual property rights
IQ	Intelligence quotient
IQAS	*International Qualifications Assessment Service*
IR	Inspection report
IRD	Inspection requirements document (e.g., PO Section 4)
IRP	Interest rate parity theory
ISN®	ISNetworld®

Abbreviation	Meaning
ISO	*International Organization for Standardization*
ISOS	In-situ oil sands
ISPM	*International Standard for Phytosanitary Measures*
IT	Information technology
ITP	Inspection and test plan
ITR	Independent team review
ITSM	Inspection and Technical Services Manitoba
ITTO	Inputs tools techniques outputs
IWT	Industrial wind turbine
JDR	Judicial dispute resolution
JHA	Job hazard analysis
JI	Job instruction
JIS	Japanese Industrial Standards
JIT	Just-in-time
JITR	Just-in-time resourcing
JMP	Journey management plan
JOA	Joint operating agreement
JOE	Justify one's existence
JSA	Job safety analysis
JV	Joint venture
KIITG	Keep it in the ground
KISS	Keep it simple stupid, keep it stupid simple, keep it simple silly, keep it short and simple, keep it simple and straightforward, keep it small and simple
KK	Key knowledge
KOM	Kick-off meeting
kPa	KiloPascal
KPI	Kept promise indicator, key performance indicator
KT	Knowledge transfer
KT Project	Knowledge Transfer Project (a tradename)
KT Project	Knowledge Transfer (KT) Project
KTP	Knowledge Transfer Project, KT Project
LGBTQ2IA+	Lesbian gay bisexual transgender queer questioning two-spirit intersex asexual and other identities
LCAM	Life cycle asset management
LCC	Life cycle cost
LCE	Low-carbon economy

Abbreviation	Meaning
LD	Light duty
LDAR	Leak detection and repair
LED	Light-emitting diode
LEED	Leadership in Energy and Environmental Design
Legit	Legitimate
LFFE	Low-fossil-fuel economy
LFL	Like-for-like
LH_2	Liquid hydrogen
LIW	Load indicating washer
LL	Lessons learned, lunch and learn
LLE	Long lead equipment
LMRA	Last minute risk analysis, last minute risk assessment
LNG	Liquefied natural gas
LOC	Letter of conformance or letter of compliance
LOHC	Liquid organic hydrogen carrier
LOMP	Lots of moving parts
LOS	Laws of simplicity, lube (lubrication) oil skid
Low hy	Low hydrogen (welding electrode)
LOX	Liquified oxygen
LPG	Liquid petroleum gas, liquefied petroleum gas
LPI	Liquid penetrant inspection, see also PT
LRC	Linear responsibility chart
LRNG	Liquids-rich natural gas
LSFO	Low sulphur fuel oil
LSP	Logistics service provider
LT	Liquid penetrant test, see also PT
LTL	Less than load (partial shipment)
LTO	Light tight oil
LV	Low voltage
m	metre
M&A	Merger and acquisition
M2M	Machine to machine
MaaS	Marketing as a service
MAEWP	Maximum allowable external working pressure
MAOP	Maximum allowable operating pressure
MAWP	Maximum allowable working pressure

Abbreviation	Meaning
MB	Megabyte (10^6 or 1,000,000 bytes)
MBA	Master of business administration
MBL	Minimum breaking load
MBO	Management by objectives
MBR	Management by results, minimum backup requirement
MBS	Minimum breaking strength
MBTI®	Myers-Briggs Type Indicator®
MC	Mechanical completion, mic controller, master of ceremonies
McaaS	Managed content as a service
MCAW	Metal-cored arc welding
MCC	Motor control centre
mcf	Million cubic feet
MDGs	Millennium development goals
MDR	Manufacturer's data report
MECE	Mutually exclusive and completely exhaustive
MEL	Mechanical equipment list
MFAT	Module factory acceptance test
MI	Mechanical integrity
Mic	Mirophone
MICC	Main instrument and control contractor
micron	One millionth of a metre (0.000001 m)
MIG/MAG	Metal inert gas/metal active gas, see GMAW
mil	one thousandth of an inch (0.001 in.)
MIS	Management information system
MIT	Most important task
MLD	Most likely development
MLV	Mainline valve
mm	millimetre
mmbbl	Millions of barrels
MMS	Multimedia messaging service
MMT	Modern monetary theory
MMTCDE	Million metric tonnes of carbon dioxide equivalents
MOC	Management of change, materials of construction, master of ceremonies
MODER	Misunderstanding omission oversight delay error and redundancy
MOE	Margin of error, manufacturer original equipment, maintain one's existence
MOM	Minutes of meeting

Abbreviation	Meaning
MOOC	Management of organizational change
MOP	Maximum operating pressure
MoSCoW	Must have, should have, could have, will not have
MOV	Motor operated valve
MPI	Magnetic particle inspection, see also MT
MPQT	Manufacturing procedure qualification test
MPS	Manufacturing procedure specification
MR	Material requisition, mixed reality
MRB	Manufacturing record book
MRM	Materials responsibility matrix
MRO	Maintenance repair and operating, maintenance repair and overhaul
MRP	Material requirements planning, material requisition for purchase
MRQ	Material requisition for quotation
MS	Microsoft
MS Teams	Microsoft Teams (software)
MSA	Master service agreement
MSD	Major structural discontinuity
MSM	Mainstream media
MSP	Managed service provider
MSS	*Manufacturer's Standardization Society*
MSSA	Master service and supply agreement
MSSP	Managed security services provider
MSU	Make shit up, make stuff up
MT	Magnetic particle test, megaton
MTBM	Microtunnel boring machine
MTBO	Mean time between overhaul
MTR	Material test report or mill test report
MUC	Multiple use contract
MV	Medium voltage
MVLP	Minimum viable loveable product
MVP	Material verification program, minimum viable product
MW	Megawatt
MWh	Megawatt-hour
n.d.	No date (available)
N/A	Not applicable, not available
N/R	Not required

Abbreviation	Meaning
N_2O	Nitrous oxide
NaaS	Network as a service
NACE	*NACE International* (formerly *National Association of Corrosion Engineers*), see AMPP
NAFTA	*North American Free Trade Agreement* (Canada-United States-Mexico)
NAFTA2	The second *North American Free Trade Agreement* (Canada-United States-Mexico)
NASA	*National Aeronautics and Space Administration*
NATO	*North Atlantic Treaty Organization*
NB	National Board
NBCC	*National Building Code of Canada*
NBS	Nature-based solution
NCG	Non-condensable gas
NCR	Non-conformance report
NCV	No commercial value
NDA	Non-disclosure agreement
NDE	Non-destructive examination
NDT	Non-destructive testing
NEB	*National Energy Board* (Canada)
NEWS	North east west south, notable events weather and sports
NG	Natural gas
NGO	Non-governmental organization
NGR	Neutral grounding resistor
NIMBY	Not in my back yard
NIST	*National Institute of Standards and Technology*
NLM	Nonlinear management
NOC	National oil company, national occupation classification
Nom	Nominal
NOP	Non-owner participant, not operational
NORM	Naturally occurring radioactive materials
NPS	Nominal pipe size
NPSH	Net positive suction head
NPT	National pipe thread
NRCan	*Natural Resources Canada*
n-Tier	All tiers of a supply chain
NTP	Normal temperature and pressure
NWL	Normal working load
NWT	Nominal wall thickness, Northwest Territories

Abbreviation	Meaning
NZE	Net-zero emissions
O&M	Operation and Maintenance
OBaS	On budget and schedule
OCI	Organizational conflict of interest
OCM	Operations control management
OCTG	Oil country tubular goods
OD	Outside diameter
OE	Operational excellence, organizational excellence, original equipment
OECD	*Organization for Economic Cooperation and Development*
OEM	Original equipment manufacturer
OEMS	Operational excellence management system, organizational excellence management system
OES	Optical emission spectroscopy, see PMI
Off spec	Off specification
OFI	Opportunity for improvement
OFSE	Oil-field services and equipment
OHS	*Occupational Health and Safety*
OJT	On-the-job training
OK, O.K., Ok, ok	All correct
Okay	All correct
OKRs	Objective and key results
OMEC	Organization of Mineral Exporting Countries
OOO	Out of office
OOT	Out of tolerance
OPEC	Organization of the Petroleum Exporting Countries
OPEC+	Organization of the Petroleum Exporting Countries plus 10 Non-OPEC nations
OpEx	Operating expense, operational excellence
OPM	Other people's money, online program manager
OPP	Ore processing plant
OPPSD	Over pressure protection by system design
OPR	*Onshore Pipeline Regulations*
ORID	Objective reflective interpretive decisional
OQ	Operator qualification
ORA	Overpressure risk assessment
Org Chart	Organizational chart
OSAG	Oil Sands Advisory Group
OSHA	*Occupational Safety and Health Administration*

Abbreviation	Meaning
OT	Operation technology, operational technology, operations technology, overtime
OTD	On-time delivery
OTF	Over the fence
OTSG	Once through steam generator
Out of spec	Out of specification
P&ID	Piping and instrumentation diagram
P&L	Profit and loss
P&U	Power and utilities
P.Eng.	Professional engineer
P2G	Power to gas
P2P	Peer to peer, point to point, procure-to-pay, purchase-to-pay
P2X	Power-to-X
P3	Public-private partnership, projects programs and portfolios
Pa	Pascal
PA	Public announcement
PAA	Project alliance agreement
PaaS	Platform as a service
PACE	Playfulness acceptance curiosity empathy
PAL	Problem action log
PAT	Panel acceptance test
PAUT	Phased array ultrasonic testing
PBHT	Post bend heat treatment
PBO	Project-based organization
PCM	Project cost management
PCN	Project change notice
PCO	Potential change order
PCS	Process control system, pieces
PCV	Pressure control valve
PDCA	Plan–do–check–act
PDF	Portable document format
PDH	Propane dehydrogenation
PDRI	Project definition rating index
PDS	Product data sheet
PDU	Professional development unit
PE	Project engineer, professional engineer, polyethylene
PEA	Preliminary economic analysis

Abbreviation	Meaning
PED	*Pressure Equipment Directive* (Europe)
PEIMS	Pressure equipment integrity management system
PEM	Polymer electrolyte membrane, proton exchange membrane
PEMFC	Polymer electrolyte membrane fuel cell
PEP	Project execution plan, procurement execution plan
PER	*Pressure Equipment Regulations* (United Kingdom)
PESL	*Pressure Equipment Safety Legislation*
PESR	*Pressure Equipment Safety Regulation, Alberta Regulation 49/2006*
PEST	Political economic social technological
PFAS	Perfluoroalkyl substances, polyfluoroalkyl substances
PFF	Pipe fitting and flanges
PFM	Prefabrication meeting
PFP	Passive fire protection
PHA	Process hazard analysis
PHE	Plate heat exchanger
PHIT	Passion happiness innovation teamwork
PHMSA	*Pipeline and Hazardous Materials Safety Administration* (USA)
Photo-op	Photograph opportunity
PI	Process intensification, project instruction
PIA	Proprietary information agreement
pig	Pipeline inspection gauge
PIM	Pre-inspection meeting
PJ	Pipe jacking
PJP	Partial joint penetration
PL, P/L	Pipeline
PLA	*Public Land Act*
PLC	Programmable logic controller *or* programmable logic circuit
PLM	Product life cycle management
PLT	Project leadership team
PM	Project manager
PMBOK®	Project management body of knowledge
PMC	Project managing contractor
PMI	Positive material identification
PMO	Project management office
PMOC	Project management of change
PMP	Project management professional

Abbreviation	Meaning
PO	Purchase order
POC	Point of contact, People of Colour
POCO	Purchase order change order
POF	Probability of failure
POP	Point of purchase
Porn	Pornography
POS	Point of sale
pp	Pages
PP	Polypropylene
PPA	Purchase power agreement
PPE	Personal protective equipment, property plant and equipment
ppm	Parts per million
PPM	Project procurement manager, project portfolio management
PPP	Pre-project planning, project procurement plan, polypropylene plant, profit people planet, purchasing power parity (theory)
PPVC	Prefabricated prefinished volumetric construction
PQ	Precision questioning
PQer	Precision questioning practitioner
PQO	Professional regulatory organization
PQR	Procedure qualification record
PR	Pipe ramming, public relations
Prequal	Pre-qualification
PRV	Pressure relief valve
PSA	Preferred supplier agreement, probabilistic schedule assessment, professional services automation, public service announcement
PSD	Pressure safety device
PSDM	Problem solving and decision making
psf	Pounds per square foot
psi	Pounds per square inch
psia	Pounds per square inch absolute
psig	Pounds per square inch gauge
PSM	Process safety management
PSSR	Pre start-up safety review
PT	(Liquid) penetrant test
PTech, P.Tech.	Professional technologist
PTFE	Polytetrafluoroethylene or Teflon®

Abbreviation	Meaning
PTP	Point to point
PTSD	Post-traumatic stress disorder
PUB	Partial upgrading of bitumen
PVC	Polyvinyl chloride
PVFF	Pipe, valves, fittings, and flanges
PWHT	Post weld heat treatment
PWP	Professional work product
Q&A	Question and answer
Q1	First quarter
Q2	Second quarter
Q3	Third quarter
Q4	Fourth quarter
QA	Quality assurance
QA/QC	Quality assurance/quality control
QAP	Quality assurance program
QC	Quality control
QMP	Quality management plan
QMS	Quality management system
QO	Qualified operator (now CEDO)
QOC	Quick opening closure
QR Code	Quick response code
QS	Quality surveillance
QSP	Quality surveillance plan
QSR	Quality surveillance representative or quality surveillance report
Qty	Quantity
QVP	Quality verification point
R&D	Research and development
RA	Return authorization
RACI	Responsible, accountable, consulted, and informed
RAGAGEP	Recognized and generally accepted good engineering practices
RAID	Risks assumptions issues and dependencies
RAIL	Rolling action item log
RAM	Responsibility assignment matrix
RAS	Required at site (date)
RBI	Risk-based inspection, report by inclusion
RCA	Root cause analysis

Abbreviation	Meaning
RCM	Reliability-centred maintenance
RE	Responsible engineer
REC	Record, renewable energy certificate
REE	Rare earth element
REGO	Renewable energy guarantees of origin
REM	Rare-earth mineral
RET	Registered Engineering Technologist
RFI	Request for information
RFID	Radio frequency identification
RFO	Ready for operations
RFP	Request for proposal, request for purchase
RFQ	Request for quotation, request for qualifications
RG	Renewable gas
RGA	Return goods authorization
RH_2	Renewable hydrogen
RIK	Repair in-kind or replacement in-kind
RIPA	Robotic industrial process automation
RLS	(Inspection) release
RMA	Return merchandise authorization
RMBI	Reliability based mechanical integrity
RMCP	Resource management certified professional
RMM	Remote management and monitoring
RMO	Resource management office
RNG	Renewable natural gas
ROC	Rapid opening closure, Republic of China, rest of Canada
ROCE	Return on capital employed
ROCK	Retention of critical knowledge
ROE	Return on expectations, right of entry
ROI	Return on investment
ROS	Required on site (date)
ROUS	Rodent of unusual size
ROW	Right-of-way
RP	Recommended practice
RPA	Robotic process automation
RPT	(Inspection) report

Abbreviation	Meaning
RRIMR	Risk assessment, root cause analysis, installation, maintenance and monitoring, and removal confirmation date
RT	Radiographic test
RTD	Resistance temperature detector, research and technical (or technological) development
S&P 500 Index	Standard & Poor's 500 Index
S&P Global	Standard & Poor's Global
S/D	Shutdown
S/N	Signal-to-noise ratio
S2P	Source-to-pay
SA	Situational awareness, situation awareness, secrecy agreement, standards association
SaaS	Software as a service, security as a service
SAD	Seasonal affective disorder
SAE	*SAE International* (formerly *Society of Automotive Engineers*)
SAGD	Steam-assisted gravity drainage
SAID	Systems approach to infrastructure delivery
SAP	SAP® Business Suite
SARA	*Species at Risk Act*
SAT	Site acceptance test *or* software acceptance test
SAW	Submerged-arc welding
SCA	*Safety Codes Act*
SCADA	Supervisory control and data acquisition
SCC	*Safety Codes Council*, Stress Corrosion Cracking
SCH	Schedule (pipe size)
SCM	Supply chain management
SCO	Safety codes officer *or* synthetic crude oil
SCP	Supply chain partner
SDDR	Supplier drawing and data requirements
SDG	Sustainable development goals
SDI	Supplier document index
SDR	Supplier document requirements (e.g., PO Section 3) *or* supplier data register
SDRL	Supplier data requirements list
SECaaS	Security as a service
SEO	Search engine optimization
Seven Ps	Purpose, passion, process, product, partner, protect, performance
SF	Safety factor
SFA	Supplier frame agreement

Abbreviation	Meaning
SHK	School of hard knocks
SI	International System of Units or Système international d'unités, strategic intelligence
SIFT	System integration and functionality test
SIPOC	Supplier inputs processes outputs customer
SIR	Serial interface report
SIS	Safety instrumented system
SIT	Safety integration test
SJW	Social justice warrior
SKD	Shutdown key
SKU	Stock keeping unit (number)
SLAPP	Strategic lawsuit against public participation
SLAPP-back	Strategic lawsuit against public participation-back
SLMS	Safety loss management system
SMA	Subject matter area
SMART	Specific measurable attainable relevant and time-bound
SMAW	Shielded metal arc welding
SMB	Small- and medium-sized business
SME	Small- to medium-sized enterprise, small to medium enterprise, small and medium employer, subject matter expert
SMLS	Seamless
SMOG	Slow moving and obsolete goods, simplified measure of gobbledygook, smoke and fog
SMR	Small modular reactor, steam methane reformer
SMS	Short message service
SNG	Synthetic natural gas, substitute natural gas, sustainable natural gas
SNR	Signal-to-noise ratio
SOARA	Situation objective action results aftermath
SOC	Supplier of choice (agreement)
SOE	Standard operating environment
SOJT	Structured on-the-job training
SOL	Shit out of luck
SOP	Standard operating procedure
SOQ	Statement of qualifications
SOS	Scope of supply (e.g., PO Section 1)
SOTA	State of art analysis
SOW	Scope of work
Spec	Specification

Abbreviation	Meaning
SPG	Small projects group
SPOC	Single point of contact
SPOF	Single point of failure
Sponcon	Sponsored content
SPT	Site performance test
SQR	Supplier quality requirements (e.g., PO Section 4), supplier quality representative
SQS	Supplier quality surveillance
SRED, SR&ED	Scientific Research and Experimental Development (Canadian tax incentive program)
SSHE	Safety, security, health, and environment
SSO	Single sign on
SSPC	Steel Structures Painting Council, see AMPP
STAR	Situation task action result
STARR	Situation task action result reflection
STD	Standard
STEG	Solar thermoelectric generator
STEM	Science technology engineering and mathematics
STP	Standard temperature and pressure
SVP	Senior vice president
SW	Southwest
SWB	Solar wind battery
SWL	Safe working load
SWOT	Strengths, weaknesses, opportunities, and threats
SWP	Safe work practice, safe work procedure
SWPS	Standard welding procedure specification
SWUT	Short wave ultrasonic test
Syngas	Synthesis gas
T/A	Turnaround
TABP	Type A behavior pattern
TAR	Task action results, turnaround
TBE	Technical bid evaluation
TBL	Triple bottom line
TBM	Tunnel boring machine
TCO	Total cost of ownership
TDC	Technical document control
TDG	Transport of dangerous goods
TDL	Technical documents list (e.g., PO Section 2)

Abbreviation	Meaning
TDN	Technical deviation notice
TCP	Transmission Control Protocol
Textrix	Textrix Inc.
THINK	(Is it) True? Helpful? Inspiring? Necessary? Kind?
This document	This *Glossary of Common Industry and Project Terminology* document
Three Ps	People, planet, profit
TIA	Time-impact analysis
TIC	Total installed cost
TIG	Tungsten inert gas, see GTAW
TKY	T shape K shape or Y shape
TL;DNR	Too long; did not read
TLC	Tender loving care. technical learning and compliance
™	Trademark
TMI	Thickness measurement inspection
TML	Thickness monitoring location
TN	Technical notes
TOC	Table of contents, theory of constraints
TOEFL	Test of English as a foreign language
TOFD	Time-of-flight diffraction, see UT
TOJ	Training on-the-job
ToP	Top of pyramid
TOR	Term(s) of reference
TOW	Time of wetness
TP	Technologue Professionnel
TPCL	Technical package contents list
TPI	Third-party inspector
TQ	Technology quotient
TQM	Total quality management
TR	Technical requisition
TRC	Tradable renewable certificate
TRL	Triple random length, technology readiness level
TROC	The rest of Canada
Ts & Cs	Terms and conditions
TSASK	*Technical Safety Authority of Saskatchewan*
TSBC	*Technical Safety BC* (formerly *BC Safety Authority*)
TSM	Towards Sustainable Mining®

Abbreviation	Meaning
TSSA	*Technical Standards and Safety Authority* (Ontario)
TTR	Time to recover
TTT	Through the trees
TWPV	Thin wall pressure vessel (*thick wall* is not generally abbreviated)
UAV	Unmanned aerial vehicle
UCD	User-centred design
UFR	Unbiased forward rate theory
UG, U/G	Underground
UHK	University of hard knocks
UHP	Ultra high pressure
UHS	Underground hydrogen storage
UI	User interface
UK	United Kingdom
ULC	*Underwriters Laboratories of Canada*
UNDRIP	United Nations Declaration on the Rights of Indigenous Peoples
UNFCCC	United Nations Framework Convention on Climate Change
UNS	Unified numbering system
UPS	Uninterruptable power supply
URL	Uniform resource locator
US	United States of America
USA	United States of America
USD	United States dollars
USGBC	United States Green Building Council
USMCA	*US-Mexico-Canada Agreement*
USSR	Union of Soviet Socialist Republics
UST	Underground storage tank
UT	Ultrasonic test
UV	Ultraviolet
UVP	Unique value proposition
UX	User experience
UXD	User experience design
V1G	Smart charging
V2G	Vehicle-to-grid
VA	Virtual assistant
VBA	Visual Basic for Applications
VBDM	Value-based decision making

Abbreviation	Meaning
VCM	Voluntary carbon market
VDB	Vendor data book
VDC	Virtual design and construction
VDDR	Vendor drawing and data requirements
VDI	Vendor document index
VDR	Vendor document requirements (e.g., PO Section 3)
VFD	Variable frequency drive
VIP	Value improving practice
VLSFO	Very low sulfur fuel oil
VOC	Voice of the customer
VP	Vice president
VPN	Virtual private network
VRE	Variable renewable energy
vRES	Variable renewables
vs.	Versus
VT	Visual test
VUCA	Volatility uncertainty complexity ambiguity
W5H	What why when where who how
WATNA	Worst alternative to a negotiated agreement
WBS	Work breakdown structure
WCS	Western Canadian select
WCSB	Western Canadian Sedimentary Basin
WET	Welding Engineering Technology
WFH	Work from home
WFMT	Wet fluorescent magnetic particle test
WFT	Wet film thickness
WFX	Work from anywhere
WHMIS	Workplace Hazardous Materials Information System
WI	Work instruction
WIIFM	What is in it for me?, What's in it for me?
WIP	Work in progress
WMS	Warehouse management system
WOG	Water oil and gas
WOM	Word of mouth
WOMM	Word of mouth marketing
WOV	Window of vulnerability

Abbreviation	Meaning
WPDS	Welding procedure data sheet
WPS	Welding procedure specification
WQR	Welder qualification record
WRI	*World Resources Institute*
WSIBY	Why should I believe you?
WSIC	Why should I care?
WT	Wall thickness
WTI	West Texas intermediate
WTL	Waste-to-liquids
WWS	Wind water storage
WWTP	Wastewater treatment plant
WWW	World Wide Web
WYSIATI	What you see is all there is
XaaS	Anything as a service, everything as a service
XBX	Ten basic exercises (for women)
XH	Extra heavy, see also XS
XHD	Extra heavy duty
XP	Explosion-proof
XPM	Extreme project management
XRF	X-ray fluorescence, see PMI
XS	Extra strong, see also XH
XTL	Feed-to-liquids
XXH	Double extra heavy, see also XXS
XXS	Double extra strong, see also XXH
ZD	Zero defects
ZEV	Zero-emission vehicle
ZOPA	Zone of possible agreement
μ	Micro
μm	Micrometre
Ω	Ohm

4 Glossary of Common Industry and Project Phrases and Terms

Table 4 defines phrases and terms used in this document, other KT Project documents and industry.

Table 4: Glossary of Common Industry and Project Phrases and Terms

1 A B C D E F G H I J K L M N O P Q R S T U V W X Y Z

Phrase or Term	Definition
1	
1-Way and 2-Way Door Decision-making, One-way and Two-way Door Decision-making	*1-way and 2-way door decision-making* or *one-way and two-way door decision-making* relates to the size of a decision and if the decision is reversible after implementation. A bigger decision may be less reversible and commit more resources than a smaller decision, so it would have increased risk. Regarding reversibility, a: • 1-way door decision is irreversible. After the decision is made, it may not be cancelled; and, • 2-way door decision is reversible (even if it seems irreversible). After the decision is made, the decision may be changed. If a decision: • Is not easily reversible, then it is considered a 1-way door. Thus, fewer resources would be invested or used, or a higher-level approval would be required to protect the downside risk. For example, purchasing a high-capacity multifunction printer; and, • Is reversible, then the resources invested or used would be less restricted, or lower-level approval would be allowed, because there is less downside risk. For example, to leasing or renting a high-capacity multifunction printer. For a business to be innovative and effective, its leaders need to make more 2-way door decisions. See the next examples: 1. Example 1 In 1984, Virgin Atlantic's (Virgin's) approach to business was to either succeed within the first year or exit the market. Virgin negotiated a 2-way door decision with airplane manufacturer Boeing that permitted the one airplane sold to Virgin to be returned to Boeing (within the first year) at no charge if Virgin's business failed within one year. By making a 2-way door decision, Virgin successfully avoided costly risk associated with failing within the first year. Virgin succeeded; thus, it did not need to return the airplane. 2. Example 2 The decision to accept a position with another company is a 1-way door decision because it is unlikely that the previous employer would re-hire the same person. By contrast, accepting a different role within a company could be a 2-way door decision, especially if it is role switching is reversible. See also, *downside* and *second-order effect*.

Phrase or Term	Definition
20/20 Vision	*20/20 vision* denotes vision of normal sharpness. Visual acuity is usually measured with a *Snellen Chart*. Normal vision is 20/20, which means that a person sees with the same sharpness as a person with normal vision sees at the same distance. See also, *hindsight is 20/20* and *Snellen Chart*.
24 Character Strengths	The *24 character strengths* comprise universally-applicable six categories that describe what is best about someone's personality. Everyone possesses these character strengths to different degrees, which gives an individual a distinct *character strengths* profile. The six categories and component character strengths are: • Courage: o Bravery; o Honesty; o Perseverance; and, o Zest. • Humanity: o Kindness; o Love; and, o Social intelligence. • Justice: o Fairness; o leadership; and, o Teamwork. • Temperance: o Forgiveness; o Humility; o Prudence; and, o Self-regulation. • Transcendence o Appreciation of beauty and excellence; o Gratitude; o Hope; o Humor; and, o Spirituality. • Wisdom o Creativity; o Curiosity; o Judgement; o Love of learning; and, o Perspective. This concept is described in the article *The 24 Character Strengths* by the VIA Institute on Character. See also, *character*.

Phrase or Term	Definition
24/7, 24-7, 24x7	Commerce and industry use the phrase *24/7*, *24-7*, or *24x7* (pronounced *twenty-four seven*) to refer to service that is available at all times of the day and every day (24-hours-a-day and 7-days-a-week). Synonyms are *continuous operations*, *nonstop service*, and *round-the-clock service*.
2G, 3G, 4G, 5G	*2G*, *3G*, *4G*, and *5G* are standards for wireless communications: • 3G is about 250 times faster than 2G; • 4G is about 10 times faster than 3G; and, • 5G is about 10 times faster than 4G.
5-D Cycle	The *5-D cycle* is the five distinct phases of *appreciative inquiry*, which are: 1. Definition. Select the right subject; 2. Discovery. Explore the best of what is, identify best practices, peak performance, sources of excellence, and strengths; 3. Dream. Envision the future that is desired, fully engaged and successful around a core purpose and strategic objectives; 4. Design. Leverage the best of what is and a vision for the future, and design high-impact strategies to move creatively and decisively in the right direction; and, 5. Destiny/Deploy/Delivery. Put strategies into action, innovate and improvise fresh solutions. This is also known as the 4-D cycle, without the definition phase. See also, *appreciative inquiry*.
7-38-55 Percent Rule	The *7-38-55 percent rule* states that when one forms an opinion about someone else, it is based on: • 7% on words; • 38% on tone of voice; and, • 55% on body language and facial expressions. This rule was created by American professor Albert Mehrabian.
9 to 5, Nine-to-Five	The expression *9 to 5* or *nine-to-five* describes the normal work schedule at many workplaces (i.e., for daytime office work, where shift work is not required). This phrase also describes normal or routine work that is boring or unfulfilling. For example, I enjoyed the holidays but now I must get back to the 9 to 5. See also, *rut*.

Phrase or Term	Definition
A to B, A:B	Going from *A to B* describes: • A journey in general, without disclosing the route of the journey. For example, determine the fastest method to get from A to B. • The beginning and end of a journey, where A is the starting location and B is the ending location. For example, your A to B is Toronto and New York, respectively. • The most direct route from one location to another. For example, the fastest way to get from A to B is by highway. • Following a logical sequence or direct course of action. For example, this plan will take the company from A to B by boosting production to support increased sales during the busy season. The expression *A:B* is a relational mathematical expression that compares the number of one thing to the number of another. For example, a bowl containing three apples and four oranges has a ratio of 3:4. See also, *A to Z*, *apples to apples comparison*, *apples to oranges comparison*, and *as the crow flies*.
A Feature Not a Bug	*A feature not a bug* is an expression often used by a software developer to convince a user that a flaw in a program is actually a beneficial, deliberately provided, software feature. This expression is used to misdirect and misinform users about a software deficiency.
A to Z	*A to Z* is an expression used to describe: • A full range of services offering all possible options or solutions (e.g., a carpet cleaning company that can clean any carpet at any time); • Something in all aspects, in every way, or that is comprehensive (e.g., a law library with an extremely comprehensive collection of legal articles); or, • The full range of knowledge related to an activity or subject (e.g., arranged it in alphabetical order). A to Z covers the timeframe of beginning to end. See also, *A to B* and *life cycle*.
A/B Testing, Bucket Testing, Split Testing, Split-run Testing	*A/B testing*, *bucket testing*, *split testing*, or *split-run testing* is a research method that uses two variants: A and B. Two versions of a single variable are compared, typically by testing a subject's response to variant A against variant B, to determine which is more effective or preferred. For example, A/B testing is used for website optimization to determine which variation of a webpage element improves conversion rates the most. Because of the high frequency of testing in the marketing industry, marketers joke that A/B testing means *always be testing*. See also, *conversion rate*.
A3 Problem Solving	*A3 problem solving* is a structured problem-solving and continuous-improvement approach first employed at Toyota and typically used by lean manufacturing practitioners. It provides a simple and strict procedure that guides problem solving performed by personnel. The approach typically uses a single sheet of ISO A3-size paper, which forms its namesake. See also, *lean manufacturing*.

Phrase or Term	Definition
Abbreviation	An *abbreviation* is a shortened form of a written phrase or word. For example, *qty* is an abbreviation for *quantity*. See also, *abbreviation rule, acronym, initialism, and mnemonic.*
Abbreviation Rule	The *abbreviation rule* is a writing convention or editing rule that dictates how abbreviations (acronyms and initialisms) are managed in a document. The abbreviation rule typically states: 1. Spell out the first instance of an abbreviation in a document, followed by the associated parenthetical abbreviation. Example: The variable frequency drive (VFD) was installed in 2019. 2. Thereafter, use the non-parenthetical abbreviation. Example: The VFD will be inspected annually. See also: • https://en.wikipedia.org/wiki/Abbreviation • https://www.btb.termiumplus.gc.ca/tcdnstyl-chap?lang=eng&lettr=chapsect1&info0=1 See also, *abbreviation*.
Abductive Logic, Abductive Reasoning	*Abductive logic* or *abductive reasoning* does not involve deductive-inductive logic or reasoning. Instead, it uses incomplete observations, educated guesses, or recommendations to generate conclusions. For example, doctors use abductive logic when providing a diagnosis based on patient test results. Another example, jurors use abductive logic to reach a verdict based on presented evidence (because they consider only an incomplete set of observations). Abductive logic makes and tests hypotheses using the best existing information available at the time. This approach often entails making an educated guess after observing a phenomenon for which there is no logical or obvious explanation. See also, *deductive logic*, *inductive logic*, and *scientific method*.
Aboveground Storage Tank	An *aboveground storage tank* stores product above grade (e.g., chemicals, fertilizer, fuel, oil, or water). See also, *above grade*, *underground storage tank*, and *tank*.
Aboveground, Above Grade	To be *aboveground* or *above grade* is to be situated above the ground. See also, *underground*.
Abrasive	An *abrasive* is a material used to finish or shape materials and products by abrading or rubbing. This material is used as a media for buffing, blasting, cleaning, grinding, honing, polishing, preparing, and sharpening operations. See also, *abrasive blasting*.
Abrasive Blasting	*Abrasive blasting* is a method used to clean or prepare a substrate or surface for cleaning, coating, or other use. This method uses high pressure compressed air or water, or mechanical force, to propel a stream of abrasive material against a surface to: • Remove surface coating or contaminants; • Roughen a smooth surface; • Smooth a rough surface; or, • Shape a surface. Types of blasting abrasive media are numerous and varied, and include dry ice, glass or metal beads, metallic grit or slag, silica (sand), and walnut shells. See also, *abrasive*, *coating* and *substrate*.

Phrase or Term	Definition
Absolute Majority	An *absolute majority* is a voting requirement to obtain more than half of the votes of: • Qualified voters; or, • Votes cast. See also, *first-past-the-post*, *simple majority*, and *supermajority*.
Abundance Mindset	An *abundance mindset* is a way of thinking that involves a deep inner sense of personal security and worth, and the belief that there is more than enough (e.g., health, intelligence, judgment, knowledge, money, relationships, time, resources, and willpower) for everyone. This helps relieve anxiety and fear in uncertain times. Abundance affects feelings and thinking by orienting the mind automatically towards adequacy, that there is enough for everyone, and there always will be enough for everyone. This results in a focus on long-term and positive feelings towards others. It also results in the sharing of decision-making, prestige, profits, recognition, and opens alternatives, creativity, options, and possibilities. Seven signs of an abundance mindset are believing: 1. Big things are possible (e.g., challenging goals); 2. There are plenty of resources in the world; 3. In optimism and being genuinely happy for the success of others; 4. Change is an integral part of life that leads to positive outcomes; 5. In taking a proactive approach to life; 6. In growth and learning to endlessly acquire knowledge and develop skills; and, 7. In the limitless possibilities in the world with a focus is on *what is working*. This mindset sees life with an abundance of resources. If life is a pie, there is enough for everyone to enjoy a piece. Further, if someone needs more pie, the solution is to create a bigger pie. With this mindset, life is an ocean whose abundance is nearly infinite. See also, *goal*, *scarcity mindset*, and *thinking mode*.
Academic	An *academic* is: • A member of an institution of learning (e.g., a professor at a university); or, • Someone who is learned and knowledgeable but inexperienced in practical matters (i.e., book smart). To be *academic* means to be: • Associated with, of, or relating to an academy or school, especially one of higher learning (e.g., a university); • Of or relating to artistic or literary rather than professional or technical studies; or, • Theoretical or speculative, with no practical application or useful significance (e.g., the design of a solar panel that orbits the earth and beams energy to earth). See also, *academy*, *book smart*, *self-taught* and *technicality*.
Academy	An *academy* is: • A place of study or training in a specialized field, skill, or subject (e.g., college, high school, or university); or, • An institution or society of academics or scholars that aims to maintain and promote standards in a specialized field, skill, or subject (e.g., art, literature, or science). See also, *academic*.

Phrase or Term	Definition
Accelerator	An *accelerator* is a dedicated multifunctional team that works with a portfolio of digital solutions to organize scaling issues, promote change management, determine deployment pathways, and measure value. An accelerator may be corporate, customer, or internal. See also, *agile* and *cluster*.
Accelerometer	An *accelerometer* is an instrument used for measuring acceleration, vibration, or impact, and recording transportation data. It has many uses, including monitoring shipments of sensitive packages and containers (e.g., by airplane, ship, train, or truck).
Accept As-Is	*Accept as-is* is a disposition that indicates no corrective action is necessary. This may be a deviation from the original specifications; hence, additional documentation may be required by the project for acceptance. See also, *use as-is*.
Access	*Access* is Microsoft database management system software that combines the Microsoft Jet Database Engine with a graphical user interface and software development tools. This software is part of the Microsoft Office suite of software. It is designed to import-export, link directly to, and manipulate data stored within it, and within other compatible applications and databases. It houses the macro programming language Visual Basic for Applications
Accountability	*Accountability* is the state of being responsible, answerable, or liable. In the context of a project, accountability may be assigned to an individual (e.g., director or manager) or group (e.g., company or team) for an activity, decision, or task. See also, *delegation* and *responsibility*.
Accountant	An *accountant* is responsible to analyze, audit, and archive financial matters and accounts. To be officially recognized, many of these professionals need certification by a professional association, such as (in Canada) the *Society of Management Accountants of Canada* and *Certified General Accountants of Canada*. See also, *banker* and *financial officer*.
Accusation Audit	An *accusation audit* is a list of the worst things that a negotiation counterpart could potentially say. This prepares a negotiator to address or deflect negative dynamics early and forcefully. Because accusations may sound exaggerated when spoken, speaking about them encourages counterparts to claim the opposite is true. This concept was developed by Chris Voss in the book *Never Split The Difference: Negotiating As If Your Life Depended On It* (with Tahl Raz). See also, *label*.

Phrase or Term	Definition
Ackerman Model	The *Ackerman Model* is a counter-offer method and haggling system for negotiations that was developed by Mike Ackerman. It has a predictable result for the negotiators to meet in the middle using an extreme anchor, loss aversion, and reciprocity in these simple steps: 1. Set the target price (i.e., the goal); 2. Set the first offer at 65% of the target price; 3. Calculate the three raises of decreased increments to 85, then 95, and finally 100% of the target price; and, 4. Use empathy and say no indirectly to the counterpart offers before any increase is offered. When calculating the final offer, use a precise non-round number to give it more credibility and weight (e.g., CAD$ 5,749). With the final offer, include also a non-monetary item – even if the counterpart does not want it (e.g., offer cash) to infer that this is the final offer. See also, *extreme anchor*, *loss aversion*, and *reciprocity*.
Acid Test	An *acid test* has many meanings: • Generally, it is a conclusive or severe test of the success or true value of something (e.g., a person's character or the performance of a product). • In accounting, it is a comparison of a company's most short-term assets to its most short-term liabilities to determine if it has enough cash to pay its immediate liabilities (e.g., short-term debt). This test disregards current assets that are difficult to quickly liquidate (e.g., inventory). This is also called the *quick ratio test*. • In metallurgy, it is a qualitative chemical or metallurgical analysis that uses acid. This test was developed in the late 18th century when it was discovered that nitric acid dissolves other metals more readily than gold.
Acronym	An *acronym* is a word formed from the initial letter or letters of each of the successive parts or major parts of a compound term. For example, OPEC is an acronym for Organization of the Petroleum Exporting Countries. Generally, all acronyms may also be called abbreviations. This document uses this convention. See also, *abbreviation*, *initialism, and mnemonic*.
Action Item	An *action item list* is an issue, concern, deficiency, or non-conformance on an action item list. See also, *action item list* and *punch list*.
Action Item List	An *action item list* is an inspection form issued with the inspection report that lists all issues, concerns, deficiencies, and non-conformances that require action (e.g., incomplete, incorrect, or indeterminate materials or work). The form is an alternative to listing issues in the inspection report itself for improved tracking. See also, *action item* and *punch list*.

Phrase or Term	Definition
Active Learning	*Active learning* is an instruction method that involves engaging students with the course material through: • Case-based learning; • Discussion or debate; • Peer teaching; • Problem solving; • Role playing; • Simulations; or, • Teamwork. Active learning places a greater degree of responsibility on the learner than passive approaches. This method of learning is student-centred. It better enables students to apply and transfer knowledge by engaging the three learning domains of attitudes: 1. Affective, cognitive, and psychomotor 2. Knowledge; and, 3. Skills. It engages higher-order thinking tasks such as analysis, evaluation, and synthesis. Learning activities may range from a few minutes, and to one or multiple class sessions. See also, *affective*, *Bloom's Taxonomy*, *cognitive*, *cone of learning*, *knowledge transfer*, *metacognative thinking*, *passive learning*, *psychomotor*, and *unlearning*.
Active Listening	*Active listening* requires that the listener fully concentrate, understand, and remember what a speaker says, and then respond thoughtfully. It is a highly valued interpersonal communication skill that ensures engagement and later recall of specific details without the need for information to be repeated. Active listening uses non-verbal and verbal techniques to keep attention on the speaker. This supports the ability to focus and helps ensure that the speaker can see the listener is engaged. The listener is not mentally rehearsing or thinking about what they will say when the speaker is done speaking. See also, *dialogue of the deaf*, *reflective questioning*, and *tone-deaf*.
Activism	*Activism* is a doctrine or practice that emphasizes direct and observable action in support of or opposition to one side of a controversial issue. It may comprise efforts to direct, impede, intervene in, or promote economic, environmental, political, or social reforms that are perceived to benefit society. See also, *activist*, *hacktivism*, and *social justice*.
Activist	An *activist* is someone who performs activism. See also, *activism*, *hacktivist*, and *social justice warrior*.
Activity	An *activity* is a specific action or function.
Actuated Valve	An *actuated valve* has a positioning device, known as an actuator, that causes it to operate automatically or remotely (e.g., for opening and closing). See also, *actuator*, *control valve*, and *manual valve*.

Phrase or Term	Definition
Actuator	An *actuator* is a positioning device that causes a machine or other device to operate automatically or remotely (e.g., for opening and closing a valve). Actuators can act as shutdown mechanisms in case of an emergency, where human interaction can be dangerous. Digital or *smart* positioning devices use a microprocessor to position an actuator while also monitoring and recording data. See also, *actuated valve* and *fail position*.
Ad Hoc Activity	An *ad hoc activity* is an unplanned event happens only for a particular purpose or when needed (e.g., an ad hoc meeting).
Ad Hominem Argument	An *ad hominem argument* is an argumentative device that deliberately ignores the convention of arguing against or rebutting the ideas an opponent espouses – and instead attacks the opponent's character. *Ad hominem argument* is Latin for *to the man* or *to the person*. Five reasons to avoid an ad hominem argument: 1. It can get a debater in trouble when words used to attack others may later become unacceptable in public discourse. For example, *Eskimo* once described indigenous Inuit people. The term *Eskimo* is now considered pejorative (and even racist) and thus is avoided; 2. It destroys civil discourse because using inflammatory or negative rhetoric has a corrosive effect on society; 3. It ends fruitful discussion of ideas because a healthy exchange of ideas requires listening and the respectful exchange ideas, even when arguments are challenged and when disagreement occurs; 4. It is cruel because words that attack someone's personal traits (e.g., accent, heritage, appearance, or weight) are hurtful; and, 5. It makes a debater look uneducated or uninformed because they cannot counter their opponent using intelligence and logic. Christopher Hitchens is an English-American intellectual and author who stated: *it is a sign of victory whenever an opponent moves to the use of an ad hominem argument*. This is because when a debater attacks an opponent's character it means that the debator has lost the argument: the debator is unable to use intelligence and logic to defeat the opponent's argument, and instead must resort to less rational approaches (such as character assassination). See also, *Hitchens' razor*, *Indigenous Peoples*, and *rhetoric*.
Adaptability Quotient	The *adaptability quotient* is the ability to overcome challenges by making a conscious effort to change, discarding obsolete knowledge, and understanding what is most relevant.
Adaptive Project Framework	*Adaptive project framework* is an approach to project management that rejects traditional, linear project management and instead accepts changing requirements and allows projects to be affected by external business environments. This approach stresses flexibility and focuses on evaluating and performing project work in stages to allow room for re-planning due to changing business goals, objectives, and requirements.
Addendum, Appendix	An *addendum* or *appendix* is ancillary content or a supplement that accompanies and supports a main or central document or volume of documents. For example, see the appendices in this document.

Phrase or Term	Definition
Additive Manufacturing	*Additive manufacturing* is the industrial production name for 3D printing. It describes the technologies used build precise geometric shapes by adding layer-upon-layer of a layering material (e.g., concrete, liquid, metal, plastic, or powder). The components commonly used include a computer, 3D modelling software (i.e., computer-aided design), and machine equipment. Once a design is produced, the equipment reads data from the design file to lay down or add successive layers, in a layer-upon-layer fashion, to fabricate a 3D object. The term encompasses many technologies including subsets such as additive fabrication, direct digital manufacturing, layered manufacturing, and rapid prototyping. See also, *digital project delivery*.
Administrative Theory	*Administrative theory* looks at an organization from the top down, and sets out steps for managers to acquire the best from personnel and run business efficiently. This concept was developed in the early 19th century by French engineer Henri Fayo, which later became known as administrative theory. Then, most organizations promoted the best technical personnel. He recognized, however, the skills that made technical personnel produce at their jobs would not necessarily make them good managers. See also, *functions of management*, *management*, and *principles of management*.
Advanced Non-destructive Examination	Advanced *non-destructive examination* is a test that examines an object (i.e., a structure, material, or component) without damaging or destroying the object to ensure object (product) reliability and quality. This test uses automation and instrumentation technology, software, and data analysis and digital storage. See also, *conventional non-destructive examination* and *non-destructive examination*.
Adversarial Collaboration	*Adversarial collaboration* is an experiment conducted by two groups of experimenters with competing hypotheses. The aim is to construct and implement an experimental design that contains no obvious biases or weaknesses (as determined by both groups).
Advertising	*Advertising* is communication that employs an openly-sponsored, non-personal message to promote or sell an idea, product, or service to existing and future customers or clients. See also, *attention interest desire and action principle*, *brand*, *copyrighting*, *marketing*, *market share*, *mindshare*, *native advertising*, *positioning*, and *sponsored content*.
Affective	*Affective* is something caused by, influenced by, or related to attitudes, emotions, feelings, or moods. Affective is an adjective formed from the noun *affect*. For example, lack of exposure to sunlight during the day contributes to seasonal affective disorder. See also, *effective* and *seasonal affective disorder*.
Aftermarket Product	An *aftermarket product* is a consumable that enhances or provides additional care for a main product. For example, a spoiler added to a vehicle after the vehicle is purchased is an aftermarket product. Aftermarket products include accessories, components, and spare parts, especially for vehicles. See also, *customer perceived value*.
Ageism	*Ageism* is antagonism, discrimination, prejudice, or stereotyping based on a person's age (e.g., the belief that a certain age is intrinsically inferior or superior to another). Age is to ageism as race is to racism. See also, *Black Indigenous and People of Colour*, *discrimination*, *intersectionality*, *overqualified*, *racism*, *racist*, and *sexism*.

Phrase or Term	Definition
Agent	An *agent* is: • An active force or substance producing an effect; someone or something that makes something happen. • A company or person who legally acts or does business on behalf of another company, group, or person (e.g., acting as an agent); or, • A representative of a government agency. See also, *engineering and procurement*.
Agglomerate	An *agglomerate* is a particulate material comprised of large particles formed by the joining or binding together of primary or smaller particles and whose original identity can still be visible in the final agglomerate form (e.g., cement). See also, *flocculant*.
Agile, Agile Method	*Agile* or the *agile method* is a software development technique involving iterative development where requirements and solutions evolve through collaboration between multifunctional teams. See also, *accelerator* and *scrum*.
Agnostic	An *agnostic* is someone who: • Believes that anything beyond material phenomena, such as an ultimate reality (e.g., a God), is unknown and probably unknowable; • Has neither faith nor disbelief in God; or, • Is unwilling to commit to an opinion or position. See also, *apolitical* and *political*.
Aha!, Aha Moment, Eureka Effect, Eureka Moment	An *aha!*, *aha moment*, *eureka effect*, or *eureka moment* is an instance of sudden comprehension, insight, inspiration, realization, or recognition. Etymology indicates the source of this expression is from the story of the ancient Greek polymath Archimedes, who observed water displacement when his body sank into his bathtub. His eureka moment came when he realized the volume of water displaced equalled the volume of his body immersed in the water. Upon this realization, Archimedes allegedly shouted εὕρηκα (*eureka*, *I have found it*!). This discovery made it possible to accurately measure the volume of irregularly-shaped objects.
Air Bridge, Engineered Floating Mat System	An *air bridge* or *engineered floating mat system* is crossing bridge typically constructed from hardwood that is used to move heavy construction or transportation equipment across wetlands, over underground pipelines, or both. This crossing bridge disperses loads over a large area and protects underground pipelines and the environment from damage. See also, *hardwood* and *wetlands*.
Air Cooled Heat Exchanger	An *air-cooled heat exchanger* is typically used where a process system generates heat that must be removed, for which there is no local use (e.g., waste heat). Air is used as the cooling medium because either water is not available or not the desired process. A fan driven by an electric motor moves air across the finned tubes that are used to increase the heat transfer surface. See also, *heat exchanger*, *finned tube*, *shell and tube heat exchanger*, and *tube*.
Alberta Boilers Safety Association	The *Alberta Boilers Safety Association* is the pressure equipment safety authority, authorized by the Alberta Government for the administration and delivery of safety programs related to boilers, pressure vessels, and pressure piping systems in Alberta (formerly the *Boilers Branch*).

Phrase or Term	Definition
Alberta's Industrial Heartland, Heartland, Upgrader Alley, Refinery Row	*Alberta's Industrial Heartland, Heartland,* or *Upgrader Alley* is the largest industrial area in Western Canada and one of the largest petrochemical processing regions in Canada. It is a joint land-use development and planning initiative between five municipalities in the Edmonton Capital Region. It comprises over 40 petrochemical companies who are active in chemical, oil, and gas, and petrochemical industries. Refinery Row is the concentration of oil refineries east of the city of Edmonton, Alberta. The two main refineries are: • The Strathcona Refinery (owned by Imperial Oil Limited); and, • The Suncor Edmonton Refinery (owned by Suncor Energy Inc.). Other main refineries in the Edmonton area are concentrated near Scotford, Alberta. See also, *petrochemical*.
Aleatory Uncertainty, Statistical Uncertainty	*Aleatory uncertainty* or *statistical uncertainty* is uncertainty generated from a random process. It is representative of unknowns that differ each time the same experiment is performed. For example, flipping a coin generates aleatory uncertainty in the form of the uncertainty about if the coin will land heads or tails side up. Aleatory uncertainty involves uncertainty observed as part of a natural process (e.g., the falling and spinning of an object towards earth as dictated by gravity and other natural influences, such as wind). See also, *epistemic uncertainty*.
Algorithm	An *algorithm* is a set of rules or protocols used to solve a problem in a finite number of steps.
Algorithm	An *algorithm* is a procedure or set of rules used to accomplish an end or solve a problem (e.g., perform automated reasoning, calculations, or data processing). Algorithms are essential to how computers process data. For example, a search engine such as Bing, Google, or Yahoo has a proprietary computation that ranks websites for keywords or keyword combinations. Algorithms have been used by mathematicians since ancient times. These methods were documented and formalized by several mathematicians like Alan Turing in the early 20th century. See also, *search engine* and *search engine optimization*.
Allophone	An *allophone* is someone whose first language is neither English nor French. Native speakers of *Indigenous Peoples* languages are not considered allophone. See also, *Anglophone*, *Francophone*, and *Indigenous Peoples*.
Alloy	An *alloy* is metal made by combining two or more metallic elements to improve its properties (e.g., toughness, strength, or corrosion resistance). For example, steel is an alloy comprised of iron, carbon, and many other elements. Stainless steel is a steel alloy comprising a minimum of 10.5% chromium. See also, *base metal*, *carbon steel*, *metal*, *mineral*, *stainless steel*, and *steel*.
Altcoin	*Altcoin* refers to all cryptocurrencies other than Bitcoin. Altcoin typically targets a perceived limitation of Bitcoin. As of 2020, it was estimated that more than 5,000 cryptocurrencies exist. Altcoin is a portmanteau of the words alternate and coin. See also, *bitcoin*, *cryptocurrency*, and *legal tender*.
Alteration	An *alteration* is any change to pressure equipment as described in the original manufacturer's data report, which also requires a change to design calculations or otherwise affects its pressure-containing capability.

<table>
<tr><th>Phrase or Term</th><th>Definition</th></tr>
<tr><td>Alternative Dispute Resolution</td><td>Alternative dispute resolution is a process that uses arbitration, mediation, or both to expedite resolution, minimize conflict, and reduce costs between two or more parties. When successful, this approach bypasses traditional legal and courts – which are time consuming, wasteful, expensive, and often less effective. This approach results in a binding or non-binding resolution:<ul><li>A binding resolution is all parties are legally committed to specific objectives, instead of relying on a handshake or vague verbal agreement.</li><li>A non-binding resolution is when all parties agree to objectives but the resolution to achieve them may not be enforced through legal means.</li></ul>Most resolutions involve two or multiple parties:<ul><li>A two-party resolution model is when two parties collaborate and compromise directly to acquire a common objective. Failure of this approach may lead to a multiple-party resolution process.</li><li>A multiple-party resolution model is when three or more parties enlist the aid of an impartial third-party mediator, in a formal or informal capacity, to obtain a common resolution. The parties may also be accompanied by other professionals (e.g., accountants and lawyers) to assist with a resolution.</li></ul>See also, arbitration and judicial dispute resolution.</td></tr>
<tr><td>Alternative Media, Niche Media</td><td>Alternative media or niche media refers collectively to media sources that differ from dominant or established types of media (i.e., mainstream media) in terms of their content, distribution, or production. Their audience is typically smaller and specialized. For example, South Front, Signs of the Times, and Zero Hedge. See also, mainstream media and the press.</td></tr>
<tr><td>Alternative Safety Approach</td><td>An alternative safety approach provides a variance to existing codes, regulations, or standards by approving a safety management plan or system that details alternative methods to achieve an equivalent or superior level of safety.</td></tr>
<tr><td>American Petroleum Institute Gravity, API Gravity</td><td>American Petroleum Institute gravity or API gravity is a measure of how heavy or light a petroleum liquid is compared to water. If its API gravity is:<ul><li>Greater than 10, it is lighter than, and floats on, water; and,</li><li>Less than 10, it is heavier than, and sinks in, water.</li></ul></td></tr>
<tr><td>Analog</td><td>Analog is being, of, or relating to, a device or mechanism in which information is represented by continuously variable physical quantities. For example, a watch that displays the time using non-digital hands or pointers. The opposite of analog is digital. See also, digital.</td></tr>
<tr><td>Analysis</td><td>Analysis is a mechanical and reductionist worldview with the dissection of complexity into manageable components or broken down into smaller parts. See also, synthesis.</td></tr>
</table>

<table>
<tr><th>Phrase or Term</th><th>Definition</th></tr>
<tr><td>Analysis Paralysis</td><td>Analysis paralysis is an expression that describes the stoppage (paralysis) of decision-making that occurs when a challenge or issue is over-analyzed. Because no decision is made, no action is taken; thus, no solution is reached. This may occur when a:<ul><li>Very complicated challenge needs resolution but personnel resist making a decision because the complex nature of the challenge may cause unforeseen, unpredicted and potentially costly or unwanted results; or,</li><li>Superior solution is available, but the analysis of the solution is an ongoing pursuit that yields diminishing returns.</li></ul>Over-analyzing a challenge can be just as deleterious to a project as under-analyzing a challenge. See also, diminishing returns and perfectionism.</td></tr>
<tr><td>Anchoring Effect</td><td>The anchoring effect is a cognitive bias that describes the tendency to rely heavily on the first piece of information received (i.e., the anchor) when making a decision. This means that the first information forms a baseline for all subsequent judgements (e.g., for an estimate or offer). It establishes a reference point for a negotiation. For example, the sticker price for a new vehicle sets the standard for subsequent negotiations. Thus, a price lower than the sticker price may seem more reasonable, even if it is still higher than the vehicle's worth. See also, extreme anchor.</td></tr>
<tr><td>Ancillary Equipment, Auxiliary Equipment, Line Replacement Equipment, Support Equipment</td><td>Ancillary equipment, auxiliary equipment, line replacement equipment, and support equipment describe:<ul><li>Items that are purchased as part of a package from a supplier that will be assembled and operated as a complete unit (e.g., a pump with motor, gearbox, lube oil console, or control system);</li><li>Items that are part of a vital system that do not produce marketable product but support equipment that does; or,</li><li>Machines and other technical items that complement main equipment, creating a complete system.</li></ul>Ancillary, auxiliary, or other support equipment are usually supplemental (e.g., do not serve a critical function or are somewhat unimportant).</td></tr>
<tr><td>Anecdote</td><td>An anecdote is a short narrative of an amusing, biographical, or interesting incident. For example, before beginning a tutoring session, a tutor tells her student how she struggled with the subject matter in the past and how she managed to grow past these difficulties. See also, anecdotal and story.</td></tr>
<tr><td>Anecdotal</td><td>Anecdotal is information, an account, or a depiction of something that is:<ul><li>Based on or consisting of an observation or report of casual (e.g., subject to, occurring by, or resulting from chance) or unscientific observation; or,</li><li>Not necessarily true or reliable. Anecdotal information is based on personal accounts rather than careful study, verifiable facts, or research.</li></ul>For example, a successful person tells a story about how they overcame a challenge. The story can be interesting and instructive, but it is anecdotal, because their method may or may not apply to the success of others. See also, anecdote, empirical, logical, and paradigm.</td></tr>
<tr><td>Anglophone</td><td>An Anglophone is someone whose first language is English. See also, allophone and Francophone.</td></tr>
</table>

Phrase or Term	Definition
Anode	An *anode* is the electrode through which electrical current flows into the device. See also, *anodic protection*, *cathode*, and *sacrificial anode*.
Anodic Protection	*Anodic protection* is a technique used to control the corrosion of a metal surface by making it the anode and controlling the electrode potential in a zone where the metal is passive. It is used where cathodic protection is not suitable due to very high current requirements. See also, *anode*, *cathode*, and *cathodic protection*.
Anti-dumping duty	*Anti-dumping duties* are applied to imported products when it is believed they are being sold at a significantly lower cost compared to the supplying country's market price, or even below the cost of production. This *dumping* of products poses a significant threat to the economic competitiveness of domestic producers of these products. See also, *countervailing duties*.
Anti-strategic Lawsuit Against Public Participation Law, Anti-SLAPP Law, Strategic Lawsuit Against Public Participation-back Law, SLAPP-Back Law	An *anti-strategic lawsuit against public participation law, Anti-SLAPP Law, strategic lawsuit against public participation-back law*, or *SLAPP-back Law* allows a defendant to file a motion to dismiss or strike a case, or both, because it involves protected speech on a matter of public concern. The plaintiff then has the burden to show a probability that they will prevail. If the plaintiff fails to do so, the plaintiff's claim is rejected and they may be required to pay a penalty (e.g., the defendant's costs) for filing the case. See also, *lawfare* and *strategic lawsuit against public participation*.
Antithesis	An *antithesis* is a: • Figure of speech in which ideas are contrasted to express a parallelism of words that are the opposed to each other. For example, from Neil Armstrong when he stepped onto the surface of the moon in 1969: *that's one small step for (a) man, one giant leap for mankind*; or, • Person or thing that is the direct opposite of someone or something else. For example, preferring to wake early but having a partner that prefers to sleep late. See also, *letter of the law* and *spirit of the law*.
Antitrust Laws and Regulations	*Antitrust laws and regulations* are rules that discourage monopolies and prohibit practices that restrain business. These rules: • Permit small companies to enter a market and operate; • Preserve competition; and, • Protect consumers. See also, *monopoly*.
A-Number	An *A-Number* identifies the deposited weld metal chemical composition of base metal. This number is assigned by ASME *Boiler and Pressure Vessel Code*. See also, *P-Number*.
Anything as a Service, Everything as a Service, X as a Service	*Anything as a service*, everything as a service, and *X as a service* refers to the applications (e.g., products, tools, and technologies) and services that are available on demand over the Internet via cloud computing and remote access (i.e., instead of being provided or used locally or on-site). This means that virtually anything can be purchased on an *as-needed* basis. See also, *cloud computing, content as a service, infrastructure as a service, platform as a service, software as a service*, and *security as a service*.

<table>
<tr><th>Phrase or Term</th><th>Definition</th></tr>
<tr><td>Apolitical</td><td>Someone who is apolitical:
• Has an aversion to political matters;
• Is not interested or involved in political matters; or,
• Is neutral or unbiased with regard to political matters.
Something that is apolitical has no political interest or significance (e.g., the Prime Minister vacationed in the Rocky Mountains — it was an apolitical trip). See also, agnostic, apoliticism, and political.</td></tr>
<tr><td>Apoliticism</td><td>Apoliticism is antipathy or apathy toward a political affiliation. See also, apolitical.</td></tr>
<tr><td>Apples to Apples Comparison</td><td>The expression apples to apples comparison means that two things that are similar are being compared. It implies that items can reasonably be compared. For example, comparing the effects of wind gusts on skyscrapers versus tall buildings. An apples to apples comparison is logical and useful. See also, A to B and apples to oranges comparison.</td></tr>
<tr><td>Apples to Elephants Comparison</td><td>The expression apples to elephants comparison means that two things that are absurdly dissimilar are being compared. It implies that items cannot reasonably be compared. For example, comparing the effects of wind gusts on horses versus skyscrapers. An apples to elephants comparison is illogical and not useful. See also, A to B, apples to apples comparison, apples to oranges comparison, and false equivalence.</td></tr>
<tr><td>Apples to Oceans Comparison</td><td>The expression apples to oceans comparison is synonymous with apples to elephants comparison. See also, apples to oranges comparison.</td></tr>
<tr><td>Apples to Oranges Comparison</td><td>An apples to oranges comparison describes comparing things that may have similarities (e.g., apples and oranges are fruit) but are not alike (e.g., a citrus fruit grown in a tropical climate and a pome fruit grown in a semi-arid climate). Hence, this expression implies that, even though items may appear to be similar, they cannot be reasonably compared (e.g., being unable to decide between buying a condominium or a house because the benefits and features are so different). An apples to oranges comparison is illogical and not useful; whereas, and apples to apples comparison is useful. See also, A to B, apples to apples comparison, apples to elephants comparison, and false equivalence.</td></tr>
<tr><td>Applicant Tracking System</td><td>An applicant tracking system (system) is a human resources software application that enables electronic management of recruitment – from receiving resumes to hiring personnel. The system may automatically filter applicants' resumes based on criteria such as certifications, education, former employers, keywords, skills, and years of experience. The advent of applicant tracking systems has prompted job candidates to use resume optimization techniques when writing resumes to exploit systems' search engines. For example, an applicant may incorporate keywords in a career or job advertisement into their resume. Alternatively, an applicant may not want to re-write their resume to accommodate keywords. Instead, they may include keywords at the end of the resume using white font (on a white background) to make them invisible. This random keyword verbiage will not be noticed by a recruiter but will be found by the search feature of the applicant tracking system. The system may then identify the resume-applicant as suitable for further consideration (e.g., to be interviewed or hired).</td></tr>
</table>

Phrase or Term	Definition
Application Programming Interface	An *application programming interface* is a set of re-useable software (e.g., protocols, routines, and tools) that allow different systems to interact with each other. For example, Microsoft has a core set of application programming interfaces available in Windows operating systems, which are informally known as WinAPI. One of these is Win32. Another example is the Apple application programming interface that is used to detect touchscreen interactions.
Applied Engineer	An *applied engineer* is a professional who is trained in a branch or discipline of applied engineering. For example, an applied engineer could be involved with the application, design, and technical components related to the development of a new product. For example, an engineer that develops the design of an improved hydrogen fuel cell. See also, *applied engineering*, *engineer*, *engineering technologist*, and *fuel cell*.
Applied Engineering	*Applied engineering* is a professional discipline in which trained personnel apply the mathematical and scientific principles of engineering to the design and management of systems, develop new product designs, improve manufacturing processes, and manage physical or technical functions of an organization (e.g., using basic engineering principles, industrial processes, project management, production and operations management, quality control, statistics, and systems integration and control). Trained personnel in this discipline may be called an *applied engineer* or *engineering technologist*. See also, *applied engineer*, *engineering*, and *engineering technologist*.
Appreciation	In finances, *appreciation* is an increase in the value of an asset over time. In relationships, appreciation is gratitude. See also, *investment*.
Appreciative Inquiry	*Appreciative inquiry* is an approach to organizational management that emphasizes working from strengths to find new directions for growth, rather than focusing on problems to be solved or weaknesses to be improved. It seeks to engage stakeholders in self-determined change. It advocates collective inquiry into the best of what is, to imagine what could be, followed by collective design of a desired future state that is compelling and thus does not require the use of incentives, coercion, or persuasion for planned change to occur. For example, how does an artist paint a landscape? Does an artist complain that a cloud is too grey, a rock too small, or a tree too crooked? No! An artist will instead focus on beautiful and positive aspects to create an artistic work, the result of which may or not closely resemble the actual landscape. Some argue that excessive focus on dysfunctions or problem-solving can worsen problems or even fail to resolve the issue at all. Answers or solutions are shaped by the questions people ask. Appreciative inquiry, therefore, attempts to use ways of asking questions and envisioning the future to foster positive relationships and build on the present potential of a given organization, person, or situation. See also, *5-D Cycle*, *management*, and *positive change*.
Apprentice	An *apprentice* is a person who works for another to learn a trade. In most industries, an apprentice will enroll in an apprenticeship program. Upon successful completion, an apprentice achieves journeyman or other professional certification. See also, *apprenticeship*, *trades person*, and *journeyman*.
Apprenticeship	An *apprenticeship* is a system of training trades personnel with on-the-job training and study (classroom and laboratory work). See also, *skilled trade*, *trades person*, and *unskilled*.

Phrase or Term	Definition
Approved Manufacturer's List	Not a preferred phrase – see *approved supplier list*.
Approved Supplier List	An *approved supplier list* is a list of suppliers that are approved by an owner, engineering and procurement company, or project, for the procurement of materials, equipment, and services.
Appurtenance	An *appurtenance* is a device (fitting, trim, or mounting) attached to a: • Boiler (e.g., safety valve, pressure gauge, gauge glass, blowdown valve, or stop valve); • Pressure vessel (e.g., valve, gauge, regulating or controlling device, flange, elbow, reducer, or tee); or, • Pipeline (e.g., corrosion control device, isolating valve, or weighted section).
Architecturally-exposed Structural Steel	*Architecturally-exposed structural steel* is steel that supports the primary structural requirements of a building (including canopies or other structures) and is also exposed to view (and therefore forms a significant part of the architectural design of the building). The design, detailing and finishing requirements of this kind of steel typically exceed that of standard structural steel, which is normally concealed by other finishing materials (e.g., drywall). See also, *drywall*.
Archrival	An *archrival* is a chief rival or opponent. Archrivals are two organizations (e.g., businesses, persons, or teams) that are both competitors and leaders in their field (e.g., Coca-Cola and Pepsi). See also, *competitor*, *frenemy*, *nemesis*, and *rival*.
Armchair Critic, Armchair Quarterback, Armchair General	An *armchair critic*, *armchair quarterback*, or *armchair general* is a someone who: • Offers comment or criticism on a subject of which they have little or no hands-on experience or knowledge; or, • Gives advice based on theory rather than practice. An *armchair critic* typically opines about a topic (e.g., a baseball game or the MBA) from the comfort of a chair, and has no responsibility for, or authority in, the outcome. Armchair critic and armchair quarterback are frequently used in sports. Armchair general is frequently used in the military. See also, *back seat driver*, *book smart*, *hands-on*, *hindsight is 20/20*, and *second guess*.
Artificial Construct	*Artificial construct* refers to something that is not physically real, at least not in terms of the cause or reason for the construct, but which has a meaning that can be seen in the mind. For example, the idea of a work-life balance is a modern-day artificial construct because the concept did not exist in earlier times. That phrase *work-life balance* was first used in the 1970s in the United Kingdom and, later, in the 1980s in the United States. Humans can recognize non-physical structures through reasoning. For example, a mathematical formula, like the equation for the area of a square (length × width). Even though a construct is not physically real, it exists, nonetheless. Logic, philosophy, and even linguistics rely upon artificial constructs to define subjects that have no real physical existence. For example, grammar rules are an artificial construct. There is no physical cause for grammar rules to be the way they are, and the rules have no physical reality. Another example is the hierarchies and roles that people recognize and assume when in social settings, even though there is no real physical reason or even any real cause for obeying the constructs. See also, *construct* and *work-life balance*.

Phrase or Term	Definition
Artificial General Intelligence	*Artificial general intelligence* is a machine that can understand or learn a human intellectual task. This is sometimes referred to as full or strong *artificial intelligence*. See also, *artificial intelligence* and *augmented intelligence*.
Artificial Intelligence	*Artificial intelligence* is non-biological intelligence that performs human-like tasks such as decision-making, speech recognition, translation, and visual perception. For example, the software that make driverless cars drive autonomously. See also, *artificial general intelligence*, *augmented intelligence*, *chatbot*, *strong artificial general intelligence*, and *weak artificial general intelligence*.
As the Crow Flies	*As the crow flies* is an expression that means the most direct or shortest route between two points (i.e., a straight line). Flying from point A to point B like a crow (or bird) is the shortest route, and shorter than the route required if walking, driving a vehicle, taking a cab, or using public transportation. For example, if we could get there *as the crow flies*, we would be there by now. This expression has been in use since the 16th century. Its origins are unknown. See also, *A to B*.
As-built	*As-built* is when a dimension, material, or another element of a finished product is different than what was required by a drawing or specification. This may occur as a result of an out of tolerance condition or a change. For example, a steel plate is specified as 19 mm (3/4 in.), but is unavailable. So, 25 mm (1.0 in.) is used instead. See also, *as-built drawing* and *out of tolerance*.
As-built drawing	An *as-built drawing* is a prepared upon completion of a project. It documents all changes made to specifications and working drawings during fabrication, manufacturing, or construction and identifies the final, exact dimensions, geometry, and location of all elements of the completed work. See also, *as-built* and *red-line drawing*.
Assembly	An *assembly* is two or more components or parts joined by bonding, bolting, brazing, screwing, soldering, or welding. See also, *sub-assembly*.
Asset	An *asset* is a resource with economic value that a corporation, country, or person controls or owns with the expectation that it will provide a future benefit. It is shown on the owner's balance sheet and is usually classified according to the ease with which it can be converted into cash. It something that, in the future, can generate cash flow, reduce expenses, or improve sales that may be: • Something that an entity has acquired or purchased, and that has money value (its cost, book value, market value, or residual value). It can be: o A right (e.g., copyright, patent, or trademark); o An assumption (e.g., goodwill); o An enforceable claim against others (e.g., accounts receivable); or, o Physical property (e.g., building, cash, inventory, machinery, or materials): or, • Something valuable that an entity owns, benefits from, or has use of, in generating income. See also, *total cost of ownership*.
Asset Integrity Management	*Asset integrity management* is an approach to the reliable and safe operation of assets, applicable throughout the life cycle of the assets. Asset integrity management also applies to aging assets. See also, *asset, enterprise asset management system*, *integrity assessment*, *inspectioneering*, and *life cycle*.

Phrase or Term	Definition
Asset Performance Management	*Asset performance management* refers to the use of equipment to realize business-specific goals (e.g., use asset and operational data to improve performance, make mission critical decisions, and reduce costs). This helps asset-intensive organizations (e.g., manufacturing, oil and gas, and utilities) to acquire the greatest possible benefit from (costly) equipment budgets. Organizations employ various strategies to maximize profits and reduce risk. It often involves historical and real-time data analysis, capture, integration, and visualization. It also includes condition monitoring, predictive forecasting, and reliability-centred maintenance. See also, *management*, and *reliability-centred maintenance*.
Assignment Instruction	An *assignment instruction* is issued to a quality surveillance representative to identify purchase-order-specific information, inspection requirements, quality surveillance levels and quality verification points. See also, *sub-assignment*.
Assumption	An *assumption* is a factor deemed to be true during a planning process, though proof of validity is not available. Assumptions can affect its risks and outcomes, so these must be considered carefully.
Atmospheric Pressure	*Atmospheric pressure* is the pressure exerted by the weight of the atmosphere, which at sea level has a mean value of 101.3 kPa (14.6 psi). See also, *internal pressure* and *external pressure*.
Attendee	An *attendee* is someone who attends a meeting or witnesses an activity (e.g., witness a factory acceptance test).
Attention Interest Desire and Action Principle, AIDA Principle	The *attention interest desire and action principle* or *AIDA principle* explains how an advertising or marketing communications message engages and involves a consumer in brand choices. It describes how advertising messages need to accomplish several tasks to move the consumer through a series of steps from brand awareness through to action for purchase and consumption. This concept was developed in 1898 by the American advertising pioneer Elias. St. Elmo Lewis. See also, *advertising*, *brand*, and *marketing*.
Audit	An *audit* is an on-site verification activity, such as the review, inspection, or examination of a process or quality system, to ensure compliance to requirements. An audit can apply to an entire organization, a specific function, process, or step. There are significant differences between how audits are planned, performed, and reported if an audit is made per these requirements: • Project quality surveillance; • Supplier regulatory quality management system (e.g., electrical or pressure equipment); and, • Supplier quality management system (e.g., ISO 9001, *Quality Management System)*.
Audit Point	An *audit point* is a verification activity, such as inspection or examination of a process or quality system, to ensure compliance with requirements.

<table>
<tr><th>Phrase or Term</th><th>Definition</th></tr>
<tr><td>Augmented Intelligence, Intelligence Amplification</td><td>Augmented intelligence or intelligence amplification is the effective use of information technology to add to, improve, or increase human intelligence. For example, 80% of cyberattacks are driven by highly organized crime rings that freely exchange data, tricks of the trade, and tools. Cybersecurity experts cannot keep up, so by augmenting their expertise, using systems that are learning how to monitor unstructured data to detect risks before they emerge. This is also referred to as cognitive augmentation, machine augmented intelligence, and enhanced intelligence. See also, artificial intelligence, artificial general intelligence, augmented reality, and tricks of the trade.</td></tr>
<tr><td>Augmented Reality</td><td>Augmented reality is computer generated graphics or images superimposed on a view of the real world. For example, Google Sky Map. See also, augmented intelligence, digital reality, mixed reality, and virtual reality.</td></tr>
<tr><td>Authentication (Digital and Physical), Sealing, Stamping</td><td>Authentication (digital and physical), sealing, or stamping means that a licensed professional has completed or reviewed and validated the work (i.e., as an accurate, professional work product) and accepts professional responsibility for the engineering or geoscience involved in the work. This may be digital (for electronic copies) or physical (for original hard copies). See also, digital signature, electronic signature, hard copy, professional work product, technical document, and wet signature.</td></tr>
<tr><td>Authorization for Expenditure</td><td>Authorization for expenditure is a budgeting and approval form used during the project planning process that includes an estimate of total costs with breakdowns. This completed form is used for the approval of a large capital or operational expenditure related to exploration, development, and production. It represents:
1. A budget for the project against which actual expenditures are compared; and,
2. A form to evidence agreement by the joint interests and owners to participate in a project.</td></tr>
<tr><td>Authorized Inspection Agency</td><td>An authorized inspection agency is authorized by a regulatory authority to perform inspections required under the Safety Codes Act (see CSA B51-14, Clause 3).</td></tr>
<tr><td>Authorized Inspector</td><td>An authorized inspector is a person regularly employed by an ASME accredited authorized inspection agency. See also, safety codes officer.</td></tr>
<tr><td>Autistic Savant</td><td>An autistic savant is someone who is highly knowledgeable or gifted in one subject, but who knows little about, or has difficulties with, virtually everything else.
For example, Kim Peek is an autistic savant made famous for being the inspiration behind Raymond Babbitt – the savant leading character in the Hollywood movie Rain Man. See also, reverse Darwinism and savant.</td></tr>
<tr><td>Automagically</td><td>Automagically is something done:
• Automatically and in a way that is ingenious, inexplicable, or seems magical (especially for the operation of a computer process);
• Automatically, but in a way that someone does not want to explain because it is too complicated, trivial, or inelegant); or,
• With such ease and speed, or in a way that is hidden from, or not understood by, the user such that it appears to be magic.
Automagically is a portmanteau of automatic and magic.</td></tr>
</table>

Phrase or Term	Definition
Automation	*Automation* is the use of computer software, machines, or other technology to complete tasks that would otherwise be performed by humans. There are many types of automation: from fully-mechanical to fully-virtual, and from very simple to very complex. The two basic types of automation are: 1. Industrial automation Controlling physical processes using control systems and physical machines to automate tasks within an industrial process (e.g., welding of doors to vehicles on a vehicle assembly line). 2. Software automation Using software to perform tasks that personnel need a computer program to perform (e.g., moving and renaming thousands of files and sorting them into folders). Automation may or may not make use of physical robots. See also, *robotics*.
Auxiliary Equipment Test	An *auxiliary equipment test* is completed for auxiliary equipment at the sub-supplier's facility to ensure it is working as specified by the purchase order (e.g., a motor, gearbox, lube oil console, or control system). See also, *complete unit test*.
Award	*Award* is the creation and issuance of a purchase order to a supplier. It is the result of successful bid or offer submitted by the supplier in response to a buyer's request for quotation. After the award is issued, the first phase of a production can begin. See also, *bid*, *buyer*, *purchase order*, and *request for quotation*.

Phrase or Term	Definition
B	1 A **B** C D E F G H I J K L M N O P Q R S T U V W X Y Z
Baader-Meinhof Phenomenon, Frequency Illusion	The *Baader-Meinhof phenomenon* or *frequency illusion* is when a word or thing that has recently come to one's attention suddenly seems to appear with improbable frequency shortly afterwards. It is named after the Baader–Meinhof Group, when a contributor to the Bulletin Board column in the St. Paul Pioneer Press reported hearing the group's name repeatedly after learning about them for the first time. See also, *bias*, *law of attraction*, *recency illusion* and *selection bias*.
Back Charge	A *back charge* is a bill to collect an unanticipated incurred expense from: • A credit for expenses incurred by the project from an unplanned event that was caused by others (e.g., supply of product that is incomplete, incorrect, or late); • A lack of payment by the recipient of products or services; • An expense amount adjustment to correct an accounting or other errors; or, • An expense amount adjustment to collect an expense that was not billable or could only be determined in the future.
Back Office	The *back office* refers to the area where administration and support functions of a business, that are not client-facing, occur. Back-office functions include accounting, information technology, operations, record maintenance, and regulatory compliance. The phrase originated from traditional office design, where companies designed offices such that the front portion was designed for interaction with customers and the back portion was designed for personnel who have no interaction with customers (e.g., accounting clerks). The phrase *back office dealings* refers to activities that customers do not see (e.g., determining profit margins for the sale of a product). See also, *front office* and *middle office*.
Backfire	*Backfire* is: • A fire that is set intentionally to arrest the progress of an approaching fire by creating a burned area in its path, thus depriving the approaching fire of fuel; • A firearm malfunction when a bullet does not fire when the trigger is pulled (i.e., the backward escape of gases or cartridge fragments when a gun is fired); • A mistimed explosion in the cylinder or exhaust of an internal combustion engine; • An action or plan that rebounds adversely on the originator and has the opposite intended effect (e.g., a comedian tells jokes but is met with boos and jeers from the audience); • An attempt to fix something that makes the situation worse (e.g., attempting to fix a malfunctioning automatic garage door only to inadvertently lock the door such that it cannot be open at all (automatically or manually); or, • To fail or to stop being successful. See also, *backfire effect* and *kickback*.

Phrase or Term	Definition
Backfire Effect	The *backfire effect* is when, in the face of contradictory evidence, established beliefs do not change and instead become entrenched and more pronounced. For example, presenting hyperlinks, facts, figures, and quotes can instill even more confidence in an opponent's psyche at the end of a of debate than was experienced at the beginning. As proponents and opponents argue, each debater's position is cemented and entrenched further. Neither debater is able to persuade the other and both walk away with even more firmly-held beliefs. See also, *Rapoport's Rules*.
Backwards Compatibility, Downward Compatibility	*Backwards compatibility* or *downward compatibility* is a property of a product, system, or technology that allows for interoperability with an older legacy system, or with input designed for such a system, especially in computing and telecommunications. For example, the ability of a computer program that translates a newer version of a programming language to accept data or programs from a previous version. Modifying a system in a way that does not allow backward compatibility is referred to as *breaking* backward compatibility. In programming jargon, the concept may be referred to as hysterical reasons, a homophone for historical reasons. See also, *forward compatibility*.
Back Seat Driver	A *back seat driver* is a: • Passenger in a vehicle who gives the driver unwanted advice or directions; or, • Person who attempts to control something that is someone else's responsibility or gives unwanted advice or direction. A back seat driver is someone who opines from the comfort of the back seat of a vehicle without being responsible for the vehicle. See also, *armchair critic* and *second guess*.
Backstory, Back-story, Background Story, Background	A *backstory*, *back-story*, *background story*, or *background* is the narrative that explains what leads up to or precedes the primary story or plot (i.e., fiction and non-fiction). The backstory includes events directly tied to the essential development of the plot or story form. For example, the parents of the story's main character dreamed of becoming business owners, but without any resources – they could only work as employees. The background provides the character's individual, historical context as an aspect of the *storytelling*. For example, before finally establishing a successful business, the main character overcame many challenges and suffered many failures. See also, *founder story*, *story*, *story form*, and *storytelling*.
Bad Faith, Mala Fides	*Bad faith* or *mala fides* (Latin) is to act with the intent to deceive or a lack of honesty when dealing with others in personal, business, or legal matters. For example, signing a contract with no intent to fulfill its terms. It may also involve self-deception (e.g., one's perceived intentions are directly contradicted by facts). For example, signing a contract knowing that the terms will not be fulfilled because of a pending bankruptcy. Bad faith is often associated with breaches of contract or trust, fraud, hypocrisy, lip service, and unfulfilled obligations. For example, signing a contract to earn initial payments but knowingly being unable to fulfill all terms and earn final payments. The existence of bad faith can minimize or nullify a claim in a lawsuit. For example, if signing a contract knowing that the company is virtually bankrupt and will be unable to provide the products and services as agreed. When the business partners learn that the contract cannot be fulfilled, they can cancel it based on bad faith. See also, *bankruptcy*, *good faith*, and *lip service*.

<table>
<tr><th>Phrase or Term</th><th>Definition</th></tr>
<tr><td>Bailout, Bail Out</td><td>A bailout (noun) is the act of giving financial assistance, resources, or both to a failing business or government. This act helps prevent the consequences of a potential economic collapse, which may include bankruptcy and default on financial obligations. A bailout may be given in the form of cash, a loan, or in bonds or stock. The bailout terms may require the recused party to repay all or some of the financial assistance. For example, in Canada, the Ontario provincial government provided General Motors and Chrysler a CAD$ 4.6 billion bailout in 2009, of which $3.6 billion was repaid. See also, corporate welfare and crony capitalism.
To bail out (verb) means to:
• Abandon a difficult or harmful situation (e.g., divest from the stock market when it drops precipitously);
• Empty water from a boat by hand (e.g., with a bucket);
• Help someone with a problem (e.g., deliver gasoline to a friend whose vehicle ran out of fuel on the highway);
• Parachute from an aircraft; or,
• Pay money to a court to facilitate suspect's release from incarceration while they await trial.
See also, divest.</td></tr>
<tr><td>Bait and Switch</td><td>Bait and switch is:
• A logical fallacy that occurs when someone presents a partial appealing truth while hiding an unappealing falsehood;
• A ploy to offer a person or people something desirable to gain favour (i.e., political support or votes) and then frustrating their expectations with something different or less desirable. For example, a politician proposes making basic legislative changes to gain initial support. Then, once support is gained, the specific details are identified – which may be different or the opposite of what was initially proposed;
• A tactic used in an authoritarian setting or cult to conceal doctrine from outsiders and new inductees. For example, nonsensical or objectionable doctrine is deliberately hidden and newcomers are not allowed to study it independently; or,
• An illegal or unethical sales tactic where a product is advertised as a bargain with the intent of substituting an inferior product or persuading the customer to purchase a more expensive product at the point of sale, instead. For example, the advertised product (i.e., the bait) suddenly becomes sold out or unavailable and then is replaced with an inferior or more expensive product (i.e., the switch).
See also, circular reasoning, cognitive bias, either/or, ethical, false equivalence, logical, logical fallacy, red herring, righteousness fallacy, slippery slope, and straw man argument.</td></tr>
</table>

Phrase or Term	Definition
Band-aid, BAND-AID®	A *band-aid* or *BAND-AID®* is a portmanteau of *band* and *aid* that describes a: • Brand of adhesive, sterile, medical bandages invented in 1902 and distributed by the Johnson & Johnson company. North Americans commonly use this brand name to describe the equivalent, non-branded phrase: adhesive bandage; and, • Quick, temporary, or superficial solution that hides symptoms of the problem but leaves the root cause of the problem unaddressed or unresolved. See also, *quick fix*.
Bandwidth	*Bandwidth* has multiple meanings. Euphemistically, bandwidth is the energy or mental capacity required to deal with a situation. Technically, bandwidth is: • A range of frequencies within a given band used for transmitting a signal (e.g., radio signal); or, • The capacity of a wired or wireless network communications link to transmit data from one point to another over a path (i.e., computer network or Internet connection) in a given time. Synonymous with capacity, bandwidth describes the data transfer rate; it is not a measure of network speed, which is a common misconception. See also, *bench strength*, *data transfer rate*, and *horsepower*.
Banker	A *banker* engages in the business of banking (i.e., financial activities to accept deposits, lend money, and manage withdrawals). See also, *accountant* and *financial officer*.
Bankruptcy	*Bankruptcy* is a legal process undertaken when an individual or business cannot repay outstanding debts within an agreed-upon pre-established period. Ultimately, bankruptcy relieves a debtor of debts. The bankruptcy process begins with a petition filed by the debtor or on behalf of creditors. During the process, all debtor's assets are measured and evaluated. Assets are typically used to repay some or all of the outstanding debt. See also, *receivership*.
Barcode, Bar Code	A *barcode* or *bar code* is a machine-readable, visible label that contains embedded data in the form of a an image or matrix. Barcodes may be scanned by an optical device (e.g., a barcode reader or camera) to retrieve information. For example, a grocery store clerk will scan products using a barcode reader to quickly determine the cost of each product, the total cost of all products purchased, and for merchant inventory tracking and restocking purposes. A barcode may use a: • One dimensional or a linear code (e.g., stripes); or, • Two-dimensional or a matrix code (e.g., dots, hexagons, rectangles, or other shapes). Barcodes became commercially successful in the early 1970s when they were used to automate supermarket checkout systems. See also, *quick response code* and *universal product code*.
Barred Tee, Scrapper Tee	A *barred tee* or *scrapper tee* is a tee or fitting with a branch connection that contains restriction bar(s) that are welded to the interior of a pipe. The bar(s) prevent a piping instrument gauge or pig from traversing the branch connection. See also, *branch connection*, *fitting*, and *pipeline instrument gauge*.

<table>
<tr><th>Phrase or Term</th><th>Definition</th></tr>
<tr><td>Barrel of Oil Equivalent</td><td>A barrel of oil equivalent is a unit of energy based on the approximate energy released by burning one barrel (160 L or 42 US gal) of crude oil. The measure is used in oil and gas company financial statements to combine oil and natural gas reserves and production into a single measure. This energy equivalence does not account for the lower financial value of energy in the form of gas.</td></tr>
<tr><td>Base Metal</td><td>Base metal is material to be cut, brazed, soldered, or welded. In the context of this document, a base metal is not a precious or noble metal (e.g., gold, silver, or platinum, which resists chemical action, does not corrode, and is not easily affected by acids). See also, electrode, metal, mineral, non-ferrous, non-metallic, and weldment.</td></tr>
<tr><td>Baseline</td><td>A baseline is the original or starting point often used for comparison or reference purposes. A baseline may be used to:<ul><li>Acquire an electrical signal from a sensor when no measured variable is present (i.e., output at no-load condition);</li><li>Calculate or compare data;</li><li>Compare project plans vs. project performance (e.g., scope, cost and schedule).</li><li>Configure software, hardware, or a process;</li><li>Identify material;</li><li>Inspect for verification of code compliance;</li><li>Measure lines for construction; and,</li><li>Measure material thickness for corrosion monitoring.</li></ul></td></tr>
<tr><td>Baseline Ultrasonic Thickness Survey</td><td>A baseline ultrasonic thickness survey uses ultrasonic testing to measure the actual thickness of material at or after construction, fabrication, or manufacturing. This information is later compared to the thicknesses measured with another thickness survey, after a period in-service, to determine the corrosion rate. See also, in-service, and corrosion rate.</td></tr>
<tr><td>Baseload Plant</td><td>A baseload plant is a power generation facility that provides a continuous supply of electricity, over 24-hours and throughout the year. It supplies a minimum power generation requirement to meet fundamental consumer demand. Power is commonly produced using energy sources such as coal, geothermal, natural gas, and nuclear. This type of facility does not change power output quickly. See also, baseload, peak load plant, and power output.</td></tr>
<tr><td>Baseload, Base Load</td><td>Baseload or base load is the minimum level of demand on an electrical grid over a period (e.g., one week). This demand can be met by a collection of smaller intermittent energy sources, dispatchable generation, and unvarying power plants. The approach to meet demand depends upon which approach has the best mix of availability, high reliability, and low cost in a market. The remainder of demand, varying throughout a day, is met by dispatchable generation – which can be turned up or down quickly – using energy storage, load following power plants, or peaking power plants. Power plants that do not change power output quickly, such as coal or nuclear plants, are generally called baseload power plants. See also, baseload plant, peak shaving, load shifting, and power output.</td></tr>
<tr><td>Baseplate</td><td>A baseplate is rigid steel frame used to mount and maintain shaft alignment between two or more pieces of equipment (e.g., a driver, gearbox, and pump). See also, skid</td></tr>
</table>

Phrase or Term	Definition
Basic Data, Basic Data Package	*Basic data* or a *basic data package* is the body of scientific information that governs and underlies the design of an asset (e.g., a facility or pipeline). For a megaproject, this is a document comprising thousands of pages. Because different functional groups are often responsible for developing the design, those who use this information may not be aware of how the basic data was produced or its quality. Erroneous and incomplete basic data leads to project failure. Basic data is used for design to identify: • Dangerous items; • How to sequence operations; • How to size equipment; and, • What materials to use.
Basis of Design	Not a preferred phrase – see *design basis*.
Beam	A *beam* is a horizontal structural member that supports loads from a floor or roof system onto columns or walls. See also, *column*.
Bean Counter	A *bean counter* is someone with financial interests (e.g., an accountant, banker, bureaucrat, or financial officer) who is perceived by others as placing excessive emphasis on controlling budgets, and insignificant or trivial expenditures. They may be reluctant to spend money or quibble over small things to save money. For example, *my bean counter boss prohibits me from spending more that $15 for a business meal.* This expression is mildly derogatory; it asserts that beans are an inexpensive commodity and that counting them is wasteful and unnecessary. See also, *accountant*, *banker*, *bureaucrat*, and *financial officer*.
Bear Market	A *bear market* is a general decline in the stock market over time (e.g., a price decline of 20% or more over at least a two-month period). See also, *bull market*, *market bottom*, and *market trend*.
Beg the Question, Begging the Question, Assume the Conclusion	To *beg the question* or *begging the question* means to: • Modern meaning (also considered a misuse of the original meaning) Invite an obvious or specific question as a reaction or response. For example, the statement the government needs to raise taxes to increase revenue begs the questions who shall pay higher taxes? • Original meaning Ignore a question with the assumption it has already been answered or settled (i.e., make an argument that assumes the truth of the conclusion instead of supporting it). For example, the Apple iPhone is the best smartphone because there is no better smartphone. Beg the question is a 16th century mistranslation of Aristotle's statement (in Latin), *petitio principii*, which means *assume the conclusion*. See also, *Catch-22*, *circular reasoning*, and *vicious circle*.

Phrase or Term	Definition
Behavioral Change Stairway Model	The *behavioral change stairway model* is a five-stage method that enables a crisis negotiator to go from listening to influencing behaviour. It is: 1. Active listening; 2. Empathy; 3. Rapport; 4. Influence; and, 5. Behavioral change. This method is described by Chris Voss in the book *Never Split The Difference: Negotiating As If Your Life Depended On It* (with Tahl Raz). This method was developed by the United States Federal Bureau of Investigation's Crisis Negotiation Unit. See also, *active listening*.
Belief-driven Buyers	*Belief-driven buyers* choose to boycott, purchase, or switch from or to a brand based on the brand's stance on social issues (i.e., chooses principle over product). This practice is common among young consumers who advocate for a brand while building loyalty around a cause. For example, the TOMS shoe company, in partnership with humanitarian organizations, donates one pair of shoes to a child in need for every pair of shoes that is purchased. The TOMS shoe company advertises these actions to attract belief-driven buyers. See also, *brand*.
Bell Curve, Gaussian Distribution	A *bell curve or Gaussian distribution* is a graph with a large, rounded peak tapering away at each end. In mathematics, a Gaussian function – often simply referred to as a Gaussian – is named after the mathematician Carl Friedrich Gauss.
Bells and Whistles	*Bells and whistles* describes the addition of attractive, but non-essential features. A non-essential feature provides no benefit for product functionality or service. For example, requiring documentation such as material test reports for parts (e.g., gauges or valves) that will be discarded and replaced instead of being repaired if damaged or worn-out This phrase was originally used to describe bells and whistles that decorate fairground musical organs. See also, *bits and pieces*, and *gold plating*.

<table>
<tr><th>Phrase or Term</th><th>Definition</th></tr>
<tr><td>Bench Strength</td><td>Bench strength is:<ul><li>The ability and skill possessed by an organization's personnel; or,</li><li>The capability and number of an organization's personnel able to move into positions of greater or leadership responsibility when required.</li></ul>Bench strength is used to measure competency management, human resources, and succession planning, including:<ul><li>Competency depth – the number of personnel with critical competencies;</li><li>Internal placement rate – the percentage of key positions that are filled by an internal candidate;</li><li>Key position coverage – the percentage of key positions with at least one personnel qualified to immediately assume the role if required;</li><li>Key position readiness – the average readiness score for personnel to assume a critical role;</li><li>Key position volatility – the percentage of key positions vacated yearly;</li><li>Promotion rate – the percentage of high performers who are promoted in a year; and,</li><li>Talent pool age – the number of days qualified personnel wait for a promotion or to fill a vacated position.</li></ul>See also, bandwidth and horsepower.</td></tr>
<tr><td>Benchmark</td><td>A benchmark is:<ul><li>A level of quality, point of reference, or a standard against which performance or things may be assessed or compared;</li><li>A geodetic mark or a survey mark or monument (i.e., a physical mark on a permanent object) indicating elevation that used as a reference point for survey measurements; or,</li><li>To evaluate or check (something) by comparison with a standard.</li></ul>See also, standard of workmanship and workmanship.</td></tr>
<tr><td>Benefit</td><td>In business, a benefit may be classified as follows:<ol><li>General
An advantage, financial reimbursement (such as that made under an insurance policy, medical plan, or pension plan), privilege, or right;</li><li>Finance
A desirable and measurable outcome or result from an action, investment, project, resource, or technology; or,</li><li>Marketing
A desirable attribute of a product or service that a customer will obtain from a purchase. For example, while a supplier sells a feature (e.g., a high-speed tungsten-carbide drill bit), the buyer seeks the benefit (e.g., the holes the drill bit will render). That is, most buyers only care if the drill bit drills the holes needed for the job and are much less concerned about the drill bit's properties.</li></ol>See also, feature.</td></tr>
</table>

Phrase or Term	Definition
Bespoke	*Bespoke* describes something that is custom made for a customer or user. For example, hand-made shoes or a tailored suit. In contemporary usage, it refers to a general branding and marketing concept implying exclusivity and limited production runs.
Best Alternative to a Negotiated Agreement	The *best alternative to a negotiated agreement* is the most advantageous alternative agreement if a negotiation fails and an initial agreement is not reached (e.g., appeal to a higher authority, suspension of negotiations, or transition to another negotiating partner). When the first agreement is not reached, a successful negotiator will acquire this alternative agreement. Generally, anything less than the best alternative to a negotiated agreement is not acceptable. This concept was developed in the book *Getting to Yes* by Roger Fisher and William Ury. See also, *wimp-win*, *win-win*, *worst alternative to a negotiated agreement*.
Best-for-project	*Best-for-project* is when project alliance decision-making is directed toward the collective, shared objectives and vision of the project (e.g., cost, quality, reputation, and time), rather than toward narrow self-serving goals (e.g., to maximum profit). These are actions and decisions that: • Align with principles developed by the participants and were incorporated into the alliance agreement. • Drive the delivery of outstanding outcomes for all objectives; and, • Reflect project commitments found in the alliance agreement. See also, *project alliance* and *project alliance agreement*.
Best in Class	*Best in class* is the highest current performance level in an industry. This designation is a benchmark, goal, or standard to be equaled or exceeded by others in the industry. See also, *best practice* and *word class*.
Best Practice	A *best practice* is a method or technique that has been generally accepted over all alternatives because it produces results that are superior to those achieved by other means (e.g., the most efficient) or because it has become a standard way of doing things (e.g., to meet certification, ethical, or legal requirements). For example, using templates to standardize documentation. Best practices are used to maintain quality and can be based on benchmarking or self-assessment. It is a feature of accredited management standards such as ISO 9000. A key talent required when applying a best practice is the ability to balance the unique qualities of an operation, organization, or system with the practices that it has in common with others. See also, *best in class*, *good practice*, *smart practice*, and *world class*.
Best-in-class Performance Management	*Best-in-class performance management* is a system that: 1. Sets direction and context; 2. Establishes clear accountabilities and metrics; 3. Creates realistic budgets, plan, and targets; 4. Tracks performance effectively; 5. Holds robust performance dialogues; and, 6. Endures action, consequences, and rewards. This concept is described in the article *Deliverology: From idea to Implementation* by Michael Barber, Paul Kihn, and Andy Moffit. See also, *deliverology* and *state of the art*.

Phrase or Term	Definition
Beta Test	A *beta test* is a trial of machinery, software, or other product, in the final stages of its development, that is carried out by a party unconnected with its development. A beta test is the second phase of testing in which a sampling of the intended audience uses the product. Beta is the second letter of the Greek alphabet (β, Β). See also, *beta-tester* and *perpetual beta*.
Beta-tester	A *beta-tester* is someone who tests an early version of a product to identify technical errors and determine usability. By providing feedback to the creator of the product, a beta tester improves the product before the product goes to production-market. Most beta testers are paid little or nothing. See also, *beta test*.
Bevel	A *bevel* is a: • Slope from the horizontal or vertical axis; or, • Weld joint preparation. See also, *bevelled end*.
Bevelled End, Bevelled Joint	*Bevelled end* or *bevelled joint* is a pipe end preparation with a bevel cut or a weld joint that is bevelled. See also, *bevel*, *plain end*, and *threaded end*.
Bias	A *bias* is disproportionate weight against or in favour of an idea or thing, which may be in a way that is closed-minded, prejudicial, or unfair. A bias can be innate or learned, and can be: • A concern with, or interest in, one thing more than others; • A systematic error introduced into sampling or testing by encouraging or selecting one answer or outcome over others; • An inclination of outlook or temperament (e.g., personal judgment, preference, or prejudice), in favour of, or against, a group, person, or thing compared with another, usually in a way considered to be unfair or unreasonable; or, • An influence in favour of a certain choice. See also, *blind spot*, *cognitive bias*, *cognitive dissonance*, *cognitive flexibility*, *confirmation bias*, *conflict of interest*, *critical thinking*, *Dunning-Kruger effect*, *hindsight bias*, *ideological bias*, *independent forecast*, *independent project review*, *lucky fool syndrome*, *objective*, *optimism bias*, *Pollyanna principle*, *recency illusion*, *reference-class forecasting*, *selection bias*, *self-serving bias*, *thinking mode*, *unconscious bias*, *uniqueness bias*, and *zero-sum bias*.
Bid	A *bid* is an offer submitted by a seller in response to a buyer's request for quotation. A bid may compete with other bids for the supply products, services or both, within a prescribed time, and at a specified cost. The buyer may: • Reject the bid; • Make a counter-offer; or, • Accept the bid and create or award a purchase order for the seller. See also, *award*, *bid-rigging*, *bidder buyer*, *request for quotation*, and *three bids and a buy*.

<table>
<tr><th>Phrase or Term</th><th>Definition</th></tr>
<tr><td>Bid Clarification Meeting</td><td>A bid clarification meeting is held by the buyer, before purchase order award, to review a bidder's proposal with the bidder, for clarification of commercial or technical requirements, and to discuss alternatives and options (as required). See also, bid, bidder, buyer, clarification, and request for quotation.</td></tr>
<tr><td>Bid Evaluation, Bid Review</td><td>A bid evaluation or bid review is an analysis of one or more bids received in response to a request for quotation, to determine the successful bidder and award a purchase order. A bid may be analyzed commercially, technically or both. See also, bid, bid clarification meeting, bidder, and purchase order.</td></tr>
<tr><td>Bidder</td><td>A bidder is a company with a request for quotation for the supply of materials or services to a project. A bidder becomes a supplier after the bidder is awarded a purchase order. See also, bid, bidder, buyer, and request for quotation.</td></tr>
<tr><td>Bid-rigging</td><td>Bid-rigging is an illegal practice in which competing parties collude to establish the winner of a bidding process. This is a form of anticompetitive collusion where two or more parties agree that, in response to a call for a bid, one or more of them will:<ul><li>Not submit a bid;</li><li>Submit a bid or bids prepared by agreement (e.g., with artificially increased prices for products, services, or both); or,</li><li>Withdraw a bid.</li></ul>Bid-rigging is a serious crime that eliminates competition among service providers and suppliers, and increases costs with fabricated prices. Regardless of where this occurs (e.g., on government or private projects), increased costs are ultimately absorbed by consumers and taxpayers without their consent. See also, bid.</td></tr>
<tr><td>Big Business</td><td>A big business is defined in different ways by different organizations. Some definitions are:<ul><li>Canadian Bankers Association
Any business with a credit limit over CAD$ 1 million.</li><li>Export Development Corporation
Any busines with over CAD$ 1 million in annual export sales.</li><li>Industry Canada
Any business with over 500 employees.</li></ul>See also, medium-sized business, micro-business, and small business.</td></tr>
<tr><td>Big Data</td><td>Big data is unstructured data in extremely large data sets. See also, data lake and large data.</td></tr>
<tr><td>Big Mac Index, Burgernomics</td><td>The Big Mac Index or Burgernomics is a theory that currency exchange rates should move towards the rate that would equalize the prices of identical products and services in any two countries (e.g., a hamburger). It takes its name from the Big Mac, a hamburger sold at McDonald's restaurants. This concept was invented by The Economist in 1986 as a lighthearted guide to if currencies are at their correct levels. The Big Mac index is not a sophisticated measure for analyzing currency rate risk; however, the burger concept makes exchange-rate theory easier to understand than the theories of exchange rates. See also, exchange rate, random walk theory, and theories of exchange rates.</td></tr>
</table>

Phrase or Term	Definition
Big Picture, Bird's Eye View, 30,000 ft View	The *big picture* is an account of the entire perspective, long-range view, or overall view of an issue or matter. This is a summative, macro viewpoint – not a detailed or micro viewpoint. For example, the big picture indicates the economy is improving because overall sales this year are greater than last year. This is also referred to as a *bird's eye view* or a *30,000 ft view* (or a 10,000 or 20,000 ft view). Management can typically see the big picture (all sales); workers typically cannot (they only see and are responsible for individual financial transactions).
Bill of Material, Bill of Materials	A *bill of material or bill of materials* is a parts list, or assembly and installation list. It is the final accounting of what materials, parts, sub-assemblies, and quantities are needed to assemble a product, or to be delivered for construction. There are different types of bill of materials depending on the business need and use for which it is intended. The labour, materials, and resources used to produce the items listed can be analyzed by category and cost. These data can also be used for traceability. See also, *job order* and *traceability*.
Binary Opposition, Binary System	*Binary opposition* or *binary system* is a pair of related concepts or terms that are opposite in meaning. It is language and thought, binary opposition are two theoretical opposites, as strictly defined and set off against one another. For example, backward and forward, left and right, on and off. See also, *black-and-white thinking*, *false dichotomy*, *middle ground*, and *thinking mode*.
Biobreak, Bio Break, Biological Break	A *biobreak*, *bio break*, or *biological break* is a euphemism meaning to take a break from a meeting or work to: • Eat, drink, or a combination (e.g., get coffee) • Re-hydrate (i.e., drink water); • Relax (e.g., pause); • Stretch (e.g., move, stand, or walk); or • Use the restroom (i.e., toilet).
Biomass	*Biomass* is organic matter used as a fuel (e.g., agricultural by-products, wood, wood by-products, or waste), especially at a power station for the generation of electricity.
Bitcoin	*Bitcoin* is a digital currency created in January 2009 following the ideas set out in the whitepaper *Bitcoin: A Peer-to-Peer Electronic Cash System* by Satoshi Nakamoto (a pseudonym). The identity of the person or persons who created this concept and its technology are unknown. There are no physical bitcoins, only balances kept on a public ledger. Despite not being legal tender, Bitcoin triggered the launch of hundreds of other virtual currencies collectively referred to as altcoin. See also, *altcoin*, *cryptocurrency*, *grantcoin*, and *legal tender*.
Bitmap	A *bitmap* is a spreadsheet with serial address information that is developed into a serial interface report by instrumentation and controls. Also known as a Modbus data map.
Bits and Pieces, Bits and Bobs	*Bits and pieces* or *bits and bobs* are: • An assortment of small items; or, • Assorted objects. See also, *bells and whistles*.

Phrase or Term	Definition
Bitumen	*Bitumen* is a naturally occurring viscous mixture of hydrocarbons that contains high levels of sulphur and nitrogen compounds. In its natural state, it is not recoverable through a well because its high viscosity inhibits flow. Bitumen typically makes up about 10% of oil sands (by weight). Saturation varies. Bitumen is typically refined into gasoline, diesel fuel, and heating oil, which release energy upon combustion. See also, *bitumen beyond combustion* and *oil sands*.
Bitumen Beyond Combustion	The 2018 report *Bitumen Beyond Combustion*, by Stantec Consulting Ltd. For Alberta Innovates, examines the potential uses of bitumen to meet emerging global needs excluding combustion products. The report identifies 14 different products. The report recommends focusing on these four areas for detailed analysis and further study: 1. Carbon fibres and combination products (e.g., steel, cement, and wood); 2. Polymers; 3. Quinone and Vanadium flow batteries; and, 4. Solid asphalt transportation. See also, *future fit hydrocarbons* and *bitumen*.
Black and White	*Black and white* refers to: • An image, illustration, movie, photograph, or television program that uses tones and shades of black and white only (no colour); • An issue that seems simple (e.g., with an absolute judgement or clear boundary) and is therefore easily resolved. For example, when driving a motor vehicle, it is always illegal to exceed the posted speed limit; • An opinion or situation involving clearly defined opposing issues or principles (e.g., either all good or all bad). For example, fund the police vs. defund the police; or, • Something that is sharply defined or clear (e.g., the arsonist started the fire). See also, *grey area* and *thinking mode*.
Black Box	A *black box* is a device, system, or object that can be viewed in terms of its inputs and outputs, without any knowledge of its internal workings. Its implementation is conceptual and opaque. Almost anything might be referred to as a black box (e.g., an algorithm, an inspection, and test plan). See also, *algorithm*.
Black Ceiling	A *black ceiling* is a barrier to professional advancement that prevents black (e.g., African-American) professionals from advancing their careers or being promoted to senior-level positions within an organization. See also, *glass ceiling*.
Black Hat Marketing	*Black hat marketing* is the use of deceptive and manipulative search engine optimization techniques (e.g., illegal or unethical strategies to promote online business). For example, in 2006, Google blacklisted BMW's website BMW.de, which received a PageRank of 0 because of artificially inflated competitive keyword rankings. Black hat marketers promise immediate results, and often obtain them; however, such results can cost a business its reputation. They may regard their strategies as simply brilliant or normal business practices, but using these methods can have dire consequences such as being blacklisted by Google, negative responses from other businesses, or significant fines. See also, *blacklist*, *ethical*, *search engine* optimization, and *white hat marketing*.

Phrase or Term	Definition
Black Indigenous and People of Colour, People of Colour	*Black Indigenous and People of Colour* and *People of Colour* refers to non-Caucasian people (i.e., non-white peoples). See also, *ageism*, *discrimination, intersectionality, racism*, and *sexism*.
Black Swan Theory, Theory of Black Swan Events	The *black swan theory* or *theory of black swan events* is a metaphor that describes an event that comes as a surprise (i.e., was highly improbable or inconceivable, or an unknown unknown), has a major effect, and is often inappropriately rationalized after the fact with the benefit of hindsight. In the second century, Roman poet Juvenal first coined this phrase. This was a time when the black swan was presumed not to exist. The importance of the metaphor lies in its analogy to the fragility of any system of thought or belief. A set of conclusions is potentially undone after any of its fundamental postulates is disproved. Juvenal's phrase was common in 16th century London and indicated *impossibility*. In 1697, after Dutch explorers discovered black swans in Western Australia, the term metamorphosed to mean that a perceived impossibility might actually be possible. See also, *green swan*, *grey swan*, *kinds of information* (*unknown unknown*), *law of cause and effect*, and *perfect storm*.
Blacklist	A *blacklist* is a record of companies, personnel, or things that are regarded as unacceptable or untrustworthy, and should be regularly avoided, boycotted, or excluded. See also, *cancel culture*, *deplatforming*, *doxing*, and *shadow banning*.
Blackout	A *blackout* is: • A period of darkness caused by a complete voltage drop, forcing the failure of electrical power (e.g., in a city or part of a city); • A period of darkness enforced as a precaution against an air raid (i.e., attack by armed airplanes on a surface target); • A prohibition or restriction of broadcasting (e.g., a sports event) or trading (e.g., restrictions on the purchase or sale of stocks); • A temporary loss of a communication or radio signal; • A time during which a special commercial offer (e.g., discount) is not valid; • A transient dulling or loss of consciousness, memory, or vision; or, • An action to blot out, erase, obliterate, suppress, or wipe out something (e.g., censor or redact information). See also, *brownout*, *outage*, *rolling blackout*, and *whiteout*.
Blanket Purchase Order	A *blanket purchase order* is a long-term agreement between a company and a supplier to deliver products or services with a set price on a recurring basis over a specified time period. In addition to price, quantity, and time frame, a blanket purchase orders may include specifications. See also, *purchase order*.
Bleeding Heart	*Bleeding heart* describes someone who is *extremely* soft-hearted or excessively sympathetic for another's misfortune. A bleeding heart is not, however, necessarily a left wing individual who supports environmental or social causes. The phrase is sometimes used to criticize those with politically liberal views who favour government spending on social programs. This phrase was originally coined by Westbrook Pegler in 1938. In 1970, Ronald Reagan, then the newly elected governor of California, said *I was quite the bleeding-heart liberal*. By then he was known as a *former bleeding heart*. See also, *left wing*.

<table>
<tr><th>Phrase or Term</th><th>Definition</th></tr>
<tr><td>Blending</td><td>Blending, in the petroleum industry, is when two similar but chemically different products are mixed to create a third product. For example, blending high-sulphur diesel with zero-sulphur diesel produces low-sulphur diesel.</td></tr>
<tr><td>Blind Spot</td><td>A blind spot can apply to:<ul><li>Thinking
A blind spot in thinking is an area in which one fails to exercise discrimination or judgment.</li><li>Vision
A blind spot in vision is the specific scotoma in the visual field that corresponds to the lack of light-detecting photoreceptor cells on the optic disc (also known as the physiological blind spot).</li><li>A vehicle
A blind spot in a vehicle are areas located outside of a vehicle that cannot be seen while looking forward, backward or through optical aids (e.g., using a camera or mirror).</li></ul>See also, bias.</td></tr>
<tr><td>Blockchain</td><td>Blockchain is an innovative technology that combines distributed computing with encryption to enable trust-less business transactions. For example, it is the record-keeping technology behind cryptocurrency bitcoin. See also, digital project delivery.</td></tr>
<tr><td>Blog, Web Log</td><td>The word blog is a shortened form of the phrase web blog. A blog or web log is a website or web page that contains writing from a writer or group of writers about writer opinions, observations, and experiences.</td></tr>
</table>

Phrase or Term	Definition
Blooms Taxonomy of Knowledge, Bloom's Taxonomy	*Blooms Taxonomy of Knowledge* or *Bloom's Taxonomy* is a categorization of educational goals used by generations of instructors and teachers at colleges, schools, and universities. It consists of these categories: 1. Knowledge; 2. Comprehension; 3. Application; 4. Analysis; 5. Synthesis; and, 6. Evaluation. This concept was developed by Benjamin Bloom with Max Englehart, Edward Furst, Walter Hill, and David Krathwohl in their book *Taxonomy of Educational Objectives: The classification of educational goals*. A revised version, *A Taxonomy for Teaching, Learning, and Assessment*, was published by Lorin W. and David R. Krathwohl. It applied a dynamic concept of classification using verbs instead of the original notion of static objectives using nouns. It comprises these categories: 1. Remember; 2. Understand; 3. Apply; 4. Analyze; 5. Evaluate; and, 6. Create. See also, *active learning*, *cone of learning*, *knowledge transfer*, *metacognative thinking*, *passive learning*, and *unlearning*.
Blue Collar	*Blue collar* constitutes or relates to the class of personnel whose duties, generally: • Are performed outside an office setting e.g., in a facility, factory, or the field); • Earn less than white collar wage earners; • Require a basic education (e.g., an apprenticeship or high school diploma); • Require manual labour, particularly with equipment and tools; and, • Require protective or work clothing. Blue-collar workers traditionally wear sturdy clothing that does not show dirt easily, such as a blue denim shirt. The phrase was first used to describe trades work in 1924, in an Alden, Iowa newspaper. See also, *facility*, *field*, *pink collar*, and *white collar*.
Blue-Collar Crime	A *blue-collar crime* is a petty or violent crime that may be committed as a reaction or in response to an opportunity by an individual or a group of individuals. Blue-collar crime has no official legal classification, but it is used to describe a group of crimes such as assault, burglary, drug possession, murder, robbery, and theft. See also, *blue collar* and *white-collar crime*.

Phrase or Term	Definition
Blue Ocean Strategy	*Blue ocean strategy* is the simultaneous pursuit of differentiation and low cost, in a new uncontested or unexplored market space, to make the competition irrelevant. It is based on the view that market boundaries and industry structure can be reconstructed by actions and beliefs with value innovation. This theory was developed in the book *Blue Ocean Strategy: How to Create Uncontested Market Space and Make The Competition Irrelevant* by W. Chan Kim and Renée Mauborgne. Blue ocean is an analogy that describes the deeper, wider potential of market space that is not in existence or is unexplored. In a *blue ocean*, new demand is created rather than competing in an existing market, so there is ample opportunity for profitable and rapid growth. For example, Netflix did this with video streaming (over the Internet) instead of competing directly with Blockbuster's brick-and-mortar video rental model. Because the rules of the game are waiting to be established, the competition is irrelevant because it has not formed yet. Another example ... before the advent of the mass production of vehicles in America, America's 500 automakers built custom novelty vehicles that were exclusively owned and used by the wealthy. The vehicle came only in black and only one model was available. However, it was reliable, durable, and easy to fix. In 1908, The Ford Model T changed everything by revising the way automobiles were built. It was built to be affordable to most Americans: • In 1908 it was USD$ 850 (or about half the price of existing automobiles at the time); • In 1909 it was USD$ 609; and, • By 1924, it was only USD$ 240. Consequently, Ford's market share surged from 9% (1908) to 61% (1921). See also, *brick-and-mortar business, disruptive innovation, market share, red ocean strategy* and *value innovation*.
Board	A *board* is a: • Composite or wood, where the length greatly exceeds its width; or, • Group of personnel with responsibilities for advice, consultation, direction, examination, investigation, management, supervision, or combination. See also, *composite* and *shadow board*.

Phrase or Term	Definition
Body of Knowledge	A *body of knowledge* is the complete set of concepts, terms and activities that make up a professional domain, as defined by the relevant learned society or professional association. It is more than simply a collection of information or terms, a description of professional functions, a professional library or reading list, or a website or collection of websites. It may be: • A set of accepted and agreed upon nomenclatures and standards pertaining to a field or profession; • A set of knowledge within a profession or subject area, which is generally agreed as both generally known and essential; • The structured knowledge that is used to guide members' practice or work and that an individual is expected to have mastered to be certified or considered as a practitioner; or, • The systematic collection of activities and outcomes in terms of their constructs, instantiations, models, principles, and values, which arise from continuous discovery and validation by the members of the profession. See also, *body of work* and *project management body of knowledge*.
Body of Work	A *body of work* is the entirety or a substantial part of the academic, creative, or professional output of a person or group. For example, the body of work put forth by the Jane Goodall Institute about great apes has become the cornerstone of modern primatology. *See also, body of knowledge.*
Body Language	*Body language* is subtle nonverbal communication through conscious or unconscious physical movements (e.g., breathing, pointing, posture, pulse, shrugging shoulders, and shaking) and gestures (e.g., blushing, facial expressions, mannerisms, and perspiration). For example, frowning and lowering one's head could indicate distress or sadness. This communication may be obvious (e.g., turning your back to a speaker mid-sentence to indicate disagreement) or subtle (e.g., a micro-expression such as a brief eye roll to indicate disagreement) that may be noticed, subconsciously noticed, or unnoticed. Body language strongly influences one's perception of others. It provides cues about how to interpret mood, motivation, and receptiveness. In face-to-face settings, body language is used more frequently than verbal communication. See also, *emotional intelligence*, *mirroring*, *paraverbal communication*, and *strong presence*.
Boilerplate Text	*Boilerplate text* is standardized text (e.g., PO notes) that includes detailed commercial and legal information for terms and conditions, and provides instructions for expediting, invoicing, and other subjects, including SQS (the fine print). Like legalese, this text is deliberately prepared with reuse in mind, such that all or most text does not need to be changed when it is used on a project. Boilerplate text is purposefully general, making it applicable to virtually any project.
Boiling Frog Syndrome	The *boiling frog syndrome* is a fable that describes a frog being slowly boiled alive. The premise is that if a frog is put into boiling water, it will jump out. If the frog is put in tepid water, which is then slowly brought to a boil the frog will not perceive the danger and will be boiled to death. This is used as a metaphor for the inability or unwillingness of people to be aware of, or react to, sinister threats that arise gradually instead of suddenly. In his novel *The Story of B*, environmentalist Daniel Quinn uses this metaphor to describe human history. See also, *ostrich effect* and *slippery slope*.
Bonder	A *bonder* performs a manual or semi-automatic bonding operation. See also, *bonding*.

Phrase or Term	Definition
Bonding	*Bonding* is to produce a permanently-fused joint in non-metallic piping by applying: • Adhesive or solvent cement; • Electrofusion; • Heat and pressure, with or without filler material; or, • Reinforcing fabric and resin. See also, *bonder*, *bonding operator*, *bonding procedure specification*, *filler metal*, and *fusion*.
Bonding Operator	A *bonding operator* operates automatic bonding equipment or machinery. See also, *bonding*.
Bonding Procedure Specification	A *bonding procedure specification* is a document that lists parameters to be used for the production of bonded joints per the code of construction. See also, *bonding*.
Book Smart	To be *book smart* is to be intelligent and have knowledge gained from education, with a good understanding of academic and theoretical matters, but without practical experience or street smarts. See also, *academic*, *armchair critic*, *easy way*, *hands on*, *hindsight is 20/20*, *on-the-job learning*, *hard way*, *school of hard knocks*, and *street smart*.

Boom	A *boom* is a:
	• Deep, loud, resonant sound;
	• Rapid expansion or increase such as:
	  o A development or settlement of an area or city (e.g., during the 1970s Calgary, Alberta, Canada was a boom town);
	  o Widespread expansion of economic activity (e.g., a gold rush); or,
	  o An upsurge in activity, interest, or popularity. For example, a surge in the popularity of soccer (football) when the Fédération Internationale de Football Association's World Cup competition is held.
	• Spar or outrigger (e.g., used for aircraft or sail boat structures);
	• Supporting arm or brace (e.g., to hold a microphone or cymbal); or,
	• Temporary floating barrier used to contain a chemical or oil spill on water.
	See also, *boon* and *bust*.

Boon	A *boon* is a: • A favour or request; or, • Something that makes life better or easier (e.g., a benefit or blessing). See also, *boom* and *boondoggle*.

Phrase or Term	Definition
Boondoggle	A *boondoggle* is an impractical or wasteful activity or project that may be politically motivated or involve white-collar crime. For example, when the Montréal–Mirabel International Airport at Mirabel, Quebec opened in 1975, it was the largest airfield ever developed. It comprised nearly 400 square kilometres of expropriated farmland and cost CAD$ 500 million (almost CAD$ 2.5 billion in 2021 dollars). It was doomed to fail because it was located a one-hour drive outside of Montreal, Quebec, the light rail transit system connecting it to Montreal was never built, and travel by supersonic passenger jet never materialized. The last commercial passenger flight departed in 2004. Today it is used primarily for cargo, general aviation, and MEDEVAC flights, and by Airbus Canada and Bombardier Aerospace for manufacturing. See also, *boon*, *bust*, and *white elephant*.
Boot Camp	A *boot camp* is: • A military basic training program with strict discipline, especially for new recruits. For example, a 10-week training program that teaches basic military skills, ethical values, first aid, physical fitness, and weapons handling; or, • An intense and short training program for basic or initial instruction, or to quickly improve physical fitness or technical skills. For example, an intensive 2-week training program for newly-hired salespeople that quickly teaches them about an organization's products, services, sales strategies, and systems). See also, *boot camp*, *continuing education*, *cross-train*, *facility*, *lifelong learning*, *lunch and learn*, *on-the-job training*, *professional development unit*, and *workshop*.
Bootleg Document	A *bootleg document* or *bootleg copy* is an unofficial, interim draft of a document produced on an ad hoc basis that facilitates informal, advanced stakeholder reviews of the draft. Obtaining a bootleg copy is quicker and easier than officially obtaining a copy through the Document Control department. Moreover, the bootleg copy is typically more up-to-date than the official copy held by Document Control (unless Document Control holds the final copy). Typically, a *bootleg copy* is superseded by a final copy (per Document Control). Etymologically, the term *bootleg* originally referred to transporting alcohol (unlawfully made, and without registration or paying taxes). The term *bootleg* is American slang derived from concealing a flask of alcohol down the leg of a high boot.
Borescope Examination	A *borescope examination* is a visual inspection aided by a mechanical or electro-mechanical device that examines the inside of inaccessible equipment or surfaces (e.g., a pipe weld).
Bottleneck	A *bottleneck* is a pinch point or delay in work caused by a component of the process (e.g., department, facility, machine, or resource that is already working at its full capacity) that can handle only a finite amount of additional work. A bottleneck limits the throughput of a process to those that will perform the next steps of the process (e.g., a baker can only bake bread if the dough makers provide the dough). If a bottleneck is not addressed, the process will slow to the pace at which the bottleneck can produce work, chronically slowing the entire process and affecting the bottom line. See also, *bottom line*, *debottleneck*, and *theory of constraints*.

Phrase or Term	Definition
Bottom of the Market	The *bottom of the market* refers to low-end products that are the cheapest or least advanced in a company's product range or the market. For example, a vehicle manufacturer introduces a new, basic, and inexpensive model that will be affordable for many customers (e.g., the Honda Civic). See also, *market trend*, *top of the market*, and *niche market*.
Bottom Line	The *bottom line* is the essential or salient point; the primary or most important consideration. It is also the line at the bottom of a financial report that shows the net profit or loss. See also, *triple bottom line*.
Bottom Line Up Front	*Bottom line up front* means to summarize the main point of a correspondence at the beginning of the correspondence (i.e., at the top of an email or letter). This method of communication conveys main points immediately and focuses the reader on the topic (allowing quick reader response). With this method, readers are not forced to review all details; instead, they understand the email or letter quickly. See also, *bottom line*.
Bottom Up Analysis	*Bottom up analysis* is a method of market analysis that calculates a total sales figure or market share by estimating potential sales. This includes an evaluation of where products can be sold, the sales of comparable products, and the sales that can be achieved. This method takes more effort than a top down analysis, but the result is usually more accurate. See also, *marketing* and *top down analysis*.
Box Breathing, Battle/Bomb Breathing, Four-square Breathing, Resetting Your Breath, Square Breathing, One-Minute Breath	*Box breathing, battle/bomb breathing, four-square breathing, resetting your breath, or square breathing* is a breathing technique that uses slow, deep breaths to: • Heighten concentration and performance; and, • Relieve stress. This technique can be employed several times a day. To employ this technique: 1. Find a stress-free, quiet environment where you can focus on your breathing; 2. Sit upright in a comfortable chair, with your feet flat on the floor, and hands relaxed on your lap with palms facing up; 3. Take deep breaths and then: a. Slowly exhale through your mouth, expelling all the oxygen out of your lungs; b. Slowly and deeply inhale through your nose (count to 4); c. Hold your breath (count to 4); d. Exhale through your mouth (count to 4); e. Hold your breath (count to 4); and, f. Repeat Step 3 as required (e.g., 4 to 10 times). During steps 3b to 3e, count to 4 slowly. Also, consider using a longer count (e.g., 6 to 10). The *one-minute breath* is a similar breathing technique that works like this: 1. Sit in a meditative posture; 2. Inhale for 20 seconds; 3. Hold for 20 seconds; and, 4. Exhale for 20 seconds. If required to build stamina, start breathing using 10-second, 15-second, and finally 20-second intervals. See also, *tricks of the trade*.

Phrase or Term	Definition
Brain Drain, Human Capital Flight, Brain Gain	*Brain drain* or *human capital flight* refers to the substantial emigration of educated or intelligent personnel (e.g., academics, business people, engineers, healthcare professionals, scientists, and others) from one geographical area, industry, or organization to another. This occurs when better benefits and salary, living conditions, or other factors, or a combination, are available elsewhere. Consequently, personnel migrate to: • Escape crime, oppression, poverty, turmoil, unrest, or a combination; • Exploit more favorable business or professional opportunities; or, • Obtain a higher standard of living for themselves and their families. Brain drain has pros and cons. Con: emigration harms the place of origin because the loss of talent emigrant diminishes supply of expertise; Pro: the migrant and the new place may realize greater economic benefits. The immigration of educated personnel may be referred to as a *brain gain* and is the opposite of brain drain.
Brain Dump	A *brain dump* is the act or an instance of comprehensively and uncritically expressing and recording one's ideas and thoughts about a topic (e.g., action items or a to-do list, concerns or plans, negative or positive thoughts, etc.). See also, *brainstorm*.
Brain Fart, Maladaptive Brain Activity Change	A *brain fart* or *maladaptive brain activity change* is: • A confusing thought or mental lapse (e.g., temporarily forgetting how to set your alarm clock); • A human error or temporary failure to reason correctly or remember something (e.g., forgetting the first name of someone you have known for years); • A poorly thought out or regrettable action or choice (e.g., having too many drinks after work with colleagues); or, • Something done or said impulsively (e.g., taking the day off work without giving notice). *Brain fart* is slang for the more formal *maladaptive brain activity change*, which was coined by Tom Eichele and a team of researchers in the research article *Prediction of Human Errors by Maladaptive Changes in Event-Related Brain Networks*. It is an honest or stupid mistake that can be made for no apparent reason that can have myriad consequences. For example, it was a real *brain fart when I left my laptop at home instead of taking it to the office for the meeting*. See also, *brain freeze*, *Darwin Awards*, *fundamental laws of stupidity*, and *reverse Darwinism*.
Brain Freeze	*Brain freeze* is a: • Severe and sudden but brief cold-stimulus headache induced by the rapid consumption of a cold drink or food (e.g., eating ice cream very quickly). The scientific term for this phenomena is *sphenopalatine ganglioneuralgia*; or, • Temporary mental confusion or failure to perform. For example, being unable to recall a fact one should easily remember (e.g., the name of a colleague). See also, *brain fart*, *Darwin Awards*, *fundamental laws of stupidity*, and *Reverse Darwinism*.

Phrase or Term	Definition
Brainstorm	A *brainstorm* is an activity performed by one or more people to quickly generate many ideas for a future endeavour (e.g., corporate or project improvement). After brainstorming, ideas generated should be carefully considered and evaluated. Brainstorm is a portmanteau of *brain* and *storm*. See also, *brain dump*, *creative destruction*, *focus group*, *lessons learned*, *whiteboard session*, and *unfocus group*.
Brainwashing	*Brainwashing* is to: • Force someone to give up their political, religious, or social attitudes and beliefs to: o Allow the introduction and acceptance of contrasting, regimented or new, unwanted ideas and thoughts; and, o Reduce their ability to think critically or independently. • Make someone believe something by repeatedly telling them that it is true and preventing them from receiving other information or thinking objectively; or, • Persuasion by propaganda or salesmanship. This term was first used by Edward Hunter in 1950 to describe how the Chinese government appeared to make its citizens cooperative. The concept is recognized but the term is not generally accepted as scientific. It may involve the use of indoctrination, psychological pressure, or even torture. Brainwashing is a portmanteau of *brain* and *wash*. It is also known as *coercive persuasion*, *mind control*, *re-education*, *thought control*, and *thought reform*. See also, *bullshit*, *disinformation*, *fake news*, *false narrative*, *media bias*, *misinformation*, *propaganda*, *rhetoric*, *social engineering*, *thought control*, *thought crime*, *truth*, *truthiness*, and *untruth*.
Brake Horsepower	*Brake horsepower* is a measure of the power of an engine taken from the degree of resistance offered by a brake, which represents the useful power that an engine can sustain. The term *horsepower* is a foot-pound-second unit of power – originally adopted to compare the output of steam engines with the power of draft horses. It was later expanded to include engines, turbines, and electric motors. See also, *horsepower*.
Branch Connection	A *branch connection* is a fitting that provides an outlet from one pipe to another pipe, which is the same or smaller size. The main pipe into or onto which the branch connection is welded is usually called the run or header pipe. The connection may be at 90° (e.g., using a cross, O-let, or tee fitting) or another angle (e.g., 45° using an O-let or Y-lateral fitting). See also, *O-let* and *Y-lateral*.
Branch Line	Not a preferred phrase – see *service line*.

Phrase or Term	Definition
Brand	A *brand* is: • A design, feature, mark, name, symbol, or term that identifies a product or service as distinct when compared to other similar products or services (e.g., the Porsche logo); • A product or service provided by a company under a particular name (e.g., Apple iPhone); or, • The experience created for customers over time (i.e., reputation), which customers come to expect based on their experience with the brand. This is based on organizational values, which directs organizational behaviours and creates the customer experience. Experiences are built from in-person interactions (e.g., a customer visit a spa), online presence (e.g., Amazon), product use (e.g., the George Foreman Grill), service use (e.g., American Motor Association), and promotions (e.g., Ikea's Kitchen Event). See also, *attention interest desire and action principle*, *brand activism*, *customer centric*, *mindshare*, *market share*, *positioning*, *values*, and *win-win*.
Brand Activism, Cause Marketing	*Brand activism* or *cause marketing* is the involvement of a corporation in political discourse (e.g., to publicly take a stand for environmental or social change). This kind of advocacy is intended to benefit society and create more profit by using a new and differentiating marketing strategy. For example, campaigns by clothing manufacturer Patagonia have always trumpeted its environmental vows. Patagonia donated all profits it earned over its 2016 Black Friday Sale to local, environmental, non-governmental organizations. Further, their subversive *Don't Buy This Jacket* campaign helped raise awareness over the negative impacts of fast-fashion (i.e., inexpensive clothing produced rapidly by mass-market retailers in response to the latest trends) and the importance of recycling clothing. Companies that employ this marketing technique must be transparent (e.g., demonstrate a genuine commitment to the cause end-to-end or throughout their supply chain). Today, consumers' easy access to information serves to keep brands honest and on point about brand-touted activism. See also, *brand*, *conscious capitalism*, *crony capitalism*, *greenwashing*, *non-governmental organization*, and *woke washing*.
Brand-new	*Brand-new* is something: • Completely new and unused, or not previously owned or used by someone else. For example, a vehicle purchased from a dealership; or, • Recently introduced. For example, a brand-new Mercedes Benz model will be available next year. See also, *first-hand* and *second-hand*.
Brass Tacks	*Brass tacks* is an expression that means basic facts or realities, details of immediate and practical importance, or getting to the point. This is often expressed as *get down to brass tacks*. Brass tacks are used in the furniture trade to secure fabric because of their decorative appearance and imperviousness to rust. The etymology of this expression meanders and is circuitous. It has American roots but its origin is subject to debate.
Brazer	A *brazer* performs a manual or semi-automating brazing operation. See also, *brazing*.

<table>
<tr><th>Phrase or Term</th><th>Definition</th></tr>
<tr><td>Brazing</td><td>Brazing is a process that joins two or more base metals by melting a filler metal into the joint. Brazing does melt base metals. Brazing is possible at temperatures above 427°C (800°F). See also, filler metal, soldering, and welding.</td></tr>
<tr><td>Brazing Operator</td><td>A brazing operator operates automatic brazing equipment or machinery. See also, brazing.</td></tr>
<tr><td>Bread and Butter, Steak and Potatoes</td><td>Bread and butter is the:<ul><li>Activity or work that provides the main part of an organization or a person's income (e.g., consulting with banks is my bread and butter); or,</li><li>Issues or matters that are important to people who are affected personally (e.g., the local hardware store is concerned with Home Depot starting a business across the street because selling construction materials and tools is the hardware store's bread and butter).</li></ul>The expression steak and potatoes is synonymous with bread and butter. See also, brass tacks, and nuts and bolts.</td></tr>
<tr><td>Breadcrumbing</td><td>Breadcrumbing is the act of leading someone along using continuous or periodic communication (e.g., emails, phone calls, social media messages, or texts). For:<ul><li>Business, this is intended to:<ul><li>Get a candidate to participate in a lengthy hiring process (e.g., subsequent interviews) even if they are unlikely to get an offer. For example, in the hope that a better candidate will apply during the hiring process; or,</li><li>An employer provide just enough compensation to motivate personnel to keep working (i.e., prevent personnel from quitting) but not enough to adequately reward them for extra effort, hard work, or initiative like a good employer should.</li></ul></li><li>Matters of romance, this is intended to keep a potential mate interested (i.e., through flirting). This is true even if the potential mate is not attracted to, or interested in, the other person. For example, in the hope that a better romantic partner will be found.</li></ul>Someone who breadcrumbs misleads others to accomplish ulterior motives. See also, gaslighting and ulterior motive.</td></tr>
<tr><td>Break-fix Model</td><td>The break-fix model for megaprojects asserts that organizations and personnel do not know how to deliver successful projects or do not have the incentive to do so. Projects break when reality catches up with manipulated or optimistic estimates of benefits, costs, and/or schedule. Projects are then paused to refinance, reorganize or both to fix the problems and deliver the initially planned project with at least some success. See also, megaproject and megaproject paradox.</td></tr>
<tr><td>Breakthrough</td><td>A breakthrough is:<ul><li>A highly dramatic significant invention or improvement in performance, achieved through consistent, focused, and synergistic effort;</li><li>A dramatic, important and sudden advancement, development, or discovery;</li><li>An act of moving through or beyond an obstacle; or,</li><li>An instance of achieving success in an activity or an area of expertise or interest.</li></ul></td></tr>
</table>

Phrase or Term	Definition
Bribe	A *bribe* is offering, giving, receiving, or soliciting an item of value to influence the actions of someone (e.g., a government official, police officer, or businessman) illegally. For example, offering a project money to award a purchase order. Bribes are used to acquire advantage, emolument, goods, money, objects of rights in action, preferment, privilege, property, or value. A bribe may also be a promise to induce or influence an action or vote. A gift of money or other item of value, which is otherwise available to everyone on an equivalent basis, and not for dishonest purposes, is not a bribe. For example, offering a discount or a refund to a purchaser is a legal rebate (not a bribe). A bribe may also be called *grease* (slang) or a facilitation payment. See also, *kickback* and *white-collar crime*.
Brick-and-mortar Business	A *brick-and-mortar business* offers products and services to its customers face-to-face in a physical building (e.g., a physical shoe store). This kind of business has difficulty competing with online businesses because it has higher operating costs and does not realize the cost savings available only to online businesses. Brick-and-mortar refers to the materials that traditionally comprise buildings. The term was first used by Herman Melville in his book *Moby Dick*. See also, *web-based business*.
Broken Windows Theory	The *Broken Windows Theory* is a metaphor for disorder within neighbourhoods. It asserts that each unattended problem that occurs in an environment negatively affects attitudes toward that environment and leads to more problems. This theory links disorder and incivility within a community to subsequent serious crimes. As a corollary to this theory, when problems are solved when they arise and the environment is well-managed, it positively affects attitude and leads to continued good management and maintenance. This theory was developed in the article *Broken Windows* by James Q. Wilson and George Kelling. They proposed that serious crime is the result of a lengthier chain of events. They theorized that crime emanated from disorder; if disorder were eliminated, then serious crimes would not occur. Pre-dating this theory, scholars and police focused on serious crime (e.g., murder and robbery). Wilson and Kelling took a different view. They stated: *... consider the pavement. Some litter accumulates. Soon, more litter accumulates. Eventually, people ... leav[e] bags of refuse from take-out restaurants ... or even break into cars.* The Broken Windows Theory teaches that good management and maintenance yields order.
Brownfield	A *brownfield* project involves new construction at an operating or previously operating facility.

<table>
<tr><th>Phrase or Term</th><th>Definition</th></tr>
<tr><td>Brazing</td><td>Brazing is a process that joins two or more base metals by melting a filler metal into the joint. Brazing does melt base metals. Brazing is possible at temperatures above 427°C (800°F). See also, filler metal, soldering, and welding.</td></tr>
<tr><td>Brazing Operator</td><td>A brazing operator operates automatic brazing equipment or machinery. See also, brazing.</td></tr>
<tr><td>Bread and Butter,
Steak and Potatoes</td><td>Bread and butter is the:<ul><li>Activity or work that provides the main part of an organization or a person's income (e.g., consulting with banks is my bread and butter); or,</li><li>Issues or matters that are important to people who are affected personally (e.g., the local hardware store is concerned with Home Depot starting a business across the street because selling construction materials and tools is the hardware store's bread and butter).</li></ul>The expression steak and potatoes is synonymous with bread and butter. See also, brass tacks, and nuts and bolts.</td></tr>
<tr><td>Breadcrumbing</td><td>Breadcrumbing is the act of leading someone along using continuous or periodic communication (e.g., emails, phone calls, social media messages, or texts). For:<ul><li>Business, this is intended to:<ul><li>Get a candidate to participate in a lengthy hiring process (e.g., subsequent interviews) even if they are unlikely to get an offer. For example, in the hope that a better candidate will apply during the hiring process; or,</li><li>An employer provide just enough compensation to motivate personnel to keep working (i.e., prevent personnel from quitting) but not enough to adequately reward them for extra effort, hard work, or initiative like a good employer should.</li></ul></li><li>Matters of romance, this is intended to keep a potential mate interested (i.e., through flirting). This is true even if the potential mate is not attracted to, or interested in, the other person. For example, in the hope that a better romantic partner will be found.</li></ul>Someone who breadcrumbs misleads others to accomplish ulterior motives. See also, gaslighting and ulterior motive.</td></tr>
<tr><td>Break-fix Model</td><td>The break-fix model for megaprojects asserts that organizations and personnel do not know how to deliver successful projects or do not have the incentive to do so. Projects break when reality catches up with manipulated or optimistic estimates of benefits, costs, and/or schedule. Projects are then paused to refinance, reorganize or both to fix the problems and deliver the initially planned project with at least some success. See also, megaproject and megaproject paradox.</td></tr>
<tr><td>Breakthrough</td><td>A breakthrough is:<ul><li>A highly dramatic significant invention or improvement in performance, achieved through consistent, focused, and synergistic effort;</li><li>A dramatic, important and sudden advancement, development, or discovery;</li><li>An act of moving through or beyond an obstacle; or,</li><li>An instance of achieving success in an activity or an area of expertise or interest.</li></ul></td></tr>
</table>

<table>
<tr><th>Phrase or Term</th><th>Definition</th></tr>
<tr><td>Budget</td><td>A budget is an estimate of costs and resources to be used over a specified period. It may also estimate revenues or be a quantity of material, typically provided in writing (e.g., a list). A budget serves as a:<ul><li>Method for managing adverse situations;</li><li>Plan of action for achieving quantified objectives; and,</li><li>Standard for measuring performance.</li></ul>See also, budgeting, contingency, definitive estimate, Delphi technique, estimate at complete, and estimate to complete, and inspection cost estimate.</td></tr>
<tr><td>Budgeting</td><td>Budgeting is a discipline that determines:<ul><li>Allocation of funds and resources for each activity or project; and/or,</li><li>Which activities or projects should and should not be fully funded or resourced (i.e., where financial funding should be increased, reduced, or stopped).</li></ul>See also, budget and stop doing list.</td></tr>
<tr><td>Build Absolutely Nothing Anywhere Near Anyone, Build Absolutely Nothing Anywhere Near Anything, BANANA</td><td>The phrase build absolutely nothing anywhere near anyone, build absolutely nothing anywhere near anything, and BANANA refers to the ongoing opposition of some groups to oppose all proposed development, suggesting these groups seek an absence of growth or development, seek to impede it, or both. A similar phrase is citizens against virtually everything. See also, not in my backyard.</td></tr>
</table>

Phrase or Term	Definition
Build Back Better	1. United Nations *Build Back Better* is defined by the United Nations as the use of the recovery, rehabilitation and reconstruction phases after a disaster to increase the resilience of nations and communities through integrating disaster risk reduction measures into the restoration of physical infrastructure and societal systems, and into the revitalization of livelihoods, economies, and the environment. This concept is common sense to the Japanese, due to their historical experience recovering from disasters and preparing for the future. This phrase was introduced by the Japan International Cooperation Agency Reconstruction Team that rebuilt infrastructure in Central Java after the devastating 2006 earthquake. They reconstructed 100,000 homes with earthquake-resistant engineering and improved materials. 2. World Bank *Build Back Better* is defined by the World Bank to mean that the repaired or replaced assets are more resilient, but also that the recovery process is shorter and more efficient, and that the entire recovery process does not leave anyone behind. Further, it is the alignment of these three strategies: • Building back *stronger* to reduces losses by ensuring that reconstructed infrastructure can resist more intense events in the future • Building back *faster* to reduce disaster impacts by accelerating reconstruction through measures such as contingent reconstruction plans, pre-approved contracts, and financial arrangements; and, • Building back *more inclusively* to ensure that post-disaster support reaches all affected population groups. 3. COVID-19 *Build back better* has been proposed for worldwide recovery plans after the COVID-19 pandemic, with investments in green technologies and new environmental policies. It has also been proposed as the basis for an updated social contract that always treats people as essential, not just times of crisis – with fair and safe workplaces, higher wages, and better benefits. See also, *green new deal*.
Building Information Modelling	*Building information modelling* is a process that involves the generation and management of digital representations of functional and physical characteristics of places. Files are usually generated in proprietary formats and contain proprietary data, which can be exchanged and networked to support decision-making. See also, *enterprise resource planning*, and *virtual design and construction*.
Bulk Items	*Bulk items* are commodities purchased in large numbers (e.g., a large number of fittings or small bore valves). See also, *commodity* and *untagged equipment*.
Bull Market	*A bull market* is a general rise in prices over time (e.g., beginning when stocks rise 20% from their low and ending when stocks drawdown 20%). See also, *bear market*, *market top*, and *market trend*.

Phrase or Term	Definition
Bullshit Detector, Bullshit Filter, Blabla Meter	A *bullshit detector*, *bullshit filter*, or *blabla meter* are informal, figurative expressions that describe the ability to detect corruption, deception, dishonesty, falsity, fraud, hypocrisy, or insincerity in others. For example, when the company said their stock would increase by 500% this year, my bullshit detector went off. See also, *bullshit*, and *truth filter*.
Bullshit, Bullshite, Bullcrap, Bull, BS	*Bullshit*, *bullshite*, *bullcrap*, *bull*, or *BS* describes an exaggerated, foolish, nonsensical, stupid, or false statement that may be used to: • Deceive or mislead; or, • Impress or persuade. In his book *On Bullshit*, Harry Frankfurt characterized bullshit as a form of untruth distinct from lying. This is because a liar understands the truth, but deliberately misleads, while a bullshitter does not care about the truth and only seeks to impress. There are many comparable terms including *bullocks* (chiefly British English) and *horseshit* (a synonym). The word bullshit is considered by many to be profane. See also, *brainwashing*, *bullshit detector*, *disinformation*, *fake news*, *false narrative*, *media bias*, *misinformation*, *propaganda*, *smells fishy*, *stretch the truth*, *truth*, *truth filter*, *truthiness*, and *untruth*.
Bully	A *bully* is a predatory person who is habitually abusive, aggressive, coercive, cruel, domineering, harmful, insulting, intimidating, or threatening to others whom they perceive as smaller, vulnerable, or weaker. Ironically, it was once an adjective that meant *superb* or *wonderful*. See also, *bully pulpit*, *cyberbullying*, *flame war*, *flaming*, *office jerk*, *save face*, *toxic behaviour*, and *trolling*.
Bully Pulpit	A *bully pulpit* is a position of authority or public office that provides its occupant with an outstanding opportunity to speak and be heard by a large audience about any topic. This phrase was coined by United States President Theodore Roosevelt, who in 1902, first referred to his office as a bully pulpit. By this, he meant his office was a terrific platform from which to advocate an agenda. Where his predecessors believed that powers not granted were forbidden, Roosevelt asserted that powers not forbidden were granted. In his autobiography, he wrote that he did not *usurp* power, but instead *greatly broadened* executive authority. At this time, the word bully had positive connotations. Roosevelt used this word as an adjective, which meant *superb* or *wonderful*. See also, *bully*.
Bureaucracy	*Bureaucracy* refers to: • A government characterized by adherence to fixed rules, a hierarchy of authority, and specialization of functions; • A system of administration marked by officialism, proliferation, and red tape; or, • An administrative policy-making group or a body of non-elected government officials. *Bureau* is French for *desk*; thus, *bureaucracy* means *government by people at desks*. See also, *bureaucrat*, *Catch-22*, and *red tape*.
Bureaucrat	A *bureaucrat* is a member of a bureaucracy (e.g., an official in a government department). This term may be used as a slur about a governmental official who is perceived as being concerned with procedural correctness at the expense of public needs. See also, *bureaucracy*.

Phrase or Term	Definition
Burn Rate	The *burn rate* is the rate at which an enterprise (e.g., a company or project) spends money in excess of its financial resources or income. For example, it spends venture capital to finance overhead costs before generating positive cash flow from operations. It is a measure of negative cash flow usually measured in cash spent per month.
Burst the Bubble	To *burst the bubble* means to bring a sudden end to a very happy, profitable, or successful experience. For example, *I don't mean to burst your bubble while you are celebrating the birth of your first child, but the police are here to arrest you*! See also, *bubble* and *burst the bubble*.
Burst Someone's Bubble	To *burst someone's bubble* means to cause them to suddenly understand that something they admire, believe, or trust is not good or true. See also, *bubble* and *burst someone's bubble*.
Business	*Business* refers to: • A commercial economic activity or transaction (e.g., to buy, lease, or sell); • A commercial or industrial enterprise (e.g., a company); • A function, objective, role, or task (e.g., *recruiting is my business*); • A serious activity requiring effort and time without distractions (e.g., *get down to business*); or, • An affair, matter, personal concern, or right (e.g., *the choice is my business*).
Business Analysis	*Business analysis* is the practice of identifying and solving business problems. It focuses on creating and implementing solutions to business needs using organizational development, process reengineering, or any number of other methods.
Business Casual	*Business casual* is a dress code that is less formal than traditional business wear; yet is still professional and appropriate in an office setting. Guidelines vary; however, these are typically: • Gender-neutral Chinos, twill, or other non-denim pants with a button-down shirt, polo shirt, or sweater. • Men Dress khakis, trousers, or slacks with a button-down shirt, polo shirt, or sweater. A tie may be optional, and a jacket is not required. • Women Dress trousers or a knee- or mid-length skirt with a blouse, button-down shirt, polo shirt, or sweater. Accessories such as a belt, jewelry, shoes, and socks must be clean and compliment attire per the dress code. Dress codes may vary considerably by company, industry, and workplace. Many organizations have a dress code policy that includes a designated day (e.g., casual Friday or other occasion) when business casual or less formal attire is permitted (e.g., blue jeans and a shirt without a collar that are clean and in good condition). See also, the *three Bs rule*.

Phrase or Term	Definition
Business Continuity Planning	*Business continuity planning* is the process of creating a plan for the prevention and recovery from potential threats to a business. The plan will also enable ongoing operations before and during execution of disaster recovery. This ensures that assets and personnel are protected and able to function in the event of an unforeseen circumstance or event. The plan is prepared in advance and involves input from key stakeholders and personnel. It involves defining all business continuity risks that can affect business operations. Threats and disruptions result in increased costs and a loss of revenue, which leads to reduced profitability, and business insurance alone may not cover all the costs, Clients or customers may go to the competition or the competition may exploit the opportunity to usurp market share. See also, *business continuity risk*.
Business Continuity Risk	*Business continuity risk* is any threat to a business' operations (i.e., assets and personnel) including: • Civil disturbance, piracy, terrorism, vandalism, or war; • Dependencies not defined or documented (e.g., for processes and technology); • Failure of critical components (e.g., assets or personnel) or mission-critical systems (e.g., due to data corruption or error); • Globally-reaching diseases, including deleterious viruses, epidemics, or pandemics. • Fraud, sabotage, or theft (e.g., external or internal for assets or information); • Infrastructure outages (e.g., Internet, electricity, natural gas or other fuel, telephone, water, or other utilities); • Lack of adequately trained personnel (e.g., expertise or number); • Man-made disaster (e.g., cyberattack such as a denial of service attack, phishing attack, malware or ransomware, or an active attack exploiting network security vulnerabilities); • Natural disaster or act of god (e.g., earthquake, fire, flood, hurricane, typhoon, or another weather-related event); • Security breach (e.g., unauthorized building or system access); • Single point of failure (e.g., a single network device or a person who is the only one possessing certain specific expertise or knowledge); and, • Unavailability of a key supplier-provided products or services. *See also, business continuity planning, risk management*, and *single point of failure*
Business Development	*Business development* is the activity of pursuing strategic opportunities for a particular business or organization. For example, cultivating partnerships or other commercial relationships to win contracts or purchase orders, and identifying new markets for existing products and services.

Phrase or Term	Definition
Business Model Canvas	The *Business Model Canvas* is a strategic management template for developing a new business model or documenting an existing one. It is a visual chart with elements that describe an organization or product's customers, finances, infrastructure, and value proposition. It assists an organization to align activities by illustrating potential compromises. The *nine building blocks* of the business model design template that became the Business Model Canvas were initially developed by Alexander Osterwalder beginning in 2004. Since its release around 2008, new canvases for specific business niches have also been developed. This concept was also developed in the book *Business Model Generation: A Handbook for Visionaries, Game Changers, and Challengers* by Alexander Osterwalder and Yves Pigneur. See also, *value proposition*.
Business Objective	*A business objective* is a result that a business or company aims to achieve. It includes the strategies that will be used and may include a time frame and list available resources. See also, *objective*.
Business Process Mapping	*Business process mapping* is a graphical method of understanding and visualizing a business process. It presents graphical data in a structured way that illustrates the steps of the process, the inputs and outputs, who is responsible, and other relevant information. Four common business process mapping documents are: 1. A flowchart, which represents a process or step-by-step approach to completing a task. It illustrates the steps as text within boxes, and indicates sequence by connecting the boxes with arrows. A flowchart can be drawn by hand or by using software (e.g., MS Visio); 2. A supplier inputs processes outputs customer diagram, which is a simplified map focused on process essentials and the people involved. This diagram excludes extraneous data; 3. A swim lane diagram, which is similar to a flowchart, except it is structured in terms of who does what; and, 4. A value stream map, which is more in-depth than a flowchart and may be more difficult to analyze at a glance because it contains detailed data.
Business Process Modeling	*Business process modeling* is the representation, analysis, and evaluation of business processes to improve them.
Business to Business	*Business to business* refers to the process of businesses selling products and services to each other (e.g., distributors, manufacturers, retailers, and wholesalers). See also, *business to consumer*, *business to government*, *customer to business*, and *customer to customer*.
Business to Consumer	*Business to consumer* refers to the process of a business selling products and services directly to the consumers who are the end-users of its products or services. This includes e-commerce and selling products and services using the Internet. See also, *business to business* and *business to government*, *customer to business*, and *customer to customer*.
Business to Government	*Business to government* refers to business conducted between private businesses and governments. See also, *business to business* and *business to consumer*, *customer to business*, and *customer to customer*.

Phrase or Term	Definition
Business Wargame, Corporate Wargame	A *business wargame* or *corporate wargame* is the strategic application of a business plan to test it against likely external reactions. In business situations where incorrectness is expensive, wargames can be used to understand challenges and opportunities. Generally, there are three types of business wargames: 1. Business is a Game These games regard business transactions as a game between participants with potentially conflicting goals. 2. Business is Business These games regard business as neither a game nor a war, so competitors do not aim to defeat each other, but to satisfy customers' preferences better than others. 3. Business is War These games are a direct adaptation of military wargames that envision competitors as the *enemy* and the goal as *victory* in a market *battle*. To some, the term *wargame* is an accurate or useful depiction of business and competitive strategy; however, others find the war metaphor counterproductive, preferring the phrase *strategy game* instead. Other business wargame methods include board games, negotiation wargames, and role-play. See also, *game*, *games against nature*, *gamesmanship*, *gamification*, *lose-lose*, *war room*, *wargame*, *win-win*, and *zero-sum game.*
Bust	A *bust* (noun) is: • A complete failure; • A depression; • A sculptured representation of the upper part of the human figure including the head, neck, and part of the shoulders and breast; or, • The upper part of the human torso between the neck and waist. To *bust* (verb) is to: • Break or smash with force; • Break down; • Bring to an end or breakup; • Demote; • Go bankrupt or broke; • Make an arrest; or, • Tame (e.g., a wild horse). See also, *boom* and *depression*.

Phrase or Term	Definition
Busy Work, Busywork	*Busy work* or *busywork* is an activity undertaken to pass time that has little or no actual value. For example, to sharpen pencils that have already been sharpened. This may occur when personnel: • Need to be occupied (e.g., an instructor gives students additional exercises to fill remaining class time); • Need to be present at a workplace but do not have a task to perform; or, • Want to appear busy to avoid being asked to work (e.g., performing administrative paperwork to avoid going outside to check equipment). See also, *fool's errand* and *make-work job*.
Butt Joint or Butt Weld	A *butt joint* or *butt weld* is when two pieces of metal in the same plane are joined by welding. See also, *joint*, *lap weld* and *socket weld*.
Buyer	A *buyer* performs a purchasing role for a project, especially with the bidders before a purchase order is awarded. See also, *supplier*.
Buyer's Remorse	*Buyer's remorse* is a buyer's feeling of guilt or regret after purchasing a product or service regarded as expensive, extravagant, or unnecessary (e.g., with the most luxurious accessories, options, styles, and trims). For example, to purchase a new vehicle with a USD$ 1 million option that allows it to submerge and navigate underwater when you live in the desert. See also, *caveat emptor*.

<table>
<tr><th>Phrase or Term</th><th>Definition</th></tr>
<tr><td>Buzzword, Buzz Phrase, Buzz Term, Catch-Phrase</td><td>A buzzword, buzz phrase, buzz term, or catch-phrase is a word or phrase that may be:
<ul>
<li>Fashionable and used to impress or persuade rather than to inform. For example, the term localization describes the process of adapting a design or product to meet the specific unique requirements of an area or market, to account for local culture or preferences.</li>
<li>A technical term that:
<ul>
<li>Is used as a shortcut or shorthand method for quicker communication between personnel who are informed of the context and meaning. For example, using comms instead of communications for brevity; or,</li>
<li>Sounds authoritative, catchy, important, or memorable. For example, using domestic engineering instead of housecleaning;</li>
</ul></li>
<li>A useful term adopted into general popular speech but has lost its usefulness (e.g., becomes non-descriptive or vague). For example, the term organic may:
<ul>
<li>Be a synonym for the words authentic, natural, or unpaid;</li>
<li>Describe how marketing works;</li>
<li>Mean authentic (e.g., communication of businesses values or to keep one's promises);</li>
<li>Food produced using certain practices (e.g., resource rotation, ecological balancing, and biodiversity conservation) with restricted use of fertilizers and pesticides, and avoiding or excluding use of industrial solvents, irradiation, or synthetic food additives;</li>
<li>Growth that occurs naturally (e.g., growth of a business or trees in a forest);</li>
<li>Something that just happens or that people can easily understand (e.g., the natural flow of water in a tributary); or,</li>
<li>Word of mouth advertising (i.e., people have non-incentivized discussions about a product or service).</li>
</ul></li>
<li>An acronym, jargon, newly-coined term, or portmanteau word that becomes popular, at a certain time or in a certain context, but has lost its original meaning. For example, mind your Ps and Qs.</li>
</ul>
Buzzwords can be ubiquitous in many business, technical, and professional workplaces. For example, in many offices, a drive-by is a buzz phrase for visiting someone's office without prior planning. Some consider buzzwords to be the anti-thesis of clear communication; others feel they quickly convey meaning without the need to be verbose.

Additional examples of buzzwords abound, including actionable, conceptualized, key performance indicator (KPI), on the same page, and out of the box. See also, acronym, drive-by, gibberish, jargon, key performance indicator, localization, mind your Ps and Qs, out of the box, plain English, portmanteau word, word of mouth advertising, and word salad.</td></tr>
<tr><td>By the Book</td><td>To do something by the book means doing it correctly and exactly according to the official guidelines, instructions, norms, or rules.</td></tr>
</table>

Phrase or Term	Definition
C	1 A B **C** D E F G H I J K L M N O P Q R S T U V W X Y Z
C-suite, CxO	*C-suite* or *CxO* refers to the executive- or senior-level (i.e., most important corporate) managers within a company – those that have the word *chief* in their title. These are the corporate officers and directors. See also, *senior management*.

Cadre	A *cadre* is a:
	• Member of an activist group;
	• Small unified group or nucleus of people:
	  o Who have been chosen, organized, and trained for a specific profession or purpose; or,
	  o Within a larger organization (e.g., project team or political party).
	• Staff of key operational personnel around which an expanded organization can be built.
	See also, *project team*.

Calculated Risk	*Calculated risk* is: • A chance of failure or hazard or for which the degree of probability has been assessed or estimated before something is undertaken; • A risk considered to be worth taking because the result, if successful, will be beneficial; or, • An undertaking or the actual or possible product of an undertaking for which the chance of failure has been previously assessed or estimated. A calculated risk is the chance of exposure to injury or loss that might be undertaken after its advantages and disadvantages have been carefully considered and weighted. For example, a business takes a calculated risk when expanding business activities into a new area or market. See also, *risk* and *risk assessment*.

Phrase or Term	Definition
Calibrated Question™	A *calibrated question™* removes aggression from a conversation or negotiation by acknowledging the counterpart openly and without resistance. It allows the negotiator to introduce ideas or requests by nudging and without being pushy. It asks for help, gives the counterpart the illusion of control, and suspends unbelief. It avoids bravado and a confrontation and allows the counterpart to solve their problem in the way the negotiator wants. It educates the counterpart about what the problem is instead of causing conflict by telling them what it is. Calibrated questions are carefully designed to ease the negotiation towards the negotiator's goals while allowing the counterpart to think that is their choice to move in that direction. Calibrated questions avoid verbs or words like are, can, do, does, or is, which are closed-ended questions. Instead, the reporter's questions are used. Of these, the words *how*, *what*, and (sometimes) *why* are the most useful, as these words will inspire the counterpart to think and speak expansively. The word *why* can be considered accusatory and must be used with caution. The words *when*, *where*, and *who* are less useful because these words get the counterpart to share facts without thinking. Examples of calibrated questions include: • How am I supposed to do that? • How can we solve this problem? • What about this is important for you? • What brought us to this situation? A calibrated question's implicit request for help reduces defensiveness and triggers goodwill by creating a situation where the counterpart uses their emotional and mental resources to overcome the negotiators challenges. This guides them to design a solution that is the negotiator's solution. See also, *closed-ended question*, *defensive behaviour*, *objective reflective interpretive decisional questioning*, *precision questioning*, *reflective questioning*, *reporter's questions*, *open-ended question*, *precision questioning*, and *unbelief*.
Calibration	*Calibration* is the comparison of a measurement device (e.g., gauge or instrument) to an equal or better reference device that is known to be accurate or correct. Typically, the reference device has a higher accuracy than that of the measurement device. Calibration identifies if or how far off the measurement device is from the reference instrument's measurement. Measurement device readings can drift over time, so calibration ensures that the results are within the acceptable tolerance range, or if it needs adjustment, repair, or replacement. Calibration may be required annually, with every use, or at other intervals. See also, *out of tolerance* and *tolerance*.

Phrase or Term	Definition
Call a Spade a Spade	To *call a spade a spade* is to do one of the following, especially if it is unpleasant or uncomfortable: • Address or describe the true nature of someone or something. For example, tell your brother that he is lazy; • Call something by its correct name. For example, tell your brother he is an unemployable; or, • Speak frankly about something. For example, tell your brother that he needs to take care of his appearance and wear clean, business casual clothes to make a positive impression at his job interview. For example, *let's call a spade a spade, he is not economical with the truth—he is a liar*. A spade is a type of shovel. However, this term became a slang term and then a derogatory slur in the early 20th century when some used it to describe black people.
Canary in a Coal Mine, Sentinel Animal	A *canary in a coal mine* is a metaphor for an early indicator or warning of potential danger or failure. For example, bleached coral can warn of oceanic pollution. Another example: the financial market can warn investors and others of an upcoming recession. During the late 19th and 20th centuries, coal mines in Canada, the United Kingdom, and the United States used live canaries as a *sentinel animal* to detect the presence of carbon monoxide and other toxic gases, or a lack or oxygen. A bird is more sensitive to toxic conditions than a human because of its anatomy (e.g., high metabolism, rapid breathing rate, and small size). The canary becoming ill or dying was a warning that miners should evacuate. Scottish physiologist John Scott Haldane pioneered the use of sentinel animals (e.g., caged canaries and white mice) in mining operations in the 1890s. In the 1980s, canaries were replaced by electronic gas detection devices.
Can Do Attitude, Can-do Attitude	A *can do attitude* or *can-do attitude* is an eagerness and willingness to accept and meet challenges instead of giving up. For example, Muhammad Ali, one of the greatest boxers in history, constantly beat athletes that were bigger, faster, and stronger. His physical talent and skill contributed to his success; however, it was his positive mental attitude – his can-do attitude – that made him a champion. Someone with a can-do attitude uses failure as a learning opportunity to increase probability of future success. See also, *growth mindset*.
Canadian Central Registration Number	The *Canadian Central Registration Number* is a design registration number, allotted by a nationally recognized organization in accordance with procedures accepted by more than one province, that allows a fitting to be used in such provinces. See also, *Canadian Registration Number*.
Canadian Registration Number	A *Canadian Registration Number* is a design registration number issued by a pressure equipment jurisdiction in Canada per the requirements of CSA B51. These are issued for boiler, pressure vessel, or fitting designs. See also, *Canadian Central Registration Number*.

Phrase or Term	Definition
Cancel Culture, Call-out Culture	*Cancel culture* or *call-out culture* is a modern form of ostracism in which someone is pushed out of personal or professional circles of influence in real life, social media, or both (e.g., renowned personnel including celebrities and politicians). At its worst, it is about unaccountable groups successfully applying pressure to punish someone (e.g., by encouraging management to fire someone) for a perceived misdeed. Those who are subject to this ostracism are said to be *cancelled*. For example, an award-winning professor at the University of Southern California was placed on leave for using the Chinese word *nèi ge* (pronounced *neh-ga*), which means *that* in English, during a lecture about how different cultures use filler words. Some students thought it sounded like a racist English word, even though the professor was simply speaking Chinese. Chinese people use the word *nèi ge* as a filler word ... like the filler words *err* and *um* in English. Cancel culture is the force that causes a boycott or withdrawal of support after a company or individual has done or said something considered objectionable or offensive. See also, *blacklist*, *deplatforming*, *doxing*, *filler words*, *peer pressure*, and *shadow banning*.
Capability–expectations Gap	The *capability–expectation gap* is the difference between what is discussed or planned, and what is delivered. If the gap is to be closed, capabilities and results must be demonstrated, rather than aspirations and potential. In other words, action – not words – must close the gap. If the gap cannot be closed, the alternative is to lower expectations.
Capacity-Requirements Planning	*Capacity-requirements planning* is the process by which a company determines how much it needs to produce and if it can meet those production goals. Small businesses must conduct capacity-requirements planning regularly to keep pace with changes in supply and demand. The frequency with which capacity-requirements planning occurs depends on the industry and type of businesses. See also, *material requirements planning*.
Capital	*Capital* is wealth in the form of assets or money that are owned by a person or organization, or are available or contributed for a particular purpose (e.g., to invest, or operate or start a company). See also, *capitalism*, *investor's capital*, *return on capital employed* and *return on investment*.
Capital Expenditure, Capex	*Capital expenditure* or capex are the funds used by a company to acquire, maintain, and upgrade physical assets such as buildings, equipment, facilities, property, and technology. These funds are often used to undertake major investments or major new projects. See also, *capital* and *capital project*.
Capital Improvement Plan, Capital Improvement Program	A *capital improvement plan* or *capital improvement program* is a short-range plan (e.g., 4 to 10 years), which identifies capital projects and equipment purchases, options for financing, and a schedule for the plan. The plan provides a link between stakeholders and the comprehensive or strategic plans and the budget. See also, *capital project*.
Capital Project	A *capital project* requires a large and long-term investment to acquire, develop, improve, and/or maintain a capital asset (e.g., infrastructure or a revenue-producing production facility). See also, *capital*, *capital expenditure*, *infrastructure*, *non-capital project*, and *project*.

Phrase or Term	Definition
Capitalism	*Capitalism* is an economic system in which private individuals or businesses own capital and capital goods. The production of goods and services is based on supply and demand in the general market – known as a market economy. The United States employs capitalism. See also, *capital*, *conscious capitalism*, *crony capitalism*, *economic system*, *goods*, and *socialism*.
Car Seal, Valve Zip Tie	A *car seal* or *valve zip tie* is a simple device used to lock or seal a valve in the open or closed position, thus preventing unauthorized operation (e.g., a cable with a self-locking and breakaway key): • A self-locking seal becomes locked when the car seal cable enters the body of the valve. The cable can be pulled tight but not released; and, • A breakaway key becomes locked when the key is manually turned until it snaps off. The tip of the key is thereby forced into the cable, locking it in place. After a valve is car sealed, it can operate only if the seal is cut. By design, a cut car seal leaves evidence of unauthorized tampering or activity. See also, *lockout* and *tag out*.
Carbon	*Carbon* is a non-metallic chemical element that readily forms compounds with many other elements. It is a constituent of organic compounds in all known living tissues. The largest sources of inorganic carbon are carbon dioxide, dolomites, and limestones. Significant quantities occur in organic deposits of coal, methane hydrates, oil, and peat. It is also an alloy used for the production of steel. See also, *alloy*, *carbon steel*, *hydrocarbon*, *hydrogen*, and *low-carbon*.
Carbon Capture and Sequestration, Carbon Capture and Storage, Carbon Capture and Utilization or Storage	*Carbon capture and sequestration*, *carbon capture and storage*, or *carbon capture and utilization or storage* is a process where carbon dioxide is separated (i.e., captured) from a gas stream and diverted (i.e., sequestered) for use or buried for disposal underground.
Carbon Credit	A *carbon credit* represents the avoidance or removal of greenhouse gases equivalent to 1 tonne of carbon dioxide. Carbon credits can be bought and sold. See also, *greenhouse gas*.
Carbon Equivalent	*Carbon equivalent* is an empirical value in weight percent, relating the combined effects of different alloying elements used in the making of carbon steels, to an equivalent amount of carbon. This value can be calculated using various long or short mathematical equations. For example: Carbon Equivalent = Carbon + Manganese/6 + (Chrome + Molybdenum + Vanadium)/5 + (Nickel + Copper)/15). This value is used to determine the weldability of a carbon steel alloy: • <0.35% excellent; • 0.36 – 0.40% very good; • 0.41 – 0.45% good; • 0.46 – 0.50% fair; and, • >0.50% poor. See the code of construction or design specification to identify the formula that should be used. See also, *alloy*, *code of construction*, *design specification*, and *weldability*.

Phrase or Term	Definition
Carbon Footprint	A *carbon footprint* is the total amount of greenhouse gas produced by human activities, both directly and indirectly. It can be attributed to an individual, organization, or country, and is usually expressed in equivalent tons of carbon dioxide (e.g., 80 million metric tonnes of carbon dioxide equivalents). Human activities like driving, food production, and heating have associated carbon footprints. A carbon footprint is the sum of these emissions that were induced by human activities within a given period (e.g., usually a year). See also, *carbon credit*, *carbon intensity*, *greenhouse effect*, and *greenhouse gas*.
Carbon Intensity	*Carbon intensity* is the number of grams of carbon dioxide per unit of electricity generated. For electricity generation using coal power stations, the carbon intensity is high because carbon dioxide is produced as part of the power generation process. For electricity generation using renewable forms (e.g., hydro or solar power) almost no emissions are produced; thus, their carbon intensity is very low. The lower the carbon intensity, the greener (or more environmentally-sound) the electricity. See also, *carbon footprint*.
Carbon Leakage	*Carbon leakage* is when there is an increase in greenhouse gas emissions in one country because of an emissions reduction by a second country with a stricter climate policy. It is the increase in carbon dioxide emissions outside the countries taking domestic mitigation action divided by the reduction in the emissions of these countries. This occurs when a country's: • Emissions policy raises local costs. One consequence could be that another country with a more relaxed policy may now have a trading advantage. Production may then be moved offshore to a country with less expensive production costs and lower quality standards, and, ultimately, global emissions will not be reduced; or, • Environmental policy adds a premium to certain commodities or fuels forcing a decline in demand and a subsequence price reduction. Then, other countries that do not place a premium on those items may take up the demand and use the same supply, ultimately negating any benefits of the environmental policy. See also, *carbon*, *carbon footprint*, and *greenhouse gas emissions*.
Carbon Negative	*Carbon negative* means that an organization removes more carbon than it emits, annually. For example, in enhanced oil recovery, a field typically permanently stores more carbon than is produced in the oil. In cement, carbon-negative indicates a product that is made with, and stores more carbon than, is emitted during its creation. See also, *enhanced oil recovery*.
Carbon Pricing	*Carbon pricing* is the cost applied to carbon pollution to incent polluters to reduce the amount of greenhouse gas they emit into the atmosphere. It usually takes the form of a (carbon) tax or a requirement to purchase permits to emit carbon pollution. See also, *carbon tax* and *greenhouse gas*.
Carbon Steel	Like alloy steel (i.e., steels alloyed with other elements) and unlike stainless steel, *carbon steel* is a metal with a carbon content of up to 2.1% by weight. As the percentage of carbon content rises, steel can become harder and stronger through heat treating; however, this process also makes the steel less ductile and weldable. See also, *alloy*, *carbon*, *low-carbon*, *stainless steel*, *steel*, and *weldability*.

Phrase or Term	Definition
Carbon Tax	A *carbon tax* is a tax levied on the carbon content of fuels, generally in the transport and energy sector, that are typically paid by businesses and industries that produce carbon dioxide as a by-product of operations. Carbon taxes are meant to reduce carbon dioxide emissions by decreasing the demand for fossil fuels and increasing their price. Carbon taxes are a form of carbon pricing. See also, *carbon pricing* and *Clean Fuel Standard*.
Cargo	*Cargo* is goods or products transported on an aircraft, motor vehicle, or ship. See also, *goods* and *products*.
Carrier	A *carrier* is a person or company that undertakes the professional conveyance of goods, and to carry, hold, or convey something. See also, *goods*.
Cartel	A *cartel* is an organization (often illegal) created by a formal agreement between similar producers of a product or service (e.g., countries, manufacturers, or suppliers) to control prices and prevent competition. See also, *consortium*, *cooperative*, *joint venture*, *monopoly*, and *monopsony*.
Cascading Style Sheets	*Cascading Style Sheets* is a simple mechanism that allows users to add styles (e.g., colours, fonts, and spacing) to Web documents.
Cash Call	A *cash call* is a request for payment for anticipated future capital and operating expenditures that is sent by joint venture operators to non-operating partners. Most joint operating agreements include a provision that allows the operator to issue cash calls to non-operating partners. See also, *joint operating agreement* and *joint venture*.
Catch Phrase	Not a preferred term. See *buzzword*.
Catch-22	*Catch-22* is a problematic situation for which the only solution is denied by a circumstance inherent in the problem or by a rule. For example, losing something is typically a conventional problem; to solve it, one looks for the lost item until it is found. But if the thing lost is one's glasses, one cannot see to look for them. That is a Catch-22. This term is also used more broadly to mean a tricky problem, or a no-win or absurd situation. Catch-22 is the title of a satirical war novel by American author Joseph Heller. The title is a reference to a fictional bureaucratic stipulation which embodies forms of illogical and immoral reasoning. See also, *beg the question, bureaucracy, false dichotomy, herding cats, paradox, pushing a rope uphill*, and *vicious circle*.
Catch-all	*Catch-all* is: • A category or term that includes a variety of different possibilities. For example, the medical adhesive brand BAND-AID® is so popular with consumers, BAND-AID is often used as a catch-all for all adhesive medical bandages; or, • Something that holds or includes odds and ends or a wide variety of things.
Cathode	A *cathode* is the electrode from which an electrical current flows out of the electrical device. See also, *anode* and *cathodic protection*.
Cathodic Protection	*Cathodic protection* is a technique used to control the corrosion of a metal surface by making it a cathode. A simple method of passive protection connects the metal to be protected to a more easily corroded *sacrificial metal* to act as the anode. The sacrificial metal then corrodes instead of the protected metal. For large structures where passive cathodic protection is not adequate, an external direct current electrical power source is used to apply adequate protection. See also, *anode* and *cathode*.

Phrase or Term	Definition
Causal Relationship, Cause and Effect Relationship	A *causal relationship* or *cause and effect relationship* is where a process or state, a cause, contributes to the production of another process or state, an effect. The cause is fully or partly responsible for the effect, and the effect is fully or partly dependent on the cause. For example, when it rains (cause) streets get wet (effect). *See also, correlation*, *false assumption*, and *non-causal relationship*.
Causality	*Causality* is the relationship between cause and effect. Cause follows effect. For example, lower temperatures occur at higher altitudes; thus, increasing altitude (the cause) yields cooler temperatures (the effects). Causality is the interdependent influence of connected components in a system that requires a deeper understanding of agency, feedback loops, connections, and relationships – which are all parts of systems mapping. See also, *chicken and egg situation*, *emergence*, *feedback loop*, *interconnectedness*, *synthesis*, *systems mapping*, and *systems thinking*.
Cause-and-effect Analysis Diagram, Fishbone Diagram, Ishikawa Diagram	A *cause-and-effect analysis diagram*, *fishbone diagram*, or *Ishikawa diagram* illustrates brainstorming of a challenge or problem. Completed correctly, this diagram illustrates all possible causes of a problem (e.g., by category and cause of the undesirable effect) and is used for troubleshooting. The shape of this diagram resembles that of a fish skeleton. See also, *brainstorm* and *troubleshoot*.
Caveat Auditor, Let the Listener Beware	*Caveat auditor* (Latin) or *let the listener beware* means that, when hearing something, the listener shall undertake due diligence to determine if the content is accurate, relevant, or reliable (e.g., has inaccuracies, is biased, or both). See also, *caveat emptor* and *caveat lector*.
Caveat Emptor, Let the Buyer Beware, Buyer Beware, Sold as-Is Where-Is	*Caveat emptor* (Latin), *let the buyer beware*, or *buyer beware* summarize these principles: • The buyer is responsible to verify the quality and suitability of a product before purchasing it (e.g., the buyer must ensure it meets expectations and has no defects); or, • The buyer takes the risk that a product may be defective or unsuitable if there is no warranty (e.g., the buyer has no recourse with the seller if the product does not meet expectations or is defective). The phrase *caveat emptor* is from the Latin statement: *caveat emptor, quia ignorare non debuit quod jus alienum emit,* which means: *Let a purchaser beware, for he ought not to be ignorant of the nature of the property which he is buying from another party.* The phrase *sold as-is where-is* has a similar meaning. See also, *buyer's remorse*, *caveat auditor*, *caveat lector*, and *warranty*.
Caveat Lector, Let the Reader Beware	*Caveat lector* (Latin) or *let the reader beware* means that when reading something, the reader shall undertake due diligence to determine whether the content is accurate, relevant, or reliable (e.g., has inaccuracies or is biased). See also, *caveat auditor* and *caveat emptor*.
Cellular Concrete	*Cellular concrete* is a mixture of cement, preformed foam, and water. The purpose of the foam is to provide stable air voids in the mixture to produce a strong cellular or porous solid upon curing. Hardened cellular concrete has an oven-dry density of 0.8 t/m^3 (50 lb/ft^3) or less. See also, *cement* and *composite*.

Phrase or Term	Definition
Cement	*Cement* is a powdery substance made with calcined lime and clay that is mixed with water to form mortar or mixed with sand, gravel, and water to make concrete. Cement is a common construction material that has been in use in various forms since 6500 BCE. See also, *cellular concrete*, *composite*, and *Portland cement*.
Centrally Planned Economy, Command Economy	A *centrally planned economy* or *command economy* is an economic system in which a central authority, such as a government (i.e., a bureaucracy), makes economic decisions regarding the manufacturing and the distribution of products and services. In a centrally planned economy, production is often by state-owned or public enterprises. See also, *bureaucracy* and *market economy*.
Centreline	A *centreline* is a line that bisects something into equal parts (e.g., land, products, and material). See also, *straddle* and *two hole*.
Certificate	A *certificate* is a document that attests the truthfulness or validity of: • Information (e.g., a product warranty); • A qualification (e.g., completion of a training course or successful examination); • Compliance (i.e., a field operator completes a mandatory H_2S safety course), and that a condition or requirement has been met); • Fulfillment of requirements and the approval to practice in a field (e.g., chartered professional accountant or welders and related machine operator); • That something is owned (e.g., an asset or a debt) or someone has permission to do something (e.g., Company B is permitted to execute environmental remediation activities); or, • That something has been accomplished or happened (e.g., an academic degree is a certificate). See also, *certificate of authorization permit*, *certificate of conformance*, *degree*, *diploma*, *license*, *material test report*, and *permit*.
Certificate of Authorization Permit	A *certificate of authorization permit*, which is issued by *Alberta Boilers Safety Association*, authorizes a company to carry out activities stated on the certificate. See also, *permit*.
Certificate of Compliance	Not a preferred phrase – see *certificate of conformance*.
Certificate of Conformance	A *certificate of conformance* states that the item(s) represented have been fabricated or manufactured in accordance with the requirements of the purchase order and all requirements specified therein.
Certification Mark	A *certification mark* is an ASME symbol identifying a product as meeting code requirements. For example, an ASME PP or U stamp. See also, *certification mark stamp*.
Certification Mark Stamp	A *certification mark stamp* is a metallic stamp issued by ASME for use in impressing the certification mark. For example, an ASME PP or U stamp. See also, *certification mark*.

<table>
<tr><th>Phrase or Term</th><th>Definition</th></tr>
<tr><td>Certified Individual</td><td>A certified individual is an employee (i.e., full-time, part-time, or contractor) of the holder of or an applicant for an ASME BPVC certificate of authorization and certification mark or an organization certified to an ASME Standard (e.g., a manufacturer). The certified individual's responsibility is to provide oversight of the activities that affect the proper utilization of the ASME certification mark. See NB-383, Rules for Certified Individuals. See also, certification mark.</td></tr>
<tr><td>Certified Material Test Report</td><td>Not a preferred phrase – see material test report.</td></tr>
<tr><td>Certified Test</td><td>A certified test is certified by the supplier (e.g., with performance data or results) to demonstrate compliance with the purchase order requirements. See also, performance test.</td></tr>
<tr><td>Chainage</td><td>Chainage is an imaginary or theoretical line used to measure distance (e.g., corresponding to the centre of a pipeline or road), the distance measured along this line, or both.</td></tr>
<tr><td>Challenge</td><td>A challenge is:<ul><li>A call to take part in a contest, competition, task, or other undertaking (e.g., compete in a marathon);</li><li>An objection or query of the truth of something, often with an implicit demand for proof (e.g., question the tenements of Buddhism);</li><li>To confront, defy boldly, oppose, or protest (e.g., participate in civil disobedience);</li><li>To dispute something as being invalid, outmoded, or unjust (e.g., attend court to arbitrate a speeding ticket); or,</li><li>To question the legality, legal qualification, right, or validity of something or someone (e.g., question the constitutionality of a police arrest).</li></ul>See also, predicament and problem.</td></tr>
<tr><td>Champion</td><td>A champion makes project success a personal responsibility by encouraging the project team to work hard, liaising with stakeholders on behalf of the project, and supporting the project manager. This is typically an informal role.</td></tr>
<tr><td>Change Agent</td><td>A change agent is an individual or group with the skill and power to facilitate, and the task of instigating and managing change in an organization. They directly or indirectly influence the transformation of how business is now conducted or how the organization is managed. A change agent can be personnel who are:<ul><li>External
Talent appointed to organization (e.g., a contracted consultant) who are not bound by the organization's culture, politics, rules, regulations, policies and traditions.</li><li>Internal
Talent appointed by the organization (e.g., an employee) who understand the organization's culture, history, operating procedures, personnel, politics, regulations, policies, rules, and traditions.</li></ul></td></tr>
<tr><td>Change Notice</td><td>Not a preferred phrase – see design change notice.</td></tr>
</table>

Phrase or Term	Definition
Change Order	A *change order* is a commercial document, which is part of a *contract* or *purchase order*, with a second or later official offer that is issued by a buyer to a contractor or supplier. Information, products, and services may be added to or deleted from the original scope of work. Depending on the change, the original price, completion date, or both may or may not be altered. See also, *contract* and *purchase order*.

Character	*Character* is:
	• A feature used to separate distinguishable things into categories. For example, chairs with three legs versus chairs with four legs;
	• A graphic symbol used in writing. For example, a hieroglyph, letter, mark, or number);
	• A person in a fictional story. For example, Huckleberry Finn was a character from the book *The Adventures of Tom Sawyer* (1876), and the protagonist and narrator of its sequel: *Adventures of Huckleberry Finn* (1884), written by Samuel Clemens (Mark Twain);
	• An attribute or feature that comprises and distinguishes an individual. For example, his humour makes him a real character in the office;
	• An individual or place marked by conspicuous or unusual traits. For example, the character of the city of Calgary is marked by the foothills Rocky Mountains, Calgary Tower, Calgary Stampede, and the Old West;
	• Moral excellence and firmness. For example, the pastor has strong ethics and character;
	• Reputation. For example, by showing up early, working hard, and working late to get the job done improves a reputation and is a sign of strong character in the eyes of others;
	• The aggregate of distinctive qualities, typical of a breed, strain, or type. For example, the character of a Percheron draft horse comprises:
	  o Black or grey colouring, with white markings;
	  o Heights between 155 and 185 cm (61 and 73 in.); and,
	  o Weights between 500 and 1,200 kg (1,100 and 2,600 lb).
	• The complex of ethical and mental traits that mark and often individualize a country, group, or person. For example, Canadians have a reputation of being friendly and overly-polite because, generally, they are gregarious and apologize frequently.
	See also, *24 Character Strengths*, *jargon*, *plain English*, *story*, and *writing style*.

Charity	*Charity* is: • Kindness and understanding towards others or a lenient judgment of others; and, • The resources provided voluntarily by an organization in the form of money or time to help others (e.g., typically for the poor, sick, or underprivileged). Charity aims to relieve the pain of a societal problem. See also, *non-profit*, *philanthropy*, *voluntary sector*, and *wealth*.

<table>
<tr><th>Phrase or Term</th><th>Definition</th></tr>
<tr><td>Charter</td><td>A charter is:<ul><li>A formal document describing the aims, principles, or rights, of a group or organization. For example, a project charter that emphasizes safety as the most important consideration; or,</li><li>The grant of authority or rights where the granter retains superiority or sovereignty, and the recipient accepts an inferior or limited status within the relationship. For example, the Magna Carta Libertatum (Latin for great charter of liberties or freedoms) or Magna Carta is a royal charter of rights that protects individual freedoms, which was granted by King John of England on June 15, 1215. It altered in 1216, 1217, and 1225.</li></ul>See also, corporate instruction, guideline, knowledge transfer, plan_project charter, and project instruction.</td></tr>
<tr><td>Chatbot, Chatterbot</td><td>A chatbot or chatterbot is an artificial intelligence application used to imitate a real conversation with a user in their native language. This enables communication by audio or text on messaging applications, mobile apps, telephones, and websites without direct human contact. See also, artificial intelligence.</td></tr>
<tr><td>Chatham House Rule</td><td>The Chatham House Rule is a system for holding debates and discussion panels on controversial topics. It is named after the headquarters of the United Kingdom Royal Institute of International Affairs, based in Chatham House, London, England.
The rule is designed to increase openness of discussion of current affairs and public policy, as it allows personnel to discuss and express controversial arguments and opinions without risk to their careers (e.g., dismissals or reprimands), and with a clear separation between their discussion and expressed opinion, and the view of their employer. The rule, which was created in 1927, and then refined in 1992 and 2002, states:<blockquote>When a meeting, or part thereof, is held under the Chatham House Rule, participants are free to use the information received, but neither the identity nor the affiliation of the speaker(s), nor that of any other participant, may be revealed.</blockquote></td></tr>
</table>

Phrase or Term	Definition
Checks and Balances	*Checks and balances* are mutual restraints within a system that: • In corporate settings, prevent error, fraud, and omission. Corporations will identify separate parties (departments, groups or individual personnel) and install mechanisms such that one party cannot complete a transaction independently, forcing it to seek out other responsible entities to complete the transaction. For example, for a project to purchase equipment, three entities are involved: o Engineering. Engineering issues the material requisition; o Procurement. Procurement issues the purchase order; and, o Accounts payable. Accounts payable makes the payment. At each step above, data is checked and cross-referenced. • In governments (primarily in constitutional governments), allow each branch of government (e.g., executive, judicial, and legislative) to amend or veto acts of another branch, to prevent one branch from exerting too much power. • In organizations, counterbalances prevent improper behaviour and decrease the risk of centralization of power into the hands of a few. See also, *freedom of the press* and *the press*.
Chemical Value Chain	The *chemical value chain* categorizes the manufacturing of thousands of chemical products into these eight major categories: 1. C_1 chemicals based on syngas including ammonia- and methanol-derived chemicals; 2. C_2 chemicals based on ethylene; 3. C_3 chemicals based on propylene; 4. C_4 based chemicals including butadiene and isobutylene; 5. Benzene-based chemicals; 6. Toluene-based chemicals; 7. Xylene-based chemicals; and, 8. Chlor-alkali–based chemicals.
Cherry Pick, Cherry-pick	*Cherry pick* or *cherry-pick* is an expression that means to: • Choose and take only the most beneficial or profitable items or opportunities from what is available for personal benefit or gain, often in a way that others may consider unfair; or, • Present only select evidence to persuade others to accept a position (i.e., ignore or reject contradictory evidence). When *cherrypicking*, evidence that argues against one position is withheld. This dishonesty is amplified as the amount evidence withheld increases. • In hockey, instead of performing as a defensive team member – frequently maintaining a short distance from the hockey net to take advantage of easy and fortuitous opportunities to score. The phrase *cherry pick* derives from *selecting the best cherries in the bowl for oneself, without regard for others who may also want cherries*. See also, *handpick*, *low-hanging fruit*, and *nutpicking*.

Phrase or Term	Definition
Chicken and Egg Situation	A *chicken and egg situation* is a cause-and-effect or causality dilemma where it is impossible to determine which of two things caused the other to happen or to determine what happened first. For example, the question: *which came first, the chicken or the egg?* This question is impossible to answer because neither could exist without the other. See also, causality.
Chicken Scratch	*Chicken scratch* is illegible or difficult-to-read handwriting. For example, how am I supposed to understand this chicken scratch? See also, *penmanship*.
Chief Executive Officer Disease	*Chief executive officer disease* describes maladies that affect powerful leaders, including: • Bipolar or manic behaviours (e.g., from the energy required to propel some leaders forward, professionally); • Inter-personal dynamics between leaders and those around them (e.g., information is filtered when others or subordinates avoid negative conversations or are reluctant to disclose bad news or worst-case scenarios); or, • Self-aggrandizement (e.g., self-serving behaviours and decisions). See also, *chief executive officer*, *founder's syndrome*, and *senior management*.
Chief Inspector	A *chief inspector*: • Meets the requirements to manage an owner's pressure equipment integrity management program; or, • Is responsible for leading work site -based inspectors or managing inspection activities, and ensuring that construction complies with contract, project, and regulatory requirements (e.g., procedures and specifications). See also, *pressure equipment* and *integrity management program*.

Phrase or Term	Definition
Chunking	*Chunking* is the process of grouping information such that the reader easily understands the message. Information that is chunked effectively is consistent, logical, organized, and sequenced correctly. Chunking information into logical, manageable units makes it easy to remember and understand. Example 1. Credit card and phone numbers are typically chunked in groups of 3 or 4 numbers (not as 10 or 16 continuous numbers). By contrast, when a telephone number (or other familiar grouping) is organized (chunked) in a different format, it is difficult to remember. Example 2. The grouping of products in a grocery store such as bread and pastries in one section, and fruits and vegetables in another section. Generally, information can be chunked in three ways: 1. Physical Objects chunked according to colour, shape, size, and type allow viewers to quickly make sense of, and remember, the categories, and identify missing objects and objects that are askew. 2. Verbal Information chunked logically assists listeners to follow and remember key details or ideas. 3. Written Information chunked hierarchically gives readers a quick overview of the narrative or task and a clear understanding of where to find additional details (e.g., a table of contents). See also, *chunking down* and *chunking up*.
Chunking Down	*Chunking down* is the process of moving from a general statement, premise, principle, or law to specific details. This is helpful or useful for: 1. Negotiations and problem solving After an agreement or consensus is achieved, chunking down can make the tasks for success apparent. For example, after a fair price is agreed-upon – how and when the purchase will be made can also be agreed-upon. 2. Organization and overcoming stress By breaking a larger task into smaller tasks, one seemingly insurmountable task is no longer overwhelming because the process of chunking down organizes the work into manageable tasks. See also, *chunking*, *chunking up*, *deductive logic*, and *inductive logic*.

Phrase or Term	Definition
Chunking Up	*Chunking up* is the process of moving from specific statement, premise, principle, or law to generalities. This is helpful or useful for: 1. Creative thinking By encouraging creativity when arriving at solutions, new and better solutions may be realized. Then, by using chunking down, it is possible to find a specific solution that was not previously considered. For example, to solve a complex or difficult problem, personnel can brainstorm to identify all potential solutions (regardless of practicality). For example, personnel could consider a scenario with no monetary or time constraints. Then, by considering and narrowing all possible choices, personnel can choose the most appropriate solution. 2. Motivation Chunking up identifies the general or overall plan. This allow teams and personnel to learn how they fit in, and can contribute to, the overall work effort. These are motivational data in that they provide a goal or purpose. Further, and importantly, these data incent personnel to start with the end in mind. 3. Negotiations and problem solving It is far easier to acquire agreement and consensus with others by providing a summative, overall, general plan than with hundreds of pages of technical detail. 4. Overcoming stress By grouping smaller tasks into one large task, personnel can easily identify overall work progress instead of fixating on less important details. See also, *brainstorm*, *chunking*, *chunking down*, *deductive logic*, *inductive logic*, and *start with the end in mind*.
Circular Economy, Circularity	A *circular economy* or *circularity* is an economic system aimed at eliminating waste and the continual use of resources. These systems employ recycling, refurbishment, remanufacturing, repair, reuse, and sharing. This creates a closed-loop system, minimizing the use of resource inputs and the creation of waste, pollution and carbon emission outputs. The aim is to keep equipment, infrastructure, and products in use for longer, thus improving the productivity of these resources. All *waste* should become feedstock for another process: either a by-product or recovered resource for another industrial process, or as regenerative resources for nature (e.g., compost). See also, *economic system*, *economy of repetition*, *economy of scale*, *hydrogen economy*, *feedstock*, *knowledge economy*, *gig economy*, *linear economy*, and *low-carbon economy*.

Phrase or Term	Definition
Circular Reasoning	*Circular reasoning* is a logical fallacy in which the reasoner begins with what they are trying to end with. The components of a circular argument are often logically valid because if the premises are true, then the conclusion must be true. This is not a formal logical fallacy but a pragmatic defect in an argument whereby the premises are just as much in need of proof or evidence as the conclusion, and therefore the argument fails to persuade. Other ways to express this are that there is no reason to accept the premises unless one already believes the conclusion, or that the premises provide no independent ground or evidence for the conclusion. It is closely related to circular reasoning, and in modern usage the two generally refer to the same thing. Circular reasoning is often of the form: *A is true because B is true; B is true because A is true*. Circularity can be difficult to detect if it involves a longer chain of propositions. This is an example of a fallacious circular argument: Calgary is in Alberta; therefore, Calgary is in Alberta. Although the argument is deductively valid, it cannot prove that Calgary is in Alberta because it contains no evidence that is distinct from the conclusion. The context – that of an argument – means that the proposition does not meet the requirement of proving the statement; thus, it is a fallacy. See also, *bait and switch*, *beg the question*, *either/or*, *false equivalence*, *logical*, *logical fallacy*, *red herring*, *righteousness fallacy*, *slippery slope*, and *straw man argument*.
Cladding	*Cladding* is: 1. For insulation, a thin flexible sheet material (i.e., aluminium or stainless steel) that is used as a jacket to cover thermal insulation on piping, ducting, or equipment. See also, *insulation*. 2. For welding, not a preferred term – see *weld overlay*.
Clarification	*Clarification* is the: • Action of making a situation or statement less confusing and more comprehensible, clearer, simpler, or more straightforward by providing additional explanation, information, or both; or, • Process of making a liquid clear by clearing or removing visible impurities, such as feculent or solid matter, by separating or removing insoluble particles that prevent the liquid from being transparent. See also, *bid clarification meeting*.
Classical Economic Theory	*Classical economic theory* is a broad phrase that refers to the dominant school of thought for economics in the 18th and 19th centuries. Self-regulating democracies and capitalistic market developments form the basis of classical economics. Before the rise of classical economics, most national economies followed a top-down, command-and-control, monarchal-type of government system. Classical economics became closely associated with economic and political freedom. See also, *demand-side economics* and *supply-side economics*.

Phrase or Term	Definition
Clean Fuel Standard	The *Clean Fuel Standard* is a set of Canadian regulations that require fuel suppliers to reduce the carbon intensity of fuels they produce and sell. Suppliers, like natural gas providers and refiners, decide which combination of fuels and technologies are the most inexpensive means of compliance – through biofuels, electricity, hydrogen, natural gas, or other solution. This means that fuel used across Canada will get cleaner, year after year. It aims to reduce carbon pollution by 30 million tonnes of carbon dioxide equivalent by 2030 (i.e., equal to taking more than seven million vehicles off the road). These regulations will cover all fossil fuels used in Canada (i.e., liquid, gaseous, and solid). The regulations for the liquid fuel class are proposed to be implemented in 2022, followed by regulations for gaseous and solid fuel 12 months later in 2023. These regulations are also known, especially to Canadians, as the *second carbon tax*. See also, *carbon tax*.
Clean Hydrogen, Clean H_2	*Clean hydrogen* or *clean H_2* is produced using renewable energy or fossil fuels with substantial carbon capture and storage. It has a lower carbon intensity compared to other types of production (e.g., brown or grey hydrogen). See also, *carbon intensity*, *colours of hydrogen*, *fossil fuel*, *fossil fuel hydrogen*, *hydrogen*, *renewable hydrogen*, and *renewable energy*.
Clearing	A *clearing* occurs during a meeting in which personnel are encouraged to share topics that prevent them from performing productive work. Lululemon uses this method to improve efficiencies. It is a concept that was developed by the *Landmark Forum*, which is a personal development seminar. This simple concept is incredibly effective and allows all involved to be heard, present (i.e., engaged), and seen. Clearings foster an empathetic team whose members want to collaborate to succeed. See also, *clearing meeting* and *team*.
Clearing Meeting	A *clearing meeting* is a team gathering that serves to: • Discuss important issues of which everyone should be aware; • Encourage personnel to solve problems independently; • Focus personnel on key issues and encourage them to act. • Gauge the team's level of self-organization; • Give personnel the opportunity to discuss concerning issues; • Increase trust within the team; or, See also, *clearing*.
Client	A *client* is an organization (e.g., corporation) that procures or provides a physical location (project site), and compensates service providers (suppliers) to design and procure equipment and materials for the project. This includes acquiring competent resources for project construction and commissioning. In some cases, a client also provides funding for project maintenance and operations, and associated capital costs. See also, *customer*, *project*, *service provider* and *site*.
CliftonStrengths, StrengthsFinder	*CliftonStrengths*, formerly *StrengthsFinder*, is an online personal assessment test hosted by Gallup that outlines users' strengths. This test is the fundamental element of the self-help book *Strengthsfinder 2.0* by Tom Rath. The authors' advocate focusing on building upon strengths rather than upon weaknesses. See also, *self-help*.

Phrase or Term	Definition
Climate Change	*Climate change* is a long-term shift in average global or regional climate patterns, in particular a change apparent from the mid to late 20th century onwards, that is attributed largely to the increased levels of atmospheric carbon dioxide produced by the burning of fossil fuels. This phrase refers to both human-caused and naturally-produced warming, and the effects they have on the planet. Climate change is commonly measured as the average increase in Earth's global surface temperature. See also, *climate porn*, *global warming* and *greenhouse gas*.
Climate Porn, Climate Pornography, Fear porn, Fear Pornography	*Climate porn*, *climate pornography*, *fear porn*, and *fear pornography* is alarmist, apocalyptic, or extreme language or images used to describe the challenges and current or future effects of climate change. See also, *climate change* and *pornography*.
Close	*Close* is: • A short distance away or apart in space or time (e.g., the office is close to the warehouse); • Denoting a family member who is part of a person's immediate family (e.g., a parent or sibling); • To block against entry or passage, or deny access to (e.g., the contractor closed the gate); • To bring to an end (e.g., close an account); • To cease operation, or stop or suspend the operation of (e.g., the factory closed); • To enter or complete an agreement (e.g., close on a deal or sale); or, • To move to bar passage through something (e.g., close a valve). This term has other meanings. See also, *SPIN selling*.
Closed Data	*Closed data* is data that is private and not freely circulated or shared. These data are not in the public domain and cannot be used by third parties. See also, *open data*.
Closed-ended Question	A *closed-ended question* is a query that can be answered with a *yes* or *no* response. See also, *open-ended question* and *reporter's questions*.
Close-out	*Close-out* is the completion phase of a project for all supplier purchase orders and includes supply of all equipment, material, document deliverables, archiving, and assignments for inspection services. This also refers to the completion of a non-conformance report.
Closure	*Closure* is the procedure to complete the production of an assembly (e.g., welding a pressure vessel) wherein, after completion, the internal areas are no longer accessible for direct examination (e.g., there is no direct visibility or manhole for accessibility). See also, *manhole*.
Cloud	*Cloud* describes a global network of digital servers, each with a unique function. Instead of a physical entity and not unlike the Internet, it is a vast global network of remote servers that are linked together and to operate as a single digital ecosystem. See also, *cloud computing* and *Internet*.

<table>
<tr><th>Phrase or Term</th><th>Definition</th></tr>
<tr><td>Cloud Computing</td><td>Cloud computing is shared computer processing and data storage services available through an Internet browser. There are four methods of deploying cloud resources:
1. Community cloud
Resources are restricted to, and only shared between, certain organizations (e.g., educational or government institutions).
2. Hybrid cloud
A combination of private and public clouds to reap the advantages of both. In a hybrid cloud, applications and data can move between the private and public clouds for greater flexibility and more deployment options.
3. Private cloud
Resources (e.g., servers and storage) are used exclusively by one business or organization. These can be hosted by a third-party service provider or physically located at an on-site data centre.
4. Public cloud
Resources (e.g., servers and storage) are owned and operated by a third-party cloud service provider and delivered over the Internet.
See also, anything as a service, cloud, infrastructure as a service, platform as a service, software as a service, and security as a service.</td></tr>
<tr><td>Cloud Method</td><td>The cloud method is a document markup convention that identifies recent or important revisions by outlining the revisions with an object shaped like a cloud. This method allows readers to disregard portions of the document and focus on recent or important revisions. The cloud shape may be drawn by hand or using software, and is removable to accommodate future revisions. This method may be used by to standardize mark-ups, improve quality control, increase the predictability of how markups are delivered, and for many other review purposes (e.g., to flag or question a section, to identify a section as on hold, and to draw attention to detail). See also, clouded, on hold, and triangle method.</td></tr>
<tr><td>Clouded</td><td>A document or document section that contains markups that use the cloud method is clouded. Generally, a clouded document is a marked-up document. See also, cloud method.</td></tr>
</table>

Phrase or Term	Definition
Cluster, Ecosystem, Hub, Incubator, Farm, Node, Supercluster	*Cluster*, *ecosystem*, *hub*, *farm*, *incubator*, *node*, and *supercluster* are terms that describe an organization or program created and developed for businesses success. These organizations are formed especially for start-ups to: • Connect founders, entrepreneurs, investors, and partners; • Foster collaboration, innovation, and synergy; • Nurture research and development; • Offer coworking space and networking opportunities; • Overcome common challenges and solve problems; • Provide advice, education, mentorship, and training; • Refer resources for business development, legal, and other expertise; and, • Support business and entrepreneurial growth and success. These organizations are usually non-profits, formed by businesses, entrepreneurs, governments, industries, regions, schools, start-ups, universities, or a combination. The first start-up incubator was the Batavia Industrial Center, formed in Batavia, New York in 1959. These terms also have other meanings. See also, *accelerator*, *coworking space*, *entrepreneur*, *hydrogen valley*, *non-profit*, *start-up*, *synergy*, and *technology hub*.
Coachable	To be *coachable* is to be easily taught and trained (e.g., in the workplace). A coachable person is receptive to constructive criticism and feedback and uses it to improve performance. These personnel are adaptable and speed business growth. They are open to making changes and improvements for the betterment of themselves, other personnel, and the business. Coachability sets good personnel apart from others. Un-coachable personnel typically blame others, deflect, dislike performance audits, make excuses, and are slow to learn. See also, *coaching*.
Coaching	*Coaching* is extending traditional training methods to include focus on: • An individual's needs and accomplishments; • Close observation; and, • Impartial and non-judgmental feedback on performance. See also, *coachable* and *mentoring*.

Phrase or Term	Definition
Coal, Metallurgical Coal, Thermal Coal	*Coal* is a combustible black or dark brown rock consisting of mainly carbonized plant matter (e.g., peat deposits). It is a fossil fuel found mainly in underground deposits and is defined as carbonaceous matter more than 50% by weight or 70% by volume. Different varieties of coal result from differences in the plant material (i.e., coal type), degree of coalification (i.e., coal rank), and range of impurities (i.e., coal grade). There are several classifications for coal, however, the two most important are: • Metallurgical, met, or coking coal, which is burned at a high temperature. For example, burned at >2,000°C (>3,632°F) to produce steel; and, • Thermal or steam coal which is burned at a moderate temperature. For example, burned at >1,000°C (>1,832°F) to produce steam and generate electricity. Coal is gasified and liquefied to produce gaseous and liquid fuels. To convert coal to an alternative fuel, hydrogen must be added either directly or in the form of water. It is a chemical source from which numerous synthetic compounds (e.g., dyes, oils, pesticides, pharmaceuticals, and waxes) are produced. It See also, *coal liquefaction*, *fossil fuel*, *hydrocarbon*, *natural gas*, and *petroleum*.
Coal Liquefaction	*Coal liquefaction* is the process of converting coal into liquid hydrocarbons (i.e., liquid fuels or petrochemicals). This process is known as *Coal to X*, where X can be many different hydrocarbon-based products. However, the most common process is *coal to liquid fuels*. Direct processes are based on approaches such as carbonization, hydrogenation, and pyrolysis. Indirect processes involve gasification of coal to a mixture of carbon monoxide and hydrogen, often known syngas. See also, *gasification*, *hydrocarbon*, *petrochemical*, and *syngas*.
Coat	A *coat* is one layer of material that is applied to a substrate in a single continuous application to form a uniform film when dry (e.g., to inhibit or prevent corrosion). See also, *coating* and *film*.
Coating	A *coating* is a liquid, liquefiable, or mastic composition that, after application to a substrate, is converted into a solid adherent film for an aesthetic or protective purpose (e.g., to provide a specified finish colour, or prevent corrosion or erosion). See also, *coat*, *coating system*, *lining*, and *powder coating*.
Coating System	A *coating system* facilitates the application of the required number and types of coats, including the sequence in which coats are applied to a substrate. More broadly, this includes surface preparation, pre-treatments, dry film thickness, and manner of application. See also, *coating*, *substrate*, and *surface preparation*.

Phrase or Term	Definition
Cobra Effect, Cobra Problem	The *Cobra Effect* or *cobra problem* is an unanticipated consequence. When an attempt to solve a problem instead exacerbates or makes the problem worse – this is the Cobra Effect. The namesake is from colonial India at a time when Delhi suffered a proliferation of cobras. To reduce their numbers: 1. The Indian government placed a bounty on cobras. As people hunted cobras, cobra numbers, predictably, dropped; 2. As the cobra population fell, it became difficult to find cobras. People began raising cobras domestically for their bounty; 3. When the government realized that there were very few cobras in the city and that the same number of bounties were being paid, they cancelled the bounty program; and, 4. People raising cobras domestically released their now-worthless cobras into the streets. In the end, Delhi had a bigger cobra problem than when it started! The unanticipated consequence of the cobra bounty program was a net increase in the number of city cobras! See also, *second-order effect*, *spillover effect*, and *unanticipated consequences*.

Code Creep	*Code creep* is when a code of construction is:
	• Continuously amended to address new technologies and information that are considered obsolete, outdated, or unnecessary. If a section is deleted during this process, it may:
	  o Become needed for reference or use again later; or,
	  o The document may contain references to the deleted section, making references to it pointless and meaningless.
	• Not readily available for purchase or use because it is out of print.
	For example, the last code of construction for steam locomotives was published in 1952 by ASME. Yet, steam locomotives in operation today still need to be maintained and repaired. Thus, steam locomotive mechanics may make use of this standard with the applicable parts of a newer standard, or a newer standard alone! See also, *code of construction* and *specification creep*.

Code of Construction	The *code of construction* is the code used for construction, fabrication, and manufacturing of equipment, facilities, and materials (e.g., API, ASME, CSA, and IEEE). Codes often refer to other documents (e.g., recommended practices or standards) and are often imposed upon or governed by other documents (e.g., acts or regulations). In addition, owner specification(s) may also be applicable. See also, *construction*, *fabrication*, and *manufacturing*.
Codification	*Codification* is a type of knowledge transfer that converts knowledge into documents, images, and videos that can be used by anyone. See also, *knowledge transfer*, *personalization*, and *tribal knowledge*.
Cogeneration, Combined Heat and Power	*Cogeneration* or *combined heat and power* is the simultaneous production of electricity by recovering and re-using the heat and power (i.e., steam) generated by a process.

Phrase or Term	Definition
Cognitive	*Cognitive* is being, involved in, of, or related to conscious intellectual activity (e.g., learning, perception, reasoning, remembering, or thinking). Cognitive knowledge and skills involve the ability to acquire information. See also, *cognitive bias*, *cognitive dissonance*, and *cognitive flexibility*.
Cognitive Bias	*Cognitive bias* is a systematic error in thinking that is a deviation from normal or rational judgment. For example, the *Ostrich effect* describes the behaviour when negative financial information is avoided or ignored. The phrase is derived from the false assumption that an ostrich will bury its head in the sand to avoid danger. In cognitive bias, individuals create their own *subjective reality* based on their interpretation or perception of an input. The constructed reality, not the objective input, may then dictate their behaviour, decisions, and judgment. This occurs from problems with attribution, attention, memory, and other mental mistakes, and leads to irrationality (e.g., illogical interpretation, inaccurate judgment, or perceptual distortion). Cognitive biases are also called *bugs in the human hardware*. There are many types of cognitive biases. Listing all cognitive biases is beyond the scope of this document. See A *List of Cognitive Biases*, here: https://en.wikipedia.org/wiki/List_of_cognitive_biases. *See also*, *bias, cognitive*, *false equivalence*, *logical fallacy*, and *social engineering*.
Cognitive Dissonance	*Cognitive dissonance* is the state of having inconsistent attitudes, beliefs, or thoughts, especially relating to attitude changes and behavioural decisions. See also, *bias, cognitive*, *cognitive flexibility*, *convergent thinking*, *creative thinking*, *creativity*, *critical thinking*, *dichotomy*, *divergent thinking*, *doublethink*, *executive skills*, *groupthink*, *false dichotomy*, *flexibility*, *forward-thinking*, *lateral thinking*, *mental prison*, *loserthink*, *outside the box*, and *vertical thinking*.
Cognitive Flexibility, Flexible Thinking	*Cognitive flexibility* or *flexible thinking* is the ability to: • Switch between thinking about two different concepts or to think about multiple concepts simultaneously; or, • Take new information and apply it to information that is already known (e.g., think about something in more than one way). Cognitive flexibility is a key part of developing different strategies for problem-solving. See also, *bias, cognitive*, *cognitive dissonance*, *convergent thinking*, *creative thinking*, *creativity*, *critical thinking*, *dichotomy*, *divergent thinking*, *executive skills*, *groupthink*, *false dichotomy*, *flexibility*, *forward-thinking*, *lateral thinking*, *mental prison*, *loserthink*, *outside the box*, and *vertical thinking*.
Coin	*A coin* (noun) is a flat and typically round piece of metal with an official stamp that is used as currency (e.g., a dime is a coin that has the value of 10 cents). Similarly, Canadians use a gold-coloured coin called a loonie that has a value of 100 Canadian cents. *To coin* (verb phrase) is to be the first to use a term or phrase, especially if it becomes frequently used in vernacular. For example, Reena Walker coined the phrase *false narrative* in a Facebook group in the 1990s – this phrase is now used frequently by the media, worldwide. See also, *false narrative*.

Phrase or Term	Definition
Coincident Indicator, Concurrent indicator	A *coincident indicator* or *concurrent indicator* is a measurement of the economic and financial market that is analyzed and used as it occurs. These are key numbers that have a substantial impact on the overall economy. For example, personal income is a coincident indicator of the economic health of a country. Higher personal income numbers coincide with a stronger economy. Lower personal income numbers coincide with a weaker economy. See also, *lagging indicator* and *leading indicator*.
Cold Bend	A *cold bend* is a bend (typically in a pipe) that has been formed by bending a pipe into a specific shape (i.e., radius) in a bending machine at normal temperatures for use in a pipeline or piping system. See also, *cold bending, induction bend, and radius*.
Cold Bending	*Cold bending* is a method of bending material (e.g., typically pipe) at normal temperatures in a bending machine. See also, *induction bending*.
Cold Call	A *cold call* is unsolicited contact made with potential customer (e.g., by email, telephone, or in-person) with whom there has been no prior interaction, interest, or referral, to offer them a product or service. See also*, referral*, *sales lead*, and *warm call*.
Cold Eyes Review	A *cold eyes review* is the review of a document performed by someone who has never or has infrequently reviewed the document. This contrasts with document authors who have frequently reviewed the document. A cold eyes review often results in perspectives and concepts that may have been overlooked by document authors. This review may be used to specify requirements, and offer feedback for opportunities to improve the document. See also, *discipline review* and *squad check*.
Cold Work	*Cold work* is operations that do not involve the use of power generation or welding equipment for construction or demolition, or where there is no flammable atmosphere or material. See also, *cold work permit* and *hot work*.
Cold Work Permit	A *cold work permit* is permission to proceed with cold work. See also, *cold work*.
Collective Bargaining	*Collective bargaining* is the negotiations conducted between an employer and a labour union about terms of employment, working conditions, workplace rules (e.g., base pay, overtime pay, health care benefits, holidays, retirement benefits, shift duration, sick leave, vacation time, and work hours) for the labour force. See also, *labour union*.
Colloquial	*Colloquial* is language used for informal or ordinary conversation that is specific to a geographic region. For example, in: • Canada, it is *French fries*; • France, it is *pommes frites*; • The United Kingdom and United States, it is *chips*. See also, *dialect*, *Ebonics*, *jargon*, *Plain English*, and *vernacular*.

Phrase or Term	Definition
Colours of Hydrogen	The *colours of hydrogen* are categories that describe the source and method of hydrogen production. Hydrogen is made from many resources using various methods. Each process has pros and cons, different life cycle carbon intensities, and diverse challenges and opportunities. Hydrogen colour is used to denote its environmental credentials; that is, how carbon intensive or clean it is. There is no global standard or consensus for colour nomenclature; terminology may differ or be similar. Commonly, in order of most to least carbon intensity, hydrogen colours are generally described as: • Black hydrogen or black H2 results in a high carbon intensity. It is produced from fossil fuels (e.g., bitumen or coal) using gasification or steam methane reformation, without carbon capture and sequestration; • Grey hydrogen or grey H2 results in a higher carbon intensity. It is produced from natural gas using thermal processes (i.e., steam methane reformation); • Blue hydrogen or blue H2 results in a low-to-moderate carbon intensity. It is produced from industrial residual gases or natural gas (i.e., methane) with carbon capture-sequestration used to store remaining carbon, or carbon is repurposed; • Brown hydrogen or brown H2 results in a high carbon intensity. It is produced from fossil fuels (e.g., bitumen or coal) using gasification or steam methane reformation, without carbon capture and sequestration; • White hydrogen or white H2 results in a low to moderate carbon intensity. It is produced as a by-product of industrial processes; • Yellow hydrogen or yellow H2 results in various carbon intensities, depending upon the source of electricity. It is produced by electrolysis using grid electricity. • Green hydrogen or green H2 has zero carbon emissions; it is the gold standard of hydrogen production. It is produced by extracting hydrogen from water using electrolysis and renewable energy (e.g., hydro, solar panels, or wind turbines); • Clear hydrogen or clear H2 has zero carbon emissions; hence its environmental impact is the same as to green hydrogen. It is produced by in-situ oxidation of an oilfield (i.e., underground burning) to produce hydrogen that is extracted while the carbon remains sequestered underground, which results in a low carbon intensity.; • Pink, purple, or red hydrogen or pink, purple, or red H2 results in a low carbon intensity. It is produced from nuclear power using electrolysis; • Rainbow hydrogen or rainbow H2 is a symbol of hope for, and promise of, a zero-carbon future. It is a fuel for heating and transportation, a source of power generation, and a means of energy storage to meet decarbonization challenges faced by economies powered by fossil fuels; and, • Turquoise hydrogen or turquoise H2 is produced using molten metal pyrolysis technology, where natural gas is passed through molten metal to produce and release hydrogen gas and solid carbon. This technology is still under research and development, and may take until 2030 for a pilot plant to become feasible. Other colours (e.g., teal) may be used to categorize hydrogen. Other processes may be used to produce hydrogen (e.g., biomass gasification). World energy demands mean that all forms of hydrogen will be used for energy; however, most projects will focus on methods that have low carbon intensity (e.g., blue, green, and pink hydrogen) if feasible. See also, *carbon capture and sequestration*, *carbon intensity*, *clean hydrogen*, *fossil fuel hydrogen*, *gasification*, *hydrogen*, *pilot plant*, *renewable energy*, *renewable hydrogen*, and *underground burning*.

Phrase or Term	Definition
Column	A *column* is a load-bearing vertical structural member that supports a floor, roof or similar member. See also, *member* and *beam*.
Combustion	*Combustion* is the burning and complete oxidation of a fuel, such as coal or wood, by a process that produces carbon dioxide and heat. See also, *gasification*.
Comfort Zone	The *comfort zone* is: • A place or situation where personnel feel at ease or safe, and are free of stress; • The level at which one functions with ease and familiarity; or, • The temperature range within which one is comfortable, such as room temperature. For example, 15 to 25°C (59 to 77°F). A comfort zone is a healthy adaptation for living (e.g., to be calm and comfortable, and free from anxiety or worry). Moving out of a comfort zone is necessary to change, grow, and learn (e.g., conquer the fear of public speaking by presenting to a group of people). See also, *peer mentor* and *smart risk*.
Commercial	A *commercial* is an advertisement usually accompanying a broadcast program that is hosted by podcast, radio, television, or other media, and supported by an advertiser (e.g., a corporation such as Walmart or WestJet); or, Being commercial is: • Being a product of a specified grade (e.g., Grade AAA Alberta beef); • To be engaged in, or occupied with, business or work (e.g., a commercial business); or, • View business in terms of acquiring profit (e.g., a corporation that measures success by profit is a commercial business). Businesses listed on the New York Stock Exchange are commercial businesses. See also, *business*, *commercial bid evaluation*, *corporation*, and *grade*.
Commercial Bid Evaluation	A *commercial bid evaluation* is the comparison or review of a bid according to business and cost related items or conditions. It is performed based on criteria that include categorized and weighted values, including delivery schedule, financial stability of the bidder, price analysis, proposal terms and conditions, and terms of payment. See also, *bid* and *technical bid evaluation*.

<table>
<tr><th>Phrase or Term</th><th>Definition</th></tr>
<tr><td>Commissioning</td><td>Commissioning is a transitional phase that encompasses activities that occur after facility construction completion and before commercial operations. The following is a list of some of these activities:
<ul>
<li>Activate all utility systems;</li>
<li>Bump electric motors to ensure turning direction is correct;</li>
<li>Calibrate every instrument and control valve;</li>
<li>Check the proper action of all controllers;</li>
<li>Complete all checklists and close-out all punch lists;</li>
<li>Confirm the circuitry on all electrical switches;</li>
<li>Flush out piping with using air or fluid to remove debris;</li>
<li>Inspect all vessels to confirm proper installation of internals;</li>
<li>Load the initial charge of all reactor catalysts;</li>
<li>Prepare a master blind list and ensure all blinds are procured;</li>
<li>Roll steam turbines while disconnected to assure balance;</li>
<li>Test the programming of all programmable logic controllers;</li>
<li>Verify all pump alignments;</li>
<li>Verify that all safety devices are correct and installed;</li>
<li>Verify that all tagging is properly marked; and,</li>
<li>Walk down every pipe to ensure they:
<ul>
<li>Conform to related P&IDs; and,</li>
<li>Have necessary high point bleeds and low point drains.</li>
</ul></li>
</ul>
See also, punch list, start-up, and walk down.</td></tr>
<tr><td>Committeespeak</td><td>Committeespeak is when a document that is reviewed by many (i.e., a committee) becomes a non-useful, non-cogent, jumbled patchwork because of contradictory comments and lack of holistic editorial oversight.</td></tr>
<tr><td>Commodity</td><td>A commodity is:
<ul>
<li>An article of commerce or goods, especially when delivered for shipment;</li>
<li>A mass-produced unspecialized product (e.g., bulk item);</li>
<li>Primary product or raw material that can be bought and sold (e.g., diesel fuel or steel); or,</li>
<li>Something useful or valuable (e.g., patience, time, or water).</li>
</ul>
See also, bulk items.</td></tr>
</table>

<table>
<tr><th>Phrase or Term</th><th>Definition</th></tr>
<tr><td>Common Sense</td><td>Common sense is sound practical judgment concerning everyday matters, or a basic ability to judge, perceive, and understand what is common to, or shared by most, people. There are two types of common sense:
1. Good sense
The knack for seeing things as they are, and doing things as they ought to be done. For example, tying an untied shoelace or to visiting a physician if feeling unwell.
2. Folk wisdom
Knowledge that is not reliant on deliberative thought or specialized training. For example, the expression out of sight, out of mind, or stories (such as Aesop's fables).
These two types of common sense are intertwined, as a person who has common sense is in touch with common-sense ideas, which emerge from the lived experiences of those commonsensical enough to perceive them. It is also what one thinks that others should know as basic, common knowledge.
This can apply to things affecting all humans, or it can pertain to certain peoples and issues. For example, ensuring that when parking a vehicle – it occupies only one parking space. Common sense psychology is also myth, because what appears to be common sense is often common nonsense. Scientific evidence can refute common sense claims. For example, what could be more obvious than the flatness of the earth or that the earth is stationary? Also, yesterday's common sense is often today's common nonsense. For example, in ancient Athens, often cited as the birthplace of democracy, only adult, male citizens who owned land were permitted to vote. Women's suffrage, or the right of women to vote in elections, began in the late 19th century.</td></tr>
<tr><td>Communication is Nonverbal Rule</td><td>Communication is nonverbal rule states that most connections with and relation to other people is not verbal (i.e., what is not spoken). A pioneer researcher in the field of body language, Albert Mehrabian observed that the total impact of a message is about:
• 7% verbal (i.e., words only);
• 38% vocal (e.g., inflection, tone of voice, and other sounds); and,
• 55% (i.e., nonverbal).
This means that 93% of communication is nonverbal. Research by Albert Mehrabian and others have validated many of the ideas and observations developed by Charles Darwin in his book The Expression of the Emotions in Man and Animals. See also, body language.</td></tr>
<tr><td>Community Spread</td><td>Community spread is the spread of a contagious disease:
• Between groups who reside close to each other or gather in groups (especially on a regular basis). Community spread can be initiated by a carrier (i.e., someone who recently travelled to, and contracted a contagious disease from, an area with a disease; and,
• Within a community.
See also, COVID-19.</td></tr>
<tr><td>Comparative Advantage</td><td>Comparative advantage is an economic phrase that refers to an economy's ability to produce products and services at a lower opportunity cost than others. This gives a business or company the ability to sell products and services at a lower price than its competitors and realize stronger sales and greater profit margins.</td></tr>
</table>

<table>
<tr><th>Phrase or Term</th><th>Definition</th></tr>
<tr><td>Competent</td><td>Competent means that a person possesses at least the minimum levels of experience, knowledge, qualifications, and skills to perform the work safely per regulatory or other requirements. See also, four stages of competence.</td></tr>
<tr><td>Competitive Advantage</td><td>Competitive advantage refers to factors that allow an organization to produce better and cheaper products or services than its competitors. Factors such as branding, cost structure, customer service, distribution network, intellectual property, and quality allow greater sales, greater profit margins, or both. Jim Collins argued that the most important competitive advantage is people:<ul><li>People are not your most important asset. The right people are; and,</li><li>The real competitive advantage in any business is one word only, which is people.</li></ul>See also, competitor.</td></tr>
<tr><td>Competitive Bid</td><td>A competitive bid occurs when competing suppliers bid on a scope of supply, and are openly advised of the required scope of supply, specifications, terms and conditions, and the criteria by which the bids will be evaluated. This structure creates a transparent process that is open and fair to all bidders. The purpose of a competitive bid is for a company to obtain products and services at the lowest prices by stimulating competition, and preventing company bias or favoritism. The competitive bid process may be:<ul><li>Required by law (e.g., for a government agency that issues a bid); and,</li><li>Associated with a variety of business opportunities in which suppliers present proposals (e.g., for a purchase order or a supplier frame agreement).</li></ul>In open bidding or open competitive bidding, bids are sealed and later opened in full view of all bid opening attendees.
In closed bidding or closed competitive bidding, sealed bids are opened only in the presence of authorized personnel.
See also, bid, three bids and a buy, purchase order, scope of supply, single source, sole source, and supplier frame agreement.</td></tr>
<tr><td>Competitor</td><td>A competitor is:<ul><li>A rival that is:<ul><li>An organization (e.g., business, group, or person) that makes similar products, operates in the same industry, and targets the same customers (e.g., Coca-Cola and Pepsi); or,</li><li>Someone attempting to gain or win something over someone else (e.g., my colleague applied for the same position in the new business unit, making her my rival); or,</li></ul></li><li>Another organization (e.g., business, person, or team) that buys or sells products or services in the same market as your organization.</li></ul>See also, archrival, frenemy, nemesis, and rival.</td></tr>
<tr><td>Complete Joint Penetration</td><td>Complete joint penetration is a welded joint that is made with full penetration (where the fused joint is completely solid with no gaps after welding). See also, full penetration, incomplete penetration joint, and partial penetration.</td></tr>
</table>

<table>
<tr><th>Phrase or Term</th><th>Definition</th></tr>
<tr><td>Complete Unit Test</td><td>A complete unit test is when an equipment package is tested as a complete unit with purchased equipment (e.g., pump and driver train with job motor and all auxiliaries). This testing is completed as specified by the purchase order in addition to, or in place of, separate testing of individual components. See also, auxiliary equipment test.</td></tr>
<tr><td>Complex Supply Chain</td><td>A complex supply chain is a condition of inter-connectedness and inter-dependencies across a network where a change in one element can influence other elements. In this context, the word complex does not mean complicated. See also, supply chain</td></tr>
<tr><td>Compliance</td><td>An item or organization is said to be in compliance when it meets prescribed contract or purchase order requirements, regulations, specifications, standards or a combination.</td></tr>
<tr><td>Component</td><td>A component is an equipment part that may be:<ul><li>Pressure-boundary or pressure-retaining (i.e., built to retain pressure). For example, a fitting or valve body or casing;</li><li>Structural and provide an attachment point or support (e.g., insulation ring, skirt, or saddle); or,</li><li>Wetted and exposed to the process fluid (e.g., the internal parts of a valve).</li></ul></td></tr>
<tr><td>Composite
(Ceramic, Concrete, Fibre-reinforced Plastic, Metal Matrix, Pipe, and Wood)</td><td>A composite is a solid material composed of two or more substances having different physical characteristics, wherein each substance retains its identity while contributing desirable properties to the finished product. These products are also referred to as engineered, man-made, or manufactured. Examples include:<ul><li>Ceramic matrix composites have ceramic and metal matrices;</li><li>Concrete is made of Portland Cement, aggregates (e.g., gravel and crushed stones), and sand. Admixtures (additives) enhance the properties of concrete. Different types of binding material are used (e.g., lime for lime concrete and bitumen for asphalt concrete) to achieve desired cohesion and viscosity;</li><li>Fibre-reinforced plastics (e.g., fibre-reinforced polymer, fibre-reinforced plastic, or fibreglass) are a polymer matrix reinforced with fibres. The fibres usually aramid, basalt, carbon, or glass;</li><li>Metal matrix composites that have at least two constituent parts: one is a metal; the other may be a different metal or another material (such as a ceramic or organic compound). A hybrid composite occurs when at least three materials are present;</li><li>Pipe that is made either by the filament wound or rotary casting method. Fibreglass or other material is wrapped around or inside a mold that is then encased in a resin or epoxy solution to bind the material into a tubular product; and,</li><li>Wood derivative products are manufactured by binding or fixing the strands, particles, fibres, or veneers or boards of wood with adhesives.</li></ul>See also, cellular concrete, cement, and Portland cement.</td></tr>
</table>

Phrase or Term	Definition
Compressor	A *compressor* is device or instrument for compressing something: • A machine (of which there are many types) that increases the pressure of a gas by reducing its volume (e.g., air or natural gas, to move it through a piping system); or, • An electrical amplifier that reduces the dynamic range of a signal.
Computational Fluid Dynamics	*Computational fluid dynamics* is a branch of fluid mechanics that uses numerical analysis and data structures to analyze and solve problems that involve fluid flows. Computers perform calculations required to simulate the free-stream flow of fluid, and the interaction of the fluid (liquids and gases) with surfaces defined by boundary conditions.
Computer-aided Design, Computer-aided Design and Drafting	*Computer-aided design* or *computer-aided design and drafting* is the use of software to assist in the analyses, creation, modification, or optimization of a design. This specialized software is used to create a database, increase the productivity of the designer, improve the quality of the design, and generate documentation. Software output is often electronic files that projects use for machining, printing, or other production operations. See also, *computer-automated design* and *digital project delivery*.
Computer-automated Design, Automated Design	*Computer-automated design* or *automated design* are extensions of traditional computer-aided design that are transformed by biologically inspired machine learning. For example, the design of semiconductor chips was semi-automated as early as the mid-1970s. Semiconductors can have billions of components; thus, automation of design is the only way the work can be completed. Common applications include: • Automotive engineering; • Civil engineering; • Composite material design; • Control engineering; • Dynamic system identification and optimization; • Financial systems; • Industrial equipment; • Invention of novel systems; • Mechatronic systems; • Steel construction; and, • Structural optimization. See also, *computer-aided design*, *digital project delivery*, *machine learning*, and mechatronic *technology*.

Phrase or Term	Definition
Computerized Maintenance Management System, Computerized Maintenance Management Information System	A *computerized maintenance management system* or *computerized maintenance management information system* is software that maintains a database of information about an organization's maintenance of assets (e.g., equipment or facilities). This information is intended to control operations and improve their effectiveness (e.g., determine what requires maintenance and where spare parts are stored) and assist with analysis and decision-making (e.g., determine the cost of a breakdown repair versus preventive maintenance for each machine). This approach also tracks record keeping, and assigned and completed tasks in a cost-effective and timely manner. See also, *enterprise asset management system*.
Concealed Damage	*Concealed damage* is hidden damage in the contents of a package that cannot be seen by observing the unopened package.
Concealed Loss	*Concealed Loss* is damage to the contents of a package that cannot be determined until the package is opened.
Concession, Concession Request	Not a preferred phrase – see *design change notice* and *request for information*.
Condensate	*Condensate*, also referred to as natural gas liquid, is a mixture of extremely light hydrocarbons that may be recovered from a gas reservoir. Condensate is a diluent that reduces bitumen viscosity and allows bitumen to flow through a pipeline.
Condition-based Inspection	*Condition-based inspection* is a method of predictive maintenance that uses periodic inspections to measure the status of an operating asset over time. This provides accurate calculations of equipment-component wear rates and predictable life span estimates. With this method of inspection, an asset is: • Maintained per a predetermined schedule (e.g., regular maintenance – engine and oil filter changes, and oil level and oil pressure checks); or, • Repaired or replaced if the risk of failure exceeds a pre-determined threshold (e.g., as calculated based on information obtained from the inspection). For example, to determine the remaining thickness and corrosion rate of a pressure vessel. See also, *condition-based monitoring* and *risk-based inspection*.
Condition-based Monitoring	*Condition-based monitoring* is a method of predictive maintenance that uses sensors to measure the status of an operating asset over time. This provides accurate calculations of remaining life, establishes trends, and predicts failures. With this method of monitoring, maintenance is only performed if data shows performance is decreasing or failure is likely. See also, *condition-based inspection* and *risk-based inspection*.

Phrase or Term	Definition
Conduct of Operations, Conduct of Operations/Operational Discipline	*Conduct of operations* or *conduct of operations/operational discipline* is the execution of management and operational tasks in a deliberate and structured manner to meet safety standards and reduce operational risk. Often referred to as *formality of operations* or *operational discipline*, this concept is closely tied to organization culture. *Conduct of operations* categorizes the activities and attributes of a well-functioning operating organization, which include: • Effective factors for human performance, responsibility, and supervision; • Equipment and process monitoring; • Operating procedures; • Response to abnormal conditions; and, • Shift team communication.
Conduit	A *conduit* is pipe used as a duct for electrical wiring. It is usually supplied threaded and coupled, in 3 m (10 ft) lengths. It is normally galvanized, slightly lighter than standard weight pipe, and has a smooth interior surface.

<table>
<tr><th>Phrase or Term</th><th>Definition</th></tr>
<tr><td>Cone of Learning</td><td>
The cone of learning or pyramid of learning is a representation of the different degrees of knowledge retention from various types of learning. The next table (adapted from various sources) shows representations made by general percentages in discrete layers.
This concept was devised in the 1940s and then further developed by Edgar Dale in his book Audio-Visual Methods in Teaching. Like all great paradigms, it is illustrative and makes perfect sense. Many versions are in use. The cone model was developed by the National Training Laboratories Institute in the early 1960s. The percentages of learning retention were added later, by others.
<table>
<tr><th>Retention after Two Weeks</th><th>Activity</th><th>Nature of Learning</th><th>Learning Outcomes</th></tr>
<tr><td rowspan="3">90% of what one does and says</td><td>Prepare and perform the real thing</td><td rowspan="5">Active</td><td rowspan="5">Analyze, Create, and Evaluate</td></tr>
<tr><td>Experience, model, or simulate the real thing</td></tr>
<tr><td>Make a collaborative or dramatic presentation</td></tr>
<tr><td rowspan="2">70% of what one says</td><td>Design a lesson or make a presentation</td></tr>
<tr><td>Practice or participate in a discussion or workshop</td></tr>
<tr><td rowspan="4">50% of what one sees and hears</td><td>Watch a demonstration on-site</td><td rowspan="7">Passive</td><td rowspan="4">Apply, demonstrate, and practice</td></tr>
<tr><td>Watch a demonstration</td></tr>
<tr><td>Attend and view exhibit</td></tr>
<tr><td>Watch a movie or video</td></tr>
<tr><td>30% of what one sees</td><td>View at pictures</td><td rowspan="3">Define, describe, explain, and list</td></tr>
<tr><td>20% of what one hears</td><td>Listen to audio</td></tr>
<tr><td>10% of what one reads</td><td>Read an article or book</td></tr>
</table>
Inconsistencies have been noted between the cone of learning and its research, which was inadequate or lost, so the original methodologies, parameters, and quality of data used for its development are unknown. Will Thalheimer concluded that there is no scientific data or other data that supports the claim that retention improves with experiential levels of the pyramid. In his article Mythical Retention Data & The Corrupted Cone, he states that the numbers are nonsense and the order of potency is incorrect. The cone of learning is also known as the cone of experience, cone of retention, learning cone, learning pyramid, and pyramid of retention. See also, active learning, Bloom's Taxonomy, knowledge transfer, metacognative thinking, passive learning, and unlearning.
</td></tr>
</table>

Phrase or Term	Definition
Confirmation Bias	*Confirmation bias* is the tendency to interpret new evidence as confirmation of one's existing beliefs or theories. See also, *bias*, *optimism bias*, *selection bias*, *reference-class forecasting*, and *unlearning*.
Conflict of Interest	*Conflict of interest* is a set of circumstances that creates a risk that professional actions or judgements regarding a primary interest will be unduly influenced by a secondary interest. It is a situation where the aims or concerns of two parties (e.g., entities, individuals, or both) are incompatible because of a clash between personal or self-serving interests and professional duties or responsibilities. For example, a government official who votes about land use legislation that affects the future use or value of land in an area where the government official is also a landowner. This occurs when a company or person has a vested interest (e.g., influence, knowledge, money, relationship, reputation, or status) that puts into question whether their actions, decision-making, judgment, or a combination, can be unbiased. A conflict of interest exists even if no improper or unethical act results and when the appearance of impropriety is created, which can undermine the confidence of impartiality, objectivity and professionalism. Conflict of interest laws regulate the avoidance and disclosure of these situations. When these situations occur, the party with the conflict may be legally required or otherwise asked to recuse or remove themselves (e.g., during discussions and final decision-making or voting). This occurs in a situation in which a familial or personal benefit can be derived from actions or decisions made in their official capacity. Or, a person can exploit familial or personal gain over duties to their employer, or to an organization in which they are a stakeholder. For example, a buyer purchases products and services through their employer, from another company, of which the buyer is the director. The buyer will profit from his employer *and* from the sale as the director; thus, there is a conflict of interest if the buyer's employer is unaware of this situation. The buyer could avoid a conflict of interest by recusing himself from the purchase decision, which would allow other personnel to make an unbiased determination about the transaction that best serves the company's interests (e.g., is beneficial and suitable for the company). See also, *bias*, *ethical*, *interest*, *lobby*, and *personal agenda*.
Confidentiality Agreement, Non-disclosure Agreement, Confidential Disclosure Agreement, Proprietary Information Agreement, Secrecy Agreement	A *confidentiality agreement*, *non-disclosure agreement*, *confidential disclosure agreement*, *proprietary information agreement*, or *secrecy agreement* is a legal contract that establishes a confidential information-sharing relationship between two or more parties. Its purpose is to legally ensure that sensitive information shared between parties is not shared with anyone else. In business, these agreements are common and typically protect corporate information, intellectual property or material, and proprietary knowledge from disclosure to others. Bank-client confidentiality and lawyer-client privilege are examples of legal confidentiality agreements that do not require a written contract. See also, *intellectual property*.
Conformance	A *conformance* is an affirmative indication or judgment that a product or service has met specified requirements. It is a characteristic that renders an item acceptable for its intended use (e.g., an item that is not blemished, defective, deficient, incomplete, or imperfect). See also, *non-conformance* and *non-conformance report*.

Phrase or Term	Definition
Confusopoly, Dilbert's confusopoly	*Confusopoly* or *Dilbert's confusopoly* is marketing that is designed to be confusing so the buyer cannot make informed decisions and is forced to use emotions to make the decision instead. Scott Adams, author of *Dilbert*, defined this as *a group of companies with similar products who intentionally confuse customers instead of competing on price*. For example, mobile phone plans price products and services using different combinations of availabilities, capabilities, and services. This makes it nearly impossible for customers to make an apples to apples comparison with the competition. See also, *apples to apples* and *Dilbert*.
Conscious Capitalism	*Conscious capitalism* is a socially responsible economic and political philosophy. A conscious capitalist believes that businesses should operate ethically while pursuing profits. This means they should consider serving all stakeholders involved, including the environment, their employees, and humanity—not just management teams and shareholders. This concept was developed by Whole Foods' co-founder John Mackey and marketing professor Raj Sisodia in the *book Conscious Capitalism: Liberating the Heroic Spirit of Business*. See also, *capitalism*, *crony capitalism*, *economic system*, and *environmental social governance*.
Consequence of Failure	*Consequence of failure* is the ranking of potential outcomes to equipment, environment, and personnel in the event of equipment failure. It comprises half of the equation used to determine risk as part of the risk-based inspection methodology. See also, *probability of failure*, *risk*, *risk equation*, and *risk-based inspection*.
Conservative	*Conservative* means: • Disposed or tending to maintain existing conditions, institutions, or views (e.g., conservative plans or policies); • Marked by moderation or caution (e.g., a conservative calculation or cost estimate); and, • Marked by or relating to business-like or traditional norms of elegance, manners, style, or taste (e.g., a conservative dress or suit). See also, *reactionary* and *right-wing*.
Consolidator	A *consolidator* is responsible to review and consolidate squad check comments (e.g., the responsible engineer).
Consortium	A *consortium* is an agreement, combination, or group (e.g., of companies) formed to undertake an enterprise beyond the resources of any one member. See also, *cartel*, *cooperative*, *joint venture*, *monopoly*, and *monopsony*.
Conspiracy Theory	A *conspiracy theory* is a belief an event or set of circumstances is the result of a secret and sinister plan by an influential organization or powerful conspirators, even if other explanations are more probable. This phrase has a pejorative connotation that implies that the belief in a conspiracy is prejudiced, based on insufficient evidence, or both. These theories resist falsification and are reinforced by circular reasoning. Evidence against a conspiracy *and* an absence of evidence for it are re-interpreted as proof of its existence. The conspiracy then becomes a matter of faith (a belief) instead of something that can be disproven or proven. For example, some conspiracy theorists believe the American Apollo moon landings (between 1969 to 1972) did not occur and were instead fabricated by American movie makers.

Phrase or Term	Definition
Construct	A *construct* is something that is: • A physical object that is built, constructed, or made (e.g., building, machine, or road), where the construct's existence does not depend upon the existence of the mind; or, • An ideal object that is created by putting different parts together (e.g., an idea, system, or writing), where the construct's existence depends upon the existence of the mind. See also, *artificial construct*.
Constructability	*Constructability* is a project management technique used during the pre-construction phase, to review construction processes from start to finish. It is used to identify obstacles before a project is built to prevent or reduce cost overruns, errors, and schedule delays.
Construction	*Construction* comprises all service and supply activities performed at site for a project using contracts rather than purchase orders. See also, *lean construction*.
Construction Management Team	A *construction management team* is responsible for all construction at site, including the receipt of materials and equipment. See also, *management* and *project team*.
Constructive Confrontation	*Constructive confrontation* is a structured, systematic approach that decreases conflict and increases accountability by identifying what people want, and what an organization or others need. It is organizational development initiative wherein personnel combine in an argumentative setting to determine, examine, and plan resolutions to issues. The goal is directness, and to minimize the destructive elements of conflict and foster conflicts' constructive potential (i.e., disagree without being disagreeable). See also, *constructive criticism*.
Constructive Criticism	*Constructive criticism* is the process of offering valid and well-reasoned opinions about the work of others (negative and positive comments), ultimately to establish and implement recommendations that collaboratively improve the overall quality of products or services. When using constructive criticism, it is prudent to use a friendly and respectful tone and *not use* an oppositional tone. Constructive criticism is more likely to be accepted if it is actionable, clear, detailed, specific, timely, and improves the product or service. See also, *constructive confrontation*.
Constructor	A *constructor* is a person or company that builds, constructs, designs, erects, or makes something.
Consultant	A *consultant* is a professional who provides expert advice or knowledge. A consultant may work in an advisory capacity only and not be accountable for the outcome of the consultation. A consultant may be an external or internal service provider to a company or organization. See also, *consulting company*, *independent contractor*, *gig*, and *side hustle*.
Consulting Company, Consultancy Firm	A *consulting company* or *consultancy firm* is a business comprised of one or more consultants who provide professional advice or services to an individual or organization. A consulting company may complement their deliverables by providing the means to implement recommendations, either with the consultants or by providing other resources. For example, engineering and procurement, or construction or project management services may be provided. See also, *consultant*, and *engineering and procurement*.

Phrase or Term	Definition
Consumable	In the context of non-destructive examination, a *consumable* is a part or product used during examination operations (e.g., contrast paint, developer, film, liquid or powder magnetic particles, penetrant, processing chemicals, and ultrasonic couplant). In the context of welding, a *consumable* is a part or product that is eroded, worn, or used during welding, cutting, or gouging operations (e.g., electrodes, fluxes, inserts, nozzles, and shielding gases).
Contact Sheet	A *contact sheet* contains multiple images on a single page. A contact sheet allows many photographs to be viewed simultaneously.
Containerized Intermodal Freight Transport System	The *containerized intermodal freight transport system* involves the transportation of cargo in an intermodal container or vehicle, using multiple modes of transportation (e.g., rail, ship, and truck), without any handling of the freight itself when changing modes. This allows for faster transportation, improved security, and reduced cargo handling, damage, and loss. See also, *intermodal container*.
Content	*Content* is the subject of copy or written matter or the substance a website or other online platform (e.g., illustrations or written matter). See also, *content marketing* and *copywriting*.
Content as a Service, Managed Content as a Service	*Content as a service* or *managed content as a service* is a business model that focuses on managing structured content into online feeds that applications and users can consume. For example, content to be used with chatbot, inter-personal, social media, and website communications. See also, *anything as a service*, *chatbot*, *infrastructure as a service*, *platform as a service*, *software as a service*, *touchpoint*, and *security as a service*.
Content Marketing, Owned Marketing	*Content marketing* or *owned marketing* means: • Creating (e.g., preparing original content to share on social media including articles, audio files, e-books, information graphics, photographs, and videos); • Curating (e.g., selecting and sharing content on social media with some added value or adding thoughtful commentary or interpretation); and, • Sharing (e.g., posting links to interesting or useful content on social media); See also, *copywriting* and *marketing*.

Phrase or Term	Definition
Contingency	A *contingency* is the provision of money, resources, or time for an unforeseen circumstance or event, or an incidental expense. For an inspection assignment, the following items may increase the cost, lengthen the duration, or both, of an inspection assignment: • Complexity of the inspection assignment and availability of inspection resources; • Duration of production is subject to change or unknown; there are an unknown number of items to be produced; • Lack of synergy or multiple shipments needed (e.g., intermittent production); • Potential for action items, additional scope, issues, non-conformances, or punch lists; • Project involvement at supplier facility (e.g., witness factory acceptance testing or walk-down); • Project request for activities in addition to those in the inspection and test plan or not previously planned (e.g., witness load-out); • Supplier unknowns (e.g., inexperienced with engineering and procurement company, or owner and similar projects); and, • Supply chain complexity (e.g., original equipment manufacturers and sub-sub-suppliers). See also, *budget*, *inspection cost estimate*, *contingency plan*, *risk*, *synergy*, *and workaround*.
Contingency Plan, Fallback Plan	A *contingency plan* or *fallback plan* is an alternative or secondary plan that may be used if the primary plan fails, becomes too costly, or encounters an unforeseen circumstance or event (e.g., schedule delays). It is developed to manage identified risks. Colloquially, this is Plan B! See also, *contingency*, *measure twice and cut once*, *risk*, and *workaround*.
Continual Improvement, Continuous Improvement	*Continual improvement* or *continuous improvement* is a gradual, never-ending, or ongoing effort to improve processes, products, or services, and fulfill an organization's policies and objectives. These efforts yield incremental improvement over time or breakthrough improvement quickly. Put simply, it describes something that improves over time. Delivery and customer valued processes are constantly evaluated to improve their effectiveness, efficiency, and flexibility. Even business strategies and results, relationships (i.e., customer, employee, and supplier), and quality may be subject to continual improvement. For example, from 1908 to 2007 (99 years), British cyclists won only one gold medal at the Olympic Games and no Tour de France races. Then, by finding 1% improvements in overlooked and unexpected areas, and making hundreds of small improvements – the results came faster than imagined. From 2007 to 2017 (just 10 years), British cyclists won 178 world championships, 66 Olympic or Paralympic gold medals, and 5 Tour de France victories in what is widely regarded as the most successful run in cycling history. How did this happen? It is so easy to overestimate the importance of one defining moment and underestimate the value of making small, continual improvements. One percent better each day adds up! See also, *small steps method*.

Phrase or Term	Definition
Continuing Education, Further Education	*Continuing education* or further education is education for adults (typically undertaken after an adult has left the formal educational system) comprised of short or part-time courses. Courses may be delivered in a campus, online, or both, and include: • College or university remediation; • Degree-equivalent courses; • Formal personal enrichment courses; • Labour force training; and, • Non-degree career training. See also, *boot camp*, *cross-train*, *lifelong learning*, *lunch and learn*, *on-the-job training*, *professional development unit*, and *workshop*.
Continuity Test	A *continuity test* verifies that an electric circuit is complete and current can properly flow. This test is also used on a bundle of cables or wires to determine which two ends belong to the same cable or wire. A multimeter or other device may be used for continuity testing. The circuit under test is de-energized before testing. See also, *multimeter*.
Continuous Improvement	*Continuous improvement* is the ongoing betterment of processes, products, or services through incremental and breakthrough improvements. *Incremental* improvement is minor improvement that occurs in small steps over time; whereas *breakthrough* improvement is a significant improvement that occurs instantaneously or over a short period. See also, *Plan-Do–Check–Act* and *total quality management*.
Contra	*Contra* is Latin for *against*, *opposite*, and *opposing*.
Contract	A *contract* is a legally-binding commercial document and the first official offer issued by a buyer to a contractor. It identifies types, quantities, and agreed-upon deliveries, including prices for products, services, or both. A contract controls the purchasing of products and services. It also specifies how and what a contractor will alter, construct, or repair at the owner's specified location. By contrast, a purchase order specifies what will be assembled, fabricated, or manufactured at a supplier's facility. See also, *change order*, *contract approach*, *contractor*, *holdback*, *order of precedence*, *purchase order*, *self-enforcing agreement*, and *terms and conditions*.

Phrase or Term	Definition
Contract Approach	The *contract approach* is the method used by an owner or sponsor and the engineering and procurement or construction companies to prepare a project contract. Contracts may comprise a vast array of components for the delivery of products and services. For example, a contract may be for the whole project or a portion thereof, and may or may not contain sub-contracts. Generally, there are four project contract approaches or types: 1. Alliance or integrated project delivery All contractors have collective ownership of opportunities and responsibilities associated with delivery of the whole project or service and participate in a single compensation scheme. It is a reimbursable, incentivized agreement to share overrun or underrun costs. It is not a consortium, frame agreement, or joint venture; 2. Fixed-price or lump-sum A contractor is responsible for engineering, procurement, or construction, or a combination, and is paid a specified amount for completing work, without a cost breakdown; 3. Mixed or hybrid A contractor is responsible for engineering and procurement and has a reimbursable contract, followed by other contractors with fixed-price contracts for fabrication and construction; 4. Reimbursable or time and materials A contractor is responsible for engineering, procurement, or construction, or a combination, and is paid based on the quantity of materials and services provided. See also, *consortium*, *contract*, *contractor*, *engineering and procurement*, *joint venture*, *project*, and *supplier frame agreement*.
Contract Equipment	Not a preferred phrase – see *job equipment*.
Contractor	A *contractor* is a company that has been awarded a contract from an engineering and procurement company or their client, to provide products, services, or both for a project. This also includes all sub-contractors from another division of the same company or a separate company. A contract supply chain may also have sub-sub-suppliers. See also, *engineering and procurement*, *contract*, *supplier*, and *supply chain management*.
Contrapreneur	A *contrapreneur* is someone who sees and leverages a business opportunity in counter-revolutionary or reactionary operations. They are risk averse, create fear, and seek to inhibit innovation – which is sometimes identified as the cause of a contrapreneur's problems. For example, fear of a climate catastrophe has been the rallying call for a major global economic, environmental, and social realignment (e.g., leaving a traditional consumption-based economy, and divesting from fossil fuels and global trade). Contrapreneur is a portmanteau of *contra* and *entrepreneur*. See also, *contra*, *entrepreneur,* and *divest*.

Phrase or Term	Definition
Contrarian	A *contrarian* is someone who: • Behaves differently or the opposite of others around them; or, • Takes an opposite attitude or position, especially against the majority opinion. A contrarian contradicts others. For example, a contrarian would respond to the statement *it is a beautiful day* with *it is not a beautiful day*, even if it is a beautiful day. Contrarian thinking is used in the fields of investing, journalism, and science. See also, *counterintuitive*.
Control Room	Not a preferred phrase – see *local control station*.
Control Signal	A *control signal* is electric, hydraulic, or pneumatic energy source that is a control command as it travels through wires, pipes, or tubes. It may also be a pulse, or frequency of electricity or light that is a control command as it travels over a network, computer channel, or wirelessly. See also, *controller* and *control valve*.
Control Valve	A *control valve* controls fluid flow by varying the size of the flow passage as directed by a control signal from a controller. This allows direct control of flow rate and the consequential control of process quantities such as pressure, temperature, and liquid level. See also, *control signal*, *controller* and *fail position*.
Controlled Hydrogen Electrode	A *controlled hydrogen electrode* is an electrode subject to a diffusible hydrogen test conducted on the deposited weld metal (e.g., per AWS A4.3 or ISO 3690). See also, *diffusible hydrogen designator*, *electrode*, and *low hydrogen*.
Controlled Valve	A *controlled valve* is a valve controlled by a safety procedure to prevent inadvertent operation using a: • Car seal; • Chain lock; • Instrumented interlock or safety instrumented system; or, • Operation and maintenance procedure. See also, *car seal* and *instrumented interlock*.
Controller	A *controller* receives control signals, compares them with settings and, if needed, adjusts the positioning device to operate it as needed (e.g., open or close a control valve). See also, *control valve*, *control signal* and *programmable logic controller*.
Conventional Non-destructive Examination	*Conventional non-destructive examination is* one of the most common methods of examining an object without damaging or destroying it. Types of conventional non-destructive examination include liquid penetrant, magnetic particle, radiographic, and ultrasonic. See also, *advanced non-destructive examination* and *non-destructive examination*.
Convergent Thinking	*Convergent thinking* involves using accuracy, logic, rationality, speed, and recognizing and repurposing previously used successful techniques to arrive at the best, most effective, and correct solution or answer. And there can only be one. Tests such as multiple choice or standardized tests require convergent thinking because only one answer for each question can be correct. Convergent thinking sees the world as black or white, with no grey areas. See also, *divergent thinking*, *grey area*, *lateral thinking*, *thinking mode*, and *vertical thinking*.

Phrase or Term	Definition
Conversion Rate	A *Conversion rate* is the percentage of potential customers who take a desired action (e.g., make a purchase, or register or subscribe for offers, communication, or information). A conversion rate is typically calculated by taking the number of customer conversions and dividing it by the total number of website visitors.
Cookie-cutter Design	A *cookie-cutter design* is a standard, mass-produced product that completely lacks customized characteristics. Products that use a cookie-cutter design are less expensive that custom built products because of economies of scale. See also, *copy scope* and *identical*.
Cooler	Not a preferred phrase – see *heat exchanger*.
Cooperative, Co-operative, COOP, CO-OP	A *cooperative*, *co-operative*, *COOP*, or *CO-OP* is an autonomous association of persons united voluntarily to meet common economic, cultural, or social aspirations and needs through a jointly-owned or non-profit enterprise. This may be a: • A business or property managed and owned by the people who use the services (e.g., a consumer or housing cooperative); • A hybrid or multi-stakeholder cooperative that shares ownership between different stakeholder groups (e.g., a non-profits or investors); • A platform cooperative (e.g., a mobile application, protocol to facilitate the sale of products and services, or website with shared governance and ownership); or, • An organization managed by the people who work there (e.g., a worker cooperative). See also, *cartel*, *consortium*, *cooperative*, *joint venture*, *monopoly*, *monopsony*, and *non-profit*.
Co-opt, Co-option, Co-optation	To *co-opt*, *co-option*, or *co-optation* is to: • Include someone in something, including against their will. For example, The oil lobby co-opted the environmentalists by hiring them for large salaries and giving them many corporate perks; • Make someone a member of a group through a vote by existing members. For example, a board or committee may co-opt a new member to balance the membership in some way (e.g., age, education, gender, or profession) or provide missing expertise; • Persuade someone who criticizes or disagrees to join your group so they can no longer oppose you. For example, the World Boxing Organization hires an opponent of violent sports as an advocate to report on the safety benefits of today's modern boxing equipment; or, • Take over or use someone else's ideas. For example, appropriate or steal an idea, modify it, and call it your idea. For example, Coca-Cola's 1971 advertisement generated the hit song *(I'd Like to) Buy the World a Coke*, which was adapted into another hit song in the same year: *I'd Like to Teach the World to Sing (In Perfect Harmony)*. Co-opt is Latin for *cooptare*, which means *to choose*. See also, *board*.

Phrase or Term	Definition
Coordinate	The definition of the verb *coordinate* is *to act in harmonious combination*. In the context of projects, this means to bring together diverse elements of complex activities or organizations into efficient, harmonious, and productive relationships. Personnel with project coordination roles manage and organize activities, equipment, information, and resources. These are examples of project responsibilities that may have a coordinator: • Factory acceptance test; • Logistics; • Non-conformance report; • Request for information; or, • Supplier quality surveillance. See also, *factory acceptance test coordinator*, *interface*, *liaise*, *logistics coordinator*, *non-conformance report coordinator*, *request for information coordinator*, and *supplier quality surveillance coordinator*.
Copy	*Copy* is a document (e.g., a hard copy, an electronic document, a book, a newspaper article, etc.). In the writing profession, copy is the product of writing and editing. See also, *copywriting*.
Copy Scope	*Copy scope* means to duplicate project design and execution from previous projects (maximizing efficiencies by leveraging engineering, procurement, supplier, and construction expertise) to lower cost and risk, and reduce implementation timelines, and while maintaining quality standards. New scopes of work are modified as needed. These types of projects are often referred to as *phased* (e.g., Phase 1, Phase 2, Phase 3, etc.). See also, *cookie-cutter design* and *identical*.
Copyright	*Copyright* is the exclusive legal right, given to an originator or an assignee to print, publish, perform, film, or record the matter and form of something, and to authorize others to do the same. See also, *infringement*, *intellectual property*, *plagiarism*, and *public domain*.
Copywriting	*Copywriting* is the act or occupation of writing text for the purpose of generating copy. Copywriting generally aims to persuade or sell readers an idea or product. An organization will use copywriting to promote itself, gain or maintain market share, and generate revenue. See also, *advertising*, *content*, *copy*, *marketing*, and *writing style*.
Corporate Culture	*Corporate culture* refers to the behaviours and beliefs that determine how a company's management and personnel manage outside business transactions and interact. Often, corporate culture is implied, not expressly defined, and develops organically over time from the cumulative, collective traits of personnel that a company recruits. A corporate culture is reflected in a company's business hours, client satisfaction, dress code, hiring decisions, office setup, remuneration and benefits, treatment of clients, turnover, and all aspects of its operations. A company has no culture; rather, every company is a culture. See also, *organizational culture*.

Phrase or Term	Definition
Corporate Instruction	A *corporate instruction* is a documented, systematic method that allows personnel to perform job tasks correctly, safely, and consistently. This allows the job motions to be performed and enables an understanding of why it is important do something a specific way. A corporate instruction is not tailored to a specific client and facility (i.e., a generic guideline), but may include all the same essential elements. See also, *charter*, *guideline*, *knowledge transfer*, *plan, project charter*, and *project instruction*.
Corporate Social Responsibility	*Corporate social responsibility* (also called conscious capitalism, corporate citizenship, corporate conscience, corporate sustainability, responsible business, or sustainable business) is a type of private business self-regulation. Over time it has moved from voluntary decisions at the level of individual organizations, to mandatory schemes at national, regional, and even transnational levels. See also, *social licence*.
Corporate Welfare	*Corporate welfare* describes a government's subsidy of a busines by aid, bailout, grants, money, tax breaks, or other favourable treatment for a business or business sector. This means a benefit is provided or a payment is made for reasons other than the purchase of products or services. The definition may be restricted to direct government subsidies of large corporations and exclude tax credits, deductions, exemptions, or reductions (e.g., tax loopholes), and regulatory or trade decisions. See also, *bailout*, *crony capitalism*, and *loophole*.
Corporation	A *corporation* is a legally established business that can own assets and incur debts. See also, *joint venture*, *incorporated*, *partnership*, *sole proprietorship* and *unincorporated*.
Corrective Action	A *corrective action* is a solution meant to reduce or eliminate an identified problem. See also, *opportunity for improvement* and *preventative action*.
Corrective Action Recommendation or Request	A *corrective action recommendation or corrective action request* is a tool (typically a document) that recommends or requests improvements to an organization's processes. The purpose of this tool is to correct or eliminate causes of non-conformities and other related undesirable situations.
Correlation	A *correlation* is a connection or mutual relationship between two or more things. It is a statistical relationship that commonly refers to the degree to which a pair of variables are linearly related. For example, generally, the taller someone is, the more they will weigh. Correlations are useful because they can indicate a predictive relationship that can be exploited. For example, the increased demand for a product of limited supply can result in or warrant a higher price. See also, *causal relationship* and *non-causal relationship*.
Corrosion	*Corrosion* is the deterioration of a material, usually a metal, from a reaction with its environment (e.g., rainwater rusting steel). See also, *corrosion allowance corrosion rate* and *rust*.
Corrosion Allowance	*Corrosion allowance* is an extra thickness added to the wall thickness of a pipe (typically) or equipment after determining the thickness that meets mechanical requirements (e.g., pressure, temperature and weight of equipment). Correctly applied, corrosion allowance allows the pipe (or equipment) to compensate for expected metal loss over the life of the asset. See also, *corrosion*.

Phrase or Term	Definition
Corrosion Rate	The *corrosion rate* is the speed at which any metal in a specific environment deteriorates. It can be defined as the amount of corrosion loss per year in thickness. The speed or rate of deterioration depends on the environmental conditions and the condition and type of metal. See also, *corrosion*.
Corrosion Study	A *corrosion study* provides information to assist with the development of effective inspection plans (e.g., predictable or potential corrosion locations, flow considerations, and fouling issues). It also provides key inputs for a risk-based inspection program (e.g., corrosion and rates, and potential damage mechanisms) that allow criticality calculations. A corrosion study may be conducted as a: • Level 1 Corrosion Study This is a high-level study of simplified process flow diagrams to provide the quickest and simplest corrosion review of assets and piping systems. • Level 2 Corrosion Study This comprises all Level 1 Corrosion Study components and a report that describes the basis for determining corrosion rates and damage mechanisms, and to provides key findings and recommendations. • Level 3 Corrosion Study This is the most detailed and in-depth analysis that includes determining the details and extent of the study with the facility owner-user. This study may be used to prepare a corrosion control manual for a facility. A corrosion study goes beyond the review of an asset (e.g., fixed mechanical equipment) for corrosion and the probability of failure as affected by: • Environmental cracking and other damage mechanisms that do not exhibit metal loss (e.g., brittle fracture, creep, or high-temperature hydrogen attack); and, • External and internal metal loss. A corrosion study is an opportunity for stakeholders to provide their knowledge about corrosion processes, including from: • A corrosion consultant; • Facility personnel (e.g., inspectors, operators, and process engineers); and, • The risk-based inspection implementation team. Corrosion study and damage mechanism reviews are closely related but have been developed separately by different industries and organizations. See also, *corrosion*, *criticality*, *damage mechanism review*, *fixed mechanical equipment*, *owner-user*, *probability of failure*, and *risk-based inspection*.
Cost Adder	A *cost adder* is a required item that will increase project cost and was not included in the original purchase order (e.g., unanticipated supply of an item, material, service, or combination). This results in a change order to purchase the additional scope of supply. See also, *change order* and *purchase order*.
Cost Increase	A *cost increase* is an increase in a budgeted cost from factors such as inflation. See also, *cost overrun*.

<table>
<tr><th>Phrase or Term</th><th>Definition</th></tr>
<tr><td>Cost Overrun, Budget Overrun</td><td>A cost overrun or budget overrun involves unexpected incurred costs, increasing the actual cost and exceeding the budgeted, estimated, original, or target cost. Costs in excess of budgeted amounts typically occur by underestimating costs during budgeting. For a project, this can be measured as the ratio of the actual final cost to the estimate made at the full-funds authorization or sanction; measured in escalation-adjusted terms. See also, cost increase.</td></tr>
<tr><td>Cost Plus</td><td>Cost plus is a method of establishing a selling price in which a percentage of the cost is added to include a profit. See also, fixed costs and variable costs.</td></tr>
<tr><td>Cost-driven</td><td>A cost-driven business or project minimizes costs and expenses wherever possible. For example, using automation (e.g., automated cashiers), lowering costs or prices (e.g., moving a business online), and outsourcing (e.g., contracting work to those who will work for less). See also, schedule-driven and value-driven.</td></tr>
<tr><td>Counterintuitive</td><td>Counterintuitive is something that is contrary common-sense expectation or intuition but is nevertheless true. For example, as counterintuitive as it may seem, the universe has no centre, and it has no boundary. See also, contrarian and intuitive.</td></tr>
<tr><td>Countervailing Duties</td><td>Countervailing duties are applied to imported products when the supplying country's production, exports or both are directly or indirectly subsidized to the extent that they cause or threaten material injury to the domestic producers of these products. See also, anti-dumping.</td></tr>
<tr><td>Coupon</td><td>A coupon is a:<ul><li>Measured and pre-weighed piece of metal (e.g., strip) that is mounted inside equipment or a system (e.g., tank or piping) to accurately monitor corrosion rates. The initial weight is compared with the weight following 60, 90, or 120 days of exposure to the process; or,</li><li>Sample weld (e.g., using pipe or plate) used to determine acceptance for a welder's qualification or a welding qualification record.</li></ul></td></tr>
<tr><td>Cover Letter</td><td>A cover letter is a document that accompanies:<ul><li>And explains the contents of another document, or package of documents or items; or,</li><li>A resume, which introduces a job applicant by summarizing the applicant credentials and suitability for a job. A cover letter may also serve as an:<ul><li>Application letter
To apply for a specific advertised position.</li><li>Inquiry letter
To express interest in a position when nothing specific is advertised (i.e., an unsolicited submission).</li></ul></li></ul>See also, cover sheet and resume.</td></tr>
<tr><td>Cover Sheet</td><td>A cover sheet is a page placed at the front of a document (e.g., a data book, document package, or report) typically with the name of the author (i.e., company or person), date, document number, and title. See also, cover letter.</td></tr>
</table>

Phrase or Term	Definition
COVID-19 Virus, Novel Coronavirus	The *COVID-19 virus* or *novel coronavirus* is an infectious virus first identified in late 2019 in Wuhan, China. In 2020, it spread to become a once-in-a-century global pandemic – infecting hundreds of millions and spreading to virtually every country on earth. The COVID-19 virus outbreak has directly affected billions of people, crippled governments, fostered fear in citizenry, shut down economies (often repeatedly). See also, *community spread*.
Coworking Space	A *coworking space* is an area that personnel can gather work independently on different projects, or in groups on the same projects. It may be operated as a private business or a cooperative. These personnel may or may not work for the same company. This shared office space allows for cost savings and convenience using common infrastructure (e.g., custodian and receptionist services, equipment, and utilities) compared to a traditional corporate office space. Coworking spaces offer the same amenities as a traditional office, usually without the need to sign a long-term lease. Basic resources may be offered such as conference rooms, desks, networking, printers, private offices, and WiFi. Other resources may include 3D printers, drinks and snacks, events (e.g., professional or social), kitchen facilities, start-up resources (e.g., advisers and coaches) or studios. The advantages of coworking include: • Ability to avoid working at home where there may be a lack of resources or too many distractions; • Community with colleagues and friends that help each other succeed; • Flexibility for independent contractors, remote workers, small businesses, and start-ups; • Motivation to be productive; and, • Networking with like-minded personnel to create new opportunities and synergies. The first co-working space was founded by hackers in Berlin in 1995, with the idea that its members would share ideas, information, and workspace. A space was opened that same year by a software company in New York, with a flexible desk setting. The term coworking was first used by Bernard DeKoven, who described it as *working together as equals*. See also, *cluster*, *cooperative*, *synergy*, and *work at home*.

Phrase or Term	Definition
Cradle-to-cradle	*Cradle-to-cradle* is when products and systems are designed to be returned after their life cycle and turned into a new product of equal, if not greater, value. It aims to minimize the environmental impact of products by employing sustainable disposal, operation, and production practices, and to incorporate social responsibility into product development. This concept was developed by Michael Braungart and William McDonough in the book *Cradle to Cradle: Remaking the Way We Make Things*. They observed that cradle-to-grave (i.e., life cycle) products were typically downcycled to a lower value recycled product after use. The concept of *waste as food*, where after-use products become *food* for new and even better products, is intended to mimic how nature works. In nature, waste equals food – all products are nutrients within biological (i.e., natural) or industrial (i.e., technical) metabolisms. For example, their book is designed to look and feel like top quality paper, yet it is durable and waterproof, and printed on a *synthetic paper* made from inorganic fillers and plastic resins. Not only is it recyclable by conventional means, it can also be called a *technical nutrient*, which is an infinitely recyclable. This *treeless* book demonstrated how many other products, can be used, recycled, and used again without losing any material quality – in a cradle-to-cradle cycle. See also, *cradle-to-grave*, *downcycle*, and *life cycle*.
Cradle-to-grave	Not a preferred phrase – see *life cycle*.
Crash	In the context of finances, a *crash* is a significant and sudden decline in the value of a market. A crash often associated with an inflated stock market that suddenly deflates; however, any market can crash (e.g., the real estate market). A crash may be defined as a precipitous drop in the value of a market index. See also, *bubble*, *burst the bubble*, and *market index*.
Crashing	*Crashing* is a schedule compression technique used to speed up project work by increasing the rate at which critical path activities are completed by adding more resources — usually more personnel or more equipment. Crashing increases project costs; thus, most projects use it first on activities that can be sped up using minimum additional cost. See also, *critical path*.

Phrase or Term	Definition
Crawl Walk Run and Fly, Crawl-Walk-Run-Fly	*Crawl walk run and fly* or *crawl-walk-run-fly* is an approach to business decision-making. To achieve something significant, it may be necessary to progress the effort in stages instead of trying to make all required changes at once. It is impossible, for example, to implement a new enterprise resource planning system that immediately becomes a world class, fully automated, heavily-integrated system by simply purchasing the software and reading about it. Other examples are business start-ups, delivery of digital healthcare, and website development. This concept is not without risk and may not be appropriate for: • Strategic projects, especially when competition is more agile, serves more customers, and usurps market share; • Situations where a business potentially faces more risk from moving slowly than moving too fast (e.g., the capital, financial, and personnel resources invested in implementing a *broad crawl effort* could potentially be more effectively and smartly implemented in a narrow *run-or-fly effort*); or, • When customer needs-wants have changed or has physically moved such that the product or service is no longer available or geographically nearby. See also, *enterprise resource planning*.
Creative Accounting	*Creative accounting* is the use of unorthodox accounting practices, which follow required laws and regulations, but deviate from the spirit of those standards. Although these practices are legal, regulations are often reformed to prevent such behaviour. It is characterized by excessive complication and novel presentation of assets, income, or liabilities, with the intent to influence readers towards the desired interpretations. Typically, this involves: • Exploiting loopholes in financial regulations to gain advantage (e.g., producing misleading advertising and marketing material); or, • Presenting data using deceptive or misleading rhetoric. Unlike the expression *cooking the books*, creative accounting is generally legal. Other associated phrases include *aggressive accounting*, *earnings management*, *Enronomics*, *financial engineering*, *innovative accounting*, *orthodox*, and *rhetoric*. See also, *creative financing* and *loophole*.
Creative Destruction	*Creative destruction* is the process whereby innovations arise and cause old ways of doing things to disappear. This concept was developed and popularized by the Austrian-American economist Joseph Schumpeter. Creative destruction and disruptive innovation are closely related but have been developed separately by different industries and organizations. See also, *brainstorming*, *cognitive dissonance*, *cognitive flexibility*, *creative innovation*, *critical thinking*, *disruptive innovation*, *focus group*, *lateral thinking*, *thinking mode*, *unorthodox*, and *unfocus group*.
Creative Error	*Creative error* is the theory that ignorance or miscalculation of the true benefits and cost of an enterprise is needed to get a project launched. This has been considered as necessary for important and large projects to get built (e.g., the Panama Canal). It encourages decision-makers and promoter to go ahead and not worry about costs or other problems. See also, *Darwin Awards*, *fundamental laws of stupidity*, *hiding hand principle*, *inverted Darwinism*, and *reverse Darwinism*.

Phrase or Term	Definition

Creative Estimating	*Creative estimating* is a problem-solving approach used to:
	• Predict an outcome when there is broken analytics, limited data, no access to a subject matter expert, a non-routine calculation, or time pressure. For example, to estimate the market appeal for locally grown organic chili peppers in Berlin. Creative estimating involves these steps:
	  o Intuitively establish a hypothesis;
	  o Identify options that will resolve the problem; and,
	  o Use analogies to find or interpret data.
	• Produce a valid estimate when:
	  o It is necessary to pad or inflate values to account for contingencies with unknown risks or variables. For example, an estimate of the number of bulk flanges needed for a piping system when engineering is only 80% complete; or,
	  o It must suit pre-conceived needs. For example, a plan to exhibit at a trade show is expected to include various elements (e.g., booth, courier, gifts, meals, promotional items, literature, and travel) but costs need to conform to a defined budget.
	See also, *extrapolation*, *intuitive*, *superforecasting*, and *top-down estimating*.

Creative Financing	*Creative financing* is to: • Create an innovative capital structure, and arrange extended loan and trade credit repayment terms to achieve a level of financial leverage not ordinarily possible; or, • Use an unconventional, uncommon, or untraditional means to buy property, land, or another major item. The goal is to finance or purchase property, with the buyer or investor minimizing the use of their own money (i.e., by leveraging or using money from others, which is often referred to as *other people's money*). *See also, creative accounting.*
Creative Thinking	*Creative thinking* is a method of looking at problems or situations from a new perspective that prompts unorthodox solutions. It can be stimulated by: • An unstructured process such as brainstorming; and, • A structured process such as lateral thinking. See also, *brainstorming*, *cognitive dissonance*, *cognitive flexibility*, *creative destruction*, *critical thinking*, *disruptive innovation*, *focus group*, *lateral thinking*, *thinking mode*, *unorthodox*, and *unfocus group*.
Creativity	*Creativity* is the capacity to produce new or novel ideas, insights, and inventions that are of aesthetic, scientific, social, or technical value. In the book *The E Myth*, Michael E. Gerber explains the concept. In the book, Gerber credits German American economist Theodore Levitt with the idea that creativity *thinks* of new things, but innovation *does* new things. See also, *flexibility*, *innovation*, and *invention*.
Crew	A *crew* is a group of personnel who work: • Closely together; or, • Together on and operate an aircraft, boat, ship, train, or spacecraft (e.g., flight crew). See also, *labour force*, *main gang crew*, *poor-boy crew*, *tie-in crew*, and *repair crew*.

Phrase or Term	Definition
Crisis	A *crisis* is a dangerous, difficult, or urgent situation that may affect an individual, group, or society. It is a crucial or unstable state or time in marked by the high possibility of a highly undesirable outcome. For example, a financial disaster (i.e., bankruptcy), a natural disaster (e.g., a hurricane), or global pandemic (i.e., the bubonic plague). See also, *predicament* and *problem*.
Critical Dimension	A *critical dimension* has the smallest tolerances, affects fit or function, and has regulatory or safety affects. These dimensions affect fit-up of assembly components or the design intent and therefore the overall performance of the assembly or item. See also, *non-critical dimension*.
Critical Energy Metals	*Critical energy metals* are strategic resources used in the production, storage, and use of renewable energy. These metals include cobalt, copper, graphite, lithium, nickel, and vanadium. Electric vehicles do not use petroleum; instead, they rely on these metals for their batteries and motors. Access to copper is increasingly seen as a national security issue; hence, in 2021, Robert Friedland coined the phrase *copper is the new oil*. This is wordplay on the phrase *lithium is the white petroleum*, which has a similar meaning. See also, *Organization of Mineral Exporting Countries*, *mineral*, *rare earth element*, and *renewable energy*, and *wordplay*.
Critical Path	The *critical path* is the longest anticipated length of time it will take to complete a project. This takes into account the completion of all project tasks and the time it takes to complete them. Ultimately, the critical path gives a clear picture of a project's actual schedule. Tasks that have some schedule flexibility are not on the critical path, meaning that if a task is delayed the project can still finish on schedule. Tasks on the critical path have no tolerance for schedule flexibility and thus these tasks may not be delayed. See also, *critical path activity* and *critical path method*.
Critical Path Activity	A *critical path activity* is a scheduled activity that is part of a project's critical path. See also, *critical path*.
Critical Path Method	*Critical path method* is used to estimate the shortest length of time needed to complete a project and determine the amount of float for activities that are not part of the critical path. See also, *fast tracking*, *float*, and *critical path*.
Critical Success Factor	A *critical success factor* is an aspect (e.g., an activity, capability, or condition) of a project that is crucial to its success. It is not a measurement of success; instead, it is a something that needs to exist; without it, the project would fail. For example, correct installation of pipeline, maintenance cost reductions, efficient management of change, or effective relationship building.

Phrase or Term	Definition

Critical Thinking	*Critical thinking* is the objective analysis and evaluation of an issue or subject to form a judgment or opinion. It is essentially conscious thinking by way of:
	• Being:
	  o Clear, independent, logical, rational and skeptical;
	  o Free of biased or egocentric tendencies; and,
	  o Self-corrective, self-directed, self-disciplined, and self-monitored.
	• Evaluating factual evidence;
	• Improving thought processes by analyzing, assessing, and reconstructing how people think; and,
	• Using mindful forms of communication and problem-solving techniques.
	Critical thinking can be applied to any problem, situation, or subject. See also, *bias*, *cognitive dissonance*, *cognitive flexibility*, *creative thinking*, *mindfulness*, *lucky fool syndrome*, and *thinking mode*.

Criticality	*Criticality* is the quality, state, or degree of being of higher importance. Something with a higher *criticality* is of more crucial importance or has higher priority. See also, *corrosion study* and *criticality rating*.
Criticality Ranking, Criticality Rating	*Criticality ranking* or *criticality rating* is a measure of the significance or impact of failure of a product, process, or service against established requirements. This information may be used to establish the quality surveillance level. See also, *criticality* and *quality surveillance level*.
Crocodile Tears, Superficial Sympathy	*Crocodile tears* and *superficial sympathy* describes insincere sorrow. The expression crocodile tears originates from an ancient belief that crocodiles would shed tears while consuming their prey. While crocodiles can and do generate tears, they are not linked to emotion. The fluid from tear ducts cleans and lubricates eyes, and is most prominent in crocodiles who have been on dry land for some time. In business, a politician may be accused of shedding crocodile tears when an oil company lays-off personnel based on a decision made by the politician.

Phrase or Term	Definition
Crony Capitalism, Cronyism	*Crony capitalism* is an economic system characterized by close, mutually advantageous relationships between business leaders and government officials. This means that a business thrives not because of risk, but the nexus between the business class and the political class. Profit is made not by success in the marketplace, but by profiteering through rent-seeking behaviour using a monopoly or oligopoly. For example, a company donates money to a politician's election campaign and in return receives favourable policies and sole-sourced awards, earning them larger profits without competition. Crony capitalism distorts the economy and corrupts public servants who serve economic, political, and social ideals. These cozy relationships are typically expensive, ineffective, and unfair to society. *Cronyism*, which occurs when government favours certain producers by subsidizing them (i.e., to determine the economic winners and losers), is more closely related to socialism and is not at all capitalistic. For example, in Canada, the Ontario provincial government provided General Motors and Chrysler with a CAD$ 4.6 billion bailout in 2009 because of the 2008 financial crisis. The government recovered only CAD$ 3.6 billion and had to write-off CAD$ 1 billion of taxpayer money from the bailout. It is estimated that the total cost to Canadian taxpayers, including money that was never repaid to the Canadian federal government or its citizenry, was CAD$ 3.7 billion. See also, *award*, *bailout*, *capitalism*, *competitive bid*, *conscious capitalism*, *corporate welfare*, *economic system*, *monopoly*, *oligopoly*, *renewable energy*, *rent-seeking behaviour*, *smoke-filled room*, *socialism*, and *sole source*.
Cross-functional	*Cross-functional* denotes or relates to a system where a group of personnel with different functional expertise work together as a team to achieve a common goal. For example, this may include personnel from finance, marketing, and operations. It may include personnel from all levels of an organization and from outside the organization (e.g., consultants, customers, or suppliers). See also, *cross-train*.
Cross-generational Mutual Mentoring	*Cross-generational mutual mentoring* facilitates knowledge transfer by encouraging cross-generational mutual understanding of both new knowledge (e.g., social media and smart phone applications) and decades of acquired business experience, skills, and wisdom. Cross-generational mutual mentoring is important because, for the first time in history, the North American labour force comprises four generations, and, in some cases, five. See also, *diversity*, *generation*, *mentoring*, *peer mentoring*, *reverse mentoring*, and *shadow board*.
Cross-train	To *cross-train* is to train personnel to do more than one job, role, skill, or task. The adaptability of cross-trained personnel makes them more valuable than those who do not receive this training. See also, *boot camp*, *continuing education*, *cross-functional*, *lifelong learning*, *lunch and learn*, *on-the-job training*, *professional*, *professional development unit*, and *workshop*.
Crude Slate	*Crude slate* is the choice of crude oil used in an oil refinery and an important determinate of refinery profitability. See also, *crude*.
Crude, Crude Oil	*Crude* or *crude oil* is naturally occurring unrefined petroleum product composed of hydrocarbon deposits and other organic materials. A type of fossil fuel, crude oil can be refined to produce usable products such as gasoline, diesel, and various forms of petrochemicals. It is a non-renewable resource, See also, *fossil fuel*, *non-renewable resource*, and *petrochemical*.

Phrase or Term	Definition
Cryptocurrency	*Cryptocurrency* is a digital or virtual currency that is secured by cryptography, which makes it nearly impossible to counterfeit or double-spend. Many cryptocurrencies are decentralized networks based on blockchain technology. Secure payments are made with online transaction records that are maintained and verified by a decentralized system, rather than by a centralized authority (i.e., through a financial institution). This renders the transactions theoretically immune to government interference or manipulation. There are many cryptocurrencies; bitcoin is one of the most well known. See also, *altcoin*, *bitcoin*, *blockchain*, *fiat currency*, *grantcoin*. and *legal tender*.
Cryptography	*Cryptography* is the practice of communication that uses techniques which prevent the public or third parties from reading private messages. Various aspects include authentication, data confidentiality and integrity, and non-repudiation. Modern cryptography uses the disciplines of communication science, computer science, electrical engineering, mathematics, and physics. Applications include electronic commerce, chip-based payment cards, digital currencies, computer passwords, and business, military, or private communications. See also, *cryptocurrency*, *digital signature*, *electronic signature*, *encryption*, *microdots*, and *steganography*.
Crystal Ball Gazing, Crystal Gazing	*Crystal ball gazing* or *crystal gazing* is to: • Stare into a crystal or glass globe or a crystal to reveal future or past events (i.e., performed by a clairvoyant or fortune teller); or, • Make a determination or prediction using questionable or unscientific means (e.g., flip a coin to determine next steps). In *Crystal Gazing: Its History and Practice, with a Discussion of the Evidence for Telepathic Scrying*, a book by Northcote W. Thomas, crystal gazing was commonly practiced by the ancient peoples of China, Egypt, and Iran, and certain Indigenous peoples of North and South America. Early crystal gazers used *specularii*, preferred beryl, a sea-green mineral that could be polished into a sphere to enhance it's reflective properties. Throughout history, the crystal ball is an ever-present and powerful symbol of spiritual mysticism, which has various historically determined meanings. See also, *creative estimating*, *extrapoliation*, *forecasting*, *mineral*, and *superforecasting*.
Cult	A *cult* is a social group with: • A set of religious devotional practices that are conventional within their culture, related to a particular figure, and often associated with a particular place; • A common interest in, and great devotion to, an idea, goal, movement, object, person, or work; or, • Unusual philosophical, religious, or spiritual beliefs. A cult may range in size from a local group with few followers to an international organization with millions of followers. Use of the term cult may be controversial and subjective, and often has a derogatory or negative connotation. See also, *unorthodox*.
Curriculum Vitae, CV	Not a preferred phrase – see *resume*.
Customer	A *customer*, also referred to as a client, buyer, or purchaser, is the recipient of a product or service that is obtained from a seller or supplier in exchange for money. See also, *buyer*, *client*, *customer relationship management*, and *supplier*.

Phrase or Term	Definition
Customer Centric, Customer-focused	*Customer centric* or *customer-focused* is a way of conducting business that creates inherent synergy between the client and service provider by alignment and a long-term relationship. For example, loyalty programs benefit the brand and customer when member feedback is used for continuous improvement. It is not transactional, but rather a win-win scenario that makes it easy for the client to solve a problem or reach a goal. This, in turn, helps the service provider reach their business goals. See also, *brand*, *positioning*, *relational customer service*, *synergy*, *transactional customer service*, and *win-win*.
Customer Perceived Value	*Customer perceived value* is the customer's belief that a product or service can meet their expectations or needs. This belief can impact product demand and pricing. For example, compare the way a customer feels about a brand that is original equipment (e.g., viewed as high status or luxurious – such as a new Ferrari) to the same feeling for an aftermarket, generic, or no-name brand (e.g., viewed a common or utilitarian – such as a used Honda Civic). See also, *aftermarket* and *customer value model*.
Customer Relationship Management	*Customer relationship management* is the combination of practices, strategies, and technologies used to analyze and manage customer data and interactions throughout the customer life cycle, with the goal of assisting in customer retention to drive sales growth and improve the customer service relationship. See also, *customer*, *management*, *relational customer service*, *outside-in*, and *transactional customer service*.
Customer to Business	*Customer to business* refers to the process where consumer or end user makes a product or service that an organization uses to complete a business process or gain competitive advantage. For example, the owner of consumer-generated content and media (e.g., blogs, podcasts, social networks, videos, and websites) provides a business with a fee-based subscription to market their products or services on the consumer's media channels. See also, *business to business*, *business to customer*, *business to government*, and *customer to customer*.
Customer to Customer	*Customer to customer* refers to the process of customers trading with each other (e.g., using online auctions or classified advertisements such as Craigslist, eBay, and Kijiji). See also, *business to business*, *business to customer*, *business to government*, and *customer to business*.
Customer Value Model	A *customer value model* is a data-driven representation of the monetary worth of what a company could do or is doing for its customers. This tool is used primarily in business to business scenarios where the choice of a given product or service is based primarily upon the amount of customer value created. Customer value is defined as: Value = Benefits – Price Thus, customer benefits are quantified as product capabilities and features – which are translated into dollar amounts. See also, *customer perceived value* and *business to business*.

Phrase or Term	Definition
Customs	*Customs* is: • The department that administers and collects the duties or taxes levied by a government on imported goods; • The duties or taxes levied by a government on imported goods; or, • A physical place at an airport, frontier, or port where customs officials check incoming goods, travellers, or luggage. See also, *goods*.
Cut Sheet, Flat Sheet, Line Card, Business Line Card, Specification, Spec Sheet	A *cut sheet*, *flat sheet*, *line card*, *business line card*, *specification*, or *spec sheet* (informally) describes and provides the specifications of a product (or a property or service). It may include photographs, sketches, a list of benefits, features, options, parts, and specifications. See also, *data sheet*, *specification*, and *specifications*.
Cut-out (Weld)	A *cut-out* is a weld that is entirely removed and replaced with a new weld. A cut-out is used if a weld has a defect or defects that are too large or numerous to replace, or if the weld being repaired has had too many previous repair attempts (as determined by the design specification). This phrase is relevant to pipeline welding, piping systems, and other pressure and non-pressure equipment. See also, *design specification* and *major repair*.
Cyber	*Cyber* refers to activities that occur exclusively online (e.g., cyberbullying). See also, *bully* and *cyberbullying*, *cybercrime*, and *cyberwarfare*.
Cyberbullying	*Cyberbullying* is bullying that takes place using a digital device (e.g., a cell phone, computer, or tablet), typically using applications and text messaging, or online in gaming or social media forums where people share or view content. Cyberbullying includes posting, sending, or sharing cruel, false, harmful, or negative content about someone else. It can also include sharing personal or private information about to cause embarrassment or humiliation. Cyberbullying is illegal in some countries. See also, *bully*, *cyber*, *cybercrime*, *flame war*, *flaming*, *hacktivism*, *office jerk*, and *trolling*.
Cybercrime	*Cybercrime* is an illegal activity that takes place on the Internet or using a digital device. It is targeted premeditated attach launched with the intent to harm or victimize an entity or someone physically, psychologically, or a combination. See also, *cyber*, *cyberbullying*, *cyberwarfare*, and *hacktivism*.
Cyberwarfare	*Cyberwarfare* is an illegal activity that occurs across international borders on the Internet or using a digital device. It involves illegal acts that occur in at least two countries. See also, *cyber*, *cyberbullying*, and *cybercrime*.

Phrase or Term	Definition
D	1 A B C **D** E F G H I J K L M N O P Q R S T U V W X Y Z
Damage Mechanism Review	A *damage mechanism review* is used in the chemical processing and refining industries to determine the susceptibility of process equipment to failure. Damage mechanisms depend on many variables including the effects of contaminants, environment, materials, operating conditions, start-up and shutdown, and upset conditions. This process typically includes: • Assessment of the causes and extent of corrosion, and future corrosion rates; • Evaluation of existing and potential damage mechanisms; • Identification of strategies to mitigate degradation and reduce corrosion; and, • Review of failure analysis reports, inspection reports, maintenance records, materials of construction, process flow diagrams, and other documents. A damage mechanism review assists in the preparation of an effective mechanical integrity plan for a facility to mitigate risk from loss of containment, unplanned outages, or both. This activity may be required by regulations. Corrosion study and damage mechanism reviews are closely related but have been developed separately by different industries and organizations. See also, *corrosion study*, *mechanical integrity*, and *process equipment*.
Dark Horse	A *dark horse* is a: • Candidate or competitor about whom little is known and: o Is not expected to succeed or win: or, o Unexpectedly succeeds or wins. • Political candidate unexpectedly nominated, usually as a compromise between factions or by power brokers, in a smoke-filled room. See also, *old boy's club* and *smoke-filled room*.
Darwin Awards	The *Darwin Awards* is are accolades awarded by https://www.DarwinAwards.com to those who accidentally and spectacularly die when performing absurd, poorly-thought-out, stupid, or unintelligent activities. For example, lawyer Garry Hoy loved telling others that the windows in his office in Toronto, Ontario, Canada were bulletproof and unbreakable. One day he decided to prove this by running towards and throwing himself against a window. Shocked onlookers watched as Gary crashed through the *unbreakable* window and fell to his death on the pavement 24 floors below. The Darwin Award is meant to be humorous: it awards people for removing stupidity from the human race by means of natural self-selection (a reference to Darwinism). Gary encapsulates this. Wendy Northcutt started collecting *Darwin Award* stories in 1993. Later, she established https://www.DarwinAwards.com, the Darwin Awards, which started a popular culture phenomenon. She is also the author of many Darwin Award books. The latest is *The Darwin Awards VI: Countdown to Extinction*. See also, *brain fart*, *brain freeze*, *Darwinian theory*, *creative error*, *fundamental laws of stupidity*, *hiding hand principle*, *idiot savant*, *inverted Darwinism*, *loserthink*, *lucky fool syndrome*, *reverse Darwinism*, and *savant*.

Phrase or Term	Definition
Darwinian Theory, Darwinism	*Darwinian theory* or *Darwinism* is a theory of biological evolution developed by the English naturalist Charles Darwin and others. It states that all species of organisms arise and develop through the natural selection of small, inherited variations that increase the individual's ability to compete, survive, and reproduce. It proports that life tends to progress without intervention. Every living thing is just a random iteration of some other organism, and the best iterations survive and persist. It gained general scientific acceptance after Darwin published his book *On the Origin of Species*. See also, *Darwin Awards*, *fundamental laws of stupidity*, *inverted Darwinism*, and *reverse Darwinism*.
Dashboard	A *dashboard* is: • A graphic or webpage that provides an at-a-glance view of data or key performance indicators relevant to a business objective or process (i.e., linked to or produced from a database that allows the data to be constantly updated); • A progress report; or, • The panel facing the driver of a vehicle or the pilot of an aircraft that contains the vehicle's controls and instruments. The term *dashboard* originates from this meaning. The information presented on dashboard may be analytical, informational, operational, and/or strategic. See also, *key performance indicator*, *scorecard*, and *traffic lights*.
Data Lake	A *data lake* is a collection of data sets into a single structure to improve access to, and electronic searching of, data. See also, *big data*, and *large data*.
Data Rich Information Poor Syndrome, Data Rich But Information Poor Syndrome	*Data rich information poor syndrome* or *data rich but information poor syndrome* occurs in an organization that has an abundance of data (e.g., indicators, patterns, or statistics). However, it lacks the capacity or processes to produce meaningful and useful information that is easy to retrieve and understand. This means that the organization is unable use the data to create a competitive advantage or for timely decision-making.
Data Sheet	A *data sheet* describes detailed technical characteristics of a product. Unlike a design specification, it does not describe how to fabricate or manufacture a product. See also, *cut sheet*, *design specification*, *specification*, *specifications*, and *technical note*.
Data Transfer Rate	The *data transfer rate* is the: • Amount of digital data that is moved from one place to another in a given time (e.g., 50 G); or, • Speed of travel of digital data from one place to another (e.g., Mbps). In general, the greater the bandwidth of a path, the higher the data transfer rate. See also, *bandwidth*.
Daughter Coil, Daughter Plate	A material that is separated during pipe manufacturing is a: • *Daughter coil* for the portion of steel removed via slitting, cutting, or shearing from the mother coil, which may be used to produce one or more pieces of pipe; or, • *Daughter plate* for the portion of steel removed via slitting, cutting or shearing from the mother plate, which may be used to produce one or more pieces of pipe. See also, *mother coil*, *mother pipe*, *mother plate*, and *starting pipe*.

Phrase or Term	Definition
Daylighting	*Daylighting* is an expression that describes uncovering and exposing of underground cables, pipelines, and utilities (e.g., for assessment, construction, examination, or repair).
Debottleneck	To *debottleneck* is to remove impediments from a bottleneck that slow, impede or inhibit a process (e.g., production of a product). Debottlenecking is completed by improving or optimizing processes or revamping equipment such that overall process capacity is increased. Normally, adding production capacity to new equipment-facilities is a major capital expenditure – with significant impact to organization finances. Thus, debottlenecking is often a mission-critical activity, particularly when production approaches or needs to exceed capacity. Debottlenecking usually avoids costs associated with capital expenditures. See also, *bottleneck*, *capital expenditure*, *mission critical*, and *optimization*.
Debrief	To *debrief* is to review a completed mission or undertaking to obtain useful information. For example, management debriefed the business development team after the proposal resulted in the successful award of a project to further the business strategy.
Decision-based Evidence Making	*Decision-based evidence making* is where analyses and facts are used to support a decision that has already been made. Personnel can think they have committed to evidence-based decision making — but have instead, without realizing it, committed to decision-based evidence making. This is most prevalent whenever evidence is highly valued within an organization (that is, evidence is effectively *mandatory*) but a conflict exists between the evidence and the decision makers' strongly held beliefs. In these circumstances, decision makers cannot simply override the disconfirming evidence; instead, they are more likely to rewrite the evidence to support their beliefs. For example, Vince Kaminski's tenure as the managing director for research at Enron Corporation before the company declared bankruptcy. Each deal undertaken by Enron was mandated to have a complete risk analysis conducted by Kaminski and Enron's team of about 50 highly skilled analysts and mathematicians. However, as Enron became increasingly focused on deal volume and increasingly hostile to analysis and facts, their risk analyses became a charade. See also, *decision-making*, *evidence-based decision-making*, *evidence-based management*, *opinion-based decision-making*, and *value-based decision-making*.
Decision-making	*Decision-making* is the act or process of making choices or decisions with a group of people, especially in business or politics. Generally, six *modes* affect decision-making: 1. Conscious beliefs; 2. Inspiration; 3. Instinct; 4. Intuition; 5. Subconscious beliefs; and, 6. Values. Most people use conscious beliefs (1), instinct (3), and subconscious beliefs (5) when making decisions. See also, *value-based decision-making*, and *values*.

Phrase or Term	Definition
Decision-matrix Method, Pugh Method, Pugh Concept Selection	Invented by Stuart Pugh, the *decision-matrix method*, *Pugh method*, or *Pugh Concept Selection* is a qualitative technique used to rank the multi-dimensional options of an option set. It is frequently used in engineering for making design decisions but can also be used to rank other options or set of multidimensional entities (e.g., investments, products, or suppliers). A basic decision matrix consists of establishing a set of criteria options, which are scored and summed to gain a total score that can then be ranked. Importantly, it is not weighted to allow a quick selection process. A weighted decision matrix operates in the same way as the basic decision matrix but introduces the concept of weighting the criteria in order of importance. The resultant scores better reflect the importance to the decision maker of the criteria involved. The more important the criteria, the higher the weighting it should be given. Each of the potential options are scored and also multiplied by the weighting given to each of the criteria to produce a result. The advantages of the decision-making matrix are that subjective opinions about one alternative versus another can be made more objective, and sensitivity studies can be performed.
Decision Tree	A *decision tree* is a graphical depiction (e.g., chart or diagram that resembles a tree with branches) used to determine a course of action or show a statistical probability. Each possible decision is shown with every potential consequence of that decision. The farthest branches (i.e., nodes or splits) represent the results of a decision. This depiction is useful to explain decision-making processes and find answers to complex problems because it: • Allows the branches to be easily interpreted; • Enables personnel to understand how to make decisions by moving from branch to branch; • Has the most important features at the top; • Is an explainable model; and, • Is easily to visualize, including by non-technical personnel. A decision tree can be used to determine a course of action in business, engineering, finance, and investment and other fields. See also, *decision tree analysis* and *node*.
Decision Tree Analysis	*Decision tree analysis* is a diagrammatic technique used to illustrate a chain of decisions and examine the implications of multiple decision-making or situational outcomes. See also, *decision tree*.
Decorum	*Decorum* is good taste and propriety in appearance or conduct (e.g., behaviour that is calm, controlled, decent, polite, and respectful). For example, the plaintiff exercised good decorum by wearing a suit and speaking respecfully in court.
Deductive Logic, Deductive Reasoning	*Deductive logic* or *deductive reasoning* is a basic form of reasoning that starts with a general statement or hypothesis, and examines the possibilities to reach a specific, logical conclusion. This means that if all premises are true, including hidden assumptions, and all inferences are correctly derived, a deductive argument can guarantee its conclusion. For example, all trees are comprised of wood. An oak tree is a tree. Therefore, an oak tree is comprised of wood. See also, *abductive logic*, *chunking down*, *chunking up*, *inductive logic*, and *scientific method*.

Phrase or Term	Definition
De-energization	*De-energization* is the disconnection and isolation of a system from a source of energy to prevent the release of energy from the system. This eliminates the possibility that the system could (accidentally, inadvertently, or unintentionally) cause harm to personnel through movement, or the release of energy, (e.g., a blast of heat, light, or sound).
Defeat Snatched From the Jaws of Victory, Victory Snatched From the Jaws of Defeat	*Defeat snatched from the jaws of victory* is an expression that means to suddenly lose when winning appears to be a foregone conclusion. It is failing in an endeavour through lack of effort or skill, from making mistakes or poor judgment, or a reversal of fortune (i.e., bad luck). For example, having the winning goal scored against your team in the final seconds of a hockey game. *Victory snatched from the jaws of defeat* is an expression that means to suddenly win when losing appears to be a foregone conclusion (e.g., to be a winning underdog). It is succeeding in an endeavor through effort, good work or judgment, skill, or a reversal of fortune (i.e., good luck). For example, scoring the winning goal in the final seconds of a hockey game. See also, *underdog*.
Defect	A *defect* is a product flaw or nonfulfillment of a service, including nonfulfillment of safety considerations. See also, *flaw* and *non-conformance*.
Defensive Behaviour	*Defensive behaviour* is: • To act aggressively or submissively in response to an imagined or real threat. For example, to respond to a manager's constructive criticism by either completely accepting or rejecting their feedback, instead of collaborating and compromising to attain the best outcome or solution; or, • To act by using or overusing defensive responses. For example, to assume that a manger's frequent constructive criticisms are personal attacks caused by negative feelings (e.g., inferiority or jealousy), instead of suggestions to improve the quality workmanship. This is a complex behaviour based on attitudes, beliefs, feelings, personality, or a combination. This behaviour is not inherently negative and can be used inappropriately, or in ways that prevent effective communication. Constant defensive behaviour, instead of offering protection from aggression or attack, can create animosity or distrust. For example, by being fearful of and immediately rejecting all constructive criticism, instead of listening to a different perspective, to make improvements or increase understanding. Someone who employs defensive behaviour may: • Appear not to listen; • Blame others instead of attempting to resolve issues; • Justify or make excuses for their behaviour; • Deflect, and claim they are doing things the same way they have always been done; or, • Tell others how to feel. See also, *constructive criticism*, *defensive driving*, *defensive person*, and *narcissistic defense mechanisms*.

Phrase or Term	Definition
Defensive Driving	*Defensive driving* is training for motor vehicle drivers that goes beyond the basic mechanics of driving (e.g., accelerating, braking, and steering) and the rules of the road. It is well known that not everyone drives safely or well — but most people think they do. Defensive driving It aims is to reduce the risk of collisions by anticipating dangerous situations, despite adverse conditions or the mistakes of others to save lives, money, and time. It requires the driver to: • Adjust speed before entering a bend, to avoid applying the brakes in the middle of a bend; • Be alert and focused; • Control vehicle speed; • Drive safely and adjust for conditions, weather, or a combination; • Look ahead and expect the unexpected; • Maintain a safe following distance; • Not expect other drivers to do what one would ordinarily do; • Prepare for any action or reaction by another driver or pedestrian; and, • Respect other drivers and pedestrians. See also, *defensive behaviour*.
Defensive Person	A *defensive person* demonstrates behaviours that are devoted to preventing or resisting aggression or attack, which can present as confrontational or negative. A defensive person has difficulty: • Accepting responsibility for their actions and speech; or, • Receiving constructive criticism, which may be perceived as a threat. See also, *defensive behaviour*.
Deficiency	A *deficiency* is any item in the supplier's scope of supply (e.g., a document, material or equipment characteristic) that is incomplete or incorrect, and includes work transferred to site. See also, *deficiency list* and *type A/B/C*.
Deficiency List	A *deficiency list* is a list of incomplete or incorrect items (e.g., produced during a walk down). See also, *deficiency*.
Define Measure Analyze Design and Verify	*Define measure analyze design and verify* is a data-driven quality strategy that focuses on the development of new products or services compared to existing ones. This approach is often used when implementing new strategies because of its ability to identify success early, basis in data, and method, which requires thorough analysis. See also, *define measure analyze improve and control* and *define review identify verify and execute*.

Phrase or Term	Definition
Define Measure Analyze Improve and Control	*Define measure analyze improve and control* is a data-driven improvement cycle used for improving, optimizing, and stabilizing business designs and processes. The steps required always proceed in this order: • Define the customer (internal and external) requirements, improvement activity, opportunity for improvement, problem, or project goals; • Measure process performance (e.g., identify benchmarks for improvement) so that new data can be compared with old data, to identify the best methods or processes; • Analyze the process to determine root causes of variation and poor performance; • Improve process performance by addressing and eliminating the root causes; and, • Control the improved process and future process performance. See also, *define measure analyze design and verify* and *define review identify verify and execute*
Define Review Identify Verify and Execute	*Define review identify verify and execute* is an approach and tool used for process improvement. It helps analyze a problem from these perspectives: 1. Define the criteria by which success will be measured and the scope of the problem, then agree on the deliverables and the success factors; 2. Review the current situation to understand the background, identify and collect information about improvements, performance metrics, and problem areas, especially low-hanging fruit; 3. Identify improvements or solutions for the problem as well as necessary changes to gain and sustain the improvements; 4. Verify that the improvements will bring about benefits that meet the defined success criteria, prioritize and test or try the improvements; and, 5. Execute the implementation of the improvements and solutions, gather feedback, and review the results. See also, *define measure analyze design and verify*, *define measure analyze improve and control*, *low-hanging fruit*.
Definitive Estimate	A *definitive estimate* is the total project cost estimate made by computing cost estimates for a project. A definitive estimate is a highly accurate estimate that is usually accurate to within 10% of the actual budget. See also, *budget*, *definitive estimate*, *estimate at complete*, and *estimate to complete*.
Defund the Police	*Defund the police* is an expression that calls for: • Modest cuts to police force budgets; • Reallocation of funds from police departments towards to non-policing forms of community support and public safety (e.g., addiction treatment, education, healthcare, housing, mental health, and social services); or, • The abolishment of police forces altogether. After the murder of George Floyd by police in Minneapolis, United States, this phrase gained popularity – especially amongst protestors of Mr. Floyd's death, but also from other anti-police protestors. The idea behind the defund the police movement is to reframe the role of the police within society — particularly in marginalized (e.g., Black and Indigenous) communities.

Phrase or Term	Definition
Degree	A *degree* is: • A certificate awarded by an educational institution (e.g., a university) to document a student's successful completion of a course of study. A degree may be issued for higher education; however, qualifications vary significantly between education systems in different countries. In North America, a degree may be a: a. Foundation or associate degree (i.e., an undergraduate academic degree); b. Bachelor's degree or baccalaureate (i.e., an undergraduate academic degree); c. Master's degree (i.e., a graduate or post-graduate academic degree); or, d. Doctorate degree (i.e., a graduate or post-graduate academic degree). • A stage or step in a course, order of classification, or process; • A unit of measurement of: ○ An angle. For example, 1° or $^1/_{360th}$ of the circumference of a circle; or, ○ A temperature, often expressed using the symbol °. For example 36°C/97°F). • The amount, extent, or level to which something happens or is present (e.g., to what degree was the suspect involved in the investigation?); and, • The extent, measure, or scope of an action, condition, or relation (e.g., the degree of pipeline corrosion caused by coating failure). See also, *certificate* and *diploma*.
Delegation	*Delegation* is the assignment of any authority to another person (normally from a manager to a subordinate) to perform specific activities. The person who delegates the work remains accountable for the outcome of the delegated work. Delegation empowers a subordinate to make decisions by shifting decision-making authority from one organizational level to another. See also, *accountability*, *micromanagement*, and *responsibility*.
Deleterious	*Deleterious* is something damaging or harmful – often in a subtle or unexpected way. For example, the pipeline rupture was unexpected and deleterious to the surrounding environment and animal populations.
Deliverables	*Deliverables* are what the supplier provides to fulfill the scope of supply for a purchase order. Specifically, this is the documentation, equipment, materials, and services.

<table>
<tr><th>Phrase or Term</th><th>Definition</th></tr>
<tr><td>Deliverology</td><td>Deliverology is an approach to managing and monitoring the implementation of activities that have significant impact on outcomes. It extends and leverages the key principles of best-in-class performance management. Deliverology is useful to (public) leaders who are committed to achieving results by:
1. Evaluating their experience in setting goals;
2. Implementing new strategies; and,
3. Reflecting on why goals were not previously achieved.
Most reforms are unsuccessful due to failed implementation; the central challenge is getting it done. The book Deliverology 101: A Field Guide For Educational Leaders (co-authored by Michael Barber, Andy Moffit, and Paul Kihn) focuses on how to accomplish meaningful results.
The British civil service originally used deliverology as a light-hearted term to denigrate the process developed by the British Prime Minister's Delivery Unit. Ultimately, the Delivery Unit adopted the term and gave it a positive definition. See also, best-in-class performance management, goal, goal setting, management by objectives, and SMART.</td></tr>
<tr><td>Delphi Technique</td><td>The Delphi technique is an estimation method based on expert consensus. Experts make estimates individually and simultaneously, and then review their estimates as a group before generating another set of estimates. This process is repeated, with the pool of estimates typically becoming narrower after each round of reviews, until a consensus is reached that produces one estimate. See also, budget and poker planning.</td></tr>
<tr><td>Delta</td><td>Delta is:
• From the United States' phonetic military dictionary that denotes the letter D (e.g., Alpha denotes A, Beta denotes B, Charlie denotes C, etc.);
• A finite increment or a variation of a function or variable; or,
• The fourth letter of the Greek alphabet (i.e., Δ, δ) transliterated as d.</td></tr>
<tr><td>DeltaV™ Control System</td><td>The DeltaV™ Control System is a scalable process control system designed by Emerson Process Management to be a single solution, for the analog and discrete worlds, as a fully digital replacement for distributed control systems and programmable logic controllers.</td></tr>
<tr><td>Demagogue</td><td>A demagogue is a political leader who exploits issues and uses emotions, false claims or promises, ignorance, and prejudice instead of rational arguments to gain power. For example, Benito Mussolini, was the Prime Minister of Italy from 1922 to 1943 and Il Duce (Italian for the leader) of Italian Fascism from 1919 to 1945. He is well-known as a demagogue for his use of his cult of personality and propaganda. For example, he said believe, obey, fight (undated). See also, propaganda.</td></tr>
</table>

Phrase or Term	Definition
Demand-side Economics	*Demand-side economics* is an economic theory that the primary factor driving economic activity and short-term fluctuations is the demand for goods and services. This means that aggregate demand is composed of consumption of goods and services, government spending on public goods and services, industry investment in capital goods, and net exports (exports-imports). When there is low aggregate demand in the short term, such as during a recession or depression, government intervention can help to reduce unemployment and stimulate growth. Demand-side fiscal policy focuses on creating a better climate for consumers. See also, *classical economic theory*, *fiscal policy*, *goods*, and *supply-side economics*.
Demobilization (Demob)	*Demobilization* (*demob*) is an activity performed by a company or project to disband, discharge, remove, and stand down personnel and resources from active project service.
Demurrage	*Demurrage* is a charge for detaining a transportation asset beyond an agreed time. For example, in shipping, demurrage charges accrue when a ship is in port for loading and unloading longer than stipulated in the charter terms.
Deplatforming, No-platforming, Financial Deplatforming	*Deplatforming* or *no-platforming* is a form of political activism or prior restraint by an individual, group, or organization to shut down controversial speakers or speech, or deny access to a venue in which to express opinions. Tactics include efforts to have speakers disinvited to an event or venue, disruption of speeches, organized social media attacks, and personal harassment such as efforts to have an individual blacklisted or fired. Social media and other technology companies may selectively ban, suspend, otherwise restrict a user's access to their platform. *Financial deplatforming* is when a bank or other financial service provider denies services to an individual or organization. See also, *blacklist*, *cancel culture*, *doxing*, *shadow banning*, and *termination for cause*.
Depletion	In oil and gas, *depletion* is the exhaustion of production of a natural resource, typically in a geographic area, or in an oil field or well. See also, *natural resource* and *peak oil*.
Depression	In dimensions, a *depression* is a part or place that is lower than the surrounding area (e.g., gutters that parallel a sidewalk); In economics, a *depression* is a long and severe recession in an economy or market, especially with rising levels of unemployment (e.g., the Great Depression). In medicine, *depression* is a common and serious medical condition that negatively affects how one feels, thinks, and acts. It causes feelings of sadness, a loss of interest in activities previously enjoyed, or a combination. It can lead to a decrease a one's personal and professional well-being and a variety of emotional and physical problems; however, this condition is treatable. It is also referred to as a *major depressive disorder*. See also, *boom*, *bust*, *Great Depression*, and *recession*.
De-risk	To *de-risk* is to: • Remove risk from a process or situation; or, • Take steps to make something less likely to involve a financial loss or risk. For example, to ensure that a company has enough insurance for potential liabilities, including product liability claims (if applicable). Another example is to identify and register or otherwise protect a company's intangible and tangible assets. See also, *risk*.

Phrase or Term	Definition
Design	A *design* is: • A drawing or sketch that depicts something, shows how it will be executed or produced or both; • A plan or protocol to accomplish or complete something; • An idea or scheme in which the methods to achieve something are described; or, • The effort or work to create something that is aesthetic or functional.
Design Basis	A *design basis* (basis of design) is a document that contains the conditions, needs, and requirements that a project considers when designing a facility. It identifies applicable acts, regulations, codes, and standards, process or technical and owner's specifications, and other relevant details. The order of precedence is usually as listed in the previous sentence. A design basis is a major deliverable for the *front end engineering and design* and *detailed engineering* phases of a project. See also, *detailed engineering*, *design basis memorandum*, *front end engineering and design*, and *order of precedence*.
Design Basis Memorandum	A *design basis memorandum* is a document that defines the basic design parameters of a project. See also, *detailed engineering*, *design basis* and *front end engineering and design*.
Design Change	A *design change* is a change to a design or specification. See also, *design change notice*.
Design Change Notice	A *design change notice* is a document used to authorize and record a change to a design or specification. See also, *design specification*, *specification*, and *supplier change request*.
Design Factor	The *design factor* refers to the stress that equipment or materials are required to withstand (e.g., the pipeline design factor is 0.80). It is generally defined for an application in advance and often set by a regulation (e.g., building code or policy). It is not an actual calculation. See also, *safety factor*.
Design Specification	A *design specification* documents requirements to be satisfied by a design, material, product, or service. A supplier uses it to guide the design, fabrication and manufacture of a product. An engineering and procurement company or owner design specification may reference other codes and standards, but a design specification is product- or project-specific. It may specify more stringent requirements and the removal of optional requirements. For example, a code may allow for a variety of welding processes, but the owner's specification takes precedence, and determines which processes are permitted and prohibited. See also, *data sheet* and *specification*.

Phrase or Term	Definition
Design Thinking	*Design thinking* refers to the cognitive, practical, and strategic processes by which design concepts (e.g., proposals for buildings, machines, or new products) are developed by designers, design teams, or a combination. Many aspects and concepts of design thinking are identified through studies, across different design domains, of design cognition and activity in both laboratory and natural settings. The traditional design thinking model is to empathize – define – ideate – prototype – test. These processes help systematically apply, extract, learn, and teach human-centred techniques to solve problems in a creative and innovative way – in businesses, countries, designs, and lives. This has been criticized for oversimplifying the design process and trivializing the role of technical knowledge and skills. Some of the world's leading brands, such as Apple, Google, and Samsung have adopted the design thinking approach. It is being taught at leading universities around the world. See also, *scientific method* and *scrappy design thinking*.
Desk Instruction	Not a preferred phrase – see *project instruction*.
Desktop Assessment, Desktop Review	A *desktop assessment* or *desktop review* is a limited-scope analysis of documentation (e.g., data, directories, drawings, Google maps, equipment lists, maps, photographs, questionnaires, records, and reports) and other resources (e.g., interview or website) at a desk, without a facility or site visit. This assessment may be prepared for an audit, evaluation, investigation, qualification, valuation, or verification of capabilities, compliance, conditions, or a combination (e.g., assets, business, environmental, financial, operations, risk management, supply chain, or technical). The purpose of this assessment is to obtain basic or preliminary information for analysis, due diligence, or initial screening to: • Determine if further investigation, a site visit, or both is required; or, • Provide information cost effectively and rapidly. See also, *due diligence*.
Destructive Testing	*Destructive testing* is an assessment that deliberately destroys or causes equipment to fail to provide the project with an understanding of equipment and material behaviour, including performance under different loads. Destructive testing is most economical and suitable for mass-produced items because the cost of destroying a small number of mass-produced items is negligible. See also, *non-destructive examination*.
Detailed Engineering	*Detailed engineering* is an engineering design phase that follows the *front end engineering and design* phase. In this phase, a project develops detailed drawings, plans, and requirements for the procurement of materials and equipment needed for construction. This includes the management of suppliers, schedules, and costs. See also, *design basis*, *design basis memorandum*, and *front end engineering and design*.
Deviation	Not a preferred term – see *design change notice*.
Developmental Trauma	*Developmental trauma* describes the distress experienced by individuals exposed to early and chronic trauma during childhood such as chronic abuse or neglect. See also, *emotionally dysregulated*, *post-traumatic stress disorder*, and *playfulness acceptance curiosity empathy*.

Phrase or Term	Definition
Dew Point	The *dew point* is the temperature to which air must be cooled to become saturated with water vapour. When further cooled, the airborne water vapour will condense to form liquid water or dew. When air cools to its dew point through contact with a surface that is colder than the air, water will condense on the surface. When the temperature is below the freezing point of water, the dew point is called the frost point because frost is formed rather than dew. The measurement of the dew point is related to humidity. A higher dew point means there will be more moisture in the air.
Dewater	To *dewater* is to: • Drain a flooded or waterlogged area; or, • Remove water from material, sediment, or tailings.
Dialect	A *dialect* is a form of a language, which is peculiar to a specific: • Area or region; or, • Occupational or social group. Many non-academic references use the terms accent and dialect interchangeably. Some languages are mutually intelligible or very similar (e.g., Danish, Norwegian, and Swedish), while some dialects are not (e.g., Chinese dialects). An accent relates to pronunciation; dialect relates to grammar, pronunciation, and vocabulary. Examples of English dialects and accents include: • African-American vernacular (Ebonics), which changes words, re-arranges them, and uses rich and exotic accents; • French Canadian English, which emphasizes the last syllables of words; and, • Texas drawl, which emphasizes very specific sounds and syllables. See also, *colloquial*, *Ebonics*, *jargon*, *Plain English*, and *vernacular*.
Dialogue of the Deaf	*Dialogue of the deaf* describes a discussion in which each party is unresponsive to each of the other parties. In this discussion, no one hears anything, which makes the discussion useless. This expression is a translation from the French *dialogue de sourds*. See also, *active listening* and *tone-deaf*.
Dichotomy	A *dichotomy* is: • A contrast or division between two things that are, or are represented to be, entirely different or opposed (i.e., especially contradictory or mutually exclusive entities or groups). For example, the two sides of a coin or yin and yang; or, • Something with seemingly contradictory qualities. For example, a restored 1950s Cadillac that runs like new. Dichotomy, which presents a striking contrast between two opposite persons or things, is used in mathematics, life science, literature, linguistics, philosophy, and politics. See also, *false dichotomy*.
Dictionary	A *dictionary* is a digital or hard copy reference that lists phrases and words, typically in alphabetical order. These are presented with their meanings, or the equivalent words in a different language. Information about form, origin, pronunciation, use, and other details are often also provided. See also, *glossary*.
Diesel Test	A *diesel test* uses diesel oil to test a tank's shell-to-bottom welds for discontinuities. See API STD 650 Para. 7.2.4.

<table>
<tr><th>Phrase or Term</th><th>Definition</th></tr>
<tr><td>Diffusible Hydrogen Designator</td><td>A Diffusible hydrogen designator is a value applied to the electrode classification of controlled hydrogen weld fillers metals (i.e., H2, H4, H8, or H16). It is measured per AWS A4.3 or ISO 3690 and relates to the maximum diffusible hydrogen content in deposited weld metal (i.e., mL/100 g). See also, controlled hydrogen electrode, electrode, and low hydrogen.</td></tr>
<tr><td>Diffusion of Innovation Theory</td><td>The diffusion of innovation theory states that every market has groups of customers who differ in their readiness and willingness to adopt a new product (e.g., innovation or technology). Also, a new product diffuses or spreads through a market in successive, overlapping waves – not in a straight line. Most populations show the following pattern in the adoption of new products (shown with the percentage of the population represented):
1. Innovators
The very first group: 2%;
2. Early adopters
The first minority group uses intuition and vision to make decisions: 14%.
3. Early majority
The first majority group waits to see whether something proves to be successful: 34%.
4. Late majority
The last majority group adopts innovations only when forced to do so (e.g., new technology becomes cost-effective or widely-used): 34%.
5. Laggards
The last minority group dislikes change and accepts new things only when there are no other options (e.g., old technology becomes unavailable or unusable): 16%.
The rate of adoption or diffusion of something new to the market can vary per type of product and price. See also, early adopter, innovator, and trail blazer.</td></tr>
<tr><td>Diffusion of Responsibility</td><td>Diffusion of responsibility is a sociopsychological phenomenon whereby a one is less likely to take responsibility for action or inaction when other bystanders or witnesses are present. It is considered a form of attribution, where it is assumed that others either are responsible for acting or have already done so. For example, this can occur when a mass email is sent out. People feel a lack of accountability since the email has not been addressed to them personally. See also, herd mentality, inclusivity, groupthink, lemming effect, loserthink, tenth man, thinking mode, and tribalism.</td></tr>
<tr><td>Digital</td><td>Digital is an electronic means of rendering data or signals, expressed as series of the digits 0 and 1, that typically represent values of a physical quantities – such as voltage or magnetic polarization. For example, a digital watch shows the time by means of displayed digits. The opposite of digital is analog. See also, analog and electronic.</td></tr>
</table>

Phrase or Term	Definition
Digital Project Delivery, Digital Project Management	*Digital project delivery* or *digital project management* uses web-based software to manage project-related duties and tasks such as scheduling milestones and reporting to stakeholders. This is a holistic approach, especially for project delivery, because the collaborative use of digital technologies, such as additive manufacturing, computer-automated design, blockchain, and modular construction, optimizes delivery, improves deliverable consistency and quality, increases efficiencies, reduces schedule delays, and tracks the status of all tasks. See also, *additive manufacturing*, *blockchain*, *computer-automated design*, *modular construction*, and *project management*.
Digital Reality	*Digital reality* is the collection of augmented and digital reality technologies and tools. See also, *augmented reality*, *mixed reality*, and *virtual reality*.
Digital Signature	A *digital signature* is a cryptographic signature on a document that is regarded as the most secure type of electronic signature. This type of signature ensures that it is valid of the signatory by a certificate of authority (e.g., by Windows). See also, *authentication*, *cryptography*, *electronic signature* and *wet signature*.
Digital Twin	A *digital twin* is a virtual version of a physical asset, used for forecasting, prediction of behaviour, simulation, and training.
Dilbert	*Dilbert* is an American comic strip illustrated and written by Scott Adams that was first published in 1989. It is known for its satirical humor about a micro-managed white-collar office engineer named Dilbert. See also, *white collar*.
Diluent	*Diluent* is chemical used to reduce bitumen viscosity and allows bitumen to flow through a pipeline. See also, *condensate* and *naphtha*.
Diluted Bitumen, Dilbit	*Diluted bitumen* or *dilbit* is bitumen mixed with diluent (e.g., condensate or naphtha), which reduces bitumen viscosity and allows bitumen to flow through a pipeline. See also, *diluent*, *condensate*, and *naphtha*.
Dimensionless Designator	A *dimensionless designator* is used by the International System of Units (Systѐme International) to describe pipe sizes. For example, nominal pipe size is a dimensionless designator of pipe size. See also, *nominal pipe size*.
Diminishing Returns	*Diminishing returns* occur when the level of profit or benefit gained becomes less than the amount of money, resources or energy invested, recurrently. See also, *analysis paralysis*, *gold plating* and *scope creep*.
Diploma	A *diploma* is a certificate awarded by an educational institution (e.g., a college) to document a student's completion of a course of study A diploma may be issued for vocational education or training; however, educational system qualifications vary by country. The certificate that a Nobel laureate receives is called a diploma. See also, *certificate* and *degree*.
Direct Tension Indicating Washer, Load Indicating Washer	A *direct tension indicating washer* or *load indicating washer* is used to apply the correct tension when tightening structural bolts. Visual means verify tightness of washers, not a feeler gauge, torque wrench, or match-marking. See also, *feeler gauge*, *torque wrench*, *turn-of-nut*, and *twist-off bolt*.

Phrase or Term	Definition
Directionally Accurate, Directionally Correct	*Directionally accurate* or *directionally correct* is an analysis or conclusion that is generally accurate or correct, but not precise in terms of its numbers. This means that it is accurate to the extent that it shows the quantity to be measured as positive or negative, even though the degree to which the quantity is positive or negative may be measured inaccurately. For example, training to lose weight to reduce body fat from 25 to 15% but only achieving 20% is: • A success in terms of achieving a positive result; and, • A failure in terms of achieving the goal.
Discipline	A *discipline* is a main category of specialized knowledge and skill. For example, procurement includes purchasing, expediting, supplier quality surveillance, and logistics disciplines. For engineering, this includes chemical, civil, electrical, process, mechanical, and other disciplines. See also, *discipline engineer*, *engineering*, *technologist,* and *technology*.
Discipline Engineer	A *discipline engineer* is an engineer who works in a specific engineering discipline (e.g., electrical, instrumentation, or mechanical). A discipline engineer may or may not be the responsible engineer. See also, *discipline*, *engineer*, *engineering*, and *responsible engineer*.
Discipline Review, Inter-Discipline Review	A *discipline review* or *inter-discipline review* is performed by related internal engineering disciplines (e.g., electrical, mechanical, and process) to ensure that the design and engineering outputs, and supplier documentation meet project design requirements. See also, *cold eyes review*, *discipline review*, *peer review*, and *squad check*.
Discrimination	*Discrimination* is: • A prejudiced or prejudicial action, outlook, or treatment that is categorical rather than individual. For example, to say *older personnel cannot learn as quickly as younger personnel* is ageist discrimination; or, • The act of making or perceiving a difference. For example, to say *this grade of stainless steel is more suitable than another grade* is a discriminating judgement. Discrimination may be based on age, culture, ethnicity, income, language, mental or physical disability, race, religion, sexual orientation, or other factors. See also, *ageism*, bias, *Black Indigenous and People of Colour*, *diversification*, *intersectionality*, *gender-based analysis*, *racism*, *sexism*, and uniqueness bias.

Phrase or Term	Definition
Disinformation	*Disinformation* is false information that is deliberately created and spread, often covertly, to deceive, influence public opinion, or obscure the truth. The seven types of disinformation are: 1. Fabricated content False content used to deceive or cause harm. 2. False connection Content is not supported by the captions, headlines, or visuals. 3. False context False contextual information is mixed with genuine content. 4. Imposter content A genuine source or person is impersonated. 5. Manipulated content Genuine information or imagery is tampered with to deceive. 6. Misleading content Information used to frame an issue or individual. 7. Parodic or satirical content Information is used to fool, humour, or make interesting connections rather than to cause harm. See also, *brainwashing*, *bullshit*, *fake news*, *false narrative*, *media bias*, *misinformation*, *propaganda*, *smells fishy*, *stretch the truth*, *thought control*, *thought crime*, *truth*, *truthiness*, and *untruth*.
Disposition	A *disposition* is an approved corrective action to address a non-conformance. Common dispositions are: • Reclassify; • Repair; • Replace; • Return; • Rework (to an existing or new specification); • Scrap; or, • Use as-is (included no deficiency found). See also, *non-conformance*.
Disrespect, Dis, Diss	To *disrespect*, *dis* (informal), or *diss* (informal) is to show: • Contempt, or a lack of courtesy or respect for someone (e.g., being rude to police after being pulled over for speeding); or, • Low regard or a lack of regard for something, such as a custom, law, or monument (e.g., a smoker who smokes in an area designated as non-smoking).

Phrase or Term	Definition
Disruptive Innovation	*Disruptive innovation* is an improvement that creates a new market and value network. It eventually disrupts an existing market and value network by displacing established and market-leading alliances, firms, products, or a combination. An innovation that is disruptive allows access to a new segment of consumers or at the bottom of a market, to a product or service that was historically only accessible to consumers with money or skill, or both. Characteristics of disruptive business, at least in initial stages, may include lower gross margins, simpler products and services, and smaller target markets. This may not appear as attractive as existing solutions when compared to traditional performance metrics. Because the lower tiers of the market offer lower gross margins, they are unattractive to organizations moving upward in the market, thereby creating space at the bottom of the market for new and disruptive competitors to emerge. For example, the smartphone business model with its applications disrupted desktop and laptop computers as the primary method for how consumers access the Internet. Today, over 50% of website visits are from a smartphone. Importantly, smartphones and their applications lead to the development of many services that did not previously exist, such as Instagram, Spotify, and Whatsapp (and countless others). This phrase was coined by Clayton Christensen and described in his book *The Innovator's Dilemma: When New Technologies Cause Great Firms to Fail*. Disruptive innovation has been called the most influential business idea of the early 21st century. Creative destruction and disruptive innovation are closely related but have been developed separately by different industries and organizations. See also, *blue ocean strategy*, *bottom of the market*, *creative destruction*, *effective disruption*, *red ocean strategy*, and *value innovation*.
Dissimilar Metal Weld	A *dissimilar metal weld* involves joining two base metals that possess different chemical or mechanical properties. Thus, this type of weld may not be suited for welding without special procedures. The electrode may be like one of the base metals or different than both.
Distributed Control System	A *distributed control system* is a computerized control system with central operator supervisory control that manages an operation or a process.
Distributed Denial of Service	*Distributed denial of service* is an attack on a computer system or website consisting of millions of simultaneous requests for service, causing the receiving system to become overwhelmed and subsequently deny access to legitimate requests.
Distributed Ledger Technology	Not a preferred phrase – see *blockchain*.
Distributor	A *distributor* is a supplier that buys non-competing products or product lines and resells them to a customer, retailers, or both. A distributor does not typically fabricate or manufacture; if needed, these resources are out-sourced. A range of services may be provided (e.g., after-sales services, product information, technical support, and warehousing). Products may be provided as bulk items, off-the-shelf, and engineered or packaged equipment. See also, *bulk items*, *off-the-shelf*, *packaged equipment*, and *supplier*.

Phrase or Term	Definition
Divergent Thinking	*Divergent thinking* involves using multiple creative solutions and free-flowing, non-linear, and spontaneous ideas to arrive at multiple alternative solutions or answers. This may involve finding or creating new procedures to resolve a challenge instead of using existing, conventional solutions. Divergent thinking is incompatible with answering multiple choice or standardized test questions because these tests require one answer per question. Divergent thinking sees the world as filled with known and unknown solutions; the world is colourful and includes grey, and black-and-white – there are always multiple options. See also, *cognitive dissonance*, *cognitive flexibility*, *convergent thinking*, *lateral thinking*, *thinking mode*, and *vertical thinking*.
Diversification	*Diversification* is a risk management strategy that eliminates unnecessary risk. It is: • In business, the practice of varying operations or products (e.g., enter a new industry or market) to reduce or spread risk (e.g., to expand or exploit spare capacity); • In finance, a strategy that mixes a wide variety of investments within a portfolio, so the positive performance of some investments neutralizes the negative performance of others (e.g., an undiversified portfolio is to hold only one or few kinds of stocks; hence, a diversified portfolio will have many stocks from a variety of asset types, company sizes, and industries); and, • The act or process of diversifying something or of becoming diversified (i.e., an increase in the variety or diversity of something). For example, recruiting diverse personnel to the labour force to avoid discrimination lawsuits, build a better consumer and employee brand, and improve creativity, productivity, and return on investment. See also, *discrimination*, *diversity*, *risk*, and *risk management*.
Diversity	*Diversity* is the condition of being composed of or having differing elements. For the individual, it is about the variety of unique dimensions, characteristics, and qualities that each person possesses. These can be elements such as age, ethnicity, gender, physical abilities, political beliefs, race, religious beliefs or other ideologies, sexual orientation, or socio-economic status (i.e., a member of a minority group). Research shows that differences make it harder for people to connect and empathize with each other. See also, *discrimination*, *diversification*, *gender-based analysis*, *inclusivity*, *intersectionality*, and *risk*.
Divest	*Divest* means to: • Deprive, dispossess, or take away something (e.g., authority, property, possessions, power, quality, rights, or title); • Eliminate or sell something that is no longer required or wanted (e.g., a business interest or investment); or, • Strip or undress, especially the removal of clothing, equipment, or ornaments A business or financial divestment or divestiture may be made for many reasons, including ethical, financial, or political. Divestment is the antonym of investment. See also, *investment*.

Phrase or Term	Definition
Do it right the first time, Do It Right This Time, DIRTT	*Do it right the first time* is a managerial accounting practice related to the just-in-time inventory system and quality control that emphasizes working correctly to avoid rework. It decreases waste and increases efficiency to avoid production delays. This practice originated as a manufacturing concept but can be applied to any task in every industry. *Do It Right This Time* and *DIRTT* are similar expressions. See also, *doing the right things right the first and every time, double-check, just-in-time, measure twice and cut once,* and *quality control*.
Do Not Look a Gift horse in the Mouth, Never Look a Gift Horse in the Mouth	*Do not look a gift horse in the mouth* or *never look a gift horse in the mouth* is an expression that means the receiver of a gift should not be critical of, unappreciative of, or ungrateful for the gift. The receiver should not closely examine the gift or attempt to discern it value; instead, the receiver should express gratitude for being given the gift. It is a common tradition among horse traders to examine the teeth of a horse to determine its age and condition, and therefore – its value. This is acceptable behaviour in business. It is also discouraged with matters related to charity and gift-giving because it tends to be critical of the gift giver, ungracious, and ungrateful on the part of the receiver.
Document Code	A *document code* is a document category used by a project. A document code helps organize supplier document requirements of a purchase order, which may comprise numerous documents that may be otherwise difficult to correlate and process. Alphabetic and numeric characters may be in document codes to expedite processing by type code. Document codes may be assigned by a document control, engineering, or material requisition system, client specification, or other resource. Regardless, the intent is to effectively communicate document metadata. The next list uses an alphabetic-descending document code order to identify welding-related documents: • W-001 – welding procedure specification • W-002 – welding procedure data sheet • W-003 – procedure qualification report • W-004 – production welding test report • W-005 – welder or welding operator certification (i.e., performance qualification) • W-006 – list of qualified welders • W-007 – welding supervisor certification • W-008 – welding inspector certification • W-009 – welding engineer certification • W-010 – company welding certification • W-011 – welding certificate of compliance In the above example, the first welding procedure specification submitted to the project would be W-001-001, the second would be W-001-002, and so on. A supplier may also use their own document codes that can be corelated to the project document codes using a supplier document index. See also, *document review code, document review stamp, metadata, review cycle, supplier document index,* and *supplier document requirements*.
Document Control	*Document Control* is the department responsible for controlling all project documents, including supplier submittals.

Phrase or Term	Definition
Document Management Group	Not a preferred phrase – see *document control*.
Document Review Code	A *document review code* summarizes the results of the engineering review process for each supplier document. The intent is to effectively communicate the same information by using a code designation instead of a lengthy description. This code is assigned by a responsible engineer. The next list uses an alphabetic-descending order: • Code 1 – May proceed without changes • Code 2 – May proceed subject to making changes and re-submitting • Code 3 – May not proceed • Code 4 – Review not required, for as-built, turnover, or record only • Code 5 – Superseded • Code 6 – Void See also, Appendix F, *document code*, *document review stamp*, *review cycle*, and *supplier document requirements*.
Document Review Stamp	A *document review stamp* is applied to all documents received from a supplier. The stamp identifies the document review code. It may be ink on hard copy or electronic and included with appropriate electronic files. See also, *document review code*.
Dogma	*Dogma* is a principle or set of principles fixed or prescribed by an authority as an established or fixed opinion that is true and without any doubt. See also, *orthodox*, *outside the box*, and *unorthodox*.
Doing the Right Things Right the First and Every Time	*Doing the right things right the first and every time* is a tenant of the Six Sigma and total quality management systems to achieve quality. The organizational attitude, goal, and performance standard is to meet the customer requirements, as defined by the customer. There are other variations of this phrase such as *do it right the first time, every time* and *doing the right things right the first time*. See also, *do it right the first time*, *quality*, *Six Sigma*, and *total quality management*.
Donation	A *donation* is a contribution or gift (e.g., equipment, materials, money, services, or time) to benefit a cause. For example, giving money to People for the Ethical Treatment of Animals (PETA). Donations of products or services are also called gifts in kind. Instead of giving money to buy products and services, the items needed are given instead (e.g., clothes and food). See also, *pro bono* and *volunteering*.
Doodle	A *doodle* is an informal drawing made during a state of affective tension (e.g., while experiencing concentration, expectation, impatience, or indecision), boredom, idleness, leisure, or meditation. For example, while one's attention is occupied (e.g., during a meeting or webinar) or unoccupied (e.g., waiting for a meeting or webinar to start). A doodle may be complex or simple and comprise abstract or random lines, or depict something real. Doodling is a form of fidgeting that can aid memory, has a positive effect on comprehension, and can relieve stress. Doodling has been around for centuries; Leonardo da Vinci's notebooks contain doodles. See also, *messy people* and *webinar*.
Dossier	Not a preferred term – see *manufacturing record book* and *turnover*.

Phrase or Term	Definition
Dotted Line Reporting	*Dotted line reporting* refers to personnel that have some level of accountability to others, but not a direct reporting line (e.g., the professional relationship between a supplier quality surveillance coordinator and the engineering manager). See also, *accountability*, *organizational chart*, *reporting line*, and *solid line reporting*.
Double Block and Bleed Valve	Per API SPEC 6D, a *double block and bleed valve* is a single valve with two seating surfaces that, in the closed position, provides a seal against pressure from both ends of the valve, with a means of venting or bleeding the cavity between the seating surfaces. This valve does not provide positive double isolation when only one side is under pressure. This definition differs significantly from the one used by North American safety codes and regulations (e.g., *Alberta Occupational Health and Safety* or *US Occupational Safety and Health Administration*). Hence, a valve described as double block and bleed by a manufacturer may not comply with safety requirements. See also, *double isolation and bleed valve*.
Double Down	To *double down* is to: • Become more resolute, tenacious, or zealous in a position or undertaking; or, • Strengthen one's commitment to a particular course of action or strategy, especially if it is potentially risky. In the card game *blackjack*, it means to double the original bet in exchange for one more card.
Double Isolation and Bleed Valve	Per API SPEC 6D, a *double isolation and bleed valve* is a single valve with two seating surfaces, each of which, in the closed position, provide a seal against pressure from a single source, with a means of venting or bleeding the cavity between the seating surfaces. This feature can be provided in one or in both directions. API SPEC 6D introduced this term to describe a valve that meets the definition of safety codes and regulations (e.g., *Alberta Occupational Health and Safety* or *US Occupational Safety and Health Administration*). See also, *double block and bleed valve*.
Double Joint	A *double joint* is two pieces of pipe that are welded together at a facility to save time and increase quality. Unlike welds produced in the field (i.e., in-situ), welds produced in a facility are not exposed to the elements, which improves the quality of the weld and reduces the total number of required welds by as much as 50% for double-jointed pipe joints. See also, *joint*.
Double Random	Not a preferred phrase – see *random length*.

Phrase or Term	Definition
Double Standard	A *double standard* is a set of principles that is applied unfairly (e.g., differently, more rigorously, or both) to one situation or group of people than to another, even though they should be applied uniformly. A double standard usually indicates a bias or hypocrisy, which is viewed negatively by those who are held to a higher standard or treated differently. For example, a code of dress that applies strict standards of business attire to women and more casual standards to men is a double standard. For example, the common legal principle that all persons are equal before the law helps prevent double standards. This is important because judges should be impartial and apply the same standards to all plaintiffs and defendants regardless of age, class, ethnicity, gender, religion, sexual orientation, or other distinctions. Another example is how some musicians will sometimes deliberately audition for a position in an orchestra *unseen* such that their musical talent is the only factor that influences acceptance or rejection.
Double-check	To *double-check* means to inspect to confirm accuracy, condition, or progress of work a second time (e.g., taking a measurement once, and then double-checking the measurement by measuring a second time). Usually, a double check is performed by the person who performed the first check; however, double-checking may be performed by anyone who is qualified. See also, *do it right the first time, measure twice and cut once*
Double-dealing	*Double-dealing* is act of behaving with duplicity or secretly working to someone's disadvantage with the intention to deceive or injure (i.e., without their knowledge), while simultaneously pretending to support or care for them. See also, *two-faced*.
Double-entry Accounting	*Double-entry accounting* is a standard accounting method that records each transaction in at least two accounts, resulting in a debit to one or more accounts and a credit to one or more accounts. The total amount of the transactions in each case must balance out, ensuring that all dollars are accounted for. Debits are typically noted on the left side of the ledger; credits on the right side. With this method, assets (e.g., cash, equipment, and inventory) are always equal to the sum of liabilities (e.g., loans, salaries, and taxes) and equity (i.e., the amount invested in the company the owners, stockholders, or a combination). For example, if a company borrows $1,000 from a bank, its assets will increase by $1,000 and its liabilities will also increase by $1,000. This method serves as the most efficient way to monitor financial growth, especially as the scale of a business grows. As a business grows, the likelihood of clerical errors increases. Although double-entry accounting does not prevent errors entirely, it limits the effect errors have on overall accounts. This method of accounting is the accepted method of record keeping. The earliest record of double-entry accounting dates to 600 AD. See also, *generally accepted accounting principles*.

Phrase or Term	Definition
Doublethink	*Doublethink* is the ability to hold two contradictory beliefs simultaneously and accept both. For example, a policy that gives clean needles to drug addicts for safety and hygiene so they can inject harmful and illegal drugs, which is dangerous and can be lethal. Euphemisms, gibberish, and jargon are also a form of doublespeak because they conceal, misrepresent, or obscure the truthy. For example, using the phrase *alternative facts* instead of *lies* or *misinformation*. Doublethink was coined by George Orwell in the book *Nineteen-Eighty Four*. See also, *brainwashing*, *cognitive dissonance*, *disinformation*, *euphemism*, *gobbledygook*, *jargon*, *misinformation*, *propaganda*, *thought control*, and *thought crime*.
Double-walled Tank	A *double-walled tank* is a primary tank within a secondary tank that is designed to prevent product releases to the environment. The space between the two tanks can be monitored for leaks. See also, an *interstitial space and tank*.
Downcycle	To *downcycle* is to recycle a product after initial use into a lower-value product. For example, after initial use, high-quality recycled plastic material is used to produce parking bumpers and street curbs. See also, *upcycle*.
Downside	*Downside* means: • A downward trend (e.g., prices or stock value); or, • The negative aspect of something (e.g., the new acquisition has a costly downside). See also, *1-way and 2-way door decision-making*, *trend*, and *upside*.
Downstream	The *downstream* sector of the oil and gas industry refines crude oil, processes and purifies natural gas, and markets and distributes by-products (e.g., gasoline, jet fuel, diesel and other oils, lubricants, waxes, asphalt, natural gas, liquefied petroleum gas, and petrochemicals). See also, *midstream*, *petrochemical*, and *upstream*.
Downturn	*Downturn* is: • A decline or downward trend, especially in business, economic conditions, or fortune; or, • To turn something downward. See also, *downside*, *trend*, and *upturn*.
Doxxng, Doxxing	*Doxing* or *doxxing* is to publicly reveal personal or private information about an individual or organization (e.g., email, postal, residential, or work addresses, or telephone numbers) using the Internet. This may be done for extortion, hacktivism, online shaming, revenge, or vigilantism. Doxing has a negative connotation because it violates privacy. Doxing comes from the abbreviation *docs* for *documents*, and refers to compiling documents of personal information to be publicly releasing or shared. The term derives from the phrase *dropping dox*. Author Mat Honan attributed the origin of this tactic to the 1990s hacker culture. See also, *blacklist*, *cancel culture*, *deplatforming*, *hacktivism*., and *shadow banning*.
Drag-reducing Agents, Drag-reducing Polymers	*Drag-reducing agents* or *drag-reducing polymers* are additives used in a petroleum pipeline to increase pipeline capacity by reducing turbulence, thereby improving oil flow efficiency. The additives are injected into a pipeline in low concentrations (e.g., 5 to 25 parts per million) because of their high cost and to maintain base product specification.

<table>
<tr><th>Phrase or Term</th><th>Definition</th></tr>
<tr><td>Dressed</td><td>A pressure vessel or tank is dressed when all ladders, platforms, piping, and other items are installed. See also, erect.</td></tr>
<tr><td>Drinking the Kool-Aid</td><td>Drinking the Kool-Aid is a metaphor for misguided followership. It was coined after Jim Jones, the delusional pseudo-guru leader of the Peoples Temple cult, which led his followers to commit mass suicide in Jonestown, Guyana in 1978. Over 900 people killed themselves by drinking a grape-flavored drink laced with cyanide. The media reported that Kool-Aid (a brand that flavours water with sugar) was in the drink. Since then, the expression drinking the Kool-Aid is used to describe reckless devotion. Drinking the Kool-Aid is also used to refer to:<ul><li>A belief in a possibly dangerous or doomed idea because of perceived potential of high reward;</li><li>Accepting an idea or changing a preference due to peer pressure, persuasion, or popularity;</li><li>Extreme dedication to a cause or purpose. For example, to say that one who is dedicated to their job drank the company Kool-Aid.</li></ul>See also, cult.</td></tr>
<tr><td>Drive-by</td><td>Drive-by has many meanings. It could mean:<ul><li>To shoot a firearm from a moving vehicle (e.g., police arrested suspects of the drive by shooting);</li><li>To participate briefly, casually, cursorily, offhandedly, quickly, or superficially (e.g., skimming a 100 word paragraph in 10 seconds); or,</li><li>A brief, unplanned, informal meeting (e.g., I will drive-by your office at 3 pm today to discuss new plans).</li></ul></td></tr>
<tr><td>Driver</td><td>A driver is a driving mechanism (e.g., an electric motor, diesel engine, or gas or steam turbine) that is used to power rotating equipment (e.g., a generator, compressor, or pump). See also, turbomachinery,</td></tr>
<tr><td>Drone</td><td>A drone is an unmanned vehicle that may be aerial, earthbound, submersible, or have the ability to float, or combination.</td></tr>
<tr><td>Drum</td><td>A drum is a cylindrical container or receptacle; a type of pressure vessel or tank. See also, pressure vessel or tank.</td></tr>
<tr><td>Dry Film Thickness</td><td>Dry film thickness is the thickness of a dried film, coating, or membrane. See also, film.</td></tr>
<tr><td>Dryland</td><td>Dryland is land that is arid and lacks water. This may include cultivated land, desert, forestland, and grassland. See also, wetland.</td></tr>
<tr><td>Dry-out</td><td>To dry-out is to remove moisture (e.g., a solvent or water) from equipment or material so that it is dry (e.g., after hydrotesting or before start-up). The degree of dryness may be specified (e.g., as a specific dew point down and maximum moisture content) or stated as a requirement that the item cannot be moist or wet.</td></tr>
</table>

Phrase or Term	Definition
Drywall	*Drywall* (also known as gyprock, gypsum panel, gypsum board, plasterboard, sheet rock, or wallboard) is a panel made of gypsum, with or without additives, typically extruded between thick sheets of facer and backer paper, and used in the construction of interior building walls and ceilings. The plaster is mixed with fibre (e.g., asbestos, fibreglass, or paper, or a combination of these materials), plasticizer, foaming agent, and other additives to reduce flammability, mildew, and water absorption. Drywall construction became prevalent in North America as a time and labour-saving alternative to the traditional lath and plaster method. See also, *gypsum*.
Dual Certified	*Dual certified* refers to a material that the manufacturer has certified to meet two or more standards (e.g., for flanges, ASTM International A350 LF2 *and* CSA Z245.12 Grade 248 Category II or III). The material test report will list the applicable standards, but will not state *dual certified*. See also, *material test report*.
Due Diligence	*Due diligence* is the: • Care and caution that a reasonable business or person is expected to take before entering into an agreement or contract, or giving advice; • Concepts or processes involving the investigation of a calculation or design, or something where a certain standard of care is needed; or, • Process of disclosing or gathering relevant and reliable information about a prospective contract, investment, purchase, or sale. It can be a legal obligation, but the phrase is more commonly applied to voluntary actions.
Dumb Machine, Dumb Technology	A *dumb machine* or *dumb technology* is equipment or hardware that lacks digital features such as onboard sensors and network communication. For example, a motor or pump without digital capabilities (i.e., operation technology) that must be operated manually. See also, *digital*, *operation technology* and *smart technology*.
Dumb Risk	*Dumb risk* is risk where little can be gained from success, and where even a minor failure would be catastrophic. Dumb risk is reckless because it is not based on facts. Instead, it is blind, frivolous, or undisciplined; thus, action must be rationalized to be justified. For example, attempting to climb onto and ride a moving vehicle. See also, *calculated risk*, *smart risk* and *risk*.
Dummy Purchase Order	A *dummy purchase order* is: • In SAP® Business Suite, the business process where a purchase order is generated from a contract, which is triggered by an invoice, to receive an item; • Prepared to have a functional or legal document that will be used for items for which there is nothing yet to be invoiced or received. For example, a preliminary purchase order used to investigate or begin processing purchase without the customer's account number or material determination. See also, *contract*, *purchase order* and *SAP® Business Suite*.
Dummy Spool	A *dummy spool* is a piece of pipe, with one flange at each end, used to (temporarily) replace a valve.

Phrase or Term	Definition
Dunbar's Number	*Dunbar's number* is the maximum number of stable social relationships a person can have, which is proposed to be between 100 and 250, with 150 commonly used. Humans have lived and worked in groups of this size throughout history, dating back to Neolithic farming villages. This number is of interest in anthropology, business, military, psychology, social media, and statistics disciplines. This concept was developed by British anthropologist Robin Dunbar in the 1990s. See also, *team*, *tribalism*, and *tribe*.
Dunning-Kruger Effect	The *Dunning-Kruger effect* is a when people assess their cognitive abilities as greater than they are when they are not qualified to make this assessment in the first place. It is related to the cognitive bias of illusory superiority and is derived from the inability of people to recognize their lack of abilities. Without self-awareness, personnel cannot objectively evaluate their competence or incompetence. This means that if you are incompetent, you cannot know that you are incompetent. This is because the skills needed to produce a correct answer are exactly the skills needed to recognize what a correct answer is. This effect was first identified by David Dunning and Justin Kruger in the study *Unskilled and unaware of it: How difficulties in recognizing one's own incompetence lead to inflated self-assessments*. See also, *bias* and *cognitive bias*.
Duopoly	A *duopoly* occurs when two companies own all, or nearly all, of the market for a product or service. For example, Mastercard and Visa dominate the credit payment industry in Europe and the United States. A duopoly can have the same impact on a market as a monopoly if both companies collude (i.e., on pricing, supply, or both). See also, *monopoly* and *oligopoly*.

Phrase or Term	Definition
E	1 A B C D **E** F G H I J K L M N O P Q R S T U V W X Y Z
Early Adopter, First-mover, Lighthouse Customer	An *early adopter*, *first-mover*, or *lighthouse customer* is a business or individual that uses a newly developed product (e.g., innovation or technology) before others (i.e., as soon as the product becomes available). For any new product to be successful, it must attract early adopters. An early adopter may pay more than others, but accepts this premium if it increases market penetration, improves efficiency, raises social status, or reduces cost. Early adopters can also be motivated to purchase (e.g., with a one-time or exclusive offer or promotion). Early adopters can provide feedback about deficiencies or use and help cover the cost of research and development. Early adopters may face risk because they use a product that may not be perfected, or may not be compatible with other products or systems. The phrase *early adopter* is sometimes misspelled as *early adapter* or *early adaptor*. See also, *diffusion of innovation theory*, *innovator*, and *trail blazer*.
Early Bird, Morning Person	An *early bird* is someone who: • Arrives somewhere or does something sooner than others, especially in the morning (e.g., there are a few early birds who arrive at the office before 7:00 am); or, • Is a *morning person* or who wakes up early (e.g., 05:00 am or earlier). See also, *early bird catches the worm*, *early bird discount*, and *night owl.*
Early Bird Catches the Worm	The *early bird catches the worm* is an expression that means someone who arrives first is most likely to get the best outcome or what they want; being first improves chances of success. See also, *early bird*.
Early Bird Discount	An *early bird discount* is a price reduction available to a consumer who purchases something sooner than normal. For example, the regular cost of a webinar ticket is CAD$ 100.00. But the early bird price, or discount price if you purchase it at least one month before the webinar, is CAD$ 75.00. See also, *early bird* and *early bird catches the worm*.
Early Closure	*Early closure* is the focus on finding the *right* answer as quickly as possible, without producing alterative ideas or developing them into different possibilities or solutions.
Easy Oil and Gas, Easy-to-reach Oil and Gas	*Easy oil and gas* and *easy-to-reach* refer to inexpensive and easy-to-produce oil and gas. For example, oil and gas sourced from drilling a conventional or shallow offshore well. See also, *unconventional gas* and *unconventional oil*.
Easy Way	The *easy way* means: • Bending a steel member across the weak principle axis (e.g., bending a 2″ x 4″ rectangular tube across the 2″ axis); and, • Completing a task using minimal effort. The easy way is not always the best way of completing a task. See also, *book smart*, *hard way*, *member*, and *street smart*.

Phrase or Term	Definition
Ebonics	*Ebonics* is American black English, which is regarded as a language rather than as a dialect of standard English. For example, the Eboincs phrase *let me hold some ends; I'll hit you back on the first* means *lend me some money and I will pay you back on the first day of the next month*. The term was originally coined by African American social psychologist Robert Williams in 1973. Ebonics is a portmanteau of ebony and phonics. See also, *colloquial*, *dialect*, *jargon*, *Plain English*, and *vernacular*.
Economic System	An *economic system* is the means by which a government or society distributes and organizes available products, resources, and services to its citizens across a geographic region or country. An economic system regulates factors of production, including capital and labour, and supports all facets of an economy. See also, *capitalism*, *centrally planned economy*, *market economy*, *mixed economy*, *socialism*, and *traditional economy*.
Economy of Repetition	*Economy of repetition* is a cost advantage gained by the: • Execution of a large number of similar projects that can be managed more efficiently and effectively by using the same or similar tools and strategies. For example, the Kearl Oil Sands Project embraced this philosophy by using the motto *design one, build many* in 2012; and/or, • Use of fixed production patterns. See also, *circular economy*, *economy of scale*, *hydrogen economy*, *knowledge economy*, *gig economy*, *linear economy*, and *low-carbon economy*.
Economy of Scale	*Economy of scale* is a cost advantage (i.e., savings) gained by an increased or more efficient level of production. This can be achieved by increasing production and lowering costs, because costs are spread over a larger number of goods. Costs can be fixed or variable. See also, *circular economy*, *economy of repetition*, *hydrogen economy*, *knowledge economy*, *gig economy*, *goods*, *linear economy*, and *low-carbon economy*.
Econophysics	*Econophysics* is the application of methods and theories, originally developed by physicists, to solve problems in economics. Problems may include nonlinear dynamics and stochastic processes or uncertainty (e.g., the fluctuation or future cost of commodities such as coper and steel). This study of financial markets, also known as statistical finance, which refers to its roots in statistical physics.
Effective	*Effective* is something that: • Fulfills a specified function; or, • Successfully produces the desired or intended result. Effective is an adjective formed from the noun *effect*. For example, *the contract is effective immediately*. See also, *affective*.

Phrase or Term	Definition
Effective Disruption, Positive Disruption	*Effective disruption* or *positive disruption* is an incremental effort to jostle apathy, complacency, or the status quo, to effect improvements in an organization or society and the human journey. It challenges existing habits or methods that are inadequate or outdated and works to find positive and productive alternatives. For example, someone seeks to improve workplace culture by building a friendship with each colleague. It: • Embraces adaptability, creativity, innovation, surprise, uncertainty, and vulnerability; • Identifies closed mindedness or opposition and reduces resistance to change • Is a search for questions rather than answers, to avoid binary opposition; • Is comfortable not knowing the answers or how to find them; • Is not a race, but progressive advancement that raises standards; • Requires honesty about one's own biases and the biases of others; • Seeks to create efficiencies and win-win solutions; • Tells others what the questions are, that the answers are unknown, and their help is needed to find them; and, • Uses empathy instead of judgement and listening more than talking. This famous quote from Albert Einstein made a strong case for effective or positive disruption: *we cannot solve our problems with the same thinking we used when we created them*. See also, *bias*, *binary opposition*, *disruptive innovation*, *status quo*, *vulnerable*, and *win-win*.
Egocentric	To be *egocentric* is to be self-centered, selfish, and to think only of oneself and one's own wants without consideration or regard for others. For example, cutting in line at the bank.
Either/or	*Either/or* is a logical fallacy where a conclusion oversimplifies an argument by reducing it to only two choices or sides. For example, this statement: *we can* either *stop using airplanes or destroy the earth's atmosphere*. See also, *bait and switch*, *circular reasoning*, *false equivalence*, *logical*, *logical fallacy*, *red herring*, *righteousness fallacy*, *slippery slope*, and *straw man argument*.
Electrical Building, E-building, Electrical House, E-house	An *electrical building*, *e-building*, *electrical house*, or *e-house* is a shop-fabricated modular structure that is dedicated to electrical equipment. Its size is usually proportional to the size of the electrical equipment. Equipment may include communications and electrical control and distribution infrastructure such as: • Access control and intrusion and/or fire or other alarms; • Distribution with emergency, standby, and/or uninterruptable power systems; • Intercommunications and/or data and voice telecommunications systems; and, • Metering for ongoing evaluation and improvement of energy performance.
Electrical Code Inspection, Electrical Regulatory Inspection	*Electrical Code Inspection* or *Electrical Regulatory Inspection* is performed by an inspection body per the inspection body's terms of accreditation with the *Standards Council of Canada*. This is not performed by a supplier quality representative.
Electrification	*Electrification* is the process of powering by electricity.

Phrase or Term	Definition
Electrode	In arc welding, an *electrode* is used to conduct electrical current to fuse two parts or pieces of base metal together. Depending on the process, the electrode is either consumable or non-consumable. For example, the tungsten electrode used for the gas tungsten arc welding process is not consumed to become part of the weld metal; it is a *consumable*. In electricity, the electrical conductor is used to contact the non-metallic part of a circuit. See also, anode, *base metal*, *cathode*, *consumable*, *electrode oven*, and *low hydrogen electrode*.
Electrode Oven	An *electrode oven* is a heated quiver or storage cabinet used to store certain types of coated electrodes. For example, the CSA E4918 electrode (which is the same as the AWS E7018) is coated with a basic (i.e., alkaline) low hydrogen iron powder that also contains calcium carbonate (i.e., limestone), calcium fluoride, and other chemicals. Low hydrogen electrode coatings that have absorbed moisture may result in welds prone to hydrogen-induced cracking, particularly in steels with a yield strength of 550 MPa (80,000 psi) and higher. Coated electrodes absorb moisture from the atmosphere. Thus, they must be dried to maintain or restore their ability to deposit quality weld metal. Unopened hermetically-sealed containers provide suitable protection in good storage conditions (e.g., at room temperature). Opened containers, or the electrodes themselves, are typically stored between 120 and 150°C (250 and 300 F). Low hydrogen electrodes may be removed from the electrode oven for welding and may be stored, unheated, for a maximum of two to nine hours before being discarded or re-dried (i.e., re-conditioned) for one hour between 340 and 430 C (650 and 800 F). The electrodes may be re-baked (i.e., re-dried or reconditioned) a maximum number of times, after which they must be discarded (e.g., one to three). See also, *electrode*, *low hydrogen electrode*, and *weldability*.
Electro-mechanical Equipment	*Electro-mechanical equipment* is a device or machine that is electrically operated (e.g., a pump package). See also, *mechatronic technology*, *packaged equipment*, and *rotating equipment*.
Electron	An *electron* is a stable subatomic particle with a charge of negative electricity found in all atoms that acts as the primary carrier of electricity in solids. See also, *electronic*.
Electronic	*Electronic* is something being, of, or relating to: • Devices constructed or working by the methods or principles of electronics, or implemented on or by using a computer; • Electrons (i.e., stable subatomic particles); • Information that is transmitted electronically (e.g., by a radio or television); or, • Music that consists electronically generated or modified sounds. See also, *digital*, *electron*, and *electronics*.

Phrase or Term	Definition
Electronic Business, e-Business, Online Business	*Electronic business*, *e-Business*, or *online business* is any business or commercial transaction that includes sharing information over the Internet. It refers to all aspects of operating an online business, in addition to buying and selling products or services, which includes: • Customer service; • Logistics; • Manufacturing and operations; and, • Marketing and sales. See also, *electronic commerce*.
Electronic commerce, e-Commerce, eCommerce, Internet Commerce	*Electronic commerce*, *e-commerce*, *eCommerce*, or *Internet commerce* is a business model that allows businesses and individuals to buy and sell products and services via the Internet. Also, it is the transfer of data and money to support these transactions. It is often used to refer to online sales, but it can also describe any commercial transaction made over the Internet. E-commerce allows businesses to establish a wider market presence by providing cheaper, more efficient, and online distribution channels. See also, *24/7*, *electronic business*, *procure-to-pay*, and *source-to-pay*.
Electronic Data Management System	An *electronic data management system* is a technology (typically software) that comprehensively manages the creation, capture, indexing, storage, archiving, retrieval, and disposition of an organization's digital records and information assets. See also, *management*.
Electronic Signature	An *electronic signature* confirms that a document, message, or transaction was digitally received by a party (often making the document legally binding). Examples include: • A signature that is captured digitally when written on a touch device, (e.g., a tablet or smartphone); • A handwritten signature on a document that is then transmitted by email or fax; • A personal identification number (e.g., used for banking machines); • The *agree* or *disagree* button on an electronic contract; or, • The typed name at the bottom of an email or on any document created electronically. See also, *authentication*, *cryptography*, *digital signature*, and *wet signature*.
Electronics	*Electronics* is: • A branch of physics that deals with the emission, behaviour, and effects of electrons and with electronic devices; and, • Electronic components, devices, or equipment. See also, *electronic*.
Electroplating	*Electroplating* is a process that uses an electric current to form a thin coating on metal parts. See also, *galvanizing*.

Phrase or Term	Definition
Elephant in the Room	The *elephant in the room* is an expression that describes an obvious issue (e.g., controversial subject, important topic, or major problem) that most personnel will avoid acknowledging or discussing. This issue makes all or some personnel uncomfortable because the subject is personally, politically, or socially controversial, embarrassing, inflammatory, or unsafe. For example, when a manager hires a family member – even though this is clearly against company policy. This hire would become the *elephant in the room* because personnel know that company policy was violated by management but few wish to discuss it out of fear of losing their jobs. This expression is based on the concept that something as conspicuous as an elephant cannot ignored or overlooked, so everyone in the room will know about it or recognize it is there (even if they do not acknowledge it).
Emergence	*Emergence* is the natural outcome of things coming together and the synergies of the parts. It is about the non-linearity and self-organization of how things interact. For example, how a caterpillar becomes a butterfly. See also, *causality*, *feedback loop*, *interconnectedness*, *synthesis*, *systems mapping*, and *systems thinking*.
Emotional Intelligence	*Emotional intelligence refers to the ability to:* • Discern between different feelings and label them appropriately; • Manage emotions, adjust emotions, or both to adapt to environments or achieve goals; • Recognize one's own emotions and those of others; and, • Use emotional information to guide thinking and behaviour. Emotional intelligence generally includes these three skills: 1. Emotional awareness, or the ability to identify and name one's own emotions; 2. The ability to harness one's emotions and apply them to tasks like thinking and problem solving; and, 3. The ability to manage emotions, which includes both regulating one's own emotions when necessary and helping others to do the same. This concept gained popularity in the book *Emotional Intelligence: Why It Can Matter More Than IQ*, by Daniel Goleman, but appeared earlier in *The Communication of Emotional Meaning*, a book by Robert Davitz and Michael Beldoch. There is no validated psychometric test or scale for emotional intelligence. This is also referred to as emotional leadership, emotional quotient, and emotional intelligence quotient. See also, *body language* and *intelligence quotient*.
Emotionally Dysregulated	*Emotionally dysregulated* refers to a poor ability to keep responses or manage them within an acceptable or typical range of emotive response. This can occur because of a brain injury, chronic maltreatment, developmental trauma, psychological trauma, or psychiatric disorder. It can result in a wide range of emotions including anger, irritability, frustration, and sadness. People who are emotionally dysregulated can have difficultly socially because their emotions may be capricious. See also, *developmental trauma* and *playfulness acceptance curiosity empathy*.

Phrase or Term	Definition
Empirical	*Empirical* is: • Knowledge or perspective that relies on experience or observation alone, often without regard for system and theory; • Something based on, concerned with, or verifiable by experience or observation rather than purely logic or theory; or, • Something capable of being disproved or verified by experiment or observation. For example, consider this statement: *Aluminum will melt (i.e., change from solid to liquid state) almost immediately after it reaches 660.32°C (1,220.58°F).* If metal is heated above this temperature but does not melt, then: o The statement is false; or, o The metal is not aluminium. See also, *anecdotal*, *informative*, *logical*, and *normative*.
Emulation	*Emulation* is a procedure that dynamically checks software code before installation using a programmable logic controller to control a model of a real system. See also, *programmable logic controller*.
Enclosure	An *enclosure* is: • An area that is sealed off by an artificial or natural barrier; • Something that closes or encloses (e.g., with a door or lid) such as a: o Building that is too small for human occupancy, also referred to as a doghouse; o Device to contain an explosion or pressure; o Cabinet to contain controls, instrumentation, or both; or, o Container to protect its contents (e.g., from vermin, weather, or both). See also, *engineered pressure enclosure*, *explosion-proof enclosure*, and *winterization*.
Encryption	*Encryption* is the encoding of information by conversion of the original representation of the information (i.e., plaintext) into an alternative form (i.e., ciphertext). Only authorized parties can decipher a ciphertext back to plaintext and access the original information. Encryption is a method of scrambling data such that it is indecipherable by unauthorized personnel. See also, *cryptography*, *microdots*, and *steganography*.
Endothermic	An *Endothermic* process or reaction is accompanied by, or requires the absorption of, heat but does not produce heat. For example, the melting of ice cubes. See also, *exothermic*.
End-to-end	*End-to-end* means from the very beginning to the very end. It is a process that comprises all of the work or cross-functional processes to deliver a complete functional solution – usually without needing to procure anything from a third party. This term is most used in the information technology sector. See also, *life cycle*.

<table>
<tr><th>Phrase or Term</th><th>Definition</th></tr>
<tr><td>Energy</td><td>In physics, energy is the capacity for doing work, which is measured in joules. It may exist in chemical, electrical, kinetic, nuclear, potential, thermal, or other forms. It is also heat and work (i.e., energy in the process of transfer from one body to another). After energy has been transferred, it is always designated according to its nature. Hence, heat transferred may become thermal energy, while work done may become mechanical energy. The base unit of energy is the joule. This means that a task, like lifting a box, requires a certain number of joules regardless of how quickly or slowly the box is lifted. See also, power.</td></tr>
<tr><td>Energy Density</td><td>Energy density is the amount of energy that can be stored in the mass of a material (i.e., substance or system). The higher the energy density, the greater the amount of energy stored in the mass. Energy can be stored in many different materials, which can release energy in these types of reactions:<ul><li>Chemical;</li><li>Electrical;</li><li>Electrochemical; and,</li><li>Nuclear.</li></ul>Energy density is expressed as:<ul><li>Gravimetric energy density or specific energy
How much energy a system contains in comparison to its mass, typically expressed in watt-hours per kilogram (Wh/kg) or Megajoules per kilogram (MJ/kg).</li><li>Volumetric energy density
How much energy a system contains in comparison to its volume; typically expressed in watt-hours per Litre (Wh/L) or Megajoules per Litre (MJ/L), which is the most common expression.</li></ul>When calculating the amount of energy in a system, usually only the useful or extractable energy is measured. Energy density does not give information about how quickly the energy can be used. See also, nameplate capacity, power density, and power output.</td></tr>
<tr><td>Energy Literacy</td><td>Energy literacy is an understanding of the nature and role of energy in daily life and the world, with the ability to apply this understanding to answer questions and solve problems. It includes an understanding of the impacts of energy consumption and production on the environment and energy-saving behaviours and methods. An energy-literate person can:<ul><li>Assess the credibility of information about energy;</li><li>Communicate about energy and energy use in meaningful ways;</li><li>Identify the flow of energy and think in terms of energy systems;</li><li>Make informed energy use decisions based on an understanding of consequences and impacts; and,</li><li>Understand how much energy they use, for what purpose, and where the energy is sourced.</li></ul></td></tr>
</table>

Phrase or Term	Definition
Energy Return on the Amount of Energy Expended, Energy Return on the Amount of Energy Invested	*Energy return on the amount of energy expended* or *energy return on the amount of energy invested* expresses the amount of effort or resources needed to produce energy in a usable form. The shorter phrase *energy return on investment* is not preferred because it may be interpreted to refer only to the return on the financial investment. For example, hydrogen requires outside energy at each step of its production: for its compression or liquefaction for storage, and then for its delivery and during ultimate use (e.g., energy is lost when hydrogen is converted to electricity by a fuel cell). See also, *hydrogen* and *return on investment*.
Energy Transition	*Energy transition* refers to the global energy sector's shift from fossil-based systems of energy consumption and production (e.g., coal, natural gas, and oil) to renewable energy sources (e.g., batteries, hydro, solar, and wind). It is shorthand for the future of energy since the 2015 Paris agreement. An energy transition is a significant structural change in an energy system, which has historically been driven by the availability of and demand for different fuels. The current transition to renewable energy is driven by a recognition that global carbon emissions must be brought to zero. Since fossil fuels are the largest single source of carbon emissions, energy systems worldwide must be changed to replace fossil fuels. See also, *fossil fuel*, *Paris agreement*, and *renewable energy*.
Energy-isolating Device	An *energy-isolating device* is a device that prevents energization of a system such as a: • Block or line valve; • Circuit breaker; or, • Manually operated disconnect switch. Circuit control switches, push buttons, and selection switches or other similar devices that cannot be locked-out are *not* considered energy-isolating devices. See also, *lockout* and *lockout device*.
Engineer	An *engineer* is a: • Member of a military group devoted to engineering work; • Person who directs an enterprise by artful contrivance, scheming, or skill; • Person who follows as a profession or is trained in a branch or discipline of engineering; or, • Person who builds, designs, runs, or supervises an engine or machine. See also, *applied engineer*, *discipline engineer*, *engineering*, *iron ring*, *professional engineer*, and *technologist*.
Engineered Lift, Planned Engineered Lift	An *engineered lift* or *planned engineered lift* is the activity of lifting equipment or materials with a crane that exceeds the rated crane capacity. Exceeding the crane load rating introduces potentially harmful risks, including: • Damage or failure of critical mechanical and electrical crane components; and, • Dropped loads, which may result in property damage, damage to equipment and materials, personal injury, and even death. ASME B30.2-3.2.1.1 considers special situations when a lift in excess of the rated load may be required (e.g., major repairs or new construction).

Phrase or Term	Definition
Engineered Pressure Enclosure	An *engineered pressure enclosure* is a device used for containing a leak or reinforcing existing pressure equipment (e.g., pipeline or pressure piping) for a limited time, until a proper alteration or repair is completed. This is also referred to as on-line leak sealing or a leak box. See also, *enclosure*, *risk assessment*, *root cause analysis*, *installation maintenance and monitoring and removal confirmation date procedure*.
Engineering	*Engineering* is a professional discipline in which trained engineers apply chemistry, mathematics, physics, and science to matter and sources of energy in nature to make them useful to people (e.g., to design and build bridges, dams, machines, and other items). It is advising or reporting on, designing, evaluating, preparing plans and specifications for, or directing the construction, maintenance or operation, or technical inspection of any process, structure, or work that: • Is aimed at the development, discovery, or use of energy, materials, or matter or in any other way designed for the convenience and use of humans; and, • Requires the professional application of the principles of chemistry, mathematics, physics, or any related applied subject be applied for the purpose of advising, directing, designing, evaluating, preparation or reporting. See also, *applied engineering*, *discipline*, *engineering technology*, *imagineering*, and *technology*.
Engineering and Procurement	*Engineering and procurement* are services provided for projects by a consulting company. The engineering and procurement company may also be the client's agent. These services may be provided together with construction or construction management services (e.g., engineering procurement and construction, or engineering procurement and construction management). See also, *agent* and *consulting company*.
Engineering Change Notice	Not a preferred phrase – see *design change notice*.
Engineering Data Book	An *engineering data book* comprises purchase order and equipment-specific documentation, which was compiled by the supplier or engineering and procurement company for turnover to the client. Contents may include a manufacturing record book or other specified supplier-reviewed documents.
Engineering House	Not a preferred phrase – see *consulting company*.
Engineering Technology	*Engineering technology* is the practical application of engineering and science to a wide range of real-world problems. Engineering technology education is more applied and less theoretical than a conventional engineering education. For example, it is used in various fields and industries such as manufacturing, maintenance, operations, product development, quality control, and technical sales. See also, *information technology*, *operation technology*, and *technology*.
Engineer-to-order, Design-build Construction, Project-driven Manufacturing	*Engineer-to-order*, *design-build construction*, or *project-driven manufacturing* is when the service provider receives functional requirements from a client and then collaborates to create a finished design. There is often an overlap between the design and engineering processes and subsequent processes such as procurement of long lead products and construction at site. This allows the client to manage all work using one contract, while fostering a culture of collaboration and a single point of responsibility. See also, *long lead*.

Phrase or Term	Definition
Enhanced Oil Recovery, Tertiary Recovery	*Enhanced oil recovery* or *tertiary recovery* is a process by which usable petroleum is drawn out from beneath the earth's surface. During this stage, crude oil is recovered that cannot be extracted otherwise. The result is 30 to 60% or more of a reservoir's oil can be extracted, compared to only 20 to 40% using primary and secondary recovery. The techniques for enhanced oil recovery include: • Chemical injection; • Gas injection; and, • Thermal applications (e.g., steam-assisted gravity drainage and underground burning). See also, *primary recovery*, *secondary recovery*, *steam-assisted gravity drainage*, and *underground burning*.
Enterprise Asset Management System	An *enterprise asset management system* is software that is used to manage the maintenance of assets (e.g., high-value equipment or facilities) throughout their life cycles. This includes the design, construction, commissioning, operation, maintenance, and decommissioning or replacement of assets. It is used to execute, optimize, and plan maintenance activities with the information, materials, priorities, resources, and tools that are needed. It is an extension of computerized maintenance management systems. *Enterprise* refers to the scope of the assets in a business across various departments, facilities, locations, and supporting business functions. See also, *asset integrity management* and *computerized maintenance management system*.
Enterprise Performance Management, Corporate Performance Management	*Enterprise performance management* or *corporate performance management* is a methodology to enhance business performance and enable management to better respond to challenges and opportunities. The domains in this method include: • Business forecasting and planning; • Financial management; • Strategy formulation; and, • Supply chain effectiveness. The six stages of this method are: 1. Strategy development; 2. Strategy translation; 3. Organization alignment 4. Operations planning; 5. Learning and monitoring; and, 6. Testing and adaptation. See also, *forecasting* and *supply chain management*.
Enterprise Resource Planning	*Enterprise resource planning* is the integrated management of main business processes using electronic data, often in real-time, and mediated by software and technology. Usually referred to as a category of business management software, it typically comprises a suite of integrated software applications used by an organization to collect, interpret, manage, and store data from numerous business activities. See also, *building information modelling* and *management*.

Phrase or Term	Definition
Entrepreneur	An *entrepreneur* manages, operates, and organizes an enterprise, especially a business, usually with considerable initiative and greater-than-normal risk. An entrepreneur is an innovator who is perceived by most as a source of new business, ideas, procedures, products, and services. Entrepreneurs seek to develop something new, find a market for it, and then profit by its sale in marketplace. Entrepreneurs are motivated by autonomy, financial success, flexibility, legacy, and purpose. For example, Elon Musk of Tesla and Space X is a well-known entrepreneur. See also, *contrapreneur*, *early adopter*, *entrepreneurship*, *solopreneur*, and *trail blazer*.
Entrepreneurship	The essence of *entrepreneurship* is to define a need, invest resources, and then develop a product or service to meet that need. See also, *entrepreneur*.
Environmental Assessment, Environmental Evaluation, Environmental Impact Assessment	An *environmental assessment*, *environmental evaluation*, or *environmental impact assessment* is a study used to predict the long-term and short-term environmental effects of a historical, existing, or proposed development, program, or project. It identifies ways to eliminate, minimize, or mitigate the effects, compensate for their impact, or a combination. An environmental impact assessment is produced as a result and used to make commitments and decisions.
Environmental Social Governance	*Environmental social governance* refers to the three key factors used to measure the societal and sustainability impact of an investment in a company or business. These criteria help to determine the future financial performance of return on investment and risk. These factors signify a commitment to accountable and sustainable capitalism. See also, *conscious capitalism* and *environmental social governance with Indigenous partnerships and participation*.
Environmental Social Governance with Indigenous Partnership and Participation	*Environmental social governance with Indigenous partnership and participation* is the effort to increase Indigenous Peoples' participation in the economy. This requires the building and maintaining of positive relationships between Indigenous Peoples and all other Canadians through the development of common goals, and mutual respect and understanding. It offers win-win financial and social outcomes such as: • Corporate social responsibility; • Diversity and employment equity; • Business and employment opportunities; and, • Talent to address labour shortages. See also, *environmental social governance*, *Indigenous Peoples*, and *win-win*.
Environmentalism, Environmental Rights	*Environmentalism* or *environmental rights* is a broad ideology, philosophy, and social movement. It concerns the improvement, protection, and restoration of the natural environment including animals, humans, non-living matter, and plants. See also, *sustainability*.
Epistemic Uncertainty, Systematic Uncertainty	*Epistemic uncertainty* or *systematic uncertainty* is an uncertainty that is due to things one could in principle know but in practice do not. A lack of knowledge can occur from: • Imprecise evaluation of characteristics; • Inadequate understanding of the underlying processes; or, • Incomplete knowledge of the phenomena. In other words, how something works is unknown – so there is uncertainty about its operation. See also, *aleatory uncertainty*.

Phrase or Term	Definition
Equipment	*Equipment* is items fabricated or manufactured by a supplier as complete assemblies, sub-assemblies, components or packages. Equipment can also be apparatus, gear, hardware, or kit (chiefly British English). See also, *electro-mechanical equipment*.
Erect	To *erect* something is to construct it or raise it to a vertical position (e.g., a building, pressure vessel, or tank).
Eroom's Law	*Eroom's Law*, which is Moore's Law spelled backwards, observes that the number of new drugs approved pre-billion dollars spent on research and development has halved every nine years since 1950. This means that drug discovery is becoming more expensive and slower over time, despite improvements in technology. See also, *Moore's Law*.
Erosion	*Erosion* is the progressive loss of material from a solid surface due to mechanical interaction between that surface and a solid or fluid, a multicomponent fluid, or solid particles in a fluid.
Error Propagation, Propagation of Uncertainty	*Error propagation* or *propagation of uncertainty* is what happens to measurement errors when an uncertain measurement is used to calculate something else. This means that, at a given stage of a calculation, part of the error arises from an error introduced at a previous stage. For example, when an incorrect length is used to calculate an area, the area calculation will also be incorrect. Unfavorable error propagation can seriously affect the results of a calculation because the error will grow much more quickly than the sum of individual errors. See also, *rounding error*.
Esoteric	*Esoteric* is something that is intended for, or likely to be appreciated, known, and understood by only a small number of people with specialized interest or knowledge (e.g., cigar aficionados' knowledge of cigars or a Formula 1 driver's knowledge of performance sports vehicles).
Estimate at Complete, Estimate at Completion	*Estimate at complete* or *estimate at completion* is the forecasted cost of the project as the project progresses. The most common way to determine this cost is a *bottoms-up* formula where the actual costs are added to the forecasted remaining spend – the estimate to complete. See also, *budget*, *definitive estimate*, and *estimate to complete*.
Estimate to Complete	*Estimate to complete* is a forecasted cost of how much more money will need to be spent to complete a project. This can be determined by: • Building a *bottoms-up* estimate where the actual costs are added to the forecasted remaining spend (e.g., by asking the project team, suppliers, or both for revised estimates); or, • Deducting actual costs from the estimate at complete. See also, *budget*, *definitive estimate*, and *estimate at complete*.
Estimating Funnel	*Estimating funnel* is a metaphor for the increased accuracy in estimation made possible as a project progresses.
Ethics	*Ethics* are the moral principles that govern human behaviour and activities. For example, the vehicle dealership acted with good ethics by first disclosing to its customer that the *new* vehicle had minor damage and then offering to repair the damage at no charge. See also, *ethical*, *honest*, and the *golden rule*.

Phrase or Term	Definition
Ethical	To be *ethical* is to: • Conform to accepted standards of conduct; or, • Express or involve moral approval or disapproval. See also, *ethics, honest* and *the golden rule*.
Ethylene Glycol, Glycol	*Ethylene glycol* or *glycol* is an organic compound used mainly as antifreeze and a raw material to manufacture polyester fibres. It is an odorless, colourless, sweet-tasting, viscous, toxic liquid. See also, *heat tracing*.
Etiology	*Etiology* is the study of the cause or origin of a given phenomenon (e.g., disease, philosophy, or physics). See also, *teleology*.
Euphemism	A *euphemism* is the substitution of expression that may be considered too blunt, harsh, or offensive, or is disagreeable, embarrassing, or unpleasant – with one that is more casual, indirect, inoffensive, kind, or mild. For example, euphemisms for *death* include *kick the bucket*, *pass away*, and *rest in peace*. See also, *doublethink*.
Evergreen	An *evergreen* document is dynamic, outdating previously printed or produced drafts. An evergeen document is typically digital and regularly updated with new information (e.g., a non-conformance report log, supplier quality surveillance log, or a ski report). See also, *perpetual beta*.
Everybody Knows Principle	The *everybody knows principle* is when one uses another's perception of reality instead of their own. For example, everybody knows that the Japanese make higher-quality vehicles than Americans. Thus, consumer's vehicle buying decisions can be made based on this. This concept was developed in the book *The 22 Immutable Laws of Marketing* by Al Ries and Jack Trout. See also, *perception is reality*.
Everything Goes According to Plan	The *everything goes according to plan* principle applies to a plan that is overly optimistic and unrealistic It assumes that an activity or project will be completed without hindrance or challenge, and exactly as planned. In this fictitious scenario, there are no: • Adverse geological developments; • Broken commitments or promises; • Contractual issues; • Management challenges; • Schedule delays; • Significant conflicts; • Specification changes; • Technical modifications; • Unexpected costs; and, • Unforeseen events or circumstances. This above scenario has never happened and will never happen. See also, *most likely development* and *worst-case scenario*.

Phrase or Term	Definition
Evidence-based Decision-making, Fact-based Decision-making	*Evidence-based decision-making* or *fact-based decision-making* helps personnel make well informed decisions about policies, programs, and projects by putting the best available evidence from research at the centre of policy development and implementation. It relies on actual experimentation to demonstrate that a plan does indeed provide a likelihood of success. The core idea is that decisions supported by hard facts and sound analysis are likely to be better than decisions made based on a subjective perspective, instinct, informal anecdotal evidence, or folklore. Using objective facts as the supporting basis for decision-making seems sensible; however, this leaves room for error. Having looked at objective data, it is easy and common to posit unproven theories to explain the data, identify causes, and predict future outcomes. Even if the data itself is reliable, how that data is used remains a key consideration. See also, *decision-based evidence making*, *decision-making*, *evidence-based management*, *opinion-based decision-making*, and *value-based decision-making*.
Evidence-based Management	*Evidence-based management* is the conscientious use of multiple sources of evidence when making management decisions. It is a direct outgrowth of evidence-based practice in medicine, public policy, criminology and other professional fields. Multiple sources means that management makes use of scientific evidence (where available), contextual information (organizational data in management and patient information in medicine), stakeholder concerns (the people affected by the decision and those who would implement it), and professional expertise or judgment. See also, *evidence-based decision making*, *evidence-based practice* and *management*.
Evidence-based Practice	*Evidence-based practice* is an interdisciplinary approach to clinical practice that has been gaining popularity following its formal introduction as evidence-based medicine in 1992. It is traditionally defined in terms of a three-legged stool that integrates these basic principles: 1. Best available research evidence bearing on whether and why a treatment works; 2. Clinical experience and judgment to rapidly identify each patient's unique health state and diagnosis, and their individual risks and benefits of potential interventions; and, 3. Client preferences and values. These principles have been applied to many disciplines, including construction safety, education, health, public policy, and social work. See also, *evidence-based management*, *management*, and *values*.
Evolution	*Evolution* is the process: • By which different kinds of living organisms are thought to have developed and diversified from earlier life forms on earth; • Of change or unfolding in a certain direction; • Of developing or working out; • Of gradual and relatively peaceful economic, political, and social advance; or, • Of gradual development of something, especially from a lower, simple, or worse-to-a-higher, more complex or better state (i.e., progress without sentient planning). See also, *Darwinism*, *intelligent design*, and *start-up*.

Phrase or Term	Definition
Excel	*Excel* is Microsoft spreadsheet software originally marketed as the Multiplan program (1982). This software is part of the Microsoft Office suite of software. Developed for Windows, macOS, Android and iOS, it features calculation and graphing tools, pivot tables, and houses the macro programming language Visual Basic for Applications. Microsoft Excel is industry standard spread-sheeting software.
Excellence	*Excellence* is a measure of consistently superior performance that surpasses expectations and requirements without demonstrating significant flaws or waste. See also, *operational excellence* and *organizational excellence*.
Excellence Model	The *excellence model* is the European Foundation for Quality Management's continuous improvement tool. It is the most widely used in the world and can be used by any organisation regardless of sector or size. It allows an organisation to evaluate and identify areas for improvement and strengths. It provides a framework to guide continuous improvement projects or transformational programs and ensures that tangible results are delivered.
Exchange Rate	An *exchange rate* is the number in an equation to determine and then convert the value of one country's currency into another. For example, how many United States dollars will one hundred Canadian dollars buy? To determine this by using an exchange rate of 0.76988, use this equation: 100 x 0.76988 = 76.99 Thus, CAD$ 100.00 will buy USD$ 76.99. Exchange rates fluctuate with market rates, currency rates, and other factors. See also, *Big Mac Index* and *theories of exchange rates*.
Executive Decision	An *executive decision* is an especially important or far-reaching determination or direction made and implemented by a person or party with executive power. For example, the company president revised the dress code policy to require that all men wear a tie while conducting company business. It is also determination or judgement regarded as domineering, trivial, or unilateral. For example, the company president revised the dress code policy to require that all men wear a tie while conducting company business and that bow ties are prohibited.
Executive Skills	*Executive skills* are five skills necessary for the cognitive control of one's behaviour: • Attentional control, or concentration; • Cognitive flexibility or being able to think about two or multiple concepts simultaneously; • Cognitive inhibition or being able to focus on the task at hand; • Inhibitory control or self-control; and, • Working memory or short-term memory. See also, *cognitive flexibility* and *nootropic*.
Exothermic	An *Exothermic* process or reaction is accompanied by the release of heat. For example, the setting of cement or concrete. See also, *endothermic*.

Phrase or Term	Definition
Expansible Fluid	*Expansible fluid* is: 1. A vapour or gaseous fluid; or, 2. A liquid under pressure and at a temperature at which the liquid changes to a gas or vapour when the pressure is reduced to atmospheric pressure or when the temperature is increased to ambient temperature. See also, *non-expansible fluid*.
Expat, Expatriate	An *expat* or *expatriate* is someone who lives or works outside of their native country.
Expediter	An *expediter* secures the timely delivery of documents, material, and equipment for a project, especially with suppliers after a purchase order is awarded. See also, *expediting*.
Expediting	Supply chain management uses *expediting* to secure the timely delivery of documents, materials, and equipment. Expediting may be used for external and internal deliverables (e.g., drawings being received and issued by the project). See also, *expediter*.
Explosion-proof Enclosure	An *explosion-proof enclosure* is a solid cabinet designed to contain an explosion and thereby minimize damage to surrounding equipment. These enclosures typically house electrical components (e.g., controls, plugs, sockets, switches, and transformers) and equipment with explosion risk. See also, *enclosure*.
Exponential Growth	*Exponential growth* occurs when the rate of escalation (e.g., of interest rates) becomes ever more rapid in proportion to the growing total number or size. It is one possible result of a reinforcing feedback loop that makes a population or system escalate by increasingly higher amounts. For example, compound interest. See also, *feedback loop*.
Extended Warranty	An *extended warranty* lengthens the coverage period of a standard warranty. It extends the time that the producer (e.g., fabricator, manufacturer, or supplier) is liable to compensate for, repair, or replace a defective item or product. This may be offered at the time of purchase or when a standard warranty nears its expiry date. See also, *warranty*.
External Pressure	*External pressure* is negative pressure that is applied to pressure equipment (i.e., a pressure inside that is less than atmospheric pressure, a *vacuum)*. See also, *atmospheric pressure*, *internal pressure*, *pressure equipment* and *vacuum*.
Externality	An *externality* is a benefit or cost caused by a producer that is not financially incurred or received by that producer. An externality can be negative or positive and can result from either the consumption or production of a product or service. For a negative example, the air pollution around a foundry that is caused by the smelting process. For a positive example, a business opens a café for its employees that is also open to the public. See also, *spillover effect*.

Phrase or Term	Definition
Extinct by Instinct	*Extinct by instinct* is an expression that describes how a group or individual can make an incorrect decision by being too quick to act or decide, or if the power or willingness to challenge choices is absent or suspended. This can occur when a decision is made based on a gut reaction or hasty judgment using groupthink, inadequate data, insufficient analysis, or a combination. For example, Coca-Cola experienced this when they created *New Coke*, a product intended to taste like Pepsi. Nationwide taste tests (in the United States) conducted before its launch in 1985 turned out to be misleading. Hence, it was abandoned (became extinct) after only a few weeks in the market and the old formula was re-introduced as *Coca-Cola Classic*. See also, *groupthink* and *paralysis by analysis*.
Extrapolation	*Extrapolation* is an estimation of a value based on extending a known sequence of facts or values beyond the area that is known with certainly. For example, estimating the global population in 2025 by extrapolating existing data from previous years (e.g., based on surveys completed during 2010, 2015, and 2020). It is the extension, expansion, or projection of known data into an area that is unknown, to produce conjectural knowledge. The application of a conclusion or method to an unknown situation assumes that existing trends will continue. See also, *creative estimating*, *forecasting*, *interpolation*, and *superforecasting*.
Extreme Anchor, Low-ball Offer	An *extreme anchor* is an aggressive (but not absurd) position for the first offer in a negotiation. The best first offers are those that fall outside the counterparts' zone of possible agreement, but not so far outside as to be discounted or ignored. An extreme anchor (one that is too high or low) may collapse a deal that could have otherwise been completed. This is also referred to as a *low-ball offer* in a negotiation to purchase something. See also, *anchoring effect*, *win-win*, and *zone of possible agreement*.
Extreme Project Management	*Extreme project management* is an approach to project management used mostly for complex projects with a high degree of uncertainty, and when project requirements are expected to change. Therefore, it focuses on flexibility instead of rigid scheduling. Traditional project management proceeds sequentially through the project management life cycle, and thus clearly defines problems, scopes, and solutions. Extreme project management accepts that problems, scopes, and solutions will change as the project proceeds, and thus emphasizes continual learning over deterministic planning. See also, *management* and *project management*.
Extrovert	An *extrovert* is a personality type characterized by being expressive, gregarious, outgoing, and unreserved. The word was popularized by C.G. Jung in the early 20th century. Extraversion is the state of primarily obtaining gratification from outside oneself. See also, *introvert*, *omnivert*, and *people person*.

Phrase or Term	Definition
F	1 A B C D E **F** G H I J K L M N O P Q R S T U V W X Y Z
Fabrication	*Fabrication* includes all preparation, assembly, and testing, or other production activities performed at a supplier's facility. The term *fabrication* may be used where products are typically custom designed or produced individually (e.g., with a serial number). See also, *fabricator* and *manufacturing*.
Fabricator	A *fabricator* is an individual or company that builds, constructs, designs, fabricates, or makes something. See also, *fabrication*.
Facilitator	A *facilitator* is an organization or person who leads one or groups of personnel through a discussion towards a meaningful conclusion. They provide indirect, unobtrusive assistance and oversight by providing guidance and supervision that allows the discussion to stay on topic and progress in a logical fashion. A facilitator assists and supports personnel in understanding common objectives and planning to reach common objectives. A facilitator makes it easier for others to implement action or processes.
Facility	*Facility* has two different meanings for projects: 1. A *facility* is any supplier premises where fabrication or manufacturing occurs. This includes factories and workshops for production, warehouses and yards for storage, and offices for administrative, engineering, and other work. 2. An owner's *facility* that is being designed and constructed by a project. This includes refineries, processing plants, storage facilities, and compressor, pump, or valve stations, and other installations. See also, *station* and *workshop*.

Phrase or Term	Definition
Factfulness	*Factfulness* is a stress-reducing habit of only having opinions for which there are strong supporting facts. It is critical thinking that enables recognition and avoidance of the most common ways information is misinterpreted. This concept is developed in the *Gapminder* article *Factfulness*, which explains that it comprises three elements: 1. Realization that one does not see the world as it is – because there is too much information, it is filtered to present an overly-dramatic and stressful impression; 2. Recognition of the 10 types of stories that trigger overdramatic interpretations; and, 3. Use of simple rules of thumb to resist overly-dramatic stories by quickly questioning the relevance of common dramatic interpretations. See more information at the Gapminder website: https://www.gapminder.org/factfulness/ In their book *Factfulness: Ten Reasons We're Wrong About the World—and Why Things Are Better Than You Think*, Swedish authors Hans Rosling, Anna Rosling Rönnlund, and Ola Rosling described how many people think the world is getting worse when, in fact – they argue – it is getting better. They proposed that the world be categorized in four levels based on income, rather than using the world model with developed and developing worlds. They observed that most of the countries in the world are categorized as Level 2 or Level 3, and only a few countries can be categorized as Level 1 or Level 4. The four levels are: • Level 1 – less than USD$ 2 a day; • Level 2 – USD$ 2 to $ 8 a day; • Level 3 – USD$ 8 to $ 32 a day; and, • Level 4 – USD$ 32+ a day. See also, *critical thinking*, *Gapminder*, *rule of thumb*, *truth*, *world model, and wealth pyramid*.
Factory Acceptance Test	A *factory acceptance test* is functional, mechanical, operational, performance and dynamic testing for equipment acceptance at the supplier's facility before release and shipment. This definition applies to all commodities including assembled and packaged equipment or systems – from motors and valves to compressors and control cabinets and panels. Equipment specifications or data sheets state testing requirements. These may describe specific tests without using the phrase *factory acceptance test*. Pressure testing and non-destructive testing are not factory acceptance tests. See also, *factory acceptance test coordinator*.
Factory Acceptance Test Coordinator (FAT Coordinator)	A *factory acceptance test coordinator* or *FAT coordinator* performs an administrative role for a project by ensuring that all factory acceptance tests are planned and executed per project requirements. Activities may be performed by a dedicated person or a supplier quality surveillance coordinator. See also, *coordinate* and *factory acceptance test*, and *supplier quality surveillance coordinator*.
Factory Acceptance Test Summary	A *factory acceptance test summary* is a document prepared to summarize the testing requirements specified in the purchase order for equipment, which may appear in the scope of supply, technical notes, data sheet, inspection requirements document, specifications, and referenced codes and standards.

Phrase or Term	Definition
Factory of the Future	A *factory of the future* is a facility that will be operated by one person and one dog. The person's job is to feed the dog. The dog's job is to ensure that the person does not adjust the machinery. This phrase is meant to be anecdotal and humorous, and is used in this context: *one day artificial intelligence and machines will operate so that the role of personnel in a factory (and elsewhere) will be usurped entirely*.
Fail Closed	*Fail closed* means that an actuated valve will close when the control signal or power is interrupted or lost. Fail closed valves might be used in a steam injection well to control dangerous steam by configuring the valve to close if the control signal or power fails, thereby maintaining a safe operating environment. This is also referred to as *spring to close* or *air to open*. See also, *actuated valve* and *fail position*.
Fail Fast	In systems design, a *fail fast* system is one that immediately reports at its interface any condition that is likely to indicate a failure. Fail-fast systems are usually designed to stop normal operation rather than attempt to continue a possibly flawed process. This phrase has been widely employed as a metaphor in business since 2001. It means that a business should undertake bold experiments to determine the long-term viability of a product or strategy, rather than proceeding cautiously and investing for years in a doomed approach. This phrase became a kind of mantra within start-up culture in the early 2000s.
Fail in Place	Not a preferred phrase – see *fail last*.
Fail Last	*Fail last* means that an actuated valve will remain as is (not react) when the control signal or power is interrupted or lost. This scenario might be used where the process cannot be shut down or where it is undesirable to do so. See also, *actuated valve* and *fail position*.
Fail Open	*Fail open* means that an actuated valve will open when the control signal or power is interrupted or lost. Fail open valves might be used to prevent overpressure in the event of a blocked line or catastrophic failure. See also, *actuated valve* and *fail position*.
Fail Position	*Fail position* describes how an actuator on an actuated valve reacts when there is a loss of power, loss of signal, or a similar event. There are three different fail scenarios, each with its own consequences, purposes, and safety effects. See also, *actuated valve*, *fail closed*, *fail last*, and *fail open*.
Fail-fast	*Fail-fast* has two meanings: • In business, it is a metaphor to undertake bold experiments and determine the long-term viability of a product or strategy, instead of proceeding cautiously over years; or, • In systems, it is a design which immediately reports at its interface any condition that is likely to indicate a failure. Fail-fast systems are usually designed to stop normal operations instead of attempt to continue a possibly flawed process.

Phrase or Term	Definition
Fail-safe	*Fail-safe* is a design feature in a system that, in the event of a specific type of failure, the system inherently responds in a way that will cause no or minimal harm to other equipment, the environment or to people. Unlike inherent safety to a particular hazard, a system that is *fail-safe* does not mean that failure is impossible or improbable, but rather that the system's design prevents or mitigates unsafe consequences. That is, if and when a fail-safe system fails, it is safe or at least no less safe than when it was operating correctly.
Failure	*A failure* is: • A shortcoming (e.g., decay, deficiency, or deterioration); • A fracturing or giving way under stress; • A lack of success; • A state of inability to perform a normal function or the abrupt cessation of normal functioning; or, • The omission of an expected or required action or occurrence (e.g., failing to perform a duty or expected action). See also, *deficiency* and *success*.
Fair	*Fair* means: • A light complexion (i.e., blonde); • Acceptable or average quality (i.e., not very good or very bad); • Conforming to the rules or standards (i.e., legitimate); • Free from favoritism, prejudice, or self-interest (i.e., unbiased); • Likely or promising (i.e., a fair chance); • Not foul or stormy (i.e., a fair day); • Open to legitimate attack, pursuit, or ridicule (i.e., fair game); • Pleasing to the eye or mind especially because of flawlessness (e.g., a fair garden); or, • Sufficient but not ample (i.e., a fair amount of food). During a negotiation, the word fair (and unfair) is an emotional term that can be used to gain concessions and put others on the defensive. This concept was developed by Chris Voss in the book *Never Split The Difference: Negotiating As If Your Life Depended On It* (with Tahl Raz). See also, *win-win*.
Fake News	*Fake news* is a phrase that describes: • False and fabricated news stories that contain no verifiable facts, citations, or authoritative sources; or, • Propaganda that is intended to influence or mislead by using falsehoods. See also, *brainwashing*, *bullshit*, *disinformation*, *false narrative*, *media bias*, *misinformation*, *propaganda*, *smells fishy*, *stretch the truth*, *truth*, *truthiness*, and *untruth*.

<table>
<tr><th>Phrase or Term</th><th>Definition</th></tr>
<tr><td>False Assumption, False Premise</td><td>A false assumption or false premise is an incorrect assumption (i.e., a premise or proposition) that forms the basis of line of reasoning or thinking. Since the assumption is not correct, the conclusion may be erroneous.
For example, this reasoning, involves a false premise:
1. If streets are wet, it has rained recently (premise);
2. The streets are wet (premise); and,
3. Therefore, it has rained recently (conclusion).
This argument is logically valid but wrong because its first premise is false. The street could be wet for another reason (e.g., flooding or washing). A simple logical analysis will not reveal the error in this argument because that analysis must accept the truth of the premise. For this reason, reasoning based on a false premise can be more difficult to discuss or refute than with a logical error.
Reasoning based on a false premise can in fact be true. In the example above, it may be true that it has recently rained and that the streets are wet. This does nothing to prove the first premise but can make the conclusion more difficult to refute. This underlies the basic problem of establishing a causal relationship.
See also, causal relationship.</td></tr>
<tr><td>False Dichotomy, False Dilemma</td><td>A false dichotomy or false dilemma is a type of informal fallacy in which something is oversimplified and falsely claimed to be an either/or situation (i.e., a dichotomy), when in fact there is at least one additional option. This can be the result of:
• A tendency to view the world with limited options, whatever the cause;
• The accidental omission of additional options (i.e., an error); or,
• The deliberate omission of additional options (i.e., a deception).
For example, the concept that religion and science are mutually exclusive and the false dichotomy that a person cannot believe in both religion and science. This is also known as bifurcation, black-and-white thinking, all-or-nothing fallacy, either-or fallacy, either-or reasoning, excluded middle, fallacy of exhaustive hypotheses, fallacy of false alternatives, fallacy of false choice, no middle ground, and polarization. See also, cognitive dissonance, cognitive flexibility, Catch-22, dichotomy, and thinking mode.</td></tr>
<tr><td>False Equivalence</td><td>False equivalence is a logical fallacy in which two things are equated based on flawed reasoning. This occurs when equivalence is drawn between two things:
• Based shared characteristics that are insignificant. For example, a bicycle and a motorcycle are two-wheeled vehicles. Regulations typically require use of a helmet to operate the former and only a driver's license to operate the latter. False equivalence would be to assume that a helmet and licence are required for both because they are both two-wheeled vehicles; or,
• Not accounting for significant differences. For example, a bicycle and a motorcycle are both two-wheeled vehicles. One can only achieve 10 to 15 kph (6.2 to 9.3 mph) on the former and can easily achieve speeds exceeding 120 kph (75 mph). False equivalence would be to assume that both vehicles are capable of similar speeds because both vehicles have two wheels.
See also, apples to oranges comparison, bait and switch, circular reasoning, cognitive bias, either/or, false equivalence, logical, logical fallacy, red herring, righteousness fallacy, slippery slope, and straw man argument.</td></tr>
</table>

Phrase or Term	Definition
False Flag	A *false flag* is a covert operation designed to deceive and misdirect others by making them believe that another country, group, or party is responsible for an activity – disguising the true identity of responsible parties. The phrase originally referred to pirate ships that would illegally fly the flag of a country as a disguise, to prevent their victims from fleeing or preparing for battle, or to disguise themselves as an ally. By flying a false flag, the blame from attacking another vessel could be assigned to the country of the false flag instead of the pirates. A false flag cyberattack is when a hacker attacks a system and attempts to fool the victims and interested parties about who is responsible, motives, or both. For example, Russian hackers attempted to disrupt the 2018 Winter Olympics in South Korea using code associated with a North Korean group. See also, *red flag* and *white flag*.
False Narrative	A *false narrative* is a story that is: • Created because of an inaccurate assessment and insufficient information but is partly true. For example, news that is inaccurate because of poor research, sloppy verification, or that is sensationalized (i.e., exaggerated for effect); • Not real but it is characterized as such by the fabrication of background information to make it appear factual. For example, inflated and overly optimistic market forecasts made by experts and promoted by the media, which push up stock prices; or, • Perceived as true but has little truth. These stories, often based on fear, are challenging to dispel, and detrimental to careers and relationships. For example, an entrepreneur thinks that there is not enough money to start a business. A false narrative is sophisticated spin with the consciously manipulated management of information to produce a predetermined political effect. It is construct, as opposed to a neutral account of actual events. This phrase was coined by Reena Walker in a Facebook group in the 1990s and then used by political producers and writers in the news media. See also, *brainwashing*, *bullshit*, *disinformation*, *fake news*, *media bias*, *misinformation*, *propaganda*, *smells fishy*, *spin*, *stretch the truth*, *truth*, *truthiness*, *and untruth*.
False Negative	A *false negative* is an error in data reporting where a test result improperly indicates the absence of a condition. For example, a cancer screening test has a negative result, yet the patient has cancer. See also, *false positive*.
False Positive	A *false positive* is an error in data reporting where a test result improperly indicates the presence of a condition. For example, if a cancer screening test has a positive result, but the person tested does not have cancer. See also, *false negative*.
Family Accepted Accounting Principles	*Family accepted accounting principles* refers to accounting practices, procedures, and standards that a family-owned business may follow when compiling financial statements. This may suit their own needs but may not use commonly-accepted methods of recording and reporting accounting information or complying with generally accepted accounting principles. See also, *generally accepted accounting principles*.
Fast Tracking	*Fast tracking* is a schedule or duration compression technique in which the duration of a critical path is shortened by performing sections of some critical path activities concurrently instead of consecutively. See also, *critical path* and *critical path method*.

<table>
<tr><th>Phrase or Term</th><th>Definition</th></tr>
<tr><td>Fastener</td><td>A fastener is a device that mechanically joins or secures something. For industrial construction, fabrication, and manufacturing, these are often a:<ul><li>Nut and bolt, or stud;</li><li>Rivet;</li><li>Screw; or,</li><li>Staple.</li></ul>Fastening devices may be used to create a:<ul><li>Non-permanent or temporary joint (e.g., shipping supports); or,</li><li>Permanent joint that will not be disassembled (e.g., structural steel for a building).</li></ul>See also, thread, turn-of-nut bolting method, and wrench.</td></tr>
<tr><td>Fatigue</td><td>Fatigue is:<ul><li>A weakness in materials, especially metals, caused by repeated variations of stress;</li><li>An attitude or state of apathy or indifference brought on by overexposure to a repeated series of similar appeals or events; or,</li><li>An extreme tiredness resulting from illness or prolonged mental or physical exertion.</li></ul></td></tr>
<tr><td>Fear of Better Options, Maximization</td><td>Fear of better options or maximization refers to the feeling or perception that all possible options must be explored for fear that one will miss out on the best choice, leading to frustration, indecision, regret, stress, unhappiness, unwanted cost, or combination. When faced with too many attractive choices, one can feel anxious not choosing the best option. Maximizers may tend to make better decisions, but they may also be less satisfied with decisions than those who make quicker decisions based on less research or satisficing.

In his book, The Paradox of Choice: Why More Is Less, Barry Schwartz demonstrates that when shoppers had to choose among many items (e.g., 6 pairs jeans or 20 types of jam), they experienced conflict and were less satisfied with their final choice. Costco employs a reverse strategy for its consumers: for example, instead of giving consumers choices of ketchup brands, they sell only one brand in one size – eliminating consumer stress associated with making the correct choice because there is only one choice. See also, fear of missing out and satisfice.</td></tr>
<tr><td>Fear of Missing Out, FOMO, Fear of Regret</td><td>Fear of missing out, FOMO, or the fear of regret refers to the feeling or perception that others are having more fun or living better lives than oneself. It is not just the sense that there might be better things that one could be doing, but the feeling that one is missing out on something important that others are experiencing right now. For example, fear of missing out on an exiting experience, profitable investment, or social event. Fear of missing out may negatively affect one's self esteem and cause envy.

This phenomenon has been exacerbated by social media that features celebrities living lavish lifestyles. This social anxiety may be exploited strategically for advertising and marketing purposes. While this concept has been around for centuries, it was first formally described by Dr. Dan Herman in the research paper Introducing short-term brands: A new branding tool for a new consumer reality. Later, this phrase was coined by Patrick J. McGinnis in 2004 and is also attributed to Joseph Reagle in 2004. See also, fear of better options and saticfising.</td></tr>
</table>

Phrase or Term	Definition
Feast or Famine	*Feast or famine* is an expression that means there is either too much (e.g., too much snow) or not enough (e.g., not enough money) of something. The cycle of too much or too little work is experienced by projects, supply chains, and seasonal contractors. Projects function in cycles of mobilizing and demobilizing personnel as the workload increases, plateaus, and then decreases. An organization may have many projects underway simultaneously that require coordinated interdepartmental synchronization to meet staffing needs. An area or industry, however, is typically unable to synchronize the needs for all organizations with projects, supply chains, and seasonal contractors. For projects, this means there is usually an abundance or shortage of personnel (i.e., during mobilization vs. demobilization, respectively). For personnel, this means there is usually an abundance or shortage of jobs (i.e., during mobilization vs. demobilization, respectively).
Feature	A *feature* is a main benefit of a product or service. It is a distinctive characteristic that distinguishes it from similar items. A feature may or may not persuade a customer to purchase product; some customers purchase products only for their intended purpose and do not consider additional features. For example, a supplier may tout that their drill bit is made of tungsten-carbide (a feature). Meanwhile, the buyer may be indifferent about this feature. That is, the buyer may only care if the drill bit drills the holes needed for the job and is much less concerned about the drill bit's properties. See also, *benefit*.
Feedback	*Feedback* is: • The corrective or evaluative information about reactions to someone's performance or a product, which is used as a basis for improvement; • The control or modification of a process or system by its results or effects (e.g., a biochemical pathway or behavioural response); or, • When the output of a system is used as input back into the system as part of a chain of cause and effect that forms a circuit or loop. The system can then be said to *feed back* into itself. See also, *feedback loop* and *feedback sandwich*.
Feedback Loop	A *feedback loop* is the part of a system in which some portion or all the system's output is used as input for future operations. Feedback loops can be either negative or positive. For example, a: • Negative feedback loop is a process whereby a company listens to customer complaints and then uses that feedback to improve their products or services. As a result of customers feeling respected and valued, the company may realize increases in customer loyalty and company profits.; and, • Positive feedback loop is a process whereby a company listens to personnel complaints and uses that feedback to improve work processes. As a result of improved workplace satisfaction, the company may increase personnel retention and company profits. See also, *exponential growth*, *feedback*, and *system thinking*.

Phrase or Term	Definition
Feedback Sandwich, Shit Sandwich	A *feedback sandwich* or *shit sandwich* describes a common method used to give negative feedback hidden, layered, or *sandwiched in* between positive feedback following this sequence: 1. Positive feedback (e.g., a compliment); 2. Negative feedback (e.g., a criticism); and, 3. Positive feedback (e.g., another compliment). This method is designed to acknowledge positive accomplishments and attributes, and identify areas for improvement. A feedback sandwich makes it easier to discuss challenges, respects the recipient's feelings, and makes criticism easier to take, especially when compared to providing only negative feedback. *See also, feedback.*
Feedstock	*Feedstock* is raw material used to fuel an industrial process or equipment (e.g., propane is the feedstock used to create polypropylene).
Feed-to-liquids, XTL	*Feed-to-liquids* or *XTL* are conversion processes that produce different and useful liquid products (e.g., synthetic crude oil) from carbon-based materials (e.g., bio-mass, coal, gas, and waste). The liquid products must have an advantage over the feedstock (e.g., a higher price) to make the transformation beneficial. These processes include: • Bio-mass-to-liquids; • Coal-to-liquids; • Gas-to-liquids; and, • Waste-to-liquids. See also, *carbon* and *feedstock*.
Feeler Gauge	A *feeler gauge* is a device comprising thin blades (usually stainless steel) calibrated to different thicknesses used for measuring clearances or narrow gaps. Blades may be calibrated with measurements in Metric units (e.g., hundredths of a millimetre) or Imperial units (e.g., thousandths of an inch).
Female Connector, Female Fitting	*A female connector is* hardware that connects to a cable, piece of hardware, or wire that has one or more shielded electrical terminals, and constructed such that a male connector can be inserted snugly into it. *A female fitting is* hardware with threads on the inside (like a nut) that screw onto male threads located on the outside (like a bolt). In electrical and mechanical manufacturing and trades, each half of a pair of mating connectors or fittings is conventionally designated male or female. The *female* connector is generally a receptacle that receives and holds the *male* connector. See also, *fitting*, *male connector*, and *male fitting*.

Phrase or Term	Definition
Fermi Estimate, Guesstimate, Order Estimation	A *Fermi estimate, guesstimate*, or *order estimation* is made using back-of-the-envelope (i.e., quick and rough) calculations and generalizations to obtain values which would otherwise require extensive analysis or experimentation to determine precisely. Personnel may use this estimate before employing a more sophisticated, precise method. This method uses unsophisticated numerical methods to estimate a solution to a problem. For example, Enrico Fermi estimated the strength of the atomic bomb that detonated at the Trinity test, based on the distance that pieces of paper that were dropped from his hand during the blast travelled. Fermi's estimate of 10 kt of Trinitrotoluene was close to the now-accepted value of 20 kt. The historical emphasis on the order-of-magnitude calculation was a result of the lack of computing power available to solve complex problems (i.e., before the use of computers and the Internet). Such approximate calculations were considered necessary to decide if an onerous and lengthy full-blown calculation was required. See also, *Fermi problem* and *Fermi Rule*.
Fermi Problem, Fermi Question, Fermi Quiz	A *Fermi problem*, *Fermi question*, or *Fermi quiz* is problem designed to teach approximation or dimensional analysis skills. This technique is named after the physicist Enrico Fermi, who was known for his ability to make good approximate calculations with little or no actual data. Fermi problems typically involve making justified guesses about quantities and their variance or lower and upper bounds. For example, how many piano tuners are there in Chicago? Fermi problems are typically restructured by breaking them up into smaller problems that are easier to approach than the original problem. See also, Fermi *estimate* and *Fermi Rule*.
Fermi Rule	The *Fermi Rule* is that it is better to be approximately right than precisely wrong. See also, *Fermi estimate* and *Fermi problem*.
Ferrous	A *ferrous* metal contains iron (e.g., cast iron, carbon steel, and stainless steel). See also, *base metal, metal, mineral, non-ferrous*, *ore*, and *non-metallic*.
Feyman Technique	The *Feyman Technique* is a communication and teaching model used to convey information using concise thoughts and simple language. It is also a simple method for learning and mastering a new concept or skill following these four steps: 1. Identify the subject. Write down everything known about it; 2. Explain the subject. Write an explanation about it in common words that a child could understand. This forces one to simplify connections and relationships between ideas and to understand at a deeper level; 3. Identify knowledge gaps. Whenever something gets complicated it identifies a need for further study. Write another explanation of the concept if it is unclear; 4. Tell a story. Practice reading the material out loud as if to a classroom of students, to hear where language stops being simple, or thoughts are incomplete. Use analogies and simple sentences to strengthen understanding. Avoid jargon or verbose language to explain the concept in as few words as possible. This concept was developed buy the American theoretical physicist Richard Feyman. See also, *jargon*.
Fiat Currency	A *fiat currency* is a government-issued currency that is not backed by a physical asset or commodity such as gold or silver bullion (e.g., bars, coins, or ingots). For example, Canada uses the Canadian dollar divided into 100 cents as its fiat currency. See also, *cryptocurrency*, *legal tender*, and *modern monetary theory*.

<table>
<tr><th>Phrase or Term</th><th>Definition</th></tr>
<tr><td>Field</td><td>A field is:
• A branch of study or sphere of activity or interest (e.g., the field of psychology);
• An area of open land, especially one planted with crops or used as a pasture (e.g., a farmer's field); or,
• The location of work or a project site (e.g., construction or operations).</td></tr>
<tr><td>Field Change Notice</td><td>A field change notice is a notification that an issued construction drawing needs to be changed. This change may be:
• Major, with cost-related implications, schedule-related implications or both; or,
• Minor, which may not need engineering or owner approval, may need to be red-lined and as-built, both.
A field change notice may or may not be subject to a management of change process or result in a design change notice. See also, as-built, design change notice, management of change, and red line.</td></tr>
<tr><td>Field Fit Weld, Field Fit-up Weld</td><td>A field fit weld or field fit-up weld is a joint that is designed to be field welded (i.e., at site). This joint is used to accommodate as-built dimensions and out-of-tolerance conditions, or where the installer (e.g., welder) determines the correct final dimension. A piping system may have more than one field fit weld so that there is a joint in each plane. The pipe component to be field fit welded will be fabricated long (i.e., with an extra length of pipe), so that the final joint can be measured and cut at the precise required location. See also, as-built, joint, field route, field weld, out of tolerance, and plane.</td></tr>
<tr><td>Field Route, Field Run</td><td>Field route or field run is when a piping system will be fabricated at site because:
• Final dimensions are:
  o Acceptable to be determined at site (e.g., by the welder);
  o Changed (e.g., due to an item that has an as-built dimension); or,
  o Not available or verified.
• Small bore piping is easily fabricated in the field (e.g., socket welded).
See also, as-built, field fit weld, field weld, and socket weld.</td></tr>
<tr><td>Field Service Management</td><td>Field service management refers to the management of a company's resources employed at, or en route to, the property of clients, rather than on company property. Examples include ensuring driver safety, locating vehicles, managing worker activity, and scheduling and dispatching work.
It includes integrating the management of such activities with accounting, billing, inventory, other back-office systems. This most commonly refers to companies who need to manage installation, repair, or service of equipment or systems. It can also refer to cloud-based and software platforms that aid in these activities.</td></tr>
<tr><td>Field Services</td><td>Field services are the required inspection, repair, and maintenance of tanks, pipelines, and oil and gas facility assets. For example, yearly cleaning of a vessel.</td></tr>
<tr><td>Field Weld, Site Weld</td><td>A field weld or site weld is a final welded joint in a piping system or structure that will be welded at site instead of at a facility. This may be required to keep item size manageable for transportation, construction handling, field fit, or a combination. See also, field fit, field route, and joint.</td></tr>
</table>

Phrase or Term	Definition
Field-erected Tank	A *field-erected tank* is a large tank that is built at site. Tank sizes range from 227,305 L (50,000 gal.) to several million Litres or gallons. These tanks must be built at site because they would be too large to transport after being built. See also, *shop-fabricated tank* and *tank*.
Fifty-fifty, 50/50	*Fifty-fifty* or *50/50* means: • Figuratively, something is uncertain or unsure (e.g., maybe); • Half favorable and half unfavorable; • Literally, there is a 50 percent chance of something happening or not happening (e.g., a tossed coin landing as heads or tails); or, • Something is assumed, borne, or shared equally (i.e., 50% each for two parties). See also, *middle ground*.
Fight-or-flight Mechanism	The *fight-or-flight mechanism* is the human body's natural defense against danger and predators. It causes the body to flood with hormones that prepare its systems to confront (i.e., fight) danger or predators or evade (i.e., flee from) them. The body produces large quantities of chemicals (i.e., cortisol, epinephrine, and norepinephrine) that trigger the following physical reactions: • Alertness; • Heightened muscle preparedness; • Increased blood pressure; and, • Sweating. See also, *stress*.
File Transfer Protocol (FTP)	*File transfer protocol* or FTP is a standard network protocol used to transfer data on a computer network.
Filibuster	A *filibuster* is a political procedure where one or more members of legislative assembly ostensibly debate proposed legislation with the real purpose being delaying or preventing a decision from being made. A filibuster obstructs progress while simultaneously not technically contravening governmental procedures. For example, if there is no rule about the duration of a member's speech, the member can prevent a vote by talking endlessly. In 1957, Senator Strom Thurmond of South Carolina set a filibuster record by speaking continuously for more than 24 hours for the purpose of delaying passage of United States President Dwight Eisenhower's *Civil Rights Bill*, which would have expanded federal supervision of integration in the southern states of the United States. His filibuster was unsuccessful in stopping the *Civil Rights Bill*.
Filler Metal, Filler Material	*Filler metal* or *filler material* is the metal (e.g., electrode) or material added when creating a joint using bonding, brazing, soldering, or welding. See also, *brazing*, *electrode*, *soldering*, and *welding*.

Phrase or Term	Definition
Filler Words	*Filler words* in: • Speech are short sounds or words used to gap the pauses that sometimes happen when deciding what to say next (e.g., ah, er, okay, uh, or um). These do not add meaning to a statement; however, they do let listeners know that the speaker is not finished speaking; and, • Writing include unnecessary adverbs (e.g., basically, just, really, or very) or empty phrases (e.g., in my opinion, needless to say, or you know) that make phrases lengthy and do not contribute to the substance of a message. When a speaker or writer includes many filler words in a message , the audience may become bored, confused, or disinterested. Occasional use of filler words is acceptable and often rarely noticed; however, frequent use of filler words often greatly impedes effective communication and annoys listeners-readers. See also, *plain English*.
Film	A *film* is a layer of coating material that may or may not be visible, and classified as a: • Thick film, >500 microns (20 mils) thick; or, • Thin film, <500 microns (20 mils) thick. This definition determines if holiday testing needs to be high voltage (for thick film) or low-voltage (for thin film). See also, *coating*, *dry film thickness*, *holiday test*, *micron*, and *wet film thickness*.
Fin Fan Cooler	Not a preferred phrase – see *air cooled heat exchanger*.
Final Data Book	Not a preferred phrase – see *manufacturing record book*.
Final Destination	The *final destination* is the location where equipment or materials will be shipped (usually to site). For example, from a sub-supplier to another sub-supplier or the prime supplier, or any combination – for internal shipments that are managed by the prime supplier. Also, from one supplier to another supplier – for external shipments that are managed by the project (i.e., free-issued). See also, *free-issued* and *site*.
Final Inspection	A *final inspection* is an examination performed at the final stage of production before issuing an inspection release. This may include visual examination, documentation review, verification of attributes, and close-out of issues (e.g., non-conformances). See also, *final-final inspection* and *kick and count*.
Final Sign-off	*Final sign-off* is the last acknowledgement or approval of something by or as if by a signature, usually acquired after a series of progressive inspections or sign-offs. See also, *sign-off*.
Final-Final Inspection	A *final-final inspection* is the very last inspection for a change order, purchase order, or scope of supply. Inspection assignments that have multiple shipments may require multiple final inspections. See also, *final inspection*.
Financial Assets	*Financial assets* are the sum of shareholders' debt and equity liabilities.
Financial Officer	A *financial officer* is the chief financial officer, controller, principal accounting officer, or treasurer of an organization. See also, *banker* and *accountant*.
Fine Tailings	*Fine tailings* are essentially muddy water produced as a result of mining extraction. Fine tailings comprise about 85% water and 15% fine clay particles (by volume). See also, *mature fine tailings* and *tailings*.

Phrase or Term	Definition
Finite Element Analysis	*Finite element analysis* is a numerical method of solving engineering challenges. Typical areas of interest include fluid flow, heat transfer, and structural analysis.
Finite Mindset	A *finite mindset* is a way of thinking understands that, in business, the goal is to be the best and beat the competition. Thus, for a business to succeed, another business must lose. Attention is focused on rivals and decisions are made based on scarcity. This leads to a decline in cooperation, innovation, and trust in an organization. This concept was developed by American author Simon Sinek. See also, *abundance mindset*, *infinite mindset*, *lose-lose*, *scarcity mindset*, and *win-win*.
Finned Tube	A *finned tube* is a tube with small fins. These fins transfer heat from the material inside the tube to the atmosphere, or vice versa. See also, *air cooled heat exchanger* and *tube*.
Fire Tetrahedron, Fire Triangle, Combustion Triangle	A *fire tetrahedron* (i.e., a pyramid with four plane faces), also known as a *fire triangle* and *combustion triangle*, is an easy-to-understand model that explains the elements required to start and sustain a fire: 1. Chemical chain reaction; 2. Fuel; 3. Ignition; and, 4. Oxidizing agent. Combustion was symbolized by the fire triangle or combustion triangle, which represented fuel, heat, and oxygen. Later research determined that a chemical chain reaction is also necessary for combustion to occur, so it is now usually symbolized by a fire tetrahedron. Nonetheless, the simpler fire triangle remains in use and may be found in older materials. It explains the oxidation process and the basics of fire; but not the entire fire process.
Fire Safe Valve	A *fire safe valve*, per API STD 607 and SPEC 6FA, is a valve that has been proven by prototype or type testing to be capable of passing a defined fire test with no more than a defined leakage rate at the conclusion of the test. These vales are typically soft seated with metal-to-metal backup seating. See also, *fireproof*.
Fireproof (Value and Actuator Assembly)	A *fireproof valve and actuator assembly* with accessories can withstand temperatures of 1,093°C (2,000°F) under test conditions of UL 1709 for 15 minutes and continue to function properly. API RP 2218 has additional requirements for fireproofing than can be applied with this specification. See also, *fire safe*.
First Article Inspection	Not a preferred phrase – see *first-off*.
First Law of Usability	The *First Law of Usability* is *don't make me think*, an adage from the book *Don't Make Me Think, Revisited: A Common Sense Approach to Web Usability* by Steve Krug. It describes making a user experience, such as on a website, as intuitive, obvious, and self explanatory as possible.

Phrase or Term	Definition
First-hand	*First-hand* describes a situation where: • Experience or information is gained from direct personal experience or an original investigation conducted in-person. For example, *I saw the thief steal the money first-hand*; or, • Something is not a derivative. For example, *this is a new, original idea; nobody thought of doing it that way before*. See also, *brand-new*, *hand*, *second-hand*, and *third-hand*.
First-off	A *first-off* occurs during the first part of a sequence (e.g., witnessing a performance test). A *first-off* may be used to specify requirements for certain equipment rather than all equipment (e.g., testing or witnessing for the first production of each size or type).
First-past-the-post	*First-past-the-post* is a voting requirement where the candidate that receives the most votes wins, irrespective of their vote share or a that a simple majority is not achieved. See also, *absolute majority*, *simple majority*, and *supermajority*.
Fiscal Policy	A *fiscal policy* is documented actions taken by a government (e.g., spending and taxation) to influence economic conditions (e.g., demand for products and services, economic growth, employment, and inflation). See also, *modern monetary theory* and *monetary policy*.
Fishing Expedition	A *fishing expedition* is an investigation that: • Seeks to uncover incriminating evidence or newsworthy information; • Seeks damaging information about another party, including through self-incrimination. It is a pejorative phrase likened to fishing because the investigating party only obtains (i.e., finds out) whatever they happen to catch; or, • Is a legal examination or interrogation designed to discover or reveal unknown information for future legal proceedings. See also, *fool's errand*, *goose chase*, and *witch hunt*.
Fitness for Service	*Fitness for service* is a best practice or standard used to determine the adequacy of an in-service asset for continued service. This standard serves as a rational basis for defining flaw acceptance limits, which allows personnel to distinguish between acceptable and unacceptable damage or flaws based on recognized and generally accepted good engineering practices. See also, *asset*, *inspectioneering*, and *recognized and generally accepted good engineering practices*.
Fitting	A *fitting* is a valve, gauge, regulating or controlling device, flange, bend, elbow, tee, reducer or Y-lateral, or any other appurtenance that is attached to, or forms part of, a boiler, pressure vessel, fired-heater pressure coil, thermal liquid heating system or pressure piping system.
Fit-up Test	A *fit-up test* ensures that the critical equipment dimensions are correct, match-marking and tagging is completed, or both, and verifies that accurate and efficient re-assembly may be performed at site. For a *fit-up test*, a supplier assembles and aligns project equipment at the supplier's facility. See also, *match-marking*.

Phrase or Term	Definition
Functions of Management	The *functions of management* were developed in the early 19th century by the French engineer Henri Fayol. These functions are: 1. Planning Assess and prepare for the future with a flexible plan that considers future market trends, resources, and work in progress. 2. Organizing Establish lines of authority and responsibility for personnel including activity coordination, duty definition, and recruitment and training. 3. Commanding Learn about and tap into personnel skills to obtain the best productivity and set a good example. 4. Coordinating Ensure that teams understand their responsibilities and obligations and the needs of other teams. 5. Controlling Continually verify that plans, processes, and rules function as they should. See also, *administrative theory* and *management*.
Five Learning Disciplines	The *five learning disciplines* are a method to develop the core learning capabilities needed to create a learning organization that can foster aspiration, develop reflective conversation, and understand complexity. These methods are: 1. A shared vision Unearthing shared pictures of the future that foster genuine commitment and enrollment rather than compliance. 2. Mental models The deeply ingrained assumptions, generalizations, or images that influence how one understands the world and takes action. 3. Team learning Dialogue or the capacity of team members to suspend assumptions and genuinely think together. 4. Personal mastery Continually clarifying and deepening our personal vision, of focusing our energies, developing patience, and seeing reality objectively. 5. System thinking The fifth discipline that integrates the first four disciplines. This concept was developed in the book *The Fifth Discipline: The Art and Practice of the Learning Organization* by Peter Senge. It focuses on group problem solving using the systems thinking method to convert companies into learning organizations. See also, *learning organization*, *the fifth discipline*, *reflective questioning*, and *system thinking*.

Phrase or Term	Definition
Five Levels of Leadership Concept, Level 5 Hierarchy of Executive Capabilities	The *five levels of leadership concept* or *Level 5 hierarchy of executive capabilities* was developed in the book *Good to Great: Why Some Companies Make the Leap ... and Others Don't* by Jim Collins. It describes these five levels of executive leadership (on an ascending scale): 1. Highly capable individual – makes productive contributions; 2. Contributing team member – contributes to the achievement of group objectives; 3. Competent manager – effectively and efficiently organizes resources; 4. Effective leader – catalyzes commitment to higher standards and a vision; and, 5. Level 5 executive – builds enduring greatness. The concept was also developed in the book *The 5 Levels of Leadership* by John C. Maxwell. It describes these five levels of leadership (on an ascending scale): 1. Position – the lowest entry level, where others follow only if persuaded; 2. Permission – based on relationship, where the leader is given permission to lead; 3. Production – to motivate others, to get things done; 4. People development – to develop other leaders; and, 5. Pinnacle – to develop organizations and create opportunities or a legacy that other leaders do not. Their leadership gains a positive reputation and these leaders often transcend their positions and organizations, and sometimes their industries. For example, Colman Mockler, CEO of Gillette (1975-1991) and David Maxwell, CEO of Fannie Mae (1981-1991) were Level 5 leaders. See also, *Level 5 leader concept* and *leadership*.
Five Rs of Materials Management, 5 Rs of Materials Management	The *five Rs of materials management* or *5 Rs of materials management* is the acquisition of materials and services: 1. At the right price; 2. At the right time; 3. From the right source; 4. In the right quantity; and, 5. Of the right quality. See also, *management* and *materials management*.
Five S Methodology, 5S Methodology	*Five S methodology or 5S methodology* is a workplace organization method that uses this list of five words: 1. Sort; 2. Set in order; 3. Shine; 4. Standardize; and, 5. Sustain. These are translated from the Japanese words seiri, seiton, seisō, seiketsu, and shitsuke, respectively. The list describes how to organize an efficient and effective workspace by identifying and storing the items used, maintaining the area and items, and sustaining the new order. The decision-making process usually comes from a dialog about standardization, which builds understanding among employees about how to perform the work. In some businesses, 5S has become 6S, with safety as the sixth element.

Phrase or Term	Definition
Five Whys, 5 Whys	The *five whys* or *5 whys* originated during the Japanese Industrial Revolution in the 1930s. The method is simple: when a problem arises, determine the cause by asking *why* five times. It sounds unsophisticated, but it should not be underestimated. Its simplicity is what makes it so helpful.
Fixed Costs	*Fixed costs* are expenses that remain the same regardless of the volume of products or services (e.g., maintenance, rent, and overhead salaries). See also, *cost plus* and *variable costs*.
Fixed Equipment Mechanical Integrity	*Fixed equipment mechanical integrity* is the creation, implementation, and maintenance of management systems to avoid loss of containment in chemical processing, oil and gas, and petrochemical industries. See also, *fixed equipment*, *mechanical integrity*, and *pressure equipment integrity management system*.
Fixed Equipment, Static Equipment	*Fixed equipment* or *static equipment* describes stationary equipment (e.g., boilers, heat exchangers, piping systems, pressure vessels, tanks, and valves). This does not include electrical equipment or instrumentation, even though these typically do not move. See also, *electro-mechanical equipment*, *fixed equipment mechanical integrity*, *process equipment*, and *rotating equipment*.
Fixed Mindset	A *fixed mindset* is the belief that qualities, like intelligence or talent, are fixed traits that can only be revealed, and not created and developed. Therefore, talent alone does not create success. Effort is also needed. Personnel with this mindset spend their time documenting, instead of developing, their intelligence or talents because they believe that they either posses or do not possess a specific talent (i.e., they cannot improve their performance or skill if they lack talent). See also, *growth mindset*.
Flame War	A *flame war* is a lengthy exchange of hostile or offensive messages by participants over the Internet for amusement or another specific purpose (e.g., social engineering). See also, *bully*, *cyberbullying*, *flaming*, *social engineering*, and *trolling*.
Flaming	*Flaming* is the act of posting or sending hostile or offensive messages over the Internet. These messages contain insults, offensive language, profanity, or a combination, and may be posted in online groups or forums, on social media, or sent by email or instant messaging applications. Flaming may or may not be intentional because a user may misunderstand the intent of the original message or post. For example, a user may take offense to a comment which was intended to be sarcastic. See also, *bully*, *cyberbullying*, *flame war*, and *trolling*.

Phrase or Term	Definition
Flange	A *flange* is an external or internal ridge, or rim (lip), that is used for: • Attachment of one object to another; • Joints (fittings) to connect pipes and pressure or other equipment. These are available with various: o Designs (e.g., blind, lap joint, slip-on, socket weld, threaded, and weld neck); o Face dimensions or styles (e.g., flat face, raised face, ring joint, or tongue and groove); o Pressure classes (e.g., with pressure and temperature ratings for different materials); and, o Standards (e.g., ASME B16.5 or ASME B16.47, DIN EN 1092-1, and MSS SP-44). • Rail car or tram wheels with a flange on one side to keep the wheels from running off the rails; and, • Strength, as the flange of an iron beam such as an I-beam or T-beam.
Flash Corrosion, Flash Rust, Rust Bloom	*Flash corrosion*, *flash rust*, or *rust bloom* is the general and rapid corrosion that instantly occurs when a metal is exposed to a corrosive environment. For example, bare steel substrate exposed to oxygen and water will rust. This corrosion can begin in minutes. Initial indicators of metal corrosion are when metal becomes covered by, or spotted with, bright orange discolouration. See also, *corrosion*, *rust*, and *substrate*.
Flash Report	A *flash report* is a communication that is made by a quality surveillance representative to the supplier quality surveillance coordinator (e.g., a telephone call, or email or text with photographs) about a significant or urgent issue. This is done during the inspection visit, from the supplier's facility, or as soon as possible thereafter. This information shall also be included in the inspection report when it is issued. See also, *inspection report*.
Flash Research	*Flash research* is quick investigation (e.g., within one or two minutes) that is made using Google or YouTube to learn if something exists, or if it is complicated or possible. For example, does a truth filter exist, who developed the concept, and how does it work? Or, can I easily change the fuel filter in my vehicle myself? See also, *truth filter*.
Flat Database	A *flat database* is a database design where data is stored in a simple table with rows and columns (e.g., a spreadsheet).
Flatearther	A *Flatearther* is someone who believes that the Earth is flat and repudiates science that proves the Earth's sphericity. Devout flatearthers are often motivated by conspiracy theories, non-scientific guidance, pseudoscience, religion, or a combination.
Flaw	A *flaw* is a mark, fault, or other imperfection that mars a substance or object.
Flexibility	*Flexibility* is the capacity to challenge old ways of thinking or working and the willingness to adapt to new ways of thinking and working using creativity and strategy. See also, *cognitive dissonance*, *cognitive flexibility*, *creativity* and *thinking mode*.

Phrase or Term	Definition
Flip-flop	A *flip-flop* is a: • Circuit or electronic device (e.g., a computer) capable of assuming either of two stable states to store information; • Rubber sandal loosely fastened to the foot by a thong; or, • Sudden reversal to act or think the opposite (e.g., an opinion, policy, or strategy). For example, a politician votes for something and then votes against it. This can be perceived as calculating and unprincipled but is sometimes necessary. The key to avoiding criticism of indecisiveness is to market or present it as a genuine change of heart (e.g., because of convincing or new information received or reconsideration of alternatives or facts).
Float	A *float* is a measure of schedule task flexibility. For example, the duration of the fabrication schedule is 10 weeks, including a float of 1 week for contingency (the unforeseen). Thus, the float allows the project to sustain itself for 11 weeks total. See also, *critical path method*, *moving to the left*, and *moving to the right*.
Floating Production Storage and Offloading Vessel, Floating Storage and Offloading Vessel	A *floating production storage and offloading vessel* is used by the offshore oil and gas industry to produce, processes, and store hydrocarbons. This type of vessel receives hydrocarbons produced by itself (i.e., by drilling) or from nearby facilities, to be processed and stored until offloaded to a tanker or transported through a pipeline. A *floating storage and offloading vessel* processes and stores hydrocarbons but does not produce them.
Flocculant, Flocculating Agent	*Flocculant* or a *flocculating agent* is a chemical substance added to a liquid to enhance aggregation or clumping of suspended particles typically used to treat industrial or municipal wastewater. It is sometimes erroneously confused with flocculent. See also, *agglomerate and flocculent*.
Flocculent	*Flocculent* describes the texture of loosely clumped material. It is sometimes erroneously confused with flocculant. See also, *agglomerate and flocculant*.
Flotation	*Flotation* is the process of separating small particles of various materials by treatment with chemicals in water, to make some particles adhere to air bubbles and rise to the surface for removal while others remain in the water.
Flow	*Flow* is the efficient progression of tasks thorough production and along the value stream. This means that a product proceeds from design to launch, order to delivery, and raw to finished material in the hands of the customer, with no backflow, delay, or scrap.
Flow-through	*Flow-through* is when incurred expenses are charged with no markup. See also, *markup*.
Focus Group	A *focus group* is a diverse group of people whose reactions to something are studied to determine the reactions that could be expected from a larger population. It is a research method used to collect data through interactive and guided or open discussions about a new product or service, or political analysis. For example, having a focus group view an early draft of a movie to determine the level of empathy viewers have for the lead characters. See also, *brainstorm*, *creative destruction*, *creative thinking*, *focus group*, *whiteboard session*, and *unfocus group*.

Phrase or Term	Definition
Follow the Money	*Follow the money* is an expression that was popularized by *All the President's Men*, a film about the American Watergate scandal. In the movie, a secret informant – known by the pseudonym Deep Throat – suggests that political corruption can be explained and understood by examining money transfers made between parties. Money is a powerful motivator and is often the prime motivation. This means that financial transactions or gaps in financial transactions, which are difficult to hide, can be investigated as evidence of a crime, criminal network, or suspicious activity. The origin of this expression is attributed to Henry Peterson who testified at 1974 Senate Judiciary Committee hearings on the nomination of Earl J. Silbert to become United States Attorney. This expression can be used to: • Apply systems thinking (e.g., use big data about financial transactions to understand money laundering trends); • Denote a lifestyle choice (e.g., use money to acquire a yacht); • Designate a pathway (e.g., only treat diseases if the remedy is profitable); • Give guidance (e.g., control power by imposing transparency); or, • Work to increase income (e.g., take the most profitable career path). See also, *big data*, *money laundering*, *systems thinking*, and *follow the science*.

Phrase or Term	Definition
Follow the Science	*Follow the science* is an expression that means science is essential to making appropriate decisions. This is a correct statement. The expression also implies that science makes policy decisions clear, which is dangerously incorrect. This is because people tend to favour evidence that supports their arguments or beliefs and dismiss evidence that does not. Scientific evidence can be a sound justification for action or inaction, but the relationship between politics, science, and society is far more complex than this. Science alone will never be sufficient to guide choices and compromises because it cannot determine policy or make value judgments. Policy is a human endeavour that combines science with priorities and values. Science can help quantify the level of risk but it cannot decide what to do. Science is a necessary tool but insufficient to manage risk by itself. Thinking otherwise usurps political power from the people and instead gives it to scientists under false pretenses. Public policy is a pragmatic science. It is what happens when science steps out of the laboratory and shakes hands with real life. Misconceptions about following the science include: 1. Credentials are not science – it is not academic degrees, level of education, or the academy in which training occurred; 2. Science is not dogma – it demands testable, falsifiable hypotheses. When there are competing ideas, studies will decide which is correct; 3. Science is not censoring – it is relentlessly challenging ideas that must be confronted, debated, discussed, and refuted. Using force to proclaim truth is antithetical; 4. Science is not a popularity contest or a belief held by the majority – it is a process and method to make sense of the world; 5. Science is not convincing others that faulty or weak evidence proves anything. It is applying criticism equally, fairly, impartially, and objectively; 6. Observational studies on topical subjects can be self-fulfilling prophecies – randomized trials need to address controversial subjects objectively; and, 7. What works in theory does not always work in practice – what works in an ideal setting, what works in the real world, and communication of a recommendation is part of its implementation. *Follow the science* is an adaptation of the expression *follow the money*. Its original source is unknown. This concept was further developed in the article *Op-Ed: What Does 'Follow the Science' Mean, Anyway?* by Vinay Prasad, MD, MPH. See also, *dogma*, *ethical*, *follow the money*, *policy*, *self-fulfilling prophecy*, and *scientific method*.
Fool's Errand	A *fool's errand* is an activity or task that has no hope of success and is needless or profitless. For example, a new apprentice mechanic is requested to fetch a *bolt stretcher* (i.e., a non-existent tool) from the shop. Experienced workers at the shop will recognize this ruse, tell the apprentice that the fictional tool is not in the shop and send the apprentice elsewhere on a wild goose chase. This exercise is intended to be humorous for experienced personnel and teach the apprentice a lesson about being thoughtful about following instructions and completing a task. See also, *busy work*, *goose chase*, *fishing expedition*, *make-work job*, *wild goose chase*, and *witch hunt*.

Phrase or Term	Definition
Fool's Gold	*Fool's gold* is the mineral chalcopyrite, iron pyrite, or pyrite. Its metallic luster and pale brass-yellow hue give it a superficial resemblance to gold. Hence, it may be mistaken for gold and is referred to as fool's gold (i.e., gold, to a fool). Fool's gold differs from real gold by brittleness, crystal form, and hardness properties. See also, *hopium*, *mineral*, *pipe dream*, *unobtainium*, and *when pigs fly*.
Force Majeure	*Force majeure* is contract clause included to remove liability for natural and unavoidable catastrophic events (e.g., an act of God) that interrupts the expected course of business and prevents participants from fulfilling contractual obligations. It is a Latin phrase that means *greater force*. Examples include an adverse weather event (e.g., , or earthquake, flood, or hurricane), labour disruption, military conflict. The International Chamber of Commerce has attempted to clarify the meaning of force majeure (although its definition is not included in the Incoterms) by applying a standard of *impracticability*, meaning that it would be, if not impossible, unreasonably burdensome and expensive, to fulfill contractual obligations. An event that brings this situation about must be external to both parties, unavoidable, and unforeseeable. It can be very difficult to prove these conditions, so most force majeure defenses fail in international tribunals. Force majeure would not cover natural and usual events, negligence or malfeasance, or circumstances included in the contract (e.g., cancellation in the event of rain or snow). See also, *Incoterms*, *labour disruption*, and *pacta sunt servanda*.
Forced Empathy	*Forced empathy* is a powerful negotiating tactic that uses a question beginning with the word *how* or *what*, which causes the other party to look at the situation from one's own perspective. This forcefully conveys to the other party that one has a problem. This appears to be an admission of uncertainty or weakness but it is a show of strength because a question can hold more power than a declaration. This tactical empathy forces or helps the other party to: 1. Consider one fairly or consider one's situation fairly; and, 2. Understand one's: a. Position; b. Issues; and, c. Constraints. For example, asking the question: *what am I supposed to do?* forces the other party to empathize. This tactic is best employed face-to-face. Online or written communication (e.g., a chat) lacks the non-verbal ques of body language. The concept of forced empathy was developed by Chris Voss in the book *Never Split The Difference: Negotiating As If Your Life Depended On It* (with Tahl Raz). Forced empathy tactics can be frustrated by whataboutism. See also, *body language*, *whataboutism*, and *win-win*.
Forecast Planning	*Forecast planning* is a method that allows organizations to make long-term plans. One challenge or opportunity (i.e., one result or scenario) is planned and then executed. For example, production capacity is expanded by undertaking a capital construction project that increases the size of the facility by adding manufacturing, office, and warehouse space. This method is used by groups such as business and project managers. See also, *forecasting*, *future-back process*, *planning*, and *scenario planning*.

Phrase or Term	Definition
Forecasting	*Forecasting* is predicting the future. It is a planning tool that helps organizations cope with uncertainty by using data from the past and present with trend analysis (e.g., statistics). The use of experience, judgment, and knowledge is also required. It can be an integral part of management decision-making and, depending on the specific application, it may be: • Short-term (e.g., to schedule personnel, production, and transportation); • Medium-term (e.g., to determine future requirements to purchase equipment, machinery, or materials, and recruit personnel); or, • Long-term (e.g., to consider environmental factors, internal resources, and market opportunities). A forecasting system may involve several approaches to predicting uncertain events, which requires the development of expertise to: • Identify forecasting problems; • Apply a range of forecasting methods; • Select appropriate methods for each problem; and, • Evaluate and refine forecasting methods over time. Forecasting is an activity that may be conducted in a personal or professional capacity. For example, planning a budget for personal finances or a capital project (respectively). By contrast, superforecasters can measurably predict events in advance. See also, *crystal ball gazing*, *enterprise performance management*, *extrapolation*, *forecast planning*, *Monte Carlo simulation*, *planning*, *scenario planning*, and *superforecasting*.
Forging	A *forging* is the product of a substantially compressive plastic working operation that consolidates material and produces the desired shape. The plastic working may be performed by a forging machine or rolls, hammer, press, ring rolling machine, and must deform the material to produce an essentially wrought structure. See also, *seamless* and *wrought*.
Form	A *form* is a: • Permanent or temporary mold (i.e., formwork) into which concrete or similar materials are poured. See also, *insulated concrete form*; or, • Prepared (e.g., printed or processed) document with blank spaces for insertion of required and requested information. See also, *form fit and function* and *template*.

<table>
<tr><th>Phrase or Term</th><th>Definition</th></tr>
<tr><td>Form Fit and Function, F3</td><td>Form fit and function or F3 is the description and identification of the major characteristics of an assembly or part. Each characteristic defines a specific aspect of the part to that meets customer demand. By standardizing and making similar the F3 of a product, changes to it minimizes documentation costs and minor design changes. Aspects of F3 include:
1. Fit
The ability of a part to connect, join, or mate with another part within an assembly. This allows the part to meet the required assembly tolerances and become useful.
2. Form
The external dimensions, size, visual appearance, and weight of an assembly or part. This element is most affected by an aesthetic choice, including the chassis, control panel, and enclosure that form the external aspects of the product.
3. Function
The criterion that is met when a part effectively and reliably performs its stated task.
See also, form.</td></tr>
<tr><td>Formula</td><td>A formula is a:
• List of ingredients for, or constituents of, something (e.g., the chemicals needed to achieve a certain reaction or make a specific product); or,
• Mathematical relationship or rule expressed in symbols (e.g., 2 x 2 = 4).</td></tr>
<tr><td>For-profit</td><td>For-profit describes a business or other organization with the primary goal of making profit for its owners or shareholders. Most companies are for-profit organizations. See also, non-profit.</td></tr>
<tr><td>Forward Compatibility</td><td>Forward compatibility is a design that has a roadmap for compatibility with future products and standards. See also, backwards compatibility.</td></tr>
<tr><td>Forward Rate</td><td>The forward rate is the future market value (i.e., as it may fall or rise) versus the spot rate (i.e., for a currency value or an interest rate applicable to a future financial transaction). See also, theories of exchange rates and spot rate.</td></tr>
<tr><td>Forward Sold</td><td>Forward sold refers to selling the product output of a project before it starts. This is very important for liquified natural gas and, sometimes, metals. Forward sold reduces some commercial risks for project sponsors but often increases project risk. This is because of the use of take-or-pay contracts that obligate the seller to provide product to the buyer regardless of if the facility is operating at the agreed-upon price. Conversely, the buyer is obligated to accept product at the agreed-upon price.</td></tr>
<tr><td>Forward-thinking, Forward-looking</td><td>To be forward-thinking or forward-looking is to:
• Favour development, innovation, progression, or vision; or:
• Plan for the future with the present in mind.
See also, thinking mode.</td></tr>
</table>

Phrase or Term	Definition
Fossil Fuel	*Fossil fuel* is a naturally-occurring source of energy (e.g., coal, natural gas, and petroleum) formed in the geological past from the remains of ancient living organisms (i.e., animals and plants). For example, of all existing global oil deposits: • 20% were formed in the Cenozoic age (65 million years ago); • 70% were formed in the Mesozoic age (252 to 66 million years ago); and, • 10% were formed in the Paleozoic age (541 to 252 million years ago). See also, *coal*, *hydrocarbon*, *natural gas*, and *petroleum*.
Fossil Fuel Hydrogen, Fossil Fuel H_2	*Fossil fuel hydrogen* or *fossil fuel* H_2 is hydrogen that is produced from fossil fuels (e.g., bitumen, coal, petroleum, or natural gas). It has a higher carbon intensity that depends upon the specific fossil fuel that is used to produce it. See also, *carbon intensity*, *clean hydrogen*, *colours of hydrogen*, *hydrogen*, and *renewable hydrogen*.
Founder Story	A *founder story* is the narrative of how a company came to be. It adds dimension to a company, becomes the early founding history and helps promote brand loyalty among the company's clients or customers. See also, *backstory* and *story*.
Founder's Syndrome, Founderitis	*Founder's syndrome* or *founderitis* is when an organization's founder maintains a disproportionate amount of influence and power following its establishment. This means that their initial charisma, creativity, passion, productivity, or a combination can become a liability instead of being an asset. This phenomenon can occur with senior management or leadership who is comfortable with wielding power and control (e.g., a chief executive officer or other high ranking corporate officer). See also, *chief executive officer disease* and *senior management*.
Four Stages of Competence, Four Stages of Learning, Conscious Competence Ladder	The *four stages of competence*, also known as the *four stages of learning* or *conscious competence ladder*, relate to the psychological states involved in the process of learning, which range from incompetence (Stage 1) to competence with a skill (Stage 4). The four stages are: 1. Unconscious incompetence – you don't know what you don't know (e.g., unconsciously unskilled or wrong intuition); 2. Conscious incompetence – you know what you don't know (e.g., consciously unskilled or wrong analysis); 3. Conscious competence – you know how to do it, but you must think your way through it (e.g., consciously skilled or right analysis); and, 4. Unconscious competence – you can do it without much effort because you know what to do (e.g., unconsciously skilled or right intuition). See also, *competence*.

Phrase or Term	Definition
Four Sublimes	The *four sublimes* are drivers that explain the increased frequency and size of megaprojects. They are: 1. Aesthetic. For example, the inherent satisfaction designers and people who love good design get from designed structures that are very large, iconic and beautiful (like the Golden Gate Bridge); 2. Economic. For example, the benefits that businesspeople (i.e., bankers, contractors, consultants, developers, lawyers, and workers) and trade unions get from generating a plethora of jobs and profit from construction of megaprojects; 3. Political. The political capital politicians generate from building monuments to themselves and their causes, and the visibility this yields with the media and public; and, 4. Technological. The progression of technical aspects of engineering disciplines as engineers and technologists examine methods of pushing the boundaries of science to accommodate megaprojects that are the fastest, largest, longest, and tallest projects ever conceived in the history of mankind. See also, *megaproject*.
Fractional C-suite Services	*Fractional C-suite services* are used by organizations to retain elite personnel who have the functional expertise and tactics required to achieve business objectives on an as-needed basis (e.g., advisory, consultant, contract, interim, part-time, or temporary basis). Their focus includes business diversification, leadership, personnel development, sales growth, and strategic execution. This service is typically provided to boards, business owners, executives, executive teams, managers, officers, or a combination, with small and mid-sized companies. See also, *C-suite*.
Francophone	A *Francophone* is someone whose first language is French. See also, *allophone* and *Anglophone*.
Free Carrier	*Free carrier* is an Incoterm where a supplier delivers goods to the carrier, as stipulated at the supplier's premises. A buyer assumes all risks and costs associated with delivery of goods to the destination, including transportation after delivery to carrier. For more information, visit the International Chamber of Commerce website: http://store.iccwbo.org/ See also, *goods*.
Free Lunch	A *free lunch* is an informal expression that indicates no expectation of repayment, service, responsibility, etc. For example: *In politics, there is no free lunch – everyone expects favours to be repaid*. See also, *free* ride and *law of cause and effect*.
Free Ride	A *free ride* is an expression that indicates: • A benefit obtained at another's expense or without the usual cost or effort; or, • Easy or soft treatment. See also, *free lunch*, *free rider problem*, and *law of cause and effect*.

Phrase or Term	Definition
Free Rider Problem	The *free rider problem* is the burden on a shared resource that is created by its use or overuse by people who are not paying their fair share for it or are not paying anything at all. This is an issue in economics that is considered an example of a market failure. That is, the inefficient distribution of goods or services that occurs when some individuals can consume more than their fair share of the shared resource or pay less than their fair share of the costs. For example, Wikipedia is a free encyclopedia that hundreds of millions of people use every month. Only a tiny fraction of the users pay to use it but all users are able to benefit from the information provided by the website. Free riding prevents the production and consumption of goods and services through conventional free-market methods. To the free rider, there is little incentive to contribute to a collective resource since they can enjoy its benefits without contributing to it. Consequently, the producer of the resource cannot be sufficiently compensated. The shared resource must therefore be subsidized in some other way, or it will not sustainable. See also, *free ride*, *game theory*, *goods*, and *Prisoner's Dilemma*.
Freedom	*Freedom* is the ability to act or change without constraint. It is the state of: • Being allowed to do something one wants to do; and, • Not being affected by something one does not want. *See also, Freedom in the World* and *freedom of the press*.
Freedom in the World	*Freedom in the World* is an annual global report on civil liberties and political rights comprised of descriptive texts and numerical ratings for every country and selected territories. It has been published regularly since 1973 by Freedom House, the oldest American organization devoted to the worldwide defense and support of democracy. For more information, visit: https://freedomhouse.org/report/freedom-world. See also, *freedom* and *freedom of the press*.
Freedom of the Press, Freedom of the Media	*Freedom of the press* or *freedom of the media* is the principle that communication and expression through various media (e.g., electronic and printed) is a right to be exercised freely as checks and balances. Such freedom implies the absence of interference (e.g., from opposed opinion or an overreaching government). Its preservation may be sought through a constitution or other legal protection or security. A government may identify information as public or protected from disclosure to the public, or subject to a sunshine law. Information may be protected from freedom of the press due to its: • Classification as sensitive or secret; or, • Relevance to protecting the national interest. See also, *checks and balances*, *freedom*, *Freedom in the World*, *the press*, and *sunshine law*.
Free-issued	*Free-issued* materials and equipment are purchased by a project, and then subsequently shipped to a supplier for assembly or use, or to site for construction. The free-issued process is reserved for specialty and large dollar value items. Materials and equipment that is not free-issued must be purchased by the supplier.

Phrase or Term	Definition
Freemium	*Freemium* describes when a supplier offers a part, product, or service for free, typically while intentionally disabling, excluding, or restricting the use of certain useful product features (e.g., downloading a free application and later paying for additional application features). Freemium is a portmanteau of *free* and *premium*.
Frenemy, Frienemy	A *frenemy* or *frienemy* is a person with whom one is friendly despite a fundamental conflict, dislike, or rivalry. The term describes commercial, geopolitical, and personal relationships among groups and individuals, or a competitive friendship. Frenemy is a portmanteau of *friend* and *enemy*. See also, *competitor*, *archrival*, *competitor*, *nemesis*, and *rival*.
Frequency	*Frequency* is the rate at which something occurs or is repeated over a period or in a given sample. For example, for a heart beating at a frequency of 120 times a minute, its period, T, — the time interval between beats—is half a second (60 seconds divided by 120 beats). See also, *probability*.
Front End Engineering and Design	*Front end engineering and design* focuses the technical requirements as well as rough investment cost for the project. The FEED can be divided into separate packages covering different portions of the project. The FEED is used as the basis for bidding the contracts for project execution (e.g., engineering and procurement) and may be used as the design basis. See also, *design basis* and *detailed engineering*.
Front End Loading	*Front end loading* refers to the work process needed to prepare a project for authorization and execution. It is generally organized into these phases: 1. Business case development; 2. Scope development; and, 3. Project definition and planning. Various industries use different terminology to define this phrase; however, the intent is similar.
Front Line Personnel	*Front line personnel* are: • Managers who interact directly with personnel; or, • Workers who interact directly with customers or are directly involved in producing a product or providing a service.
Front Office	The *front office* refers to the customer-facing functions of a business. For example, traditional customer service, experts who provide advisory services, reception, and sales. Front office functions generate most revenue for a business. See also, *back office* and *middle office*.
Fuel Cell	A *fuel cell* is a device in which chemical potential energy (i.e., energy stored in molecular bonds) is converted directly into electrical energy by the oxidation of a gaseous fuel (e.g., hydrogen) or liquid fuel (e.g., methanol). A fuel cell is used to power equipment in remote locations or electric vehicles, or as source for back-up or emergency power.
Full Penetration	A *full penetration* weld is where the fused joint is completely solid with no gaps after welding. There are many ways to accomplish this (e.g., multiple weld passes), but the final product is solid metal at the joint. This may be specified as a complete joint penetration weld. This type of weld is required for joints subjected to high stresses because this type of weld has higher strength than a partial penetration weld. See also, *partial penetration*.

Phrase or Term	Definition
Full-time Equivalent	*Full-time equivalent* is the ratio of the total number of paid hours during a period (part time, full time, and contracted personnel) to the number of working hours in that period (e.g., 40-hour work week). For example, the ratio 1:40 indicates one person working full-time for 40 hours per week. Ratio units are full-time equivalent units or equivalent personnel working full-time. In other words, one full-time equivalent is equal to one person working full-time at 40 hours per week or two personnel working part-time at 20 hours per week. See also, *labour force* and *labour hour*.
Function Test	A *function test* analyzes equipment functionality and operability, and ultimately demonstrates compliance with purchase order requirements.
Functional Lead, Functional Manager	*Functional lead* and *functional manager* are personnel who have management authority over an organizational unit—such as a department—within a business, company, or other organization. Functional roles have ongoing responsibilities, and are not usually directly affiliated with project teams, other than ensuring that goals and objectives align with the organization's overall strategy and vision. See also, *lead* and *management*.
Fundamental Laws of Stupidity	The *fundamental laws of stupidity* are: 1. Always and inevitably everyone underestimates the number of stupid individuals in circulation; 2. The probability that a certain person will be stupid is independent of any other characteristic of that person; 3. A stupid person is a person who causes losses to another person or to a group of persons while deriving no gain and even possibly incurring losses themselves; 4. Non-stupid (i.e., smart) people always underestimate the damaging power of stupid people and constantly forget that at all times and places and under any circumstances to associate or deal with stupid people always turns out to be a costly mistake; and, 5. A stupid person is the most dangerous type of person. These laws were described by Italian economic historian Carlo Cipolla in the book *Basic Laws of Human Stupidity*. See also, *brain fart*, *brain freeze*, *Darwinian theory*, *creative error*, *hiding hand principle*, *lucky fool syndrome*, and *reverse Darwinism*.
Funishment	*Funishment* is an activity meant to be fun for personnel; however, for some, it is has the opposite effect: it is a distraction from work and painful obligation. For example, personnel participate in a safety-themed game of bingo that requires participants to identify examples of safe behaviour in their workplace (which may be at home). Funishment is a portmanteau of fun and punishment.
Fusion	*Fusion* is the process or result of joining two or more things together to form a single entity, by, or as if by, melting (e.g., a pipeline comprises the fusion of many joints of pipe). See also, *bonding* and *welding*.

Phrase or Term	Definition
Future Back Process, Future-back Process	The *future back process* or *future-back process* is a view of a desirable future state (e.g., what a business will achieve, what markets will be opened, or what moonshot will be attempted). This uses a *dual transformation* or an iterative approach to extend the lifespan of the current core business while identifying and seizing the new growth opportunities for the future. This is when vision is converted into action, by defining a future state and working backwards or translating the future state into a *to-do list*. A great future back strategy goes beyond the one-to three-year planning horizon that typifies most corporate strategies. This concept was developed in the *book Competing for the Future* by Gary Hamel and C.K. Prahalad. See also, *forecast planning*, *moonshot*, *present forward process*, *stop doing list* and *to-do list*.
Future Fit Hydrocarbons	*Future Fit Hydrocarbons* is an initiative by the Energy Futures Lab to develop different uses for Canada's hydrocarbon resources in the global transition to a low-carbon future. See also, *Bitumen Beyond Combustion* and *low-carbon*.
Future Self	Your *future self* is the ideal version of yourself by a given future date (e.g., I want to become president or a millionaire by 2025). Thinking about the present is to stay in the present. By contrast, planning for and thinking about the future can lead to realizing it, as it is envisioned. Thinking about the future is a positive exercise that boosts: • Health and well-being, and allows better states of being and living; • Hope and optimism, and allows better recognition of the positive, even amidst negativity and stress; and, • Positive emotions, and encourages contentment, excitement, gratitude, interest, joy, and peace. The general steps that help achieve your desired future self include: 1. Select a date in the future (e.g., five years from now) and imagine you have become your best possible self; 2. Remember details about how hard you have worked hard to accomplish goals, deepen relationships, and realize dreams. These things must be attainable (e.g., I want to become a lawyer) and not unreasonable (e.g., I want to become a space cowboy); 3. Write down details about your best possible self. It is very helpful to document mini goals that support and ultimately help accomplish the overall goal. Documenting goals is enormously helpful in reaching them. Without a destination, the journey meanders and does not end at a meaningful location. 4. Write a list of character strengths that your desired future self should have; and, 5. Think about what is required to become your desired future self. For example, if I want to become a competing body builder, I should have a plan for lifting weights. This concept was developed in the article *How to Meet Your Future Self* by Dr. Ryan Niemiec. See also, *24 character strengths*, *character*, and *goal*.

Phrase or Term	Definition
G	1 A B C D E F **G** H I J K L M N O P Q R S T U V W X Y Z
Galvanic Anode	*A galvanic anode* is the main component of a cathodic protection system used to protect buried or submerged metal structures from corrosion. The loss or sacrifice of the anode material gives a galvanic anode the alternative name *sacrificial anode*. See also, *anode* and *cathode*.
Galvanizing	*Galvanizing* is the process of applying a protective zinc coating to steel parts to prevent corrosion. With hot-dip galvanizing, the parts are submerged in a bath of molten zinc. Electroplating may also be used. See also, *electroplating*.
Game	A *game* is: • An activity that involves a procedure, strategy, or tactic. The goal of most games is to gain advantage over other game players and win the game (e.g., play a game of poker); • An area of expertise (e.g., performing complex calculations is the engineer's game); and, • An illegal or shady scheme, maneuver, or racket (e.g., blackmail). To *have game* or *be game* is to be eager, or willing and able to do something challenging or new. See also, *business wargame*, *games against nature*, *gamesmanship*, *gamification*, *lose-lose*, *sandbagging*, *win-win*, *wargame*, and *zero-sum game*.
Game Changer	A *game changer* is a new element, factor, or idea that significantly changes the current way of doing or thinking about something. For example, the Amazon online retail business model was a game changer for traditional brick-and-mortar companies because it introduced competition that previously did not exist. See also, *brick-and-mortar* and *status quo*.
Game Theory	*Game theory* is the study of interdependent decision making, or how individuals make decisions when their optimal choice depends on what others have chosen. This is a branch of mathematics that studies models of strategic interaction among rational decision-makers. A solution to a game describes the optimal decisions of the players, who may have mixed, opposed, or similar interests, and the outcomes that may result from these decisions. Game theory was originally developed by mathematician John von Neumann and economist Oskar Morgenstern in the book *The Theory of Games and Economic Behaviour*. They described economics as a game wherein players anticipate each other's moves, which they called game theory. It has applications in business, law, the military, and the social sciences. See also, *free rider problem*, *gamification*, and *Prisoner's Dilemma*.
Games Against Nature	*Games against nature* are problems that are modelled by games played against a disinterested opponent who makes moves at random and who has no strategic interest in the outcome. The effect is to add a player (Nature), whose practical role is to act as a random number generator. For example, the dealer in a game of poker assumes the role of the Nature and poker players play *against Nature*. See also, *business wargame*, *game*, *gamesmanship*, *gamification*, *lose-lose*, *sandbagging*, *win-win*, *wargame*, and *zero-sum game*.

Phrase or Term	Definition
Gamesmanship	*Gamesmanship* is the: • Art or practice of winning by using various ploys and tactics to gain a psychological advantage (e.g., making an agreement, proposal or request for proposal unnecessarily complex and difficult to analyze or comprehend, so that specific elements may be misunderstood or overlooked by others); or, • Use of ethically-dubious methods to gain an objective (e.g., making recommendations to clients, business partners, investors, or the public, while privately making different recommendations to others such as friends or specific business partners). See also, *business wargame*, *ethics*, *ethical*, *game*, *games against nature*, *gamification*, *lose-lose*, *sandbagging*, *win-win*, *wargame*, and *zero-sum game.*
Gamification	*Gamification* is the application of the engaging and fun elements of games to a business model. Online marketing techniques include the use of competition, ease of use, point scoring, rewards, and rules of play as to encourage engagement with a product or service and increase the conversion rate. For example, most LinkedIn (a networking website) features are gamified. From one-click endorsements and profile view statistics, to the profile progress bar and *Who's viewed my profile?* Component – these features evoke curiosity and delight, and drive action. Gamification has positive effects on individuals; however, contextual, and individual results vary. See also, *conversion rate*. See also, *business wargame*, *game*, *game theory*, *games against nature*, *gamesmanship*, *lose-lose*, *sandbagging*, *win-win*, *wargame*, and *zero-sum game.*
Gamma Ray	A *gamma ray* is a penetrating electromagnetic wave of high energy with a very short wavelength, produced by the radioactive nuclear decay, which can pass through many materials that are opaque to light. The French chemist and physicist Paul Villard discovered gamma radiation in 1900. See also, radiograph, radiographic test, and *X-ray*.
Gantt Chart	Henry Gantt invented the *Gantt chart* in the early 1900s. A modern Gantt chart is a bar chart that illustrates a project schedule over time, and identifies interdependent relationships between activities (past and future) and the status of the current project schedule. Used to plan projects of all sizes, a Gantt chart identifies what and when work is scheduled, and provides users the ability to view data in date ranges from one day, to the entire project timeline, and everything in between.
Gap Rod	A *gap rod* is a strip of metal used to bridge the gap in a weld joint that is poorly fit-up (e.g., too wide) or to produce a tack weld. The material used may be: • Bare welding electrode; • Coated welding electrode with the flux coating removed; or, • Metal that is the same as or like the base metal. The use of material to bridge a gap in a joint is typically prohibited by a specification or the welding procedure specification because of the likelihood of introducing defects into the weld (e.g., lack of fusion). A gap rod may be used where not specifically prohibited (e.g., non-code work). See also, *code of construction*, *joint*, *tack weld*, and *welding procedure specification*.

Phrase or Term	Definition
Gap-A-Let®	*Gap-A-Let®* is an engineered split ring device made of stainless steel that provides a pre-measured 1/16 in. gap for socket welds. See also, *socket weld*.
Gapminder	*Gapminder* is a non-profit museum on the Internet that promotes achievement of the United Nations' Millennium Development Goals and sustainable global development. It was founded in Stockholm and registered as a foundation by Ola Rosling, Anna Rosling Rönnlund, and Hans Rosling on February 25, 2005. See also, *factfulness, Millenium Development Goals, and wealth pyramid*.
Garbage in Garbage Out	*Garbage in garbage out* refers to how computers and software will unquestioningly process unintended, even nonsensical, input data (*garbage in*) and produce undesired, often nonsensical, output (garbage out). This principle highlights the need for input validation. See also, *garbage in gospel out*.
Garbage in Gospel Out	*Garbage in gospel out* is a more recent expansion of the phrase *garbage in garbage out*. Both phrases are synonymous. See also, *garbage in garbage out*.
Gasification	*Gasification* is an endothermic process that converts organic (i.e., biomass) or fossil-fuel-based carbonaceous materials into carbon dioxide, carbon monoxide, and hydrogen. This is achieved by reacting the materials at high temperatures (i.e., >400°C or >752°F), without combustion, using a controlled amount of oxygen, steam, or both. The resulting syngas or producer gas is itself a fuel. The power derived from combustion of this fuel is considered to be a source of renewable energy if the gasified compounds were obtained from biomass. See also, *coal*, *combustion*, *biomass*, *endothermic*, *fossil fuel*, *hydrogen*, *liquification*, and *syngas*.
Gasket	A *gasket* is a flat piece or ring of softer material that creates a mechanical leak-tight seal at the junction between two harder surfaces (e.g., between flanges in a piping system). Common gasket materials are graphite, rubber, paper, polytetrafluoroethylene (Teflon®), and silicone. See also, *leakproof* and *Teflon®*.
Gaslighting	*Gaslighting* describes an abusive a pattern of behaviour where the abuser uses persistent contradiction, denial, lying, and misdirection to make the victim feel unsure of their own sanity. These tactics are used to make the victim feel that they *imagined* other abuse, or simply to disorientate them. The term originated from the 1938 play *Gaslight* and its 1944 film adaptation. The protagonist's husband, Gregory, slowly manipulates Paula into believing she is going mad. The name comes from the portion of the film where Gregory uses the gas lights in an upstairs apartment, causing them to dim in his own apartment. Throughout the film version of the story, Paula sees the gaslights brighten and dim for no apparent reason. Gregory, however, convinces Paula that she is imagining the brightening and dimming of the lights. Gregory's subterfuge was switching the attic lights on and off to create gaslight flickers and ultimately manipulate Paula's belief in her own perception of reality.
Gate	A *gate* is an end-of-phase checkpoint at which decisions are made regarding whether and how to continue with a project.
Gathering Line	A *gathering line* is a pipeline, usually with a small diameter, used to gather product from the field to a central point. See also, *mainline* and *service line*.

Phrase or Term	Definition
Gender-based Analysis	*Gender-based analysis* is an assessment the *gendered* aspects of the impact of policies and programs (i.e., how impacts differ between men and women). This analysis was created by the *Status of Women Canada* to support gender equality goals. See also, *discrimination*, *diversity*, *gender-based analysis plus*, and *intersectionality*.
Gender-based Analysis-plus, GBA+	*Gender-based analysis plus* or *GBA+* is an assessment of the different aspects of the impact of policies and programs on men, non-binary people, and women. The *plus* is consideration for the intersectionality of other identity factors. See also, *gender-based analysis*.
General Arrangement Drawing	A *general arrangement drawing* is a technical illustration of the overall layout and main components of a building, equipment, or facility. The drawing may be rendered with a top view, side view, cut-out view or variation. It may include installation details, overall dimensions, mass, and references to design documents and other drawings, such as these: • Arrangement; • Assembly; • Detail; and, • Fabrication.
Generally Accepted Accounting Principles	*Generally accepted accounting principles* refer to a common set of accepted accounting practices, procedures, and standards that companies and their accountants must follow when compiling financial statements. It is a combination of authoritative standards and commonly-accepted ways of recording and reporting accounting information, to improve the clarity of the communication of financial information. See also, *double-entry accounting* and *family accepted accounting principles*.
Generation	A *generation* is: • All the people born and living at about the same time, regarded collectively. It is the average period, generally considered to be about thirty years, during which children are born, grow up, and become adults (e.g., Generation X); or, • A cohort or social generations who are the people within a delineated population who experience the same significant events within a given period (e.g., Millennials). The description and identification of the population cohorts in the North America labour force are widely, but not universally, agreed upon by demographers and market researchers. Generally, they are: 1. Baby Boomers I Born between 1946-1954 and become adults (21) between 1963-1972. 2. Baby Boomers II or Generation Jones Born between 1955-1965 and become adults (21) between 1973-1983. 3. Generation X Born between 1966-1976 and become adults (21) between 1988-1994. 4. Generation Y, echo boomers, or millenniums Born between 1977-1994 and become adults (21) between 1998-2006. 5. Generation Z Born between 1995-2012 and become adults (21) between 2013-2020.

Phrase or Term	Definition
Generator Set, Genset	A *generator set* or *genset* is the combination of a diesel or gas engine, or a gas or steam turbine as a prime mover, packaged with a generator to produce electricity. This may be a fixed or mobile unit that includes some or all of these components: • A fuel supply (e.g., a supply line or a tank); • A generator voltage regulator; • A starter motor (e.g., using a battery, compressed air, or natural gas). • An automatic starting system and transfer switch (i.e., to disconnect the load if the utility power source fails and connects it to the generator); • Auxiliary systems (e.g., cooling, exhaust, noise control, and lubrication); • Configuration with single or multiple units; • Constant engine speed regulator (governor); and, • Control and paralleling panels, and remote monitoring. A genset may provide an alternate or standby (e.g., emergency) source of electricity as single or multiple units. See also, *packaged equipment*, *prime mover,* and *standby*.
Gibberish, Gobbledygook, Jibber-jabber	*Gibberish*, *gobbledygook*, or *jibber-jabber* is language that is: • Difficult to understand (i.e., made difficult to understand) but sounds important and official (e.g., *a set of socially shared meanings the artist chooses to make visible in the space of art* actually means *an artist has an art exhibition*); • Nonsense (i.e., not actual words) or appears to be (e.g., see Lewis Carol's Jabberwocky); or, • Unintelligible and wordy jargon (e.g., the controlled double-blind and randomized trial was in fact a recipe for false positives). See also, *buzzword, doublethink*, *jabberwocky*, *jargon*, *plain English*, and *word salad*.
Gig	A *gig* is an informal term that describes a job that lasts for a limited period (e.g., play a gig in band or provide a temporary consulting service). A gig is usually short term; however, it can evolve into a long-term position provided a business or project has a sustained need for the service. All gigs are jobs (i.e., a position of paid employment) but not all jobs are gigs (i.e., a position of temporary employment is a gig, not a job). A gig may be referred to as a main gig or a side gig depending upon the frequency of performed work (e.g., full time or part time). See also, *gig economy*, *independent contractor*, *liquid labour force*, *side hustle*, and *virtual assistant*.
Gig Economy	In a *gig economy* is a labor market characterized by the prevalence of freelance work or short-term contracts instead of permanent jobs. It is predicted that in the future, more of the labour force will temporary (i.e., not salaried company employees). A gig economy undermines the traditional economy of full-time workers who rarely change positions and instead focus on a lifetime career. See also, *circular economy*, *economy of scale*, *economy of repetition*, *hydrogen economy*, *knowledge economy*, *gig*, *linear economy*, *liquid labour force*, and *low-carbon economy*.

Phrase or Term	Definition
Girder	*A girder* is a support beam used in bridge and building construction. It is the main horizontal support of a structure which supports smaller beams. Girders often have an I-beam cross-section composed of two load-bearing flanges separated by a stabilizing web that is closed or open, but may also have a box or Z shape, or another form. Girders may be: • Concrete, which is precast; • Steel, which is bolted, rolled, or welded; or, • Wood, which is composite or laminated (i.e., engineered), or solid timber. See also, *member*.
Givers Gain®	*Givers Gain®* is a philosophy based on the law of reciprocity. In the context of networking groups, people who adopt this philosophy dedicate themselves to referring business to their fellow networkers rather than making their foremost concern getting business for themselves. In doing so, others naturally become eager to reciprocate. This concept was developed by the American business networking expert Ivan Misner in his article *'Givers Gain' Is a Standard, Not a Sword*. See also, *reciprocity* and *relational customer service*.
Glass Ceiling	A *glass ceiling* is an artificial barrier to advancement that prevents people (e.g., typically minorities and women) from advancing in a profession or being elevated to senior-level position within an organization. See also, *black ceiling*.
Global Positioning System, GPS	The *Global Positioning System* or *GPS* is a satellite-based radio-navigation system owned by the United States government and operated by the United States Space Force. Other global or regional satellite navigation systems are also available. GPS uses satellite signals to provide location, velocity, and time synchronization to fix the location of a radio receiver on or above the earth's surface (i.e., with an unobstructed line of sight to four or more GPS satellites). The GPS receivers released in 2018 can pinpoint a location to within 30 cm (11.8 in). GPS receivers are now widely available (e.g., in smartphones, vehicles, and watches) and used as a common means of navigating from between destinations. The GPS radio receiver is often called a GPS.
Global Warming	*Global warming* is the long-term heating of the Earth's climate system, observed since the pre-industrial period (i.e., between 1850 and 1900). Global warming is attributed primarily to human activities such as the burning of fossil fuels, which increases greenhouse gas levels in the atmosphere. See also, *climate change*, *greenhouse gas*, *Industrial Revolution*, and *Sinn paradox*.

Phrase or Term	Definition
Globalization	*Globalization* is the process by which businesses or organizations develop worldwide influence or start operating on an international scale. It is the growing interdependence of the world's economies, cultures, and populations, brought about by cross-border trade in products and services, technology, and the flow of information, investment, and people. It is considered by some to be a form of capitalist expansion with the integration of local and national economies into a global, unregulated market economy. Globalization has many: • Cons. For example, jobs are lost and transferred to lower-cost countries, multinational corporations exploit tax havens in other countries, and the rich get richer); and, • Pros. For example, it makes companies more competitive, lowers prices for consumers, and provides economic development for poor countries. See also, *hyperglobalization*, *internationalization*, *localization*, and *market economy*.
Glossary	A *glossary* lists and defines terms and phrases. The purpose of a glossary is to define terminology. A well-constructed glossary gives clear and simple explanations that communicate the basic and essential meanings of terminology that facilitates communication and understanding. See also, *body of knowledge*, *buzzword*, *dictionary*, *jargon*, and *plain English*.
Go and See (for Yourself), Genchi Genbutsu	*Go and See (for yourself)* is more of a mindset than a technique or tool be used. *Genchi Genbutsu* literally translates from Japanese as *real location*, *real thing* and it is a key principle of the Toyota Production System. Both phrases suggest that, to truly understand a situation, one needs to observe what is happening where work takes place. This means collecting data and facts at the actual site of the work or in front of the problem. In general, most conflict is based on differing perspectives. Thus, various personnel will see a problem differently. Because of this, arguments and disagreements about solutions are likely.
Go By	A *go by* is an informal expression that means: • To base an analysis, judgement, or opinion on a specific piece of information or observation, which is stated after *by*. For example, what features do you go by when you buy a vehicle?; • To believe in, hold true or valid, or rely on some information. For example, to say I *go by* the numbers on the list; or, • To estimate (e.g., use the rough-in measurements as a *go by*). See also, *go/no go*.
Go/No Go Check, Go/No Go Test	A *go/no go check* or *go/no go test* refers to a pass/fail or yes/no principle using two boundary conditions or a binary classification. The test is passed only when the go condition is met, and the no go condition fails. The test gives no information as to the degree of conformance to, or deviation from, the stated conditions or classification. These tests can be used for production of equipment (e.g., a fit-up test) or statistical process control purposes. See also, *go by*.

Phrase or Term	Definition
Go/No Go Gauge, Pass/Fail Gauge	A *go/no go gauge* or *pass/fail gauge* measures equipment, including: • Physical dimensions (e.g., 50 ±1.0 mm or 2 ±0.04 in.); or, • Resistor value (e.g., 100 Ω ±1%). See also, *go/no go*.
Goal	A *goal* is: • In sports, a pair of posts linked by a crossbar and often with a net attached behind it, that forms a space into or over which an object (e.g., ball or puck) is to be sent to score (e.g., soccer goalposts); • The end product that will be delivered when a project or assignment is complete (e.g., a document or other product); or, • The object of a person's ambition or effort, specifically an aim or desired result (e.g., graduate from college). In general, goals are less specific and more difficult to measure than objectives. Business goals are part of a larger process that starts with the mission and vision of the organization and end with specific SMART goals and objectives, and action plans that help move the business forward. See also, *deliverology*, *goal setting*, *objective*, *program*, *SMART*, and *stretch target*.
Goal Setting	*Goal setting* is the process of creating goals (e.g., SMART goals) and setting deadlines to achieve them. See also, *deliverology*, *goal*, and *SMART*.
Going South	*Going south* is an informal expression or colloquialism that indicates a problem is worsening. For example, perpetual lack of logistical support made the construction effort *go south*.
Gold Plating	*Gold plating* is a term used to describe: 1. Working on a project or task past the point of diminishing returns; 2. The addition of any feature not considered in the original project scope or product description; or, 3. The addition of any feature that does not provide an additional benefit for product functionality or service (e.g., using a higher grade of stainless steel when lower grade steel is adequate). This type of unnecessary addition is informally known as *bells and whistles*. See also, *diminishing returns*, *bells and whistles*, and *scope creep*.
Golden Weld	Not a preferred phrase – see *tie-in weld*.
Goldilocks Zone	A *Goldilocks Zone* is a range for something that is just right (e.g., the Goldilocks Zone for diesel engine's idle speed is between about 600 to 800 rpm). In astronomy, the Goldilocks Zone is the *habitable zone*, located in the orbital region around a star, in which a planet can possess liquid water and support human life. The origins for this phrase are from Robert Southey's fairytale *Goldilocks*. In it, one of three trespassing bears in Goldilocks' home determines that one of the three bowls of porridge in the kitchen is neither too cold nor too hot – but instead *just right*.

Phrase or Term	Definition
Good Faith, Bona Fide, Bona Fides	*Good faith*, *bona fide* (Latin), or *bona fides* (Latin) is to act with integrity, honesty, and lawfully in business, legal, and personal matters – without the intent to defraud others or be malicious. It is being decent, fair, open, reasonable, transparent, and trustworthy – regardless of the outcome or potential outcome of an interaction. A demonstration of good faith is when someone makes an agreement that they can reasonably fulfill. For example, it is good faith for a contractor to enter into a contract without the required number of personnel *only if* the contractor is reasonably certain that adequate labour force resources will exist and be available to complete the work when it needs to be performed. This phrase may also have specific legal meanings, such as for compliance with reasonable commercial standards. See also, *bad faith*, *ethics*, *ethical*, *honest*, *transparent*, and *trust*.
Good is the Enemy of Great	*Good is the enemy of great* is a quote by Jim Collins. According to him: • The enemy of great is not bad, it is good; and, • If you think that good is *good* enough, you will never become great (i.e., remain only mediocre or okay). The meaning of this quote is similar to John D. Rockefeller's quote: *don't be afraid to give up the good to go for the great*.
Good Practice	A *good practice* is a good or wise thing to do. Best practices are learned through experience and documented for future use by future generations of personnel. See also, *best practise* and *smart practise*.
Goods	*Goods* are an inherently useful and tangible items (e.g., articles, commodities, materials, merchandise, products, supplies, and wares) that are produced from construction, fabrication, or manufacturing activities. See also, *cargo* and *products*.
Goose Chase, Wild Goose Chase	A *goose chase* or *wild goose chase* is: • An absurd, complicated, lengthy, or wild search for something nonexistent or unobtainable (e.g., searching the forest for a unicorn); or, • The fruitless, hopeless, or senseless pursuit of an end or object (e.g., *after my cell phone fell out of the bus, I went on a wild goose chase to find it*). These phrases are thought to originate from a futile attempt to catch a wild goose by chasing it. The origin of this phrase, oddly, is rooted in a type of 16th century horse race called a *wild goose chase*. It involves several horses racing behind the main lead rider at a predetermined distances and in a formation (like wild geese). Because all horses follow the lead horse, on an often unpredictable course, the phrase came to mean *doing something aimlessly or uselessly* (e.g., following the lead horse for no other purpose except to follow it). The phrase *wild goose chase* was first used as an idiom by playwright William Shakespeare in *Romeo and Juliet* in mid-1590s: *Nay, if thy wits run the wild-goose chase, I have done, for thou hast more of the wild-goose in one of thy wits than, I am sure, I have in my whole five.* See also, *fishing expedition*, *fool's errand*, and *witch hunt*.

Phrase or Term	Definition
Gordian Knot	A *gordian knot* is a metaphor for an extremely difficult, intricate, or involved problem, especially a problem unsolvable in its own terms (e.g., untying an impossibly tangled knot of rope). It can be solved easily, however, by finding an approach to the problem that renders the perceived constraints of the problem moot (i.e., by cutting the knot with a knife). The Gordian Knot is a legend of Phrygian Gordium, who was associated with Alexander the Great.
Gotcha Journalism, I Got You Journalism	*Gotcha journalism* or *I got you journalism* is an interview technique designed to entrap or trick an interviewee into making a statement that is nonsensical or stupid, or is damaging or discredits their cause, character, integrity, or reputation. See also, *gotcha management*, *gotcha mentality*, and *gotcha question*.
Gotcha Management, I Got You Management	*Gotcha management* or *I got you management* is a management approach that focuses on catching personnel doing something – anything – wrong. It is demoralizing management style for personnel. This management style finds and punishes errors but fails to reward correct behaviour (as described the book The One Minute Manager by Ken Blanchard and Spencer Johnson). A manager who focuses on mistakes and fails to acknowledge good work is unlikely to succeed as a coach, leader, or team organizer. See also, *gotcha journalism*, *gotcha mentality*, and *gotcha question*.
Gotcha Mentality, I Got You Mentality	A *gotcha mentality* or *I got you mentality* is a mindset that attempts to disgrace, embarrass, or expose someone caught making a mistake or with a moral put down. Someone with this mentality seeks to make themself feel and look good because of the errors or failures of others. See also, *gotcha journalism*, *gotcha management*, and *gotcha question*.
Gotcha Question, I Got You Question	A *gotcha question* or *I got you question* is a question that has no correct answer and is designed to explain, rationalize, or reveal a personal vulnerability. For example, a gotcha question often asked at job interviews is: *what is your biggest weakness*? This type of question may be used to attack someone and is commonly seen in gotcha journalism. See also, *gotcha journalism*, *gotcha management*, and *gotcha mentality*.
Government	A *government* is the: • Agency or system through which a political organization exercises authority and functions (e.g., a company or a state), typically over a district, municipality, nation, province, region, state, or other geographically-defined area; • Personnel that constitute the governing authority of a political group or organization; or, • Political customs, institutions, and laws through which the function of governing is accomplished. See also, *kleptocracy*, *meritocracy*, *non-governmental organization*, *oligarchy*, and *plutocracy*.
Government Job	A *government* job is: • Employment by a government organization (e.g., the Government of Canada); • Employment in the public sector; or, Sarcastically, a government job is work performed for one's own benefit using employer resources, while being paid for employment (e.g., to photocopy materials for an amateur sports league or repair a motorcycle fuel tank while at work). See also, *make-work job* and *public sector*.

Phrase or Term	Definition
Grade	*Grade* is: • A datum or reference level, especially ground level; • A material classification used to distinguish different types of steel based on their unique properties; • A classification. There are many grading systems in use and definitions vary per specific application, industry, product, or standard. Commercial and professional classifications are defined the same; however, these terms are used in different standards: o Commercial grade A durable product that will work better or longer with regular use, or is of average or inferior quality. For example, truck tires. o Consumer grade A product designed for use by consumers (e.g., for family, household, or personal consumption, enjoyment, or use). For example, a box of cereal. o Industrial grade A heavy duty product that is large, resilient, and tough, or has very high performance. For example, a heavy lift military helicopter. o Professional grade A durable product that will work better or longer with regular use and is superior to regular grade. For example, microphones in a professional recording studio. • The degree of inclination of a road or slope. Steels are classified by their composition and physical properties as grades. These grades were developed by many standards organizations including AISI/SAE (American), BS (British), CSA (Canadian), DIN (German), EN (European), ISO (International), JIS (Japanese), and UNS (North America)(International). See also, *material*, *material classification*, and *type*.
Grandfather	To *grandfather* is to exempt someone or something (e.g., an activity or practice) from a new law or regulation. For example, each time the electrical code is revised, existing wiring may be *grandfathered* on the assumption it was installed correctly. The regulation may only require that existing wiring to be updated if a renovation or upgrade is made.
Grant	A *grant* is an asset (e.g., land or money) given by an individual, government, or organization to others for a particular purpose. To *grant* is to: • Agree or concede that something is true; • Bestow or transfer formally; or, • Consent to allow or give something that is requested. See also, *grantcoin*.

Phrase or Term	Definition
Grantcoin	*Grantcoin* is: • New imaginary currency conjured from nowhere by the Canadian government that is used to fund programs geared towards Canada's climate pledges (e.g., The Paris Agreement). Positive examples are programs to drive down energy consumption, reduce fugitive methane emissions, or tree planting. Negative examples are programs deemed green but are uneconomic due to a low efficiency or a poor return on investment such as the Crescent Dunes solar facility. This use of the term was coined by Terry Etam in his article *From Bitcoin to Canada's new Grantcoin – the earth-shaking, energy-sucking impact of The New Money*; and, • The first cryptocurrency to be distributed and managed by a non-profit organization. It was created in 2016 and in 2017 was renamed Manna. See also, *altcoin*, *bitcoin*, *cryptocurrency*, *grant*, *green*, *green elephant*, *non-profit*, and *Paris agreement*.
Granularity, Graininess	*Granularity* or *graininess* refers to the extent to which a material or system is composed of distinguishable pieces or grains. It can either refer to the extent to which a larger entity is subdivided, or the extent to which groups of smaller indistinguishable entities have joined to become larger distinguishable entities. It is the: • Level of detail considered in a model or decision-making process; or, • Scale present in a set of data or another phenomenon. The greater the granularity, the deeper the level of detail. For example, a kilometre broken into centimetres has finer granularity than a kilometre broken into metres. By contrast, centimetres or meters grouped as kilometres reflect coarser granularity.
Graphicacy	*Graphicacy* is the ability to present and understand information presented as charts, diagrams, graphs, maps, photographs, plans, sketches, and other non-textual formats. It is also the communication of spatial information that cannot be conveyed adequately by numerical, textual, or verbal means (e.g., the blueprint plans of a city or the piling locations for a drainage network). Graphicacy is embraced by cartography, computer-graphics, graphic arts, geography, and photography. Numbers and words are still conventional methods of communication because they are easy to produce, distribute, analyze, and disseminate when compared to graphics. Advances in information and communications technology and visualization techniques continue to increase the accessibility, importance, use of graphics, and the importance of graphicacy.

Phrase or Term	Definition
Great Depression	The *Great Depression* was the greatest and longest economic recession in modern world history. It began with the United States stock market crash on October 29, 1929 (known as Black Tuesday) and did not end until 1946 after World War 2. This unprecedented depression was caused by sudden declines in economic credit and demand, and a widespread drop in world commodity prices. These factors led to rapid declines in global trade and massive unemployment. The Great Depression is commonly used as an example of how quickly the global economy can decline. Due to a crippling drought that occurred in the Prairies of Canada and the Untied States during that time, the 1930s became known as the *Dirty Thirties*. Widespread losses of jobs and savings gave rise of populist political movements, led governments to be more active in the economy, and triggered the birth of many social welfare programs. See also, *depression*, *populism*, and *recession*.
Green	*Green* refers to: • A member of an activist or political party focused on environmental and social issues (e.g., an environmentalist); • A nauseated, pale, or sickly appearance; • Being deficient in experience, knowledge, sophistication, or training, or not fully qualified for or experienced (e.g., the welder is green); • Covered by green growth or foliage; • Make less harmful or more sensitive to the environment (e.g., the company instituted a green process that emits no carbon dioxide); • Not cured (e.g., concrete), mature (e.g., personnel), or ripened (e.g., fruit); • Not fully processed or treated or not in condition for a particular use (e.g., green lumber); • Someone who is envious or jealous (i.e., green with envy); • Tending to preserve environmental quality (e.g., by being biodegradable, nonpolluting, or recyclable); or, • The colour between blue and yellow in the spectrum (e.g., coloured like grass or an emerald). See also, *environmentalism* and *greenwashing*.
Green Building, Green Construction, Sustainable Building	*Green building*, *green construction*, or *sustainable building* refers to a structure and the application of processes that are environmentally responsible and resource-efficient throughout a building's life cycle – from planning to design, construction, operation, maintenance, and renovation, to demolition. This requires close cooperation between architects, engineers, contractors, and clients at all project stages. This practice expands and complements the classical building design concerns of comfort, durability, economy, and utility. See also, *leadership in energy and environmental design*.

Phrase or Term	Definition
Green Elephant	A *green elephant* is a: • Carved jade piece of jewelry or talisman considered by many as a symbol with many positive characteristics (e.g., the ability to strengthen love, luck, or power, or to increase feng shui); or, • A green energy project that is a white elephant. For example, the CAD$ 1.34 billion (USD$ 1 billion) Crescent Dunes solar facility in the desert between Las Vegas and Reno, Nevada was designed to generate enough energy to power a city the size of Sparks, Nevada (population 100,000), but it never came close. It used concentrated solar power technology comprised of 10,000 mirrors constructed in a spiral almost 3.2 km (2 mi) wide. By the time it opened in 2015, it was already obsolete. The efficiency of the less-expensive photovoltaic panel had already surpassed its technology by producing the same amount of power at a fraction of the cost. See also, *white elephant*.
Green Energy	*Green energy* is collected from renewable resources (e.g., geothermal heat, rain, sunlight, tides, waves, and wind), which are naturally replenished. For example, energy obtained from solar panels. See also, *greenwashing*, *renewable energy*, and *renewable resource*.
Green New Deal, New Deal	The *green new deal* is a legislative approach that aims to address climate change and economic inequality in the United States. It comprises economic reforms, social reforms, and public works projects. The *new deal* was a series of financial reforms, programs, public work projects, and regulations enacted by President Franklin D. Roosevelt in the United States between 1933 and 1939. It responded to needs for reform, recovery, and relief from the Great Depression. The green new deal combines Roosevelt's economic approach with *modern* ideas for massive investments in infrastructure and renewable energy projects to decarbonize the energy sector, and make the economy fairer and more just. The phrase was first used by Pulitzer Prize-winner Thomas Friedman in January 2007. See also, *build back better* and *Great Depression*.
Green Swan	A *green swan* is a metaphor for an event that comes as a surprise and has financially disruptive effect that could cause systemic financial crisis. For example, climate change which poses unprecedented and unpredictable risks that cannot be confronted with known decision-making processes. This is because existing climate-economic models and traditional backward-looking risk assessments cannot accurately anticipate how climate-related risks will occur. This concept was developed by Patrick Bolton, et al, in the book *The Green Swan - Central banking and financial stability in the age of climate change*. See also, *black swan theory* and *grey swan*.
Greenfield	A *greenfield* project involves new construction at site where there is no existing or operational facility.
Greenhouse Effect	The *greenhouse effect* is the trapping of the sun's warmth in earth's lower atmosphere due to the greater transparency of the atmosphere to visible radiation from the sun than to infrared radiation emitted from the earth's surface. See also, *carbon footprint*, and *greenhouse gas*.

Phrase or Term	Definition
Greenhouse Gas	*Greenhouse gas* is a gas that contributes to the greenhouse effect by absorbing infrared (i.e., longwave) radiation and radiating it back, to trap heat in the atmosphere. Greenhouse gases include carbon dioxide, chlorofluorocarbons, methane, nitrous oxide, ozone, and water vapour. Water vapour may stay in the atmosphere for days before precipitating out, while other greenhouse gases may stay in the atmosphere for years or centuries. See also, *carbon credit*, *carbon footprint*, *climate change*, *global warming*, *greenhouse effect*, and *methane*.
Greenhouse Gas Emissions	*Greenhouse gas emissions* are gases and other particles that are released into the atmosphere from burning fuels and other processes. Generally, these emissions are from automobiles, industrial processes, and power generation. See also, *carbon leakage*, *greenhouse effect*, *greenhouse gas*, and *Greenhouse Gas Protocol*.
Greenhouse Gas Protocol, GHG Protocol	The *Greenhouse Gas Protocol or GHG Protocol* is a global standard, developed by the *World Resources Institute*, that informs companies and organisations how to manage, measure, and report greenhouse gas emissions. See also, *greenhouse gas*.
Greenwashing, Green Sheen	*Greenwashing* or *green sheen* is the misleading act of a company, government, industry, individual, or organization to promote false or unjustified (e.g., deceptive or exaggerated) environmentally-friendly practices, products, and services through advertising, branding, mislabelling, packaging, or public relations. This strategy is used to persuade the public that an organization's activities, policies, or products are environmentally friendly and therefore better than competitors. For example, a company involved in greenwashing might make claims that their products are made from 100% recycled materials or have huge energy-saving benefits when this is completely or partly false. Another example: in April 2017 a United States federal judge ordered German vehicle manufacturer Volkswagen (VW) to pay a criminal fine of USD$ 2.8 billion for *rigging diesel-powered vehicles to cheat government emissions tests*. As of June 2020, the scandal, known as Dieselgate, cost VW USD$ 33.3 billion in buyback costs, fines, financial settlements, and penalties. Other vehicle manufacturers have faced similar allegations in recent years, including BMW, Chevrolet, Ford, and Mercedes-Benz. Another form of greenwashing is for a supplier of fossil fuel energy to purchase renewable energy guarantees of origin certificates issued by a green energy facility (e.g., hydroelectric power station or wind farm) and then claiming their energy use is green. Greenwashing is a play on the term *whitewashing*. See also, *green*, *green energy*, *whitewashing*, *wind farm*, and *woke washing*.

Phrase or Term	Definition
Grey Area	A *grey area* is: • A topic that is not clearly one thing or the other; open to interpretation; • An ill-defined situation or field not readily conforming to a category, an existing policy, or set of rules. For example, a company makes changes to its organizational structure by adding, changing, or eliminating roles, leaving some responsibilities and tasks as unassigned or undefined. The unassigned or undefined roles become a grey area because the company did not explicitly make it clear who is responsible for the roles; or, • An area between two mutually exclusive categories or states, where the border between the two is fuzzy (e.g., not illegal or legal). For example, sovereignty of the South China Sea. See also, *black and white* and *thinking mode*.
Grey Hair	The phrase *grey hair*: • Refers to an older person with extensive hands on or practical experience and knowledge (slang); or, • Means that someone has reached a mature age or lived a long life. To *get grey hair* means that grey hair was acquired as a result of being stressed, upset, or worried about someone or something. See also, *hands on*.
Grey Swan	A *grey swan* is a metaphor for an event that comes as a surprise and has a major effect, but unlike a black swan event, it was not inconceivable. For example, the outbreak of Word War I or II, or the 9/11 terrorist attacks were preceded by dangers and events that were known and understood. See also, *black swan theory* and *green swan*.
Gross Domestic Product, GDP	*Gross domestic product* or *GDP* is a monetary measure of the market value of all the final products and services produced by a country in a specific period (e.g., one quarter or year). It includes all additions to private inventories, government payments, investments, paid construction costs, private and public consumption, and the foreign balance of trade. As a broad measure of overall domestic production, GDP functions as a comprehensive scorecard of a country's economic health. This is useful data, especially when comparing national economies. It does not reflect differences in countries' cost of living and inflation rates; thus, it is not useful for comparing living standards between nations. See also, *gross happiness product*.
Gross National Happiness, Gross Happiness Product	*Gross national happiness* or *gross happiness product* is an index used to measure the collective happiness and well-being of a population. The term was coined in 1972 by Sicco Mansholt, a founder of the European Union and the fourth President of the European Commission. The implementation of a gross national happiness index was instituted as the goal of the government of Bhutan in its constitution on July 18, 2008. See also, *gross domestic product*.
Ground	*Ground* or *earth* is the reference point in an electrical circuit from which voltages are measured, a common return path for electric current, or a direct physical connection to the earth. Electrical circuits may be connected to ground for several reasons. For example, this allows any stray or generated electrical current within the equipment to be transferred to a safe and convenient location. See also, *grounding lug*.

Phrase or Term	Definition
Ground Disturbance	A *ground disturbance* is an activity, operation, or work that results in disturbance of the earth, or a reduction in the depth-of-cover over a buried asset (e.g., a pipeline). Protecting buried assets from damage preserves the integrity of underground infrastructure, protects the environment, and promotes safety for personnel and the public. Ground disturbance is typically controlled by various acts, codes, guidelines, and regulations to ensure that excavation or exposure of buried assets, and their operation, is done responsibly and safely.
Ground Truth	*Ground truth* has many meanings, including: • Factual data provided by direct observation (i.e., empirical evidence) instead of by inference or remote sensors; • A military slang for facts that comprise an actual tactical situation as opposed to relying on military reports, reconnaissance documents, or other military documentation; • Information obtained or verified in-situ, on location, or on site (e.g., for meteorology or satellite imagery, and other techniques that gather data at a distance); • Results validated for accuracy against the real world (e.g., for machine learning and statistics); and, • The actual location of an object on the ground compared to the coordinates estimated by a location method (i.e., a geographical information system). See also, *truth*.
Grounding Lug	A *grounding lug* is a mechanical device that is fastened to equipment for the purpose of allowing it to be electrically grounded. Typically, a copper cable is fastened to the grounding lug and a grounding system (e.g., the earth). See also, *lifting lug*.
Grounding System	A *grounding system* connects equipment or specific parts of an installation with the earth's surface for safety and functional purposes. See also, *ground*.
Group of 77, G77	The *Group of 77* or *G77* is a coalition of 134 developing countries designed to create an enhanced joint negotiating capacity in the United Nations and promote its members' collective economic interests. Founded in 1964, it's namesake is from the original 77 founding members. See also, *Group of Eight*, *Group of Eight + Five*, *Group of Seven*, *Group of Six*, *Group of Twenty*, and *world model*.
Group of Eight, G8	The *Group of Eight* or *G8* was an intergovernmental political forum that existed between 1997 and 2014, when it incorporated Russia into the Group of Seven. When Russia was expelled in 2014, it reverted to its original name: Group of Seven. See also, *Group of 77*, *Group of Eight + Five*, *Group of Seven*, *Group of Six*, *Group of Twenty*, and *world model*.
Group of Eight + Five, G8+5	The *Group of Eight + Five* or *G8+5* was an intergovernmental political forum comprising all G8 nations and the governments of the five leading emerging economies of Brazil, China, India, Mexico, and South Africa. Founded in 2005, it was superseded by the G20 in 2019. See also, *Group of 77*, *Group of Eight*, *Group of Seven*, *Group of Six*, *Group of Twenty*, and *world model*.

Phrase or Term	Definition
Group of Seven, G7	The *Group of Seven* or *G7* is an intergovernmental political forum comprising all G6 countries, which was joined by Canada in 1976. The forum can refer to the member states or to the annual summit meeting of the heads of these governments. The G7 ministers also meet throughout the year (e.g., the environment, finance, and foreign ministers). See also, *Group of 77*, *Group of Eight*, *Group of Eight + Five*, Group of Six, *Group of Twenty*, and *world model*.
Group of Six or G6	The *Group of Six* or *G6* is an intergovernmental political forum comprising France, Germany, Italy, Japan, Russia, the United Kingdom, and the United States. It was formed in 1975 and, with the addition of Canada, became the G7 in 1976. See also, *Group of 77*, *Group of Eight*, *Group of Eight + Five*, *Group of Seven*, *Group of Twenty*, and *world model*.
Group of Twenty, G20	The *Group of Twenty or G20* is an international forum for governments and central bank governors from 19 countries and the European Union. Founded in 2019, the 19 countries are: Argentina, Australia, Brazil, Canada, China, France, Germany, India, Indonesia, Italy, Japan, Mexico, Russia, Saudi Arabia, South Africa, South Korea, Turkey, United Kingdom, and the United States. See also, *Group of 77*, *Group of Eight*, *Group of Eight + Five*, *Group of Seven*, *Group of Six*, and *world model*.
Groupthink	*Groupthink* is a psychological phenomenon that occurs when the desire for conformity or harmony within the group results in dysfunctional or irrational decision-making outcomes. Known as concurrence-seeking behaviour that may be instinctive or rationalized, group members minimize conflict to reach a consensus decision without critical evaluation of alternative viewpoints by actively suppressing alternative ideas and isolating themselves from outside influences. Groupthink requires that individuals avoid controversial issues and alternative solutions. Thus, with groupthink, there is loss of individual creativity and independent thinking. Furthermore, there is a severe lack of critical analysis because some ideas are deliberately excluded and not analyzed or considered at all. See also, *diffusion of responsibility*, *herd mentality*, *inclusivity*, *lemming effect*, *loserthink*, *tenth man*, *thinking mode*, and *tribalism*.
Growth Mindset	A *growth mindset* is the belief that abilities can be developed through dedication and hard work—intelligence and talent are the starting point. It is therefore believed that anyone can be good at anything, because abilities are entirely due to actions. People with this mindset love of learning because they believe that anyone can be good at anything; one's abilities are entirely because of one's actions. See also, *fixed mindset*.
Grunt	A *grunt* is: • A low, short guttural or similar sound (e.g., characteristic of a pig); • A low-ranking or unskilled worker; or, • Someone who performs grunt work. See also, *grunt work*.
Grunt Work, Donkey Work, Shit Work	*Grunt work*, *donkey work* (chiefly British English), or *shit work* is labour that is: • Boring, repetitive, or unglamourous (e.g., stacking boxes); or, • Difficult, menial, and thankless (e.g., shovelling manure); or • Simple and needs no training (e.g., picking up garbage). See also, *grunt*.

Phrase or Term	Definition
Guarantee	A *guarantee* is a formal assurance or promise, especially in writing, that: • An item or product will perform satisfactorily for a given length of time; or, • Specified conditions relating to a product, service, or transaction shall be fulfilled. See also, *warranty*.
Guideline	A *guideline* is general advice, outlines, principles, or rules that may be used to determine a course of action, which may or may not be mandatory (as directed by the owner of the guideline). A guideline provides best practices, ideas, information, and suggestions for successful strategies and techniques. A guideline may be adopted, modified, or rejected based on PO- or project-specific needs, or other instructions or requirements. See also, *charter*, *corporate instruction*, *knowledge transfer*, *plan*, *project charter*, and *project instruction*.
Gypsum	*Gypsum* is calcium sulfate dihydrate.

Phrase or Term	Definition
H	1 A B C D E F G **H** I J K L M N O P Q R S T U V W X Y Z
Hack	A *hack* (noun) is a: • Clever tip or technique for doing or improving something; • Creative or improvised solution to a computer hardware or programming limitation or problem; • Person who works only for money and with mediocre professional standards; or, • Short dry cough. To *hack* (verb) is to: • Chop or cut with heavy or rough blows; • Cope with, manage, or tolerate something; • Gain unauthorized access to a computer or telephone system; • Make chopping blows or strokes or play inexpert golf; • Work only for money and with mediocre professional standards; or, • Write computer programs for enjoyment or malicious purposes. See also, *life hack* and *hacktivism*.
Hacker	A *hacker* is someone who performs hacking. See also, *hack* and *hacktivist*.
Hacktivism, Hactivism	*Hacktivism* or *hactivism* is illegal computer hacking performed by a hacktivist over the Internet or using a digital device to promote a political or social agenda. For example, in 2014, North Korea's intelligence apparatus hacked the Sony Pictures studio to force them to abandon the release of *The Interview*, a comedic movie about a fictitious assassination plot against North Korea's leader Kim Jong-un. When Sony refused, the hackers destroyed the company's internal network and leaked private emails and studio data online. Targets may include corporate, criminal, government, or religious individuals and organizations. Hacktivism is a portmanteau of hack and activism. See also, *activism*, *cyberbullying*, *cybercrime*, *hack*, *hacker*, and *hacktivist*.
Hacktivist	A *hacktivist* is one who performs hacktivism. See also, *activist*, *hack*, *hacker*, *hacktivism*, and *social justice warrior*.

Halfway House	A *halfway house* is:
	• A place that is:
	  o Halfway through a progression (e.g., a partially-complete negotiation is a halfway house); or,
	  o Suitable to stop midway on a journey for refreshment or rest (e.g., to stop roughly halfway along a destination to eat in a restaurant).
	• A residence for those released from a hospital or (correctional) institution that facilitates reintegration into society. This may also be referred to as a recovery house or sober house; or,
	• An arrangement or thing that has some of the qualities of two different things (e.g., a sport utility vehicle is a halfway house between a sedan and a truck).

Phrase or Term	Definition
Hand Hole	A *hand hole* is a hole through which a person may insert a: • Hand to provide leverage for lifting; or, • Hand and arm for accessing an enclosed area or parts (e.g., for cleaning a pressure vessel or tank). See also, *manhole*.
Hand, At Hand, Helping Hand, Hired Hand, Lead Hand	A *hand* is: • An ability or skill (e.g., try your hand at a manual calculation to verify the computer-generated result); • An instrumental part (e.g., the manager had a hand in writing the plan); • Applause (e.g., give the presenter a hand); • Control or supervision (e.g., put a decision in the manager's hands); • One of two sides or aspects of an issue or argument (e.g., on one hand most people agree it is necessary; on the other hand it may be too expensive); • Something resembling a hand (e.g., the hands of a clock); or • The end part of a person's arm, beyond the wrist, including the palm, fingers, and thumb. A *helping hand* is aid or assistance given, especially with physical effort (e.g., please give me a hand). A *hired hand* is someone hired to do short-term manual work (e.g., the hired hand stacked hay on the farm). A *lead hand* is someone responsible for overseeing a group of workers assigned to them, to ensure tasks are completed in a safe and timely manner (e.g., the lead hand managed the group of labourers). A lead hand typically has less responsibility than a manager. *At hand* is within reach (i.e., near in time or place; the time is at hand) or currently receiving or deserving attention (e.g., the challenge at hand). This term and its variations have many other meanings. See also, *first-hand*, *hand hole*, handpick, *hands on*, *hiding hand*, *line management*, *second-hand*, and *supervisor*.
Handover	*Handover* describes a project phase where asset ownership (i.e., care, control, and custody) at site or a facility is permanently and formally given to the client. For example, commissioning turns over assets to operations for asset start-up. See also, *turnover*.
Handpick, Hand-pick	*Handpick* or *hand-pick* is an expression that means to: • Pick by hand as opposed to a machine process (e.g., pick cherries from a tree by hand); • Select carefully with a particular purpose in mind. For example, to handpick team members; or, • Select personally or for a specific purpose. See also, *cherry pick* and *low-hanging fruit*.

Phrase or Term	Definition
Hands On	*Hands on* refers to: • Having active involvement in decision-making and the management or organization of a project; • Having active participation (e.g., practical experience doing or using something) rather than having only theoretical knowledge (e.g., read, or watched); or, • Allowing handling or involving touching with the hands (e.g., training with equipment, materials, and tools). See also, *book smart*, *easy way*, *grey hair*, *hard way*, *on-the-job training*, *school of hard knocks*, *street smart*, and *tool time*.
Haphazard	*Haphazard* is: • Marked by a lack of arrangement, direction, order, organization, or planning; or, • Something that is aimless, accidental, by chance, fortuitous, random, or rare. See also, *hazard*.
Hard Copy	A *hard copy* is a document printed on paper. See also, *technical document*.
Hard Facing	Not a preferred phrase – see *weld overlay*.
Hard Way	The *hard way* means: • Bending a steel member across the strong principle axis (e.g., bending a 2" x 4" rectangular tube across the 4" axis); and, • Completing a task using more effort than necessary (e.g., painting a structure that does not need to be painted). The hard way is not the best way of completing a task because applying a solution the *hard way* requires more resources than applying conventional solutions. See also, *book smart*, *easy way*, *member*, *school of hard knocks*, and *street smart*.
Hardness Test	A *hardness test* measures the resistance of a material to deformation, indentation, or penetration by means such as abrading, drilling, impacting, scratching, wearing, or combination of these methods. Because there is no standard hardness scale, each test expresses results in a unique (and arbitrarily-defined) measure. Nonetheless, material hardness can be measured by a scale that is useful for that material. Hardness is measured by tests such as Brinell, Knoop, Rockwell, or Vickers. See this hardness conversion chart, which details 14 different hardness scales and correlates test values with tensile strength. For some metals, such as steel, the hardness and tensile strength are empirically-related. This means that there is an approximate relationship between the hardness and tensile strength of steel. Usually, the harder the steel, the higher its tensile strength (and the lower its ductility). This is very useful information because hardness can be measured readily by non-destructive examination. Tensile strength, for example, can only be measured by destructive testing. A durometer or similar instrument can measure the hardness of pliable materials, such as plastics and rubbers. See also, *destructive testing* and *non-destructive examination*.
Hardware Acceptance Test	A *hardware acceptance test* is a type of factory acceptance test that involves testing components. See also, *software acceptance test*.
Hardwood	*Hardwood* is the wood produced from a broadleaved trees such as ash, beech, or oak. See also, *softwood*.

Phrase or Term	Definition
Hashtag, #, Octothorpe	A *hashtag* or # is a metadata tag prefaced by the hash symbol (i.e., #). For example, #renewableenergy. Hashtags are used on blogging and photograph sharing services (e.g., Instagram, LinkedIn, and Twitter) as user-generated tagging that enables content that shares a subject or theme to be cross-referenced by other users. This makes content more discoverable and helps like-minded users connect with one another. A hashtag may include letters, digits, and underscores after the hash symbol, but no punctuation, spaces, or special symbols. The use of hashtags was first proposed in 2007 by Chris Messina. The symbol # is also known as an *octothorp*. Other synonymous terms include hash sign, hex, number sign, pound, and pound sign. Its origin is disputed and there is no known usage before it was adopted by Bell Labs in the late 1960s or early 1970s. Accounts about its origin are inconclusive.
Hazard	A *hazard* is an agent that has the potential to cause harm to a vulnerable target. Hazards can be both human induced and naturally caused. Some natural hazards, such as droughts and floods, can be caused by human activity. For example, droughts can be caused by over-irrigation or groundwater pollution, and floods can be caused by bad drainage facilities. The terms hazard and risk are often used interchangeably; however, in terms of risk assessment, they are two very distinct terms. A hazard is any agent that can cause harm or damage to humans, property, or the environment. A risk is the probability that exposure to a hazard will lead to a negative consequence, or more simply – a hazard poses no risk if there is no exposure to that hazard. See also, *haphazard*, *hazard and operability study*, *risk*, *risk assessment*, and *type A/B/C*.
Hazard and Operability Study, Hazop	A *hazard and operability study* or *hazop* is a structured and systematic examination of a complex operation, process or both (existing or planned) to identify risks to equipment, personnel, or both that is performed by a multi-disciplinary team of qualified personnel. The intent of a hazop is to identify design and engineering issues that may have been overlooked. A hazop categorizes a complex system into simpler, manageable sections. The team then reviews each section individually. A hazop is qualitative. It aims to stimulate the imagination of team members to consider and identify potential hazards and operability problems that were not considered previously. The team uses a list of standardized guide-words and process parameters to identify potential deviations from the design intent. Team members are expected to use good judgement, and leverage professional intuition, experience, and expertise. A hazop should foster a working environment of frank, thorough, constructive collaboration, positive thinking, and professional discussion. See also, *haphazard* and *qualitative*.

Phrase or Term	Definition

Head	A *head* is:
	• An end cap on a cylindrically shaped pressure vessel, which may be:
	  o Flanged and dished;
	  o Flat;
	  o Hemispherical; or,
	  o Semi-elliptical.
	• The height of a liquid column (e.g., in a pipe, tank, or vessel) that corresponds to a particular pressure exerted by the liquid column on the base of its container. It may also be called pressure head, static pressure head or static head – but *not* static head pressure.
	See also, *pressure vessel* and *shell*.

Headhunter	A *headhunter* is an individual or company that provides personnel recruiting services, especially at the executive level. They find talent and locate individuals who meet specific job requirements. The term may be referred to as an *executive recruiter* and the function called an *executive search*. They may have a roster of candidates for specific positions, recruit talent from competitors' personnel, or both.
Headshot	A *headshot* is a: • Bullet or gunshot that hits the head; or, • Digital image or photograph of a head and face. See also, *mug shot* and *talking head*.
Heads-up	A *heads-up* is a warning that something is going to happen (slang).
Heat	A *heat* is a batch of steel tapped from a melting furnace. See also, *heat number* and *heat analysis*.
Heat Affected Zone	A *heat affected zone* is the non-melted area of metal that has experienced changes in its material properties because of exposure to high temperatures from welding or high-heat cutting procedures. For some dissimilar metal welds, more than one type of filler metal is required. Further, there may be up to seven heat affected zones, with different material properties in each zone. See also, *weldment*.
Heat Analysis	*Heat analysis* is the chemical analysis reported by the steel producer as being representative of the heat of steel. This may or may not be representative of the product analysis because further processing may occur. See also, *heat* and *product analysis*.
Heat Balance	*Heat balance is:* • The distribution of the heat energy supplied to a thermomechanical system (e.g., a chemical or steam power plant) among the various drains upon it including useful output and losses; and, • The evaluation or record of such distribution. See also, *mass balance*.

Phrase or Term	Definition
Heat Exchanger	A *heat exchanger* is pressure equipment that transfers heat between two or more fluids, and drives cooling or heating processes. The fluids are separated by a solid wall (e.g., a tube) to prevent mixing or may be in direct contact with each other. There are many types of heat exchangers; however, these types are the most common: • Air cooled; • Plate or plate and frame; and, • Shell and tube. See also, *air cooled heat exchanger*, *heat transfer*, *plate heat exchanger*, *pressure equipment*, and *shell and tube heat exchanger*.
Heat Number	The *heat number* is a coupon identification number that is marked or stamped on a material after it is removed from the ladle and rolled at a steel mill. It is similar to a lot number, which is used to identify production runs of any other product for quality control purposes. See also, *heat* and *heat analysis*.
Heat Tracing	*Heat tracing* is a system used to maintain or raise the temperature of equipment (e.g., to maintain process temperature or protect equipment from freezing). Equipment is usually covered with thermal insulation designed to retain heat. Two media are used for heat tracing: 1. Electrical heat tracing uses electrical cables and heating elements; and, 2. Glycol or steam heat tracing uses tubing with glycol or steam. See also, *glycol*, *insulation*, and *winterization*.
Heat Transfer	*Heat transfer* is the movement of thermal energy (heat) from one thing to another thing of different temperature. It concerns the conversion, exchange, generation, and use of thermal energy and is classified by various mechanisms (e.g., conduction, convection, and radiation). See also, *heat exchanger*.
Heat Treatment	*Heat treatment* is the heating and cooling of a metal alloy to obtain specific hardening or softening properties (e.g., annealing, case hardening, precipitation strengthening, tempering, normalizing, and quenching). See also, *post weld heat treatment*.
Hedgehog	A *hedgehog* is a: • Military obstacle made from three heavy metal members (e.g., three I-beams) that may also be cast with concrete. Hedgehogs that serve as a deterrent to armoured vehicles, landing vehicles, ships, and tanks. These are also known as a Czech hedgehog; • Person who, when asked to change habits (e.g., adopt new digital technology) becomes defensive and resists adoption – not unlike a hedgehog (animal) who will curl into a defensive ball with their quills pointing out when threatened – making it difficult to engage in constructive dialogue. A human hedgehog will resist change by making excuses. For example, *digital solutions will not work here* or *it is better to do things manually*; and, • Spiny nocturnal mammal found throughout parts of Africa, Asia, Europe, and New Zealand. It rolls into a tight ball for self-defence, causing its spines to point outwards. See also, *beam*, *hedgehog concept*, and *member*.

Phrase or Term	Definition
Hedgehog Concept	The *Hedgehog Concept* was developed in the book *Good to Great: Why Some Companies Make the Leap ... and Others Don't* by Jim Collins. It is a simple concept that flows from a deep understanding about the intersection of three circles that describe what: 1. Best drives your economic or resource engine; 2. You are deeply passionate about; and, 3. You can be the best in the world at. See also, *hedgehog*.
Hedging	*Hedging* is de-risking or taking less risk by limiting or qualifying something by using conditions or exceptions. It protects against failing or losing by taking a counterbalancing action, to strike a balance between risk and safety. For example, making two bets at the same time when one pays off and the other does not. Or, it is limiting or offsetting the probability of loss from price fluctuations of commodities, currencies, or securities. For example, a foreign exchange currency hedge transfers the foreign exchange risk from the investing or trading company to a business that carries the risk, such as a bank. There is cost to the company for a hedge and the company will also give up any profit if the change in the exchange rate is favourable. See also, *de-risk* and *risk*.
Helix of Innovation Framework, Helix of Innovation Model	The *helix of innovation framework* or *helix of innovation model* refers to a set of interactions between various sectors that foster economic and social development in a knowledge economy. The innovation helical framework theory was first developed by Henry Etzkowitz and Loet Leydesdorff in the 1990s. Each sector is represented by a circle (helix), with interactions shown as overlapping sections. The three types of helix of innovation are: • Triple: 1. Academia (i.e., universities); 2. Government; and, 3. Industry. • Quadruple: 1. Academia (i.e., universities); 2. Government; 3. Industry; and, 4. Public. • Quintuple: 1. Academia (i.e., universities); 2. Environment; 3. Government; 4. Industry; and, 5. Public. Interactions between industries, governments, and universities resulted in new hybrid or intermediary institutions (e.g., technology transfer offices and science parks). See also, *knowledge economy*.

<table>
<tr><th>Phrase or Term</th><th>Definition</th></tr>
<tr><td>Helper</td><td>A helper is an unskilled labourer who assists a journeyman or trades person (e.g., a helper will buff or grind a weld joint or hand electrodes to a welder during welding). See also, journeyman and trades persons.</td></tr>
<tr><td>Herd Behaviour,
Herd Instinct,
Herd Mentality</td><td>Herd behaviour, herd instinct, herd mentality refers to:<ul><li>Animals
How each animal minimizes danger to itself (e.g., to evade a predator) by choosing a location that is as close to the centre of the group as possible. The group appears to act gregariously, – always moving and acting together, but its actions are the result of the uncoordinated behaviour of individuals acting in their own self-interest. This was described by biologist W. D. Hamilton in the paper Geometry For The Selfish Herd as the Selfish Herd Theory.</li><li>Psychology
Situations in which a group of individuals react coherently or act in the same way at the same time without there being any co-ordination between them. A collective fad or mania can have mild consequences (e.g., fashions) but others lead to mass hysteria (e.g., violent demonstrations). For example, investors may follow what they perceive other investors are doing, rather than their own analysis.</li></ul>This is also known as gang mentality, mob mentality, and pack mentality. People in such a group may be referred to as sheeple, which is a portmanteau of the words sheep and people. See also, diffusion of responsibility, groupthink and lemming effect.</td></tr>
<tr><td>Herding Cats</td><td>Herding cats is an idiom that denotes a futile attempt to control or organize a class of entities which are inherently uncontrollable – as in the difficulty of attempting to command many cats or technical writers into a group. Cats are not animals that form herds by instinct, unlike some other animals (e.g., cows). See also, Catch-22 and pushing a rope uphill.</td></tr>
<tr><td>Heroic Theory of Invention and Discovery.</td><td>The heroic theory of invention and discovery is the hypothesis that the principal authors of inventions and scientific discoveries are unique and heroic individuals (e.g., geniuses or great scientists). For example, Thomas Edison invented the motion picture camera, phonograph, and a practical electric light bulb. See also, multiple discovery.</td></tr>
<tr><td>Hidden Work</td><td>Hidden work is tasks required to complete a job that are not identified, planned, or resourced in the original plan or scope of work such as:<ul><li>Approval;</li><li>Coordination;</li><li>Decision-making;</li><li>Repairs and rework;</li><li>Rework; and,</li><li>Scope creep.</li></ul>Hidden work is often difficult to measure, overlooked, and underestimated. See also, scope creep and rework.</td></tr>
</table>

Phrase or Term	Definition
Hiding Hand Principle	The *hiding hand principle* is the idea that an invisible or hidden hand beneficially hides difficulties so that difficulties are underestimated. This is then supposedly offset by a roughly similar error in underestimating our ability to overcome difficulties, which then helps accelerate the rate at which problem-solving can occur. It encourages project decision-makers and promoters to proceed and not worry about cost overruns or other problems. See also, *creative error*, *Darwin Awards*, *fundamental laws of stupidity*, *hand*, *inverted Darwinism*, and *reverse Darwinism*.
Heighten the Smokestack	*Heighten the smokestack* describes using a taller flue gas stack to manage environmental concerns related to the disbursement of air pollution (e.g., noxious fumes and soot) caused by industrial processes (e.g., producing steam, refining processes, and smelting ores). Taller smokestacks release pollution at a higher elevation, farther away from animals, people, and plants it could harm, but does not address the source of the problem. To comply with environmental regulations during the 20th century, the height of some smokestacks reached 397 m (1,300 ft.). Building a taller flue gas stack is an easier or less expensive method of mitigating pollution compared to other methods (e.g., processing to neutralize or remove toxins); however, environmental regulations typically no longer permit this practice. An analogous example is how today some municipalities release untreated wastewater into waterways – using the heightened flue gas stack method – by building a pipeline to release raw sewage into the ocean away from shore – instead of treating wastewater at a wastewater treatment plant to address the source of the pollution.
High Capacity Factor	A facility with an output consistently near or above the nameplate capacity has a *high capacity factor*. See also, *nameplate capacity*.
High-Level	*High-level*: • Means at or of a level above that which is average or normal (e.g., he is a high-level student); • Describes being of high importance, influence, or rank (e.g., a high-level official); • Involves or relates to people of great authority or influence, or high administrative rank (e.g., the prime minister is a high-level public servant); or, • Describes ideas or operations that are abstract in nature; wherein the overall goals and systemic features are typically more concerned with the complex or wider macro system. For example, climate is a high-level description of the actions of the atmosphere and oceans. High-level also describes a project schedule that contains only general information (e.g., key activities, deliverables, and milestones). See also, *low-level*.
Highly Hazardous Chemical	A *highly hazardous chemical* is a substance classified by the American Occupational Safety and Health Administration as material that is toxic and reactive, and, if released, has a high potential for human injury.
Highly Leveraged, Over Leveraged	To be *highly leveraged* or *over leveraged* means that an asset or organization has more debt than equity. See also, *leverage*.

Phrase or Term	Definition
High-reward Project, High-value Project	A *high-reward project* or *high-value project* offers significant return on investment, value or both for stakeholders. For example, a mine that produces diamonds that are up to CAD$2,850 (USD$2,130) a carat, when the average market price is CAD$ 215 (USD$ 161) and the lowest market price is CAD$ 135 (USD$ 101). See also, *high-risk project*, *megaproject*, and *risk*.
High-risk Project, High-profile Project	A *high-risk project* or *high-profile project* has significant financial, legal and social risks. These projects have a high profile and sweeping impact inside and outside of the organization, and can pose significant threats to the ability of a project team to deliver. For example, attempting to drill an offshore ultra-deep well – 3,500 m (11,500 ft) deep. See also, *high-value project*, *megaproject risk*, and *unconventional oil and gas*.
Hindsight Bias, Knew-it-all-along Phenomenon, Creeping Determinism	*Hindsight bias*, *knew-it-all-along phenomenon*, or *creeping determinism* occurs when a person knows the outcome of something, and that knowledge skews perception of what they thought before they knew the outcome. It refers to the common tendency for personnel to perceive an event that has already occurred as having been more predictable than it was before the event took place. As a result, personnel often believe, after an event has occurred, that they would have predicted, or would have known with a high degree of certainty, what the outcome of the event would have been, before it occurred. This is a source of overconfidence regarding the ability to predict the outcome of a future event. For example, following the dot-com bubble in the late 1990s and the recession of 2008, many analysts tried to demonstrate how what seemed like trivial events at the time, were harbingers of future financial trouble. If the financial difficulty had been that obvious, it would likely have been avoided altogether. See also, *bias* and *hindsight is 20/20*.
Hindsight is 20/20, *Twenty-twenty Hindsight*	*Hindsight is 20/20* or *twenty-twenty hindsight* means that, only after an event has occurred, do participants come to an understanding of how it should have happened. For example, *I never would have accepted the promotion if I had known how much travel is required*. Using this phrase indicates that errors have occurred on a job that could only have been avoided if the personnel performing the job had previous experience on the same or similar project. See also, *20/20*, *hindsight bias*, and *second guess*.
Historic Emissions	*Historic emissions* are the total carbon emissions emitted during an organization's operational history. In 2020 Microsoft Corporation pledged to address climate change by, among other things, removing their historic emissions from the atmosphere by 2050. This was, in part, a response to Amazon's pledge in 2019 to meet the Paris Climate Change targets by 2040, 10 years ahead of schedule. See also, *Greenhouse Gas Protocol*, *net-carbon neutral*, and *scope 1, 2, and 3 emissions*.

Phrase or Term	Definition
Hitchens' Razor	*Hitchens' razor* is the dictum that what can be asserted without evidence can be dismissed without evidence. For example, a comment on a blog may be based on a conspiracy theory or misinformation that lacks credible evidence. Instead of trying to correct or refute the comment, educate the commentator, engage in a futile or nonsensical debate, or give a platform to these views, the blogger deletes the comment. This is not hypocrisy about being open to debate and welcoming an opposing opinion. It is a consideration for how and when a critical thinker can best avoid wasting time or debunk misinformation. It is a variation of the Latin proverb *quod gratis asseritur, gratis negatur*, which means *that which is easily asserted is easily negated*. See also, *ad hominem argument* and *conspiracy theory*.
Hockey Stick	A *hockey stick* is: • A long, thin implement with a curved end used to hit or direct a ball in field hockey or a puck in ice hockey; or, • A phrase used to describe a sharp increase in a variable following a period of little or no change (i.e., graphically the line connecting the data points represent a sharply upward-curving line resembling a hockey stick). This type of graph shows a dramatic shift such as in business results, global temperatures, medical results, or poverty statistics. A sudden increase may be an aberration or a permanent situation.
Hold Point	A *hold point* is a critical quality verification point where quality surveillance representative attendance is mandatory. Work may not proceed past a hold point without their attendance or acceptance in writing. The supplier shall provide advance notification for attendance to be scheduled.
Holdup	To *holdup* is to: • Call attention to, or single out, someone or something (e.g., for criticism or praise). For example, the engineering team's success was held up as an example of using creativity and collaboration to overcome challenges; • Delay a process or person, make them late, or prevent them from proceeding (e.g., the gridlock in traffic is a real holdup); or, • Rob someone at gunpoint (e.g., this is a holdup; everyone on the ground!). See also, *hold up*.

<table>
<tr><th>Phrase or Term</th><th>Definition</th></tr>
<tr><td>Holdback, Holdout</td><td>A holdback is:<ul><li>A sum of money that is not paid at the closing date to secure an incomplete or future obligation (e.g., 5 to 10% of the purchase price). This sum of money is withheld:<ul><li>Under certain conditions (e.g., if there are missing equipment and materials or incomplete turnover of documentation); and,</li><li>Until a certain condition is achieved (e.g., if construction is complete and satisfactory and if there are no liens or litigation).</li></ul>Holdbacks are common in contracts and purchase orders.</li><li>A thing serving to hold something else in place (e.g., a cable strap).</li></ul>To holdback (verb) is to retain or restrain something. For example, to hold back from expressing one's opinion during a discussion.
To be a holdback or holdout is to disagree with the common consensus of a group. For example, only two jurors are holdouts; all other jurors agree.
See also, contract, holdup, purchase order, and terms and conditions.</td></tr>
<tr><td>Hold Up</td><td>To hold up (verb) means to:<ul><li>Display something by holding it the above the head or waist (e.g., the graduates held up their certificates to celebrate their graduation);</li><li>Prevail in good state despite difficult circumstances or conditions (e.g., the runner's stamina will hold up during the marathon);</li><li>Remain true or valid, even after close examination (e.g., despite defendant objections, the prosecution's argument managed to hold up); or,.</li><li>Support something to prevent it from failing or falling (e.g., the father was able to catch and hold up his son who fell off his bicycle).</li></ul>See also, holdback and holdup.</td></tr>
<tr><td>Hold One's Horses, Hold the Horses, Hold Your Horses, Cool Your Jets</td><td>To hold one's horses, hold the horses, or hold your horses is to:<ul><li>Be more careful or patient; or,</li><li>Slow down or wait.</li></ul>Hold your horses and cool your jets are synonymous.</td></tr>
<tr><td>Holiday</td><td>A holiday is a discontinuity in a protective coating that exposes unprotected substrate to the environment. Protective coatings that prevent holidays are typically applied externally to pipelines and internally to piping systems, tanks, and vessels.</td></tr>
<tr><td>Holiday Test</td><td>A holiday test is a continuity or spark test applied to protective coatings to detect discontinuities (such as flaws, insufficient thickness, pinholes, and voids). The test involves applying voltage to the coating using a probe. Sparks at discontinuities identify the location of discontinuities. In addition, the testing device may have a red flashing indicator or an audible alarm that identifies discontinuities. See also, flaw.</td></tr>
</table>

Phrase or Term	Definition
Holistic Engineering	*Holistic engineering* is an approach that sees both curriculum or education and design or practice as an interconnected whole. It seeks to integrate the knowledge, methodologies, and skills of multiple disciplines into a collaborative effort for the good of the project. The lack of interdisciplinary co-ordination is often a cause for cost and schedule overruns. This strategy synergizes the services from various disciplines and delivers superior value by providing fully-coordinated and compatible solutions, which enable efficiencies and savings to be achieved, informed decisions to be made by all stakeholders, and performance to be enhanced. For example, in the mid-2000s, Stockholm considered ways to reduce traffic volume, including adding a bridge. However, there were already 57 bridges that connected the 14 main islands that constitute the city. Stockholm retained IBM to design, build and operate the technical operations that comprise the traffic system. IBM designed a *tax and drive* system in which vehicles were fitted with transponders and drivers charged a fee based on how long vehicles reside within the city. In the first month of operation, the system yielded a 25% reduction in traffic, removed 100,000 vehicles from roads during peak periods, and increased use of mass transit by 40,000 riders per day. Consequently, Stockholm no longer faced the costs of yet another bridge. Even better, it realized the benefits of conserving energy and reducing pollution using existing resources and infrastructure. Since then, other cities have implemented similar successful holistically-engineered systems. See also, *value-added engineering*.
Hollow Man Argument, Hollow Manning	A *hollow man argument* or *hollow manning* is a fictitious and weak argumentative position that the proponent and opponent would never encounter. These arguments frequently use ambiguous or vague phrasing, generalizations, misdirection, misleading statements, or weasel words, such as: • Intelligent and popular people believe … • Many people think; …; • Some argue …; • Some people say …; or, • There are those who feel … . This weak kind of argument may be realized through a broad, popular, or tenuous movement in general (e.g., the philosophy of *flatearthers*), and not guided by conventional logic, an individual, or organization. For example, Barack Obama stated *there are those who embrace a view that can be summarized in two words, anything goes*. See also, *flatearthers, nutpicking, steel man argument, straw man argument, weak man argument*, and *weasel words*.
Hollow Structural Steel	*Hollow structural steel* is metal shaped into a hollow and tubular form, which may be elliptical, rectangular, round, or square. This type of steel is used as members in bridges, buildings, and other structures, and for a wide variety of manufactured products. See also, *member, pipe, TKY joint*, and *tube*.
Honest, Honesty	*Honest* or *honesty* is sincerity and freedom from deceit, fraud, and untruthfulness. Telling the truth and acting accordingly is a good virtue and a measure of ethics and moral honour. Honesty is *not lying*. Honesty is key to retaining lasting success for clients, customers and personnel. See also, *ethics, ethical, transparent, the golden rule, trust*, and *vulnerable*.

Phrase or Term	Definition
Hookup	*Hookup* is a permanent or temporary connection to electrical power, a water line, or a similar service.
Hopium, Hopeium	*Hopium* or *hopeium* is: • An addiction to false hope or optimism (e.g., to hope that someone who is dead will come back to life); • Exaggerated claims about the benefits and profitability of a business opportunity with the hope to acquire money from a government or someone (e.g., acquire an investor's capital); or, • To wallow in self-pity with delusions of fame and fortune but never achieve these dreams (e.g., to romanticize struggle and hope others will save them). Hopium is a portmanteau of *hope* and *opium*. See also, *fool's gold*, *hyperbole*, *investor's capital*, *pipe dream*, *unobtainium*, and *when pigs fly*.
Horizon chart	A *horizon chart* is an alternative to ridgeline plots and small-multiple area charts that allow greater precision for a given vertical space by using coloured bands. These charts can also employ diverging colour scales to differentiate positive and negative values. See also, *ridgeline plot* and *risk cube*.
Horse Around	To *horse around* is to: • Act in a boisterous or lively way; • Engage in horseplay; or, • Fool around. See also, *horseplay*.
Horseplay	*Horseplay* is unnecessary boisterous, rough, or rowdy behaviour, such as hitting, name calling, or pushing. In the 16th century, *horse* was an adjective describing anything big, coarse, or strong – so horseplay meant *strong play*. Although it may be intended as harmless fun, it can cause anger, humiliation, hurt feelings, prevent personnel from performing their work by way of distraction, or result in unintentional physical harm or property damage. Most accidents are caused by unsafe acts – and horseplay is an unnecessary and unsafe act. Thus, horseplay is usually specifically prohibited by workplace safety rules by name. See also, *horse around*.
Horsepower	*Horsepower* is: 1. Euphemistically Effective power (e.g., computing, creative, or intellectual). 2. Technically A foot-pound-second unit of power, equivalent to 550 foot-pounds per second, or 745 Watts. See also, *bandwidth*, *bench strength*, and *brake horsepower*.
Horstman's Corollary	*Horstman's corollary* is that work contracts to fit within the time allocated for it. For example, if a report due on Friday, it will be produced by end of business day on Friday – regardless of if it can be produced earlier. See also, *Parkinson's law* and *Parkinson's law of triviality*.
Hot Bend	Not a preferred phrase – see *induction bend*.
Hot Bending	Not a preferred phrase – see *induction bending*.

Phrase or Term	Definition
Hot Microphone, Hot Mic	A *hot microphone* or *hot mic* is a microphone that broadcasts a speaker's voice when the speaker does not know that the microphone is on. A hot microphone may result in political gaffes. For example, during a sound check before a radio address to the United States in August 1984, President Ronald Reagan jokingly remarked: *My fellow Americans, I'm pleased to tell you today that I have signed legislation that will outlaw Russia forever. We begin bombing in five minutes*. This comment was not broadcast live; instead, it was recorded and later released to the public. Consequently, the Union of Soviet Socialist Republics denounced the comment and placed their armed forces on a temporary high alert. Another example is when someone uses the toilet during a video meeting with a microphone that is not muted. The same effect occurs with a live camera during a video meeting when participants can see participants' surroundings that are not intended to be seen (e.g., a pet cat walks in front of the camera or a crying toddler appears in the background).
Hot Tapping	*Hot tapping* is the process of attaching a branch connection to an in-service pipeline. See also, *branch connection*, *in-service*, and *out of service*.
Hot Work	*Hot work* is welding or other operations that involve the use of power generation or welding equipment for construction or demolition that makes use of a flammable atmosphere or material. This includes cutting, grinding, welding, and any activity that produces flames, heat, or sparks. See also, *cold work* and *hot work permit*.
Hot Work Permit	A *hot work permit* is permission to proceed with hot work. See also, *hot work*.
Hotwashing	*Hotwashing* describes the discussions and evaluations of an organization's performance following an activity, event, or exercise. It is an exchange of honest feedback about the content and processes to identify strengths and weaknesses. See also, *post-mortem*.
Hybrid Energy System, Hybrid Power, Integrated Energy System	A *hybrid energy system*, *hybrid power*, or *integrated energy system* combines electricity and heat generation, energy storage, energy conversion technologies, or a combination into a system that achieves better outcomes compared to standalone alternatives. A hybrid energy system may: • Minimize mismatch between energy generation and use; • Optimize cost of installation; • Optimize performance compared with individual technologies; • Reduce carbon emissions; and, • Reduce overall demand or waste. A hybrid energy system can facilitate the continuous production of electricity for longer periods when large-scale use of individual sources is limited by the amount of energy that can be generated. These systems are useful in remote areas because of advancements in renewable energy technologies and the cost of petroleum products. For example, a power supply comprised of a diesel generator, batteries, and solar panels is a hybrid energy system. This differs from a conventional power source (e.g., a single diesel generator is a conventional power source). See also, *integrated hybrid energy system* and *renewable energy*.

Phrase or Term	Definition
Hydraulic Fracturing	*Hydraulic fracturing* (also known as fracking, hydro fracturing, or hydro fracking) is a well stimulation technique in which underground rock is fractured by a pressurized liquid to facilitate the extraction of oil or gas.
Hydrocarbon	A *hydrocarbon* is a liquid or solid compound of carbon and hydrogen, which are the primary components of natural gas and petroleum. See also, *carbon*, *coal*, *fossil fuel*, *hydrogen*, *natural gas*, and *petroleum*.
Hydrogen, H_2	*Hydrogen* or H_2 is a colourless, flammable, odourless, and tasteless gas. It is the simplest chemical element in the periodic table. It is virtually non-existent in its free-form on earth and requires energy to liberate it from the material in which it is found. Hydrogen can be found on earth in the greatest quantity as water (chemical formula: H_2O). Hydrogen can store and deliver usable energy but it must be produced from compounds that contain it. It is therefore not an energy source, but an energy carrier that can deliver or store a tremendous amount of energy. Hydrogen can be produced from diverse domestic resources including biomass, fossil fuels, and water electrolysis with electricity. The resulting synthesis gas contains hydrogen and carbon monoxide, which then reacts with steam to separate the hydrogen. Hydrogen is used for chemical processing, fertilizer production, food processing, fuel, metal treatment, and petroleum refining. Hydrogen is often categorized by colour, based on how it is produced. The types include black, blue, green, grey, pink, teal, and turquoise. There are no established colours for hydrogen produced from biomass or varieties of grid electricity because the environmental impact of each of these production methods can vary considerably depending on the energy source, feedstock, region, technology, and type of carbon capture utilization and storage used. See also, *biomass*, *carbon*, *clean hydrogen*, *colours of hydrogen*, *fossil fuel hydrogen*, *hydrogen enriched natural gas*, *hydrogen embrittlement*, *hydrogen economy*, *hythane*, *liquid organic hydrogen carrier*, *power-to-gas technology*, *power-to-X*, *renewable energy*, *renewable hydrogen*, and *syngas*.
Hydrogen Bake-out, Out-gassing	*Hydrogen bake-out* or *out-gassing* is used to drive diffused atomic hydrogen out of steel pressure equipment that has been exposed to hydrogen (e.g., high temperature high pressure hydrogen service or as result of sour service such as wet H_2S), before welding. There are many factors that affect the duration and temperature of a hydrogen bake out (e.g., 200°C [392°F] or higher for a minimum of four hours). Trapped hydrogen can cause cracking in a weld or the base metal, or hydrogen embrittlement. See also, *base metal*, *hydrogen*, *hydrogen embrittlement*, and *sour service*.
Hydrogen Economy	The *hydrogen economy* is the use of hydrogen as a low-carbon fuel, particularly for heating, hydrogen vehicles, long-distance transport of energy, and seasonal energy storage. Governments are creating a low-carbon economy to help phase out fossil fuel consumption and limit global warming. When hydrogen is used as a fuel, its combustion releases only clean water and no carbon monoxide to the atmosphere. As of 2019, however, hydrogen was mainly used as an industrial feedstock – primarily to produce ammonia and methanol, and for petroleum refining. See also, *circular economy*, *economy of repetition*, *economy of scale*, *fossil fuel*, *gig economy*, *knowledge economy*, *gig economy*, *linear economy*, and *low-carbon economy*.

Phrase or Term	Definition
Hydrogen Embrittlement	*Hydrogen embrittlement* is the degradation of certain metals (i.e., aluminum alloys, carbon steels, stainless steels, and titanium alloys) caused by the diffusion of atomic hydrogen into the material. Some materials (e.g., aluminum, copper, and nickel alloys) have a low susceptibility hydrogen embrittlement. Hydrogen embrittlement can seriously reduce a material's ductility and load-bearing capacity, and result in catastrophic brittle failure at a stress level below the material's yield stress. This phenomenon is not completely understood, and its detection is difficult. The damage mechanisms related to hydrogen embrittlement include: • Hydrogen assisted cracking; • Hydrogen flaking; • Hydrogen-induced cracking; • Hydrogen stress cracking; and, • Stress corrosion cracking. It has been proposed that the ingress of hydrogen creates voids that become bubbles that build-up pressure within the material to propagate cracking. For hydrogen embrittlement to occur, these three conditions are required: 1. Presence and diffusion of hydrogen; 2. Stress; and, 3. Susceptible material. Hydrogen is introduced into a material during production operations (e.g., coating, cleaning, forming, plating, welding, or a combination). Hydrogen may also be introduced over time during in-service use (e.g., environmental exposure to process fluids such as chemicals or petrochemicals, soils, and water, or corrosion processes including coating corrosion, cathodic protection, and galvanic corrosion). This phenomenon was first described in 1875. See also, *hydrogen*, *low hydrogen*, *petrochemical*, and *weldability*.
Hydrogen Enriched Natural Gas	*Hydrogen enriched natural gas* is a mixture of compressed natural gas and 4 to 9% hydrogen by energy. See also, *hydrogen*, *hythane*, and *natural gas*.
Hydrogen Valley	A *hydrogen valley* is a geographical area (e.g., a city, industrial cluster, or region) where hydrogen technologies are combined to include the entire hydrogen value chain (e.g., distribution, production, storage, and use). This forms an integrated hydrogen economy or ecosystem – advantageous for projects that: 1. Produce and consume a significant amount of hydrogen; 2. Showcase technologies (e.g., a demonstration project or pilot plant); 3. De-risk projects that would be isolated or limited in scope or size; 4. Improve economical performance; and, 5. Enable technologies to be scaled up and viable. See also, *cluster*, *hydrogen*, and *pilot project*.
Hydrophile	A *hydrophile* is a molecule or molecular entity that is attracted to water molecules and tends to dissolve in water. See also, *hydrophobe*.
Hydrophobe	A *hydrophobe* is a molecule or molecular entity that seems to be repelled by water molecules. Strictly speaking, there is no repulsive force involved because a hydrophobe is the absence of attraction. See also, *hydrophile*.

Phrase or Term	Definition
Hydrostatic Test, Hydrotest	A *hydrostatic test* or *hydrotest* tests pressure equipment or systems for strength and leaks. See also, *pressure test*.
Hyperautomation	*Hyperautomation* is using a combination of automation tools, multiple machine learning, and packaged software that uses advanced technologies including artificial intelligence and machine learning, to enhance the efficiency of personnel and automate processes. It also refers to the sophistication, or steps, of automation (i.e., analyze, automate, discover, design, measure, monitor, and reassess). It is therefore an expansion of automation in both breadth and depth using existing technologies and tools to complete processes faster, more efficiently, and with fewer errors. For example, hyperautomation can automate the end-to-end multi-layered process of processing an invoice – instead of only a part of it (e.g., getting a robot to extract relevant information and enter it into an accounting system). See also, *artificial intelligence*, *automation*, *end-to-end*, and *robot*.
Hyperbole	*Hyperbole* is: • A literary technique where a speaker intentionally over-exaggerates for emphasis and effect (e.g., *when I saw my dinner bill, my wallet exploded*). • An obvious and intentional extravagant exaggeration that makes someone or something sound better or bigger than the truth. For example, my usual restaurant is *a thousand times more economical*.
Hyperglobalization	*Hyperglobalization* is a dramatic increase in the scope, size, and velocity of globalization. It includes all three main dimensions of globalization: economic, cultural, and political. See also, *globalization*.
Hypertext Markup Language, HTML, Hypertext Transfer Protocol, HTTP	*Hypertext Markup Language*, *HTML*, *Hypertext Transfer Protocol*, and *HTTP* is the standard markup language or software used to create most web browser documents. It can be assisted by technologies (such as Cascading Style Sheets) and scripting languages (such as JavaScript). It describes the structure of a web page semantically and originally included cues for the appearance of the document. Web browsers receive these documents from local storage or a web server and programmatically render them into multimedia web pages. See also, *Cascading Style Sheets* and *JavaScript*.
Hythane	*Hythane* is a mixture of hydrogen and methane (e.g., 20% hydrogen and 80% methane). Compared to methane, hythane can significantly reduce emissions of carbon monoxide and oxides of nitrogen. See also, *hydrogen*, *hydrogen enriched natural gas*, *natural gas*, and *methane*.

Phrase or Term	Definition
I	1 A B C D E F G H I J K L M N O P Q R S T U V W X Y Z
I Am Normal Paradox	The *I am normal paradox* is the perception is that the others in the world should have the same perception as you. For example, in this paradox, if you have an assertive negotiating style, you assume others will have a similar negotiating style. See also, *negotiating styles*.
Identical	Something that is *identical* has the same design, size, materials, and parts, and operates in the same service. See also, *cookie-cutter design* and *copy scope*.
Ideological Bias	An *ideological bias* is an implicit assumption or value embedded within discourse, social practices, or writing. For example, decisions by editorial staff and journalists driven by political opinion that distort factual presentation of news stories It is created by intense commitment to philosophical, political, or religious ideas. See also, *bias*.
Ideological Turing Test	An *Ideological Turing test* determines if one can state an opposing view as clearly and persuasively as the opponent. This is similar to the Turing test. See also, *steel man argument* and *Turing test*.
Ideology	An *ideology* is: • A system of beliefs, ideas, ideals, or philosophies, especially one which forms the basis of economic, political, or religious policy and theory; or, • The content or manner of thinking characteristic of a culture, group, or individual. Those who espouse an ideology may be criticized as narrow-minded and rigid. This term was coined in 1796 by French philosopher Antoine Destutt de Tracy.
Idiot Savant	Not a preferred phrase – see *autistic savant*.
Idling	*Idling* is a deliberate, temporary cessation of production (e.g., typically from a shortage of feedstock or labour disruption). While idling, a facility maintains readiness to quickly restart production. See also, *feedstock*, *labour disruption*, *mothballing*, *shutdown*, and *shuttering*.
If it ain't broke don't fix it	The phrase if it *ain't broke don't fix it* or – without contractions – *if it is not broken do not fix it* – means that something should only be changed (fixed) if there is a problem with it (it is broken). Similarly, if something is going or working well, *do not change it*. The gist is to avoid correcting, fixing, or improving something that is already adequate or performs adequately because the risk of introducing an impediment that may adversely affect performance may be greater than the benefits of improvement. This phrase suggests *it is better to just leave it alone and not make any changes that could potentially break it*. For example, an attempt to increase the speed of an operating fan that resulted in a broken fan – the status quo was preferable to no working fan. Now the fan needs to be replaced! In this situation, it would have been better to take no action. This mid-1900s folksy phrase, which originated in America, has a similar meaning to the expression *leave well enough alone*.

Phrase or Term	Definition
Iffy	*Iffy* is slang for something that is: • Comprised of one or more unknown conditions or qualities; • Doubtful, inconsistent, unreliable, of questionable quality, or may not be legal; • Not good or appropriate in some way; • Odd, strange, or uncommon, including actions and situations; • Reasonably accurate, but not precisely accurate; or, • Uncertain or unresolved.
IKEA Effect	The *IKEA Effect* suggests that when people invest time creating something at the beginning of a project, they are more vested in the outcome. This means that initiatives or projects are more successful when those involved participate in a meaningful way at the beginning of the effort. The phrase is derived from the name of Swedish manufacturer and furniture retailer IKEA, which sells furniture that requires assembly.
Imagineering	*Imagineering* is the implementation of creative ideas in a practical form. The term was coined by the American industrial Alcoa Corporation around 1940 and has been widely used in numerous disciplines and industries since then. Disney Enterprises, Inc. registered it as a trademark in 1990. This term is a portmanteau of imagination and engineering. See also, *engineering*.
Inclusivity	*Inclusivity* is the practice or policy of including people who might otherwise be excluded or marginalized, such as members of a minority group. For *the collective*, it is about creating a culture that accepts, embraces, respects, and values differences, and strives for equity. Homogeneous workplaces can be destructive to a company's labour force and profitability. This is because of groupthink, and how inclusivity can drive innovation and cultural, economic, and social vitality. Ultimately, when personnel are truly included in a workplace, they are more likely to share information, and participate in, and contribute to, decision-making. See also, *diffusion of responsibility*, *groupthink* and *diversity*.
Incomplete Joint Penetration	*Incomplete joint penetration* is a welded joint that has partial penetration. See also, *complete joint penetration*, *full penetration*, and *partial penetration*.
Incorporated	*Incorporated* means a company or other organization is a legal corporation. See also, *corporation*, *joint venture*, *partnership*, *sole proprietorship*, and *unincorporated*.

Phrase or Term	Definition
Incoterms®	*Incoterms®* are a standard set of terminology, created by the *International Chamber of Commerce*, that are used world-wide for defining key parts of freight forwarding. The Incoterms® rules feature abbreviations for terms (e.g., free on board, delivered at place, ex works, carriage and insurance paid to), which all have a very precise and universal meaning for buyers and sellers for the sale of goods around the world. Incoterms® 2020 is the most current version of the trade terms. These were previously updated in 2019, however, the 2010 edition is still in use. Although other systems for global trade exist around the world (e.g., Harmonized Tariff Schedule of the United States), Incoterms® rules are global in reach. Incoterms® rules do not include trade terms codified for national purposes (e.g., the less than truckload shipping rule of the United States). Unlike national trade policies, Incoterms® rules are universal, to provide business with clarity and predictability. The 2020 edition is available in 29 languages. For more information, visit: http://store.iccwbo.org/ See also, *goods*.
In-camera, In Camera, In Chambers	*In-camera*, *in camera*, and i*n chambers* (chiefly British English) means a private or secret meeting or proceeding in: • Legal settings The press and public are prohibited from observing a procedure or process (i.e., the opposite of an open court where all parties testify in public). • Non-legal settings A meeting or session (e.g., for business, diplomacy, or governance) that is confidential, so decisions and information are not publicly divulged or recorded in minutes (e.g., to discuss sensitive matters such as financial, personal, or business matters).
Independent Contractor, Freelancer, Self-Employed	*Independent contractor*, *freelancer*, and *self-employed* are phrases and terms that describe personnel who are employed by different employers at different times for short-term positions, as opposed to being permanently employed by one employer for a long-term position. These personnel may be represented by a temporary staffing agency or company that resells their labour, or they may work directly for the client (i.e., independently). See also, gig, *independent contractor*, *liquid labour force*, *permatemp*, *side hustle*, *sole proprietorship*, and *virtual assistant*.
Independent Forecast, Independent Review	An *independent forecast* or *independent review* is an audit (e.g., of a project budget, financial statements, plans, or a combination). An independent forecast is performed by a professional who does not have a stake in, or direct professional responsibility for, the work; thereby assuring independence and objectivity. This is based on the principle that personnel who routinely perform the work may have biases, overlook errors, or both. See also, *bias*, *cold eyes review*, *discipline review*, *independent project review*, *peer review*, and *value-added engineering*.

Phrase or Term	Definition
Independent Project Review	An *independent project review* is a project management mechanism that enables an unbiased, outside expert to assess project progress and health, and recommend corrections and opportunities for improvement. This review can be used in three ways: 1. To assess and evaluate a program or project at the earliest stages of development and ensure it is positioned for success; 2. As a program or project health check to validate status and proactively address emerging issues; or, 3. As a comprehensive assessment of a program or project, including root cause analysis of issues and the development of a plan of corrective actions. See also, *bias, independent forecast, peer review, and value-added engineering*.
Indigenous Peoples, Aboriginal Peoples	*Indigenous Peoples* or *Aboriginal Peoples* are phrases used to describe the three categories of peoples who are native to Canada: • First Nation People who primarily inhabit territories south of the Arctic. These were the first people to encounter sustained European contact, settlement, and trade. • Inuit People who primarily inhabit territories of the arctic. Their homeland, known as Inuit Nunangat, includes the ice, land, and water. • Métis People who are of mixed European and Indigenous ancestry who live mostly in the Prairie provinces and Ontario. Indigenous Peoples were erroneously described as Indian by Christopher Columbus, an Italian explorer who believed he had discovered the Far East. He gave the name *indios* (Indians) to the indigenous peoples he encountered there. Thus, the native people of the Americas today are often collectively (and erroneously!) called *Indians*.
Induction Bend	An *induction bend* is a bend (typically in pipe) that has been formed by induction bending a pipe into a specific shape (i.e., radius) in a bending machine at an elevated temperature for use in a pipeline or piping system. See also, *cold bend*, *induction bending* and *radius*.
Induction Bending	*Induction bending* is a method of bending material (e.g., typically pipe) at elevated temperatures, using electromagnetic energy (i.e., electromotive force or voltage is applied to create a changing magnetic field), in a bending machine. Also known as hot bending. See also, *cold bending*.
Inductive Logic, Inductive Reasoning	*Inductive logic* or *inductive reasoning* is a method of reasoning in which the premises are viewed as supplying some evidence for the truth of a certain conclusion. Basically, there is data, then conclusions are drawn from the data. For example, coins taken from a bag are pennies. Therefore, all the coins in the bag are pennies. Many dictionaries define inductive reasoning as the derivation of general principles from specific observations, though there are inductive arguments that do not have that form. See also, *abductive logic*, *chunking down*, *chunking up*, *deductive logic*, and *scientific method*.

Phrase or Term	Definition
Industrial Integrated Project Delivery	*Industrial integrated project delivery* is a project delivery method that uses high levels of stakeholder collaboration and integration during the design, development, and construction phases. Projects that use this method can be delivered with improved quality and a significantly shorter timeline to achieve business objectives with more certainty than projects that do not use this method. See also, *stakeholder*.
Industrial Internet, Industrial Internet of Things, Industry 4.0	The *industrial Internet*, *industrial Internet of things*, or *Industry 4.0* is the use of actuators and smart sensors to enhance industrial and manufacturing processes. It leverages the power of real-time analytics and smart machines to take advantage of the data that *dumb machines* have produced in industrial settings for years. The driving philosophy is that smart machines are not only better than humans at analyzing and capturing data in real time, but are better at communicating important information that can be used to drive business decisions faster and more accurately. The phrase Industry 4.0 (Industrie 4.0 in German) was coined in 2011 in the German government's high-tech strategy. See also, *dumb machine*, *Industrial Revolution*, and *Internet of things*.
Industrial Revolution, Technological Revolution, Digital Revolution, Information Age, Industry 4.0	The first *Industrial Revolution* (1760 to about 1820 or 1840) changed agrarian and handicraft economies to economies dominated by industry and machine manufacturing, especially with the increasing use of steam power. This revolution primarily affected Europe and the United States. The second Industrial Revolution or the Technological Revolution (1870 to 1914) was a phase of rapid industrialization that involved the standardization of manufacturing and production technology. The third Industrial Revolution or the Digital Revolution (1950 forward) marked the beginning of the Information Age, which was the change from analog electronics and mechanical technology to digital electronics. The fourth Industrial Revolution or Industry 4.0 is the ongoing automation of traditional industrial and manufacturing practices using smart technology (about 2011 forward). See also, *analog*, *digital*, *industrial Internet*, and *Information Age*.
Infinite Mindset	An *infinite mindset* is a way of thinking that understands, in business, sometimes one has a better product and sometimes the competition does. Thus, there is no such thing as being the best because there is no such thing as winning business. It is seen as an infinite game where the goal is to outlast and outdo oneself. This concept was developed by Simon Sinek in the book *The Infinite Game*. He asserts that five things are needed to lead an organization with an infinite mindset: 1. A just cause or higher purpose than making money; 2. Trusting teams where personnel dare to raise ideas and suggestions; 3. A worthy rival who can encourage you to be better; 4. A capacity for existential flexibility or willingness to change course completely if needed; and, 5. Courage to lead for the long game because success does not happen overnight. Simon believes that the ability to adopt an infinite mindset is a prerequisite for any leader who aspires to leave their organization better than they found it. See also, *abundance mindset*, *finite mindset*, *lose-lose*, *scarcity mindset*, and *win-win*.

Phrase or Term	Definition
Influencer	An *influencer* is a: • Person or thing that exerts influence on someone or something (e.g., affects or changes behaviour). For example, a manager can influence their team to exceed its annual sales targets. Influence in-person (e.g., a business unit manager or chief executive officer) does not translate directly into influence on social media (e.g., favouriting, liking, replying, retweeting, and sharing). • Person with a reputation for expertise and knowledge on a topic, who makes regular posts about that topic on social media, and generates large followings of engaged, enthusiastic people who pay close attention to their views. For example, Gary Vaynerchuk is a Belarusian-American investor and serial entrepreneur who was one of the Internet's earliest adopters of social media as a tool to build a lucrative following. He was number 1 on Forbes' first list of Top Tech/Business Influencers (2017). • Person with the ability to affect the purchasing decisions of others because of their authority, knowledge, position, or relationship to an audience on social media (i.e., to promote or recommend a product or service). For example, Portuguese professional footballer (soccer player) Cristiano Ronaldo's sponsored posts, including paid partnerships with Nike and his own clothing brand, are seen by over 240 million followers (2020). See also, *influencer marketing*.
Influencer Marketing	*Influencer Marketing* is a form of social media marketing involving endorsements and product placement from influencers, people, and organizations who have a purported expert level of knowledge or social influence in their field. Influencer content may be framed as testimonial advertising; influencers play the role of a potential buyer, or may be involved as third parties. This not a new phenomenon. Influencer marketing works because people are influenced by people, and find word-of-mouth to be the most trusted source for recommendations. Today more than ever, consumers are choosing to listen to each other over brands. This is why influencers are so powerful. They are trusted individuals with loyal followings across digital platforms. Influencers start conversations and make it easier for brands to target their audiences on a big scale. See also, *influencer*.
Infomercial	An *infomercial* is a program or video that: • Is an extended or television-show-length advertisement that often includes demonstrations, detailed discussion about a product or service, or both; or, • Promotes a product or service in an entertaining, informative and ostensibly objective way for the purpose of acquiring sales. An infomercial attempts to sell products or services. Infomercial is a portmanteau of *information* and *commercial*. See also, *infotainment*.

Phrase or Term	Definition
Information Age, Computer Age, Digital Age, Internet Age, New Media Age	The *Information Age*, *Computer Age*, *Digital Age*, *Internet Age*, or *New Media Age* (1950 forward) is a time period within which information became a commodity. With the advent of computer technology, information is now accessible, and quickly and widely disseminated. The onset of the Information Age is closely tied to the advancement and development of transistor technology, which is the fundamental building block of digital electronics. See also, *Industrial Revolution*.
Information Technology	*Information technology* is the use of computers, electronic devices, infrastructure, networking, processes, and storage to create, exchange, process, secure and store electronic data, often in the context of a business or enterprise. See also, *engineering technology* and *operation technology*.
Informative	*Informative* is descriptive, interesting, or useful information, which is provided to assist with conceptual understanding. It is additional or supplemental information, such as background, commentary, development, explanation, guidance, history, recommendations, relationships, statements (e.g., capabilities, permissions, or possibilities) and tutorials. These are descriptive ancillary requirements that do not compel compliance. See also, *empirical* and *normative*.
Infotainment	*Infotainment* is a program or video that entertains and informs but does not formally educate or instruct. Infotainment does not attempt to acquire sales directly; although, it may attempt this indirectly (e.g., a program to describe hardware and software installed in vehicles to improve drivers' experiences while driving). The term was coined during the convention of the Intercollegiate Broadcasting System in 1974. Infotainment is a portmanteau of *information* and *entertainment*. See also, *infomercial*.
Infrastructure	*Infrastructure* is capital assets (public facilities, private facilities or both, and systems) that are necessary for the economy of a city, country, or other area to function. These assets include energy distribution (e.g., oil and gas pipelines), power generation and distribution, telecommunications, transportation (e.g., bridges, railways, roads, and tunnels), and water or sewage systems. See also, *megaproject*.
Infrastructure as a Service	*Infrastructure as a service* is cloud computing service that provides virtualized computing resources over the Internet. See also, *anything as a service*, *cloud computing*, *content as a service*, *platform as a service*, *software as a service*, and *security as a service*.
Infringement	An *infringement* is: 1. The act of breaking or violating the terms of an agreement or regulation; or, 2. An encroachment or trespass on a right or privilege. For example, copyright infringement occurs when a copyrighted work is reproduced, distributed, performed, publicly displayed, or made into a derivative work, without the permission of the copyright owner. See also, *copyright*, *intellectual property*, and *plagiarism*.
In-house	Something that is done *in-house* is done or exists internally, within an organization, or without assistance from outside an organization (e.g., the project wrote the report in-house). See also, *outsourcing*.

Phrase or Term	Definition
Initial Contact Form	An *initial contact form* is used by the expediter to contact a supplier after purchase order award, to provide instructions and confirm, obtain, and/or share information including: • Confirm readiness to proceed with supply and delivery; • Document control procedures; • Job or work order numbers; • Key personnel names and contacts; • Locations for production and shipping; • No production before initial document review and project-supplier meetings; • No shipment with out inspection and/or shipping releases; and, • Scheduling and shipping dates. See also, *expediter*, *form*, and *purchase order award*.
Initial Service Test	An *initial service test* is a leak test that is performed when pressure equipment or a system is first put into operation.
Initialism	An *initialism* is an abbreviation comprised of initial letters that are pronounced separately (i.e., initial letters *do not* spell a word). For example, *CPU* as an initialism for central processing unit (i.e., a computer) because it is pronounced: SEE-PEE-YU. Generally, all initialisms may also be called abbreviations. See also, *abbreviation*, *acronym*, and *mnemonic*.
In-line Inspection	*In-line inspection* is the evaluation of pipes and pipelines with a tethered or non-tethered (smart) pipeline inspection gauge or pig. In-line inspection uses non-destructive examination techniques to detect pipe size (i.e., unequal diameter) and internal damage (e.g., corrosion or cracks). See also, *pipeline instrument gauge* and *non-destructive examination*.

Phrase or Term	Definition
Innovation	*Innovation* is the action or process of introducing something new (e.g., a device, idea, method, or product). It is often viewed as the application of a better solution that meets existing, new, or unarticulated in-demand needs (e.g., 3G replace 2G telephone networks). An innovation is something original and more effective and, therefore, may be ground-breaking and new. This results from the provision of more-effective business models, processes, products, services, or technologies for consumers (e.g., governments, individuals, markets, and society). Innovation is related to, but not the same as, creativity or invention. In the book *The E Myth*, Michael E. Gerber explains the concept. In his book, Gerber credits German American economist Theodore Levitt for the idea that creativity *thinks* of new things, but innovation *does* new things. He also explains that innovation should simplify things and make them easier; otherwise, it is complication, not innovation. Innovation is more likely to involve the practical implementation of an invention (e.g., an improved or new ability, like a supercharged engine for the Formula 1 racing industry) that makes a meaningful impact on consumers. Not all innovations require invention. For example, digital photography uses computer-based mechanical and photoelectric techniques, without the need for photographic film and paper, or wet bath chemical processing. The first digital signals were saved to magnetic tape in 1951 using the first video tape recorder. The first digital image was produced in 1957 by Russell Kirsch using a computer (an image of his son). The adoption of digital photography only became widespread when: 1. During the late 1990s, professional photographers transitioned to digital photography because digital files were needed to fulfill the demand for faster turn-around time than conventional methods would allow; and, 2. Beginning in the 2000s, the availability of cell phone cameras became widespread, particularly due to their use with messaging and social media websites. See also, *creativity* and *invention*.
In-process	*In-process* describes an inspection activity (e.g., document review or verification) that occurs during production, before completion and final inspection.
In-service	*In-service* means that the equipment or system is in operation (e.g., pressurized). See also, *out of service*.
Inside-out Thinking	*Inside-out thinking* means to look at a business with a subjective view, from the business' perspective. With this type of thinking, decisions, policies, processes, products, systems, and tools are designed and implemented based on internal thinking and intuition. Customer needs are not considered because decisions are based what is best for the business – not for the customer, and what is thought to be best suited for customer needs. For example, after years of impressive success, Toyota's focus shifted away from meeting customer needs toward the goal of beating General Motors and maximizing growth. Consequently, Toyota lost sight of its customer's needs and the quality of its products declined. See also, *customer relationship management*, *outside-in thinking*, *relational customer service*, *thinking mode*, and *transactional customer service*.
In-situ	*In-situ* refers to something (e.g., equipment, resources) situated in its original, natural, or existing place or position (e.g., the pipeline repair will be in-situ).

Phrase or Term	Definition
Inspection	*Inspection* is a quality surveillance activity performed by comparing inspection results with specified requirements to verify if conformity is achieved for a product or service (e.g., physically examining or measuring, reviewing documentation for one or more characteristics, or both).
Inspection Agency	An *inspection agency* provides personnel who perform inspection services as quality surveillance representatives per an inspection assignment. See also, *inspection agency coordinator*, *inspection assignment*, and *quality surveillance representative*.
Inspection Agency Coordinator	An *inspection agency coordinator* provides administration services for the inspection agency. They assign personnel who perform inspection services as quality surveillance representatives per an inspection assignment and issue their reports to the supplier quality surveillance coordinator. See also, *inspection agency* and *supplier quality surveillance coordinator*.
Inspection and Test Plan	An *inspection and test plan* is a quality document used to record and control all inspection and testing requirements relevant to a purchase order or scope of work. It details all inspections or tests, with referenced specifications or requirements and the acceptance criteria, and identifies the parties responsible for acceptance (e.g., signature and date for verification, witness, and hold points). This document lists the supplier's name, address, job number, relevant equipment, and project names and numbers.
Inspection Assignment	An *inspection assignment* is issued to a quality surveillance representative by the supplier quality surveillance coordinator. It provides detailed information about project expectations for inspection, and includes additional requirements that are not specified in procurement documents (e.g., schedule of visits, concerns for materials and equipment, supplier travel expenses, and reporting). An inspection assignment is typically issued for one purchase order but may cover more than one scope of supply at supplier and sub-supplier facilities if they are local.
Inspection Assignment	An *inspection assignment* is issued to a *quality surveillance representative* by the *supplier quality surveillance coordinator*. It includes a copy of the purchase order, provides detailed information about project expectations for inspection, and includes additional requirements specified in procurement documents (e.g., schedule of visits, concerns for equipment and materials, supplier travel expenses, and reporting). An inspection assignment is typically issued for one purchase order but may cover more than one scope of supply at supplier and sub-supplier facilities if they are local. The first inspection assignment for a purchase order is the *prime* assignment. Many purchase orders have only one assignment; thus, the term *prime* is used rarely. If the purchase order is multi-disciplined, has sub-suppliers or both, then sub-assignments are required. See also, *inspection assignment number* and *inspection sub-assignment*.

Phrase or Term	Definition
Inspection Assignment Number	*An inspection assignment number* is a unique identification (i.e., serial) number used by the supplier quality surveillance coordinator, inspection agency and supplier quality surveillance representative to perform and track a variety of functions (e.g., filing, invoicing, logging, and reporting). Because inspections relate to a specific purchase order, the supplier's purchase order number is often used (e.g., 2019-000-C-0001). A unique sequential log number (e.g., 1234, generated by SharePoint) may be used instead of, or together with, the purchase order number. For an inspection sub-assignment, the term *sub* may or may not be used as part of the filename. An inspection sub-assignment number may be identified using a method such as (shown with sub as part of the inspection assignment number and filename): • Alphabetical (2019-000-C-0001 Sub A, Sub B, and Sub C); • Discipline code (2019-000-C-0001 Sub C, Sub E, and Sub M, where C = coating, E = electrical, and M = mechanical); or, • Numerical (e.g., 2019-000-C-0001 Sub 1, Sub 2, Sub 3). See also, *inspection agency*, *inspection assignment*, and *inspection sub-assignment*.
Inspection Cost Estimate	An *inspection cost estimate* is a budget of costs for performing an inspection assignment. It includes: • Accommodations and meals or a per diem; • Contingency; • Mark-up of all expenses that are not flow-through (i.e., not charged at 100% of the cost); • Number of days or hours charged at a daily or hourly rate, respectively; • Other eligible expenses (e.g., communications, consumables, printing, and tools); • Transportation fares, fees, and taxes (e.g., airplane, ferry, taxi, or train); and, • Travel distance. See also, *budget*, *contingency*, *flow-through*, *markup*, and *per diem*.
Inspection Point	Not a preferred phrase – see *quality verification point*.
Inspection Release	An *inspection release* is a completed form issued by the supplier quality representative to indicate completion of inspection activities and acceptance of quality and technical requirements. After an inspection release is issued, a shipping release may also be issued. An inspection release may be for the complete or partial supply of a purchase order. See also, *shipping release*.
Inspection Report, Inspection Reporting,	An *inspection report* or *inspection reporting* is a document provided by the supplier quality representative to the project, during and inspection assignment or upon completion of inspection activities. This report may be issued daily, weekly, or at another frequency. An inspection report includes information or observations related to the assignment and purchase order, including action items, completed activities, findings, issues, photographs, progress or status, and questions and concerns. This report may also include attachments (e.g., an action item log, inspection release, quality control record, or packing or shipping slip).

Phrase or Term	Definition
Inspection Requirements Document	An *inspection requirements document* specifies the quality surveillance level and quality verification points planned for a purchase order. This document is usually contained within Section 4 of a purchase order.
Inspection Sub-assignment	An *inspection sub-assignment* is an inspection assignment for a purchase order that is multi-disciplined, has sub-suppliers, or both. See also, *inspection assignment* and *inspection assignment number*.
Inspection Test Plan, ITP	An *inspection test plan* or *ITP* is a quality document used to record and control all inspection and testing requirements relevant to a purchase order or scope of work. It details all inspections or tests, references specifications or requirements and acceptance criteria, and identifies parties responsible for acceptance (e.g., for signature and date for verification, and quality verification points). An ITP lists the supplier's name, address, job number, relevant equipment, and project names and numbers. It is also known as a quality control checklist, traveler and by other names. See also, *quality plan* and *quality verification point*.
Inspectioneering Journal	Established in 1995, the *Inspectioneering Journal* is a technical publication about mechanical integrity and reliability in the chemical and refining industries. Primary readership is inspectors and engineers engaged in the business of asset integrity management. It is published bi-monthly by Inspectioneering, LLC of Houston, Texas. See also, *asset*, *engineering*, *fitness for service*, *inspection*, and *life cycle*. See also, *asset integrity management*, and *inspectioneerng*.
Inspectioneering,	*Inspectioneering* is asset integrity management function that employs the engineering and inspection disciplines to ensure the continued reliable and safe operation of an asset throughout its life cycle (i.e., fitness for service). It includes activities such as analysis, assessment, budgeting, evaluation, examination, forecasting, logistics, planning, safety, scheduling, and testing. Inspectioneering is a portmanteau of *inspection* and *engineering*. See also, *asset*, *asset integrity management*, *engineering*, *fitness for service*, *imagineering*, *inspection*, *Inspectioneerng Journal*, and *life cycle*.
Inspector	Not a preferred term – see *quality surveillance representative* or *authorized inspector*.
Instruction	Not a preferred phrase – see *corporate instruction* and *project instruction*.
Insulate	To *insulate* is to separate conducting bodies by means of non-conductors to prevent the transfer of electricity, heat, or sound. See also, *insulate*, and *winterization*.
Insulated Concrete Form, Insulating Concrete Form	An *insulated concrete form* or *insulating concrete form* is a system of formwork for reinforced concrete made with a rigid thermal insulation that stays in place as a permanent interior and exterior substrate for walls, floors, and roofs. The forms are interlocking units that are dry-stacked, somewhat like Lego bricks. This type of construction has become commonplace for low rise commercial and residential construction as industry adopts more stringent energy efficiency and natural=disaster=resistant building codes. See also, *composite* and *form*.
Insulation	*Insulation* is a material used for insulating. See also, *cladding*, *insulate*, and *winterization*.

Phrase or Term	Definition
Integral Reinforcement	*Integral reinforcement* is where the reinforcement of an opening (e.g., made in a pipe or shell for a fitting) is inherent in the excess material thickness that is available in the attachment welds, nozzle neck, pipe, shell plate, or weld buildup. One or many pieces may make up the assembly, without any separate reinforcing elements. All parts must be butt-welded and full penetration welded. Lap-welded parts and partial penetration welds provide non-integral reinforcement. See also, *butt-weld*, *full penetration*, *lap-weld*, *O-let*, *partial penetration*, and *reinforcing pad*.
Integrated Control and Safety System	*An integrated control and safety system* is a process automation system that provides process control functionality and separate protection functions for the project facility. These two components are individually referred to as the *process control system* and *safety instrumented system*. These separate systems are integrated into a common display and alarm annunciation system with data logging to a central repository. See also, *process control system* and *safety instrumented system*.
Integrated Management System	An *integrated management system* combines all related components of a business into one system for easy management and operation. Environmental, quality, and safety management systems are often combined and managed this way. These systems are not separate systems that are later joined; instead, they are integrated with linkages so that similar processes are seamlessly managed and executed without duplication The following systems often establish as a single integrated management system: • ISO 14001, *Environmental Management Systems – Requirements with Guidance for Use*; • ISO 45001, *Occupational Health and Safety Management Systems – Requirements with Guidance for Use*; • ISO 9000, *Quality Management Systems – Fundamentals and Vocabulary*; and, • ISO 9001, *Quality Management Systems – Requirements*. See also, *integrity management program*, *operational excellence management system*, *organizational excellence management system*, *pressure equipment integrity management system*, and *quality management system*.
Integration	*Integration* is the process of assembling components and subsystems into one system and making all parts function together as an effective single system (e.g., building a computer or an instrumentation and controls package). See also, *interface test*, *packaged equipment*, *safety integration test*, and *system integration and functionality test*.

<table>
<tr><th>Phrase or Term</th><th>Definition</th></tr>
<tr><td>Integrity Assessment, Asset Condition Assessment</td><td>An integrity assessment or asset condition assessment is an evaluation to determine whether an asset is in good or bad condition, and what steps, if any, are required to improve or repair it (e.g., a bridge, equipment, facility, pipeline, or road). This report:<ul><li>Outlines how an organization can manage capital assets to improve its operations; or,</li><li>Provides a periodic evaluation of an asset using data collected through inspection and non-destructive examination of its condition to determine if preventative maintenance is needed. This may be to either:<ul><li>Identify if defects or hazards are present and if they need repair. This may be done more frequently (e.g., annually or bi-annually); or,</li><li>Provide an overall condition assessment; to determine how much useful asset life remains. This may be done less frequently (e.g., every 5 or 10 years).</li></ul></li></ul>Asset condition assessments help an organization plan its capital maintenance and renewal budgets. Assets are purchased with a given estimated useful life that, when considered with estimated maintenance costs, facilitates an estimate that outlines how much it will cost to replace the asset in the future.
Asset condition assessment may also be referred to as a facility condition assessment when it pertains to a building. See also, asset integrity management and integrity management program.</td></tr>
<tr><td>Integrity Dig or Investigative Dig</td><td>An integrity dig or investigative dig is the inspection of a section of a pipeline after physically exposing it by excavation and removing the pipeline coating. This allows for a detailed examination of the pipeline surface and subsurface using non-destructive examination methods. The pipeline may or may not be in-service during the dig. The pipeline is then repaired, if necessary, before being recoated and backfilled. See also, integrity assessment, in-service and non-destructive examination.</td></tr>
<tr><td>Integrity Management Program, Integrity Management System</td><td>An integrity management program or integrity management system is a documented, systematic plan designed to ensure the long-term integrity of an asset or assets throughout its life cycle. This over-arching, long-term program or system systematically assesses and mitigates risks to reduce likelihood and consequences of incidents per the Safety Codes Act. See also, asset, integrated management system, integrity assessment, integrity dig, life cycle, management, mechanical integrity, and Safety Codes Act.</td></tr>
<tr><td>Intellectual Property</td><td>Intellectual property is creative ideas, expressions of human mind, or knowledge that have commercial value and are protectable under copyright, patent, trademark, or trade secret laws from dilution, imitation, and infringement. Intellectual property includes artistic works, brand names, designs, discoveries, drawings, formulas, inventions, knowledge, and software. See also, copyright, infringement, patent, plagiarism, proprietary trademark, and trade secret.</td></tr>
<tr><td>Intelligence Quotient, IQ</td><td>Intelligence quotient or IQ is a total score derived from a set of standardized tests designed to assess human intelligence. The abbreviation IQ was coined by the psychologist William Stern from the German term Intelligenzquotient, his term for a scoring method for intelligence tests at University of Breslau he advocated in the German book Die psychologischen Methoden der Intelligenzprüfung: und deren Anwendung an Schulkindern. The English version The Psychological Methods of Testing Intelligence was translated by Guy Montrose Whipple. See also, emotional intelligence.</td></tr>
</table>

Phrase or Term	Definition
Intelligent Design	*Intelligent design* is a theory that certain features of living things and the universe are best explained by an intelligent cause or entity, not by an undirected process such as natural selection (i.e., evolution). Higher intelligence (e.g., works of art, business, and literature) or a system that is designed with a purposeful arrangement of parts, is therefore only surmised to be the creation of an intelligent mind. See also, *evolution*.
Interconnectedness	*Interconnectedness* is how everything is connected to, and reliant upon, something else for survival (often as a complex array of other things). For example, how animals and humans need air, food, and water to sustain life and thrive. Or how a house needs trees to provide wood for its construction. A specific mindset is needed to untangle and work with complex systems. It requires a shift from a linear, mechanical, and structured worldview, to one with chaotic, dynamic, and interconnected arrays of feedback loops and relationships. See also, *causality*, *emergence*, *feedback loop*, *interdependent*, *synthesis*, *systems mapping*, and *systems thinking*.
Interdependent	*Interdependent* means that two or more people or things are dependant upon one another for success or survival (i.e., mutually dependant). For example, globalization contributes to interdependence between countries. See also, *globalization and interconnectedness*.
Interdisciplinary Inspection	*Interdisciplinary inspection* occurs when a purchase order requires inspection from more than one discipline (e.g., any combination of coating, electrical, mechanical inspection). This scenario is typical for packaged equipment. Inspection may be arranged by either: 1. Assigning a qualified inspector who is knowledgeable of the requirements for all disciplines; or, 2. Preparing inspection assignments with a separate qualified inspector for each discipline. How this is accomplished may depend on many factors, including: • Availability of local and qualified inspection resources; • Equipment cost, complexity, criticality, and scope of supply; and, • Supplier certification, experience, and reliability.
Interest	*Interest* is: • A commitment, duty, goal, or obligation associated with a social practice or role; • A concern or feeling that accompanies or causes special attention to something or someone; • A legal share, right, or title in something; • An advantage, benefit, responsibility, or self-interest; • The fee charged regularly (e.g., daily, monthly, or annually) at a rate for the use of money that is lent, or for the delay of debt repayment; • The profit in goods or money that is made on a capital investment; or, • The state of wanting to know or learn about something or someone. See also, *conflict of interest*, *goods*, and *lobby*.

Phrase or Term	Definition
Interface	An *interface* is a: • Device or program that enables a user to communicate with a computer; or, • Point where two organizations, subjects, or systems communicate or interact and meet. For a project, an interface has the potential to negatively impact the cost and schedule by way of: • Contractual obligations between stakeholders and suppliers or contractors that limit the ability of or require parties to take action, make decisions, or both; • Interdependency with resources needed by others; • Issues with physical or electronic interaction between assets, components, or resources. • Lack of functionality with system requirements; or, • Miscommunication of information or a lack of common knowledge between stakeholders and suppliers or contractors, or between disciplines and within organizations. See also, *coordinate*, *interface management*, and *liaise*.
Interface Management	*Interface management* is the coordination, control, monitoring, and planning of the work of others, often involving complex, costly, large projects. This strategy is used for projects with engineering procurement and construction, or engineering procurement and construction management services to manage communications, deliverables, and relationships. Interface management offers increased alignment and communication that typically yields improved performance and eliminates, mitigates, or reduces risks associated with: • Budget and cost; • Joint venture agreements; • Multiple engineering centres; • Numerous stakeholders or suppliers and contractors; • Schedule; • Scope; and, • Technology. See also, *engineering and procurement*, *joint venture*, and *interface*.
Interface Test	An *interface test* is an integration test with equipment devices that communicate with the process control system. This testing may be performed as preliminary link tests or as full data-transfer tests.

Phrase or Term	Definition
Intermodal Container	An *intermodal container* is a large standardized shipping receptacle designed and built for international transport. These containers accommodate different modes of transport – from ship to rail to truck – without unloading and reloading cargo. An intermodal container is used primarily to store and transport materials and products efficiently and securely in the global containerized intermodal freight transport system. These containers are also used for regional transport. These containers are known by many names, including container, cargo or freight container, ISO container, sea van or Conex box, sea can or c can, and a sea container, shipping container, or ocean container. See also, *containerized intermodal freight transport system* and *packaging*.
Internal Pressure	*Internal pressure* is positive pressure that is applied to pressure equipment (i.e., a pressure inside that is greater than atmospheric pressure*)*. See also, *atmospheric pressure*, *external pressure*, *pressure equipment*, and *vacuum*.
Internals	*Internals* are parts installed inside equipment (e.g., a pressure vessel or tank). These may be removable (e.g., bolted) or permanent (e.g., welded). An internal may be a filter cartridge, inlet deflector, ladder, mist pad, screen, strainer, or vortex breaker.
International Oil Company	An *international oil company* is privately owned and operates globally. See also, *national oil company*.
International System of Units	*International System of Units* is from the French Système international d'unités, the modern form of the metric system.
Internationalization	*Internationalization* is the process of designing products to be easily modified, meet the needs of users in many countries, or both. For example, designing a website such that, when it is translated from English to Spanish or vice-versa, the aesthetic layout remains functional. This may be difficult to achieve as some words in Spanish have more characters than in English. Thus, more space may be needed for a page in Spanish when compared to English. See also, *globalization* and *localization*.
Internet of Things	The *Internet of things* is the interconnection via the Internet of computing devices embedded in everyday objects (e.g., cell phones), enabling them to send and receive data. See also, *Internet of things*.
Interpolation	In mathematics, *interpolation* is estimating an unknown value by analyzing a known sequence of values. For example, to estimate the world's population in 2017 by interpolating existing data (e.g., based on surveys completed in 2015 and 2020). In editing, interpolation is the text added to a citation, typically by an editor. For example, in the next citation ... The president *was certain* [the Cubans] *would comply*. ... the interpolated text "[the Cubans]". For editors, interpolated text should be used in citations sparingly and only when it adds context and meaning to a citation (as it does in the above citation). See also, *extrapolation*.

Phrase or Term	Definition
Intersectionality	*Intersectionality* is the complex and cumulative nature of social categorizations such as class, gender, and race as they apply (i.e., combine or overlap) to groups or individuals with multiple forms (i.e., interdependent and overlapping) of disadvantage and oppression, or privilege and empowerment. The term was coined by black feminist scholar Kimberlé Williams Crenshaw in 1989. See also, *ageism, Black Indigenous and People of Colour* and *People of Colour, discrimination, diversity, gender-based analysis, racism*, and *sexism*.
Interstitial Monitoring	*Interstitial monitoring* detects the presence of a gas or liquid in an intermediate space located between two surfaces (interstitial space) by electronic or visual means. See also, *double-walled tank* and *interstitial space*.
Interstitial Space	An *interstitial space* is an intermediate space located between two surfaces. See also, *double-walled tank* and *interstitial monitoring*.
Intervention Point	Not a preferred phrase – see *quality verification point*.
Intrinsically Safe	*Intrinsically safe* means that arcs or sparks in equipment have insufficient energy (heat) to ignite a vapour.
Introvert	An *introvert* is a personality type characterized by being analytical, reflective, reserved, and shy. The word was popularized by C. G. Jung in the early 20th century. Introversion is the state of being predominantly interested in one's own mental self. See also, *extrovert* and *omnivert*.
Intuition	*Intuition* is: • A mental algorithm coupled with a psycho-physiological reaction (e.g., a gut feeling based on experience for making judgement about culture, people, and strategy); or, • The ability to understand something immediately, without the need for conscious reasoning (e.g., an experienced canoer who tries kayaking for the first time). See also, *intuitive*.
Intuitive	*Intuitive* is: • Based on or use of information that one feels to be true, even without conscious reasoning or evidence (i.e., instinctive with quick and ready insight). For example, it was intuitive to the team that the estimated cost was incorrect; or, • Easy to use or readily learned and understood (e.g., computer or smart phone software). For example, to pilot an airplane, the pilot will move the control wheel or yoke by pushing to nose down, pulling to nose up, turning left to turn left, and turning right to turn right. See also, *counterintuitive* and *intuition*.
Intumescent	An *intumescent* coating or sealant swells when heated, protecting the material underneath it or sealing a gap to prevent oxygen ingress in the event of a fire.
Invention	An *invention* is a product or material that has never been made before, or the process of creating something that has never been made before (e.g., the invention of flight by the Wright brothers or intermittent windshield wipers by the Robert Kearns). See also, *creativity, innovation*, and *patent*.

Phrase or Term	Definition
Inverted Darwinism	*Inverted Darwinism* is the *survival of the unfittest*! This occurs when the project that looks best on paper is implemented (Project A) instead of a project that will work better (Project B). Project A may grossly underestimate costs and overestimate benefits. The result is that costly Project A proceeds. Worse, Project B can no longer proceed because the required resources were usurped by Project A! One forecast, business case or analysis usually does not reveal all facets of a project. Different stakeholders have different interests; outside forces are always at play. Forecasts, then, are not only products of data and mathematical modelling – but also of perspective, negotiation, power, and politics. To avoid inverted Darwinism, use due diligence, and independent forecasts and risk assessments. See also, *creative error*, *Darwinian theory*, *due diligence*, *independent forecast*, *hiding hand principle*, *looks good on paper*, *reverse Darwinism*, and *risk assessment*.
Invest	To *invest* means to: • Endow or provide someone or something with a particular attribute or quality (e.g., pay a student's university tuition); • Engage or involve, especially emotionally (e.g., invest in a joint venture); • Expend or lend money or resources with the expectation of achieving a future financial return (i.e., profit) or material result (e.g., land); or, • Make use of something for future advantage or benefit (e.g., purchase a stockpile of material that can be resold later, at a higher price, for a profit). See also, *investment* and *joint venture*.
Investment	An *investment* is an asset acquired with the goal of producing appreciation or income. When an investment is purchased (or, when investing in an investment), the intent is to create future income or wealth. An investment always requires providing something today or in the future (i.e., effort, money, or time) in the hope that the investment will grow in value over time and yield future profit. See also, *appreciation*, *divest*, and *invest*.
Investor's Capital	*Investor's capital* are assets invested to further business goals and objectives. This is often money but may also be fixed or long-term assets such as equipment, facilities, and real estate (i.e., property, plant, and equipment). See also, *capital* and *invest*.
Invoice	An *invoice* is a list of products delivered or services provided by the seller to the buyer, with a statement of the amount that is payable for these. An invoice may contain additional information about the terms of payment such as how payment shall be made and when it is due (e.g., immediately or in 30 days), and pricing (e.g., hourly or unit cost).
Iron Imperative	The *iron imperative* is a principle used by good writers that treats reader's time as more valuable than their own. This means that when writer writes a document (e.g., email, instruction, or letter), the writer takes sufficient time to maximize the clarity and efficiency of the communication to save readers the time and effort associated with understanding an imprecise, poorly-written document. Writers who adopt this principle stand out and often become candidates for career advancement. The recognition of writers' use of the iron imperative by companies is significant – successful companies reward effective communicators.

Phrase or Term	Definition
Iron Law of Megaprojects	The *iron law of megaprojects* is over budget, over time, over and over again. This law occurs in the presence of weak front-end planning and poor project management. See also, *management* and *megaproject*.
Iron Ring	An *iron ring* is a ring worn by most professional Canadian engineers. It reminds those who wear it – and those interested in the practice of engineering – of an engineer's ethical obligation to live and work to a high standard of professional conduct and public service. The ring is made from iron or stainless steel, faceted, and intended to be worn on the little finger of the working (dominant) hand. In Canada, the ring is presented to engineering graduates in a closed ceremony known as the *Ritual of the Calling of an Engineer*. The concept of the ritual and its iron rings originated from H. E. T. Haultain in 1922, with assistance from Rudyard Kipling. It is not a symbol of legal qualification and registration to practice, which is the exclusive responsibility of provincial and territorial licensing bodies. A myth persists that the first iron rings were made from the beams of a Quebec, Canada bridge that collapsed during construction in 1907 from poor planning and design by engineers. In fact, the original iron rings were produced by World War I veterans at the Christie Street Military Hospital in Toronto in 1907. In the United States, a similar program was created – the *Order of the Engineers* (1970). This organization conducts similar ceremonies, in which recipients also receive a (stainless steel) ring. Unlike the Canadian iron ring – the United States' *iron* ring can be smooth, *may not be* faceted and conventionally comprises stainless steel. See also, *engineer*, *ethics*, and *ethical*.
Is the Juice Worth the Squeeze?, Is it worth It?	*Is the juice worth the squeeze* is an expression that ponders if the hard work or trouble will be worth the compensation-results. This means that to make juice, labour must be expended to squeeze fruit. Is the effort of this labour worth the resulting juice? For example, is spending four years to acquire an academic degree (the labour) worth the expanded career opportunities that accompany the degree (the juice)? Is the juice worth the squeeze and *is it worth it* are synonymous.
ISNetworld®, ISN®	*ISNetworld®* or *ISN®* is an online contractor management database that collects health, safety, procurement, quality, and regulatory information designed to meet governmental record keeping and client requirements.

Phrase or Term	Definition
J	1 A B C D E F G H I **J** K L M N O P Q R S T U V W X Y Z
Jabberwocky	*Jabberwocky* is meaningless speech or writing. In the poem *Jabberwocky* in the book *Through the Looking-Glass and What Alice Found There*, Lewis Carroll used nonsensical words to warn readers about a frightful and mythical beast – the Jabberwock:. *Beware the Jabberwock, my son!* *The jaws that bite, the claws that catch!* *Beware the Jubjub bird, and shun* *The frumious Bandersnatch!* See also, *gibberish*.
Jargon	*Jargon* is abbreviations, expressions, and specific words used by a group or profession, which are difficult for others outside the group or profession to understand (e.g., in the pipeline construction industry, the phrase *poor-boy crew*). See also, *buzzword, colloquial, dialect, doublethink, gibberish, glossary, Ebonics, Plain English, poor-boy crew, vernacular*, and *word salad*.
JavaScript	*JavaScript* is a high-level, just-in-time compiled, object-oriented programming language that is one of the core technologies of the World Wide Web.
Jeep	A *jeep* is the sound that a holiday detector makes when it detects a holiday. The pipeline industry describes this sound as a jeep, which is a sharp and sudden arcing-electronic beep sound. See also, *holiday*.
Jerry-build	To *jerry-build* is to construct something of poor quality. For example, a roof that is not water-tight is jerry-built. See also, *jerry-rig* and *stopgap*.
Jerry-rig	To *jerry-rig* is to construct or organize something in a crude, improvised, makeshift, or temporary manner. For example, to use an orange juice carton make a gasket is to *jerry-rig* the pump. See also, *jerry-build* and *stopgap*.
Jevons Paradox, Jevons Effect	*Jevons paradox* or *Jevons effect* is when government policy, increased efficiency, or technological advancement for the use of a resource leads to a higher level of consumption, which negates the environmental benefits of increased energy efficiency. This was first described in English economist William Stanley Jevons' book *The Coal Question; An Inquiry Concerning the Progress of the Nation, and the Probable Exhaustion of Our Coal-Mines*. He argued that improvements in fuel efficiency tend to increase – rather than decrease – fuel use. Jevons observed that England's consumption of coal soared after James Watt's steam engine greatly improved the efficiency of the coal-fired steam engine. Watt's innovations made coal a more cost-effective power source, which lead to the increased use of the steam engine in a wide range of industries. This resulted in increased total coal consumption, even though the amount of coal required for any application was reduced. See also, *Khazzoom-Brookes postulate* and *rebound effect*.
Jig	A *jig* is a type of custom-made tool used to control the location, motion of parts or both, or other tools. Its primary purpose is to provide for the accuracy, interchangeability, and repeatability, of manufactured products. It is often confused with a fixture, which holds the work in a fixed location. A jig can hold the work *and* guide a tool.

Phrase or Term	Definition
Job Equipment	*Job equipment* is equipment purchased by a project to be delivered to site for installation and operation (e.g., a pump and motor). These items may or may not be used for fit-up, performance, or other testing.
Job Instruction	Not a preferred phrase – see *project instruction*.
Job Number, Job Order Number, Work Order Number	A *job number, job order number*, or *work order number* is a unique number or code that is used to identify a scope of supply or scope of work and all related items such as assemblies, components, documents, materials, requisitions, purchase orders, and timesheets. See also, *job, job order, sales order number*, and *work order*.
Job Shop Manufacturing	*Job shop manufacturing* is a method of production where a facility takes an order to produce specific products that are designed by the customer.
Job, Job Order, or Work Order	A *job, job order*, or *work order* is a task or a job for a customer that can be scheduled or assigned to a qualified person. Such an order may be from a customer request or created internally within the organization. A work order may be for products or services and include one or more of the following: • Cost estimate; • Date and time to perform the work; • Information about the location and entities or requirements to execute the work; • Instructions and forms; and, • The person to whom the work is assigned. For contracting, a work order is a written order from the customer providing specific or blanket authorization to the contractor to proceed with the performance of the contract without further instructions. For fabrication or manufacturing, a job order is an internal authorization to incur costs in the design, development, and production of a specific product. Work orders differ from job orders in breadth of scope. See also, *job number, job order number*, and *work order number*.
Johari Window	The *Johari window* is a technique used to help people better understand their relationship with others and themselves. It was created by psychologists Joseph Luft and Harrington Ingham in 1955. Johari is a portmanteau of the names Joseph and Harrington. In the exercise, the subject picks several adjectives from a list that they feel describes their personality. The subject's peers also each pick an equal number of adjectives that describe the subject's personality. These adjectives are then inserted into a two-by-two grid of four quadrants (window) for analysis. The: 1. Arena or Open quadrant includes adjectives that both the subject and peers selected; 2. Blind quadrant includes adjectives not selected by subject, but only by their peers; 3. Façade quadrant includes adjectives selected by the subject, but not by any of their peers; and, 4. Unknown quadrant includes adjectives that neither the subject nor their peers selected.

Phrase or Term	Definition
Joint	A *joint* is: • A length or piece of pipe; • A weld; or, • The point or edge where two or more pieces of material are joined together. The AWS identifies five types of welded joints: butt, corner, edge, lap, and tee. See also, *butt weld*, *lap weld*, *seam*, and *seamless*.
Joint Operating Agreement	A *joint operating agreement* is a contract between two or more organizations with mineral interests (e.g., a gas or oil lease) that collaborate to share expertise and resources. The contract governs a joint venture between those who sign the agreement, while allowing each company to retain its own identity. See also, *joint venture* and *mineral*.
Joint Venture	A *joint venture* is a business arrangement in which two or more parties pool resources for a specific task (e.g., activity or project). Through collaboration, parties share resources, profits, losses and expenses. A joint venture is a legal entity separate from the parties' other business interests. See also, *cartel*, *consortium*, *cooperative*, *corporation*, *monopoly*, *monopsony*, *partnership*, *project alliance*, and *sole proprietorship*.
Jointer	A *jointer* is two or more lengths of pipe that are welded together to produce a single length of pipe. See also, *jointer weld*.
Jointer Weld	A *jointer weld* is a weld that connects two pieces of pipe to produce a single length of pipe. See also, *jointer*.
Journeyman	A *journeyman* is a skilled worker who has successfully completed an apprenticeship in a skilled trade. Journeymen are considered competent, authorized to work in their trade, and fully qualified. See also, *apprenticeship*, *skilled trade*, *trade*, and *trades person*.
Judicial Dispute Resolution	*Judicial dispute resolution* is a confidential pre-trial settlement conference led by a judge and attended by parties in dispute. The objective is to resolve the dispute and making a trial unnecessary, or limit the trial only to issues on which the parties do not agree. See also, *alternative dispute resolution*.
Junk Science	*Junk science* is: • An expression that describes fraudulent or spurious analysis, research, or scientific data; • An unproven or untested theory disguised as scientific fact; • False, inaccurate, or misrepresented analysis or data used to create a biased opinion, or to promote or support an agenda with ideological, financial, political, or unscientific motives. See also, *pseudoscience* and *scientific method*.

Phrase or Term	Definition
Just-in-time	*Just-in-time* is a manufacturing system in which materials or components are delivered immediately before they are required to eliminate or minimize storage, preservation, and inventory requirements and costs. This system is based on the Toyota Production System, which uses an approach that protects vehicle assemblies from damage and theft, makes efficient use of personnel, and preserves valuable space (especially where space is limited). See also, *do it right the first time*, *just-in-time delivery schedule*, *lean construction*, and *lean manufacturing*.
Just-in-time Delivery Schedule	A *just-in-time delivery schedule* integrates the delivery of materials directly with the production schedule or, in an engineering and construction environment, with scheduled activities on a job site. See also, *just-in-time*.

Phrase or Term	Definition
K	1 A B C D E F G H I J **K** L M N O P Q R S T U V W X Y Z
Kanban	*Kanban* is a supply chain control system that focuses on cost reduction by implementing just-in-time inventory control and sets limits for inventory-holding. *Kanban* is Japanese for *billboard*.
Karma	*Karma* is Sanskrit for action, deed, or work. It also refers to the spiritual principle of cause and effect where actions and intent of an individual (cause) influence the future of that individual (effect). Good deeds and intent contribute to good karma, while bad deeds and intent contribute to bad karma. See also, *Givers Gain®*, *law of reciprocity*, *pay it forward*, *poetic injustice*, *poetic justice*, and *reciprocity*.
Kayak	A *kayak* is a small, narrow watercraft, which has a covered deck and one or more cockpits, each seating one paddler. It was first used by the indigenous Aleut, Inuit, Yupik and possibly Ainu hunters in subarctic regions. See also, *kayak problem*.
Kayak Problem	A *kayak problem* is when personnel experience a challenge but manage to address it immediately – without any help. It is like rolling over in a kayak; if the kayak is righted quickly enough there is no harm. See also, *kayak*.
Keep It In the Ground	*Keep it in the ground* is a philosophy that promotes the idea of keeping existing fossil fuels in the ground and not extracting them. By extension, this philosophy ultimately champions halting fossil fuel use and, as an alternative, the consumption of renewable energies (e.g., solar or wind) instead. See also, *climate change*, *fossil fuel*, and *renewable energy*.
Keep it Simple Stupid Principle, KISS principle	The *keep it simple stupid principle* or *KISS principle* states that most systems work best if kept simple and uncomplicated. Hence, simplicity should be a key goal in design; unnecessary complexity should be avoided. The phrase has been attributed to aircraft engineer Kelly Johnson, is a design principle noted by the United States Navy beginning in 1960, and has been adopted by countless organizations. KISS is also an abbreviation for keep it stupid simple, keep it simple silly, keep it short and simple, keep it simple and straightforward, keep it small and simple See also, *Occam's razor*.
Kepner Tregoe Method, KT-method	The *Kepner Tregoe method* or *KT-method* is a documented, synchronized, and well-orchestrated root cause analysis and decision-making method. This method employs a system that follows a logical process, that is well-organized and requires data to be analyzed before acting. Dr. Benjamin Tregoe and Dr. Charles Kepner, two social scientists who researched flaws in decision making in the United States Air Force, developed this method in 1958. See also, *root cause analysis*.

Phrase or Term	Definition
Key Knowledge, Knowledge Transfer Project Key Knowledge, KK	*Key knowledge* or *KK* is intelligence of crucial or paramount importance that provides an explanation or solution for project personnel. It is practical or theoretical knowledge that is explicit (i.e., a clearly developed or formulated explanation) or implicit (i.e., an expertise or practical skill). The Knowledge Transfer (KT) Project describes a series of KK topics, which are guidelines with proven rubrics for effective and profitable solutions that were developed from decades of experience. Find KT Project KK in KT Project documents. Find all KT Project key knowledge in KTP-GEN-005, *Knowledge Transfer Project Key Knowledge*. See also, *codification*, *guideline*, *Knowledge Transfer Project*, *lessons learned*, *organizational memory*, *retention of critical knowledge*, *rubric*, and *serial number*.
Key Performance Indicator	A *key performance indicator* is a measurable value that demonstrates how effectively an organization is achieving key business objectives. Key performance indicators are applicable to the personnel and processes of an organization. These are used at multiple organization levels to evaluate if or how well an organization reaches business objective targets. See also, *dashboard*, *project performance indicators*, and *scorecard*.
Keyman Insurance, Key Employee Insurance, Key Person Insurance	*Keyman insurance*, *key employee insurance*, or *key person insurance* is a form of business insurance for which there is no legal definition. In general, it is an insurance policy purchased to compensate that business or organization for financial losses that would arise from the death or extended incapacity of a critically or highly important person or keyman. See also, *keyman*.
Keyman, Key Employee, Key Person	A *keyman*, *key employee*, or *key person* is critically or highly important to the functioning or operation of a business or organization. Their creativity, inspiration, knowledge, reputation, and/or skills are essential and vital to growth and viability. Their loss may have a crippling effect. See also, *keyman insurance*.
Khazzoom-Brookes Postulate	The *Khazzoom-Brookes postulate* describes the idea that energy efficiency gains paradoxically result in increased overall energy use. For example, jet engines use less fuel than in previous decades, but travel by aircraft today is far more common largely because of inexpensive ticket prices made possible by lower fuel costs. This phrase was coined by economist Harry Saunders and is the modern-day equivalent of Jevons' paradox. He modelled energy efficiency gains and showed that the postulate is true over a wide range of assumptions. For example, efficiency gains for other factors, such as capital and labour, can also increase overall energy use. See also, *Jevons effect* and *rebound effect*.
Kick and Count	*Kick and count* is slang phrase for a basic final inspection that usually comprises only a visual verification of bulk items (e.g., fittings or valves). See also, *bulk items* and *final inspection*.

Phrase or Term	Definition
Kickback	*A kickback* (noun) is: • A sharp, violent reaction. For example, if proper cutting techniques are not used, the lightning-fast kickback of a chain saw can be very dangerous and result in serious injury; or, • An illegal payment (bribe) provided as compensation for coercion, a confidential or secret agreement, or preferential treatment. A kickback may be anything of value (e.g., credit, a gift, or money). Paying or receiving a kickback is a corrupt and often illegal practice that interferes with unbiased decision-making. For example, Practice Fusion solicited and received payments from a major, unnamed opioid company in exchange for using its electronic health record software to increase opioid pain medication prescriptions. Consequently, the company was fined CAD$ 184 million (USD$ 145 million) by the Department of Justice for the District of Vermont in the United States. *To kickback* (verb phrase) is to take a relaxed attitude or position (informal). For example, I went fishing at the lake to kickback for the weekend. See also, *backfire*, *bribe*, and *white-collar crime*.
Kick-off Meeting	A *kick-off meeting* is held after the award of a purchase order by the buyer or expediter. The purpose of this meeting is to review the commercial and technical requirements of the awarded purchase order with the supplier, to facilitate successful execution. See also, *expediter*, *expediting*, *prefabrication meeting* and *supplier*.

Phrase or Term	Definition
Kinds of Information	The three *kinds of information* about risk or available for a negotiation are: 1. Known knowns (e.g., things established and understood such as an airplane that is delayed because of adverse weather); 2. Known unknowns (e.g., things one is aware of such as an airplane that could be delayed because of adverse weather) 3. Unknown unknowns (e.g., things so unexpected they cannot be considered, such as an airplane that disappears never to be seen again – a black swan event). The three *kinds of information* used everyday are: 1. Context (i.e., the big picture that explains or informs content); 2. Content (i.e., the essential elements that prove or disprove the context); and, 3. Meaning (i.e., the impact or implementation and interpretation). The four *kinds of information* sources are: 1. Factual information presents only established or proven data; 2. Analytical information that interprets factual information; 3. Subjective information presented from only one viewpoint; and, 4. Objective information presented from multiple viewpoints. The three *kinds of information* used for research are: 1. Primary sources which are original materials (e.g., an interview transcript or written record); 2. Secondary sources which analyze or describe primary sources (e.g., an article that interprets or a reference such as a textbook); and, 3. Tertiary sources which are used to organize and locate secondary and primary sources (e.g., a database or index). See also, *big picture, black swan event,* and *risk*.
Kitchen Sink Approach	The *kitchen sink approach* is a reference to the expression *everything but the kitchen sink*. It means to make use of everything that can be conceived. For example, to make use of a very large number of things, whether needed or not, to improve results or solve a problem. See also, *more is more*.
Kleptocracy	*Kleptocracy* is a government controlled or ruled by corrupt leaders or thieves (kleptocrats) who use political power to appropriate national wealth by embezzling or misappropriating government funds. Members of a kleptocracy enrich themselves through bribes, kickbacks, and special favours, or direct state funds to their associates and themselves (e.g., secretly or otherwise). Kleptocrats often export stolen wealth to a foreign country to protect it in anticipation of losing power. See also, *bribe*, *government*, *kickback*, oligarchy, *meritocracy*, *plutocracy*, and *populism*.
Knightian Uncertainty	*Knightian uncertainty* is risk that is unpredictable. It is a lack of any quantifiable knowledge about some possible occurrence, as opposed to the presence of measurable risk (e.g., attempting to plant a garden on the planet of Mars). The concept acknowledges an essential unpredictability, fundamental degree of ignorance, and a limit to knowledge of future events. The economist Frank Knight first differentiated between risk, the possibilities that can be measured, and uncertainty, which is the unpredictable. See also, *risk*.

Phrase or Term	Definition
Knock Off, Knockoff	*Knock off* or *knockoff* means to: • Copy or imitate someone or something, especially as a cheaper version. For example, a fake Gucci handbag is a knock off; • Defeat, kill, or overcome. For example, he knocked off his opponent; • Do quickly or routinely. For example, the gymnast is knocking off a cardio session; • Reduce an amount or price to make it more affordable or attractive. For example, the merchant knocked 20% off the price; • Steal (e.g., rob a bank). For example, the thieves knocked off the bank; or, • Stop something or stop working. For example, the job was not completed but the workers knocked off work early because it was Friday. See also, *knock on effect* and *knocked down*.
Knocked Down	*Knocked down* means that an item or package is partially or entirely disassembled or taken apart (e.g., such as for shipping after trial assembly). See also, *knock off*, *knock on effect*, and *set up*.
Knock-on Effect, Knock On, Knock-on,	The *knock-on effect*, *knock on*, or *knock-on* is when an event or situation causes other events or situations, but indirectly. See also, *Cobra effect*, *knock off*, *knocked down*, *second order effect*, *spillover effect*, and *unanticipated consequences*.
Knowledge Economy	A *knowledge economy* is an economy in which growth is dependent on the accessibility, quality, and quantity of available information, rather than the means of production. It is a system of consumption and production that is based on intellectual capital, which represents a large component of all economic activity in developed countries. See also, *circular economy*, *economy of repetition*, *economy of scale*, *gig economy*, *helix of innovation framework*, *hydrogen economy*, *linear economy*, and *low-carbon economy*.
Knowledge Transfer	*Knowledge transfer* is the process by which experienced personnel share-distribute behaviours, knowledge, and skills and with colleagues, and with personnel who will replace them. See also, *active learning*, *Bloom's Taxonomy*, *charter*, *codification*, *cone of learning*, *corporate instruction*, *guideline*, *metacognative thinking*, *passive learning*, *personalization*, *plan*, *project charter*, *project instruction*, *tribal knowledge*, and *unlearning*.
Knowledge Transfer Project (KT Project)	The *Knowledge Transfer Project* or *KT Project* provides key documents (guidelines and best practices) and key knowledge to support the successful supply and service of oil and gas projects, and projects in other sectors (e.g., industrial, infrastructure, mining, and power). Knowledge transfer is the effective and efficient transfer of knowledge from subject matter experts to other professionals. KT Project documents capture decades of valuable, relevant subject matter expert knowledge and experience, and help organizations mitigate, minimize, and eliminate project challenges caused by knowledge gaps. KT Project documents save organizations significant time and money. See also, *guideline* and *key knowledge*.

Phrase or Term	Definition
L	1 A B C D E F G H I J K **L** M N O P Q R S T U V W X Y Z
Label	A *label* (noun) is a: • Descriptive or identifying word or phrase applied to a person or thing, especially if it is inaccurate or restrictive (e.g., the team members are unmotivated); or, • Piece of material (e.g., fabric, metal, paper, or plastic) attached to and providing information about an object (e.g., its contents and identification). To *label* (verb) in the context of a negotiation is a technique used to identify and influence emotions. To do this, the negotiator identifies their counterpart's feelings, defines them in words, and then calmly and respectfully describes these emotions back to them. This acknowledges and validates their feelings using tactical empathy. It is a shortcut to intimacy that is a time-saving and emotional hack that diffuses negative emotions and reinforces good ones. This is because exposing negative thoughts can make them seem less daunting or frightening. A label can be phrased as an open-ended question or statement that begins with: • It looks like ... ; • It seems like ... ; or, • It sounds like ... . A label should not use the word *I*, which indicates self-interest and takes responsibility for the words that follow. Once a label is described, the negotiator must be silent and let the counterpart answer the question. This concept was developed by Chris Voss in the book *Never Split The Difference: Negotiating As If Your Life Depended On It* (with Tahl Raz). See also, hack, *marking and labelling*, and *open-ended question*.
Labour Day	A *labour day* is the amount of work performed by the average worker in one day. This phrase is used to estimate of the total amount of uninterrupted labour required to perform a task (by day). Labour day is also an American holiday. See also, *full-time equivalent* and *labour force*.
Labour Disruption	A *labour disruption* is a labour dispute, lockout, strike, or other labour action or strife, whether direct or indirect and whether lawful or unlawful, that halts or interrupts the work of personnel. See also, *wildcat strike*.
Labour Force	A *labour force* is a group of personnel or workers. This phrase describes the human resources that perform work for employers. See also, *full-time equivalent*, *gig economy*, *independent contractor*, *labour day*, *labour hour*, *liquid labour force*, and *side hustle*.

<table>
<tr><th>Phrase or Term</th><th>Definition</th></tr>
<tr><td>Labour Force Congestion</td><td>Labour force congestion is the measure of how many personnel may safely simultaneously occupy a volume or area. Two underlying concepts determine a safe personnel-to-area ratio:<ul><li>Health Safety and Environment
Depending on the location of the work and the jurisdiction of law, local regulations outline specific conditions for egress, space, and other requirements, which limit the number of people in any space.</li><li>Productivity
Worker productivity is the most dynamic factor in labour force congestion. Even if regulations permit a high personnel-to-area ratio, the productivity at the workface will create its own limit. This limit can be increased occasionally to accommodate needed progress (at the expense of productivity).</li></ul>See also, labour force.</td></tr>
<tr><td>Labour Hour</td><td>A labour hour is the amount of work performed by the average worker in one hour. This phrase describes the estimated total amount of uninterrupted labour required to perform a task by hour. See also, full-time equivalent and labour force.</td></tr>
<tr><td>Labour Union, Labour Organization, Trade Union, Workers' Union</td><td>A labour union, labour organization, trade union, or worker's union is a labour force organization (e.g., in a company, industry, or trade) formed to advance its members' collective interests for benefits, wages, and working conditions. A labour union democratically selects representatives to negotiate with the employer through collective bargaining. A labour union may also advocate for legal, political and social interests. See also, collective bargaining.</td></tr>
<tr><td>Lacuna, Non Liquet</td><td>A lacuna or non liquet is a situation whereby no law exists to address or prohibit something. Non liquet is Latin for it is not clear. For example, if there is no sign that states no parking, it is legal to park. A loophole is a subset of lacunae. See also, loophole.</td></tr>
<tr><td>Lagging Indicator</td><td>A lagging indicator is a pattern or trend that is reactive in nature and:<ul><li>An output-oriented measurement (i.e., easy to measure after the fact, but difficult to improve or influence). For example, the number of injuries on a project in over a year; or,</li><li>In progress or has occurred. For example, a rise in a country's unemployment rate indicates poor overall economics.</li></ul>See also, coincident indicator and leading indicator.</td></tr>
<tr><td>Lamacoid</td><td>Lamacoid (also Lamacoid and Lamacoid) is a trade name for engraving material that is no longer manufactured; however, the term is still used when specifying laminated and engraving plastic name or legend plates, or tags. This material is ideal for long-term identification of buttons, switches and panels. See also, name plate.</td></tr>
</table>

Phrase or Term	Definition

Lame Duck	A *lame duck* is:
	• A politician whose term will expire or will not be continued who is:
	  o Free to make decisions that exercise the standard powers with little fear of repercussions (e.g., issuing executive orders or pardons); and,
	  o Seen as having less influence in government or with other politicians because of their limited remaining time in power.
	• An ineffectual or unsuccessful person or thing.

Language Planning, Language Engineering	*Language planning* or *language engineering* is the deliberate effort to influence the behaviour of others about status, structure, or use of language. This involves formal and informal activities within corporate and other organizations (e.g., countries and non-profit groups), to control the use of terminology within a domain (e.g., law or project management). Corporate example: A company glossary specifies that *supplier* is a preferred term and *vendor* is not a preferred term. Government example: In 1982, the *Canadian Charter of Rights and Freedoms* made English and French the official languages of Canada. Also, the Canadian Province of Quebec employs *language police*. See also, *language police*.
Language Police	*Language police* is an informal name for civil servants in the Canadian Province of Quebec who are responsible for enforcing provincial language laws, who are not part of the police force. These laws are Bill 101 and Bill 178, which require commercial signs to make French words more prominent than English ones.
Lap Joint, Lap Weld	A *lap joint* or *lap weld* is where two or more materials in different planes (i.e., with one overlapped on top of another) are joined by welding. See also, *joint*, *butt weld* and *socket weld*.
Large Data	*Large data* is well-structured data in extremely large data sets. See also, *big data* and *data lake*.

<table>
<tr><th>Phrase or Term</th><th>Definition</th></tr>
<tr><td>Lasswell's Communication Model, Lasswell's Model of Communication</td><td>Lasswell's communication model or Lasswell's model of communication describes the basic elements of communication or public speaking by answering the following questions:
<ul><li>Who said it?</li><li>What was said?</li><li>In what channel was it said?</li><li>To whom was it said?</li><li>With what effect was it said?</li></ul>
This is a linear or one-way model that organizes scientific study of the communication process as shown in the next table.
<table>
<tr><th>Question</th><th>Element</th><th>Analysis</th></tr>
<tr><td>Who?</td><td>Communicator</td><td>Control Analysis</td></tr>
<tr><td>Says What?</td><td>Message</td><td>Content Analysis</td></tr>
<tr><td>In Which Channel?</td><td>Medium</td><td>Media Analysis</td></tr>
<tr><td>To Whom?</td><td>Audience</td><td>Audience Analysis</td></tr>
<tr><td>With What Effect?</td><td>Effect</td><td>Effects Analysis</td></tr>
</table>
This concept was developed by the American political scientist Harold Lasswell in his article The Structure and Function of Communication in Society. See also, public speaking.</td></tr>
<tr><td>Last Minute Risk Assessment, Last Minute Risk Analysis</td><td>A last minute risk assessment or last minute risk analysis is used to create awareness and remove potential risks from the workplace. It may be a:
<ul><li>Brief, individual, hazard assessment of a task and the personnel, which may be completed mentally or verbally; or,</li><li>Final assessment completed by personnel (e.g., a short checklist with a few common topics such as environment, health, and safety).</li></ul></td></tr>
<tr><td>Lateral Effect</td><td>A lateral effect is a side effect that has a negative, unwanted, or unintended reaction. For example, if the side effect of a headache medicine is nausea or worse, consumers may choose not to consume it. See also, lateral thinking and second-order effect.</td></tr>
<tr><td>Lateral Thinking</td><td>Lateral thinking involves the development and exploration of alternatives and ideas to respond to a challenge by using a variety of approaches that are often unconventional. These approaches include finding wrong answers, jumping from one step to another at random, brainstorming, and considering unexpected and unusual methods. For example, knowing that many equally or more qualified applicants will apply for the same position, a lateral thinker may decide to bring their resume and a pizza to a lunch-time interview at a construction site as a method to increase the chances they will be remembered by the interviewer and get the job. As shown in the previous example, lateral thinking involves dispelling preconceived notions and creating unconventional approaches while still respecting what is acceptable in a given scenario. See also, cognitive dissonance, cognitive flexibility, convergent thinking, divergent thinking, lateral effect, thinking mode, and vertical thinking.</td></tr>
</table>

Phrase or Term	Definition
Law of Attraction	The *law of attraction* is the belief that negative or positive energy attracts the same energy (into a person's life). Through like-energy attracting like-energy, a person can improve health, personal relationships, and wealth. The law of attraction has no scientific foundation and is considered, at best, a pseudoscience. See also, *Baader-Meinhof phenomenon* and *pseudoscience*.
Law of Cause and Effect	The universal *law of cause and effect* states that for every effect there is a definite cause. Likewise for every cause – there is a definite effect. This means that nothing happens by chance and that every action has a reaction or consequence. Similarly, Newton's third law of motion states that for every action there is an equal and opposite reaction. Therefore, if one object exerts a force on a second object, the second exerts an equal and oppositely directed force on the first one. See also, *black swan theory* and *free lunch*.
Law of Reciprocity	The *law of reciprocity* is considered by many to be the most powerful law of human nature. It states *if you do something nice for me, I will feel obligated to reciprocate and do something nice for you*. See also, *Givers Gain®*, *Golden Rule*, *karma*, *poetic justice*, *reciprocity*, and *relational customer service*.
Laws of Simplicity	The *Laws of Simplicity* is a book by John Maeda that describes 10 laws for balancing complexity and simplicity in business, design, and technology. These laws are guidelines for needing less and getting more. They are: 1. Reduce The simplest way to achieve simplicity is through thoughtful reduction. 2. Organize Organization makes a system of many appear fewer. 3. Time Savings in time feels like simplicity. 4. Learn Knowledge makes everything simpler. 5. Differences Simplicity and complexity need each other. 6. Context What lies in the periphery of simplicity is not peripheral. 7. Emotion More emotions are better than less. 8. Trust In simplicity we trust. 9. Failure Some things can never be made simple. 10. The One Simplicity is about subtracting the obvious and adding the meaningful. See also, *less is more*.

Phrase or Term	Definition
Lawfare	*Lawfare* is a strategy that uses or misuses legal systems and principles as a substitute for other means (e.g., commercial, military, police, or political) to achieve an operational objective or attack an adversary. The punishment is in the process, not the verdict. This may be done using provincial/territorial (state), federal, international, or other courts. The effect may be to: • Confine political debate to a court of law; • Damage or delegitimize a party or party's reputation; • Distort the electoral process; • Drive away business partners and investment; • Encumber and waste a party's resources (i.e., money, personnel, or time); • Permanently delay a project or proposal; • Silence a party; or, • Win a public relations victory. As a military example, prior to starting operations in Afghanistan in 2001, the United States military purchased exclusive rights to all high-resolution, near real-time commercial satellite imagery of the operational area. Access to this information, of obvious military value, was thereby denied to potential opponents by using contract law. As a commercial and political example, environmentalists have used lawfare to impede and stop Canadian pipeline projects such as Energy East, Keystone XL, Northern Gateway, and Trans Mountain Expansion by using an endless series of court challenges and other means. This also involved having thousands or people demand intervenor status at public hearings. The concept of lawfare is traceable to Hugo Grotius and his Latin book *Mare Liberum, sive de jure quod Batavis competit ad Indicana commercia dissertatio*. In English, Mare Liberum means The Freedom of the Seas. Lawfare includes financial lawfare. For example, in 2019 the United States announced plans to impose a new 10% tariff on about USD$ 300 billion of imported Chinese products. Lawfare is a portmanteau of *law* and *warfare*. See also, *anti-strategic lawsuit against public participation law,* and *strategic lawsuit against public participation*.

Lawyer Up	To *lawyer up* means to:
	• Arrange matters in a way that conforms to legal advice;
	• Conduct matters according to legal formalities or to avoid legal risk;
	• Obtain legal representation, especially when refusing to answer law-enforcement officials' questions without legal representation; or,
	• Obtain legal council to:
	  o Defend oneself against a legal claim, especially while being investigated for possible wrongdoing; or,
	  o File a claim for a legal matter or threaten to do so.
	In a dispute, after one party retains legal council, the level of communication tends to decrease, ande the level of conflict and distrust between both parties tends to increase.

Phrase or Term	Definition
Laydown Area, Laydown Yard	A *laydown area* or *laydown yard* is a (usually temporary) space located at or near a construction site that is dedicated to the assembly, receipt, and storage of equipment and materials to be installed or constructed. For example, a laydown area may be used to receive components of a tank, which will be constructed in the future. A contractor or supplier may also dedicate a space at their facility, or another location, for this purpose.
Lead (Hand or Team)	A *lead hand* or *team lead* is a senior leader responsible for the work execution of a trade or discipline team, including activity coordination and timely completion of tasks (e.g., lead buyer, lead mechanical engineer, or lead welder). See also, *discipline* and *lead supplier quality surveillance coordinator*.
Lead From Behind	*Lead from behind* is an expression that describes active leadership by: • A leader who steps back when there is no need to direct or lead from the front (e.g., be the centre of attention). It requires the leader to enable intra-team leadership, encourage innovation, set clear goals, and assist only during key points to unlock the potential of personnel; or, • Leading by example. For example, a leader who asks for feedback, initiates discussions, organizes meetings, and takes action on input also encourages others to do likewise. To lead from behind does not mean to abrogate leadership responsibilities. Nelson Mandela equated being a great leader to being a shepherd – by staying behind the flock and letting the bravest lead others. During this, the flock does not realize they are being directed from behind. See also, *leadership*.
Lead Supplier Quality Surveillance Coordinator, Lead SQS Coordinator	The *lead supplier quality surveillance coordinator* or *lead SQS coordinator* is a senior leader responsible for the SQS team, and providing development services, support services, or both (e.g., for corporate and project). If there is no lead, the SQS coordinator(s) perform SQS activities. See also, *lead (hand or team)*.
Leadership	*Leadership* is the: • Action of leading a group of people or an organization, or the ability to do this; or, • Position or state of being a leader, either formally or informally. Leadership involves: 1. Establishing a clear vision; 2. Sharing that vision with others so others follow willingly; 3. Providing the information, knowledge, and methods to realize that vision; 4. Coordinating and balancing conflicting interests of all members and stakeholders; and, 5. Acting and thinking creatively in difficult situations. Leadership comprises many different leadership styles. Ultimately, a leader compels a group to follow. Strong leadership fosters a working environment in which people feel and want to be their best. Leadership is the art of getting full cooperation from those who are not forced to comply. See also, *five levels of leadership concept*, *Level 5 leader concept*, *lead from behind*, *management*, *managing the grey*, and *managing up*.

Phrase or Term	Definition
Leadership in Energy and Environmental Design	*Leadership in Energy and Environmental Design* is a worldwide ecology-oriented green building certification program. It was developed by the non-profit *United States Green Building Council* to help building owners and operators be environmentally responsible and use resources efficiently. This program provides rating systems for the design, construction, operation, and maintenance of new construction and major renovations, and existing buildings. See also, *green building*.
Leading Indicator	A *leading indicator* is a pattern or trend that is proactive in nature and: • An input-oriented measurement (i.e., hard to measure and easy to influence). For example, activities reported with the aim of preventing adverse events before they happen with safety; or, • Indicative of future events. For example, if housing starts rise, it means builders are optimistic about the demand for newly constructed homes. See also, *coincident indicator* and *lagging indicator*.
Leak Before Break, Leak Before Burst	*Leak before break* or *leak before burst* is the concept that leakage from a circumferential or girth weld will be detected well before it reaches a size that results in a rupture. This means that, although a leak could create a dangerous situation for nearby personnel, the release of pressure, product, or both will likely not be catastrophic. See also, *rupture*.
Leak Test	A *leak test is an atmospheric or pressure test of equipment or systems for strength and leaks*. See also, *pressure test*.
Leak-proof, Leak-tight	*Leak-proof* or *leak-tight* is a feature (e.g., using seal welds or threads) that ensures a vessel will not lose its liquid contents, or something that is constructed and designed to prevent leakage. See also, *gasket*, *leak test*, *seal weld, and thread*.
Lean Construction, Lean Manufacturing, Lean Production, Lean	*Lean construction*, *lean manufacturing*, *lean production*, or *lean* is a methodology focused on minimizing waste within manufacturing systems while simultaneously maximizing productivity. It is based on concepts of the Toyota Production System, which are employed by many companies today – including Toyota. The Toyota Production System automates and standardizes repetitious and tedious processes, and processes that are hazardous to humans or prone to human error (e.g., weld a frame to the chassis of a vehicle using a welding robot instead of human resources). Similarly, in a demand-based flow (pull) manufacturing system, inventory is only delivered (pulled) when it is needed to satisfy a customer's order for just-in-time delivery. Lean principles have profoundly influenced manufacturing concepts worldwide, including industries outside of manufacturing (e.g., healthcare, service industries, and software development). The benefits of lean include improved product quality, and reduced lead times, operating costs, and waste. See also, *just-in-time*, *lean construction*, *manufacturing*, and *minimum viable product*.
Learning Curve, Experience Curve	A *learning curve or experience curve* is often expressed as a graphical representation of how an increase in learning (measured on the vertical axis) comes from greater experience (measured on the horizontal axis). It is the rate at which personnel progress in gaining experience, knowledge, or new skills, where: • A body of knowledge is learned over time; or, • The same task is repeated in a series of trials. See also, *body of knowledge* and *steep learning curve*.

Phrase or Term	Definition
Learning Organization	A *learning organization* is a business or company that continuously transforms itself and facilitates the learning of its members. This phrase was coined by Peter Senge. See also, *system thinking* and *the fifth discipline*.
Left-wing	The *left-wing* is the liberal, radical, or socialist section of a political party or system. This labelling system originated from the seating pattern of the French National Assembly, which put liberals on the left, moderates in the middle, and conservatives on the right. Someone who is left-wing usually supports egalitarianism and social equality. See also, *moving to the left*, *political*, *right-wing,* and *radical*.
Legal Tender	*Legal tender* is anything recognized by law to meet a financial obligation or settle a private or public debt (e.g., exchange for products and services or payment of legal fines or taxes). The national currency is legal tender in most countries. For example, Canada uses the Canadian dollar as its legal tender. Legal tender is defined differently in different countries and does not ban the use of other forms of currency. Banknotes and coins are usually accepted as legal tender. Non-cash methods of payment (e.g., credit cards or personal cheques) are not considered legal tender but may be accepted as payment. Using legal tender as a method of payment does not compel anyone to accept it. For example, a restaurant can refuse payment for a meal if the payment is made using only coins. Some countries allow a specific foreign currency as legal tender, either as its exclusive legal tender or concurrently with its national currency (e.g., the use of the Peso *and* American dollar in Mexico). See also, *cryptocurrency* and *fiat currency*.
Lemming Effect	The *lemming effect* is the phenomenon wherein a group of people exhibit a certain kind of behaviour for no reason except that most of their peers are behaving similarly. It is named after the misconception that lemmings, small Arctic rodents, are driven to commit mass suicide by attempting to cross a body of water or jumping off cliffs during migration. For lemmings, this is not a deliberate, collective, suicidal act; instead, this behaviour is driven by strong migratory instincts that drives lemmings to search for new habitat when population density becomes too great. See also, *diffusion of responsibility*, *groupthink* and *herd behaviour*.
Less is More	*Less is more* is an expression of the view that a minimalist approach to aesthetic, artistic, or other matters is more effective than other methods. It suggests that there is value in simplicity and that more can be accomplished with less. For example, using fewer words can be more poignant and thus more effective to successfully communicate a message. Likewise, using simpler patterns to decorate can look more beautiful and elegant than overly ornate patterns. Or, using many notes and musical techniques can make it difficult for some listeners to enjoy music such as complex jazz. See also, *more is more* and *Laws of Simplicity*.
Lessons Learned	*Lessons learned* are knowledge gained from project experience with review sessions held during close-out. The purpose of lessons learned is to share and use knowledge to promote the recurrence of desirable outcomes and avoid the recurrence of undesirable outcomes. Projects typically archive lessons learned to guide future projects. See also, *brainstorm* and *key knowledge*.

Phrase or Term	Definition
Let the Buyer Beware, Caveat Emptor	*Let the buyer beware* or *caveat emptor* (Latin) means the purchaser assumes the risk that a product or service could have defects or flaws and may fail to meet their requirements. The buyer alone is responsible for performing due diligence (e.g., verifying the quality and suitability of a product or service) before purchasing the product or service. See also, *the customer is always right*.
Letter of Compliance	Not a preferred phrase – see *certificate of conformance*.
Letter of Conformance	Not a preferred phrase – see *certificate of conformance*.
Letter of the Law	To obey the *letter of the law* means to follow the literal interpretation of the words of a rule (i.e., the exact and strict force of the language) but not necessarily the intent. This phrase may be used in a legal or non-legal context. It is the antithesis of the spirit of the law. For example, a request for proposal states that bids shall be received by 12:01 pm on Friday. When a bid is received at 12:05 pm, the project rejects and discards the bid without reviewing it because the bid submitter missed the deadline. See also, antithesis, *spirit of the law*, *technicality*, and *work to rule*.
Level 5 Leader Concept	The *Level 5 leader concept* was developed in the book *Good to Great: Why Some Companies Make the Leap ... and Others Don't* by Jim Collins. It describes the highest level of leadership where a leader: • Embodies a paradoxical mix of personal humility and professional will; • Has ambitions for their organization rather than themselves; • Sets up their successor for even greater success; • Displays compelling modesty, self-effacement, and understatement; • Is fanatically driven with an incurable need to produce sustained results; • Displays workmanlike diligence of a plow horse rather than a show horse; • Attributes success to good luck and factors other than themselves; and, • Takes full responsibility for disappointing results. See also, *five levels of leadership concept*, *leadership*, and *paradox*.
Level of Analysis	The *level of analysis* is how a business or project faces challenges depending on a macro- or micro-level orientation or perspective (no clear line of demarcation separates these perspectives): • Macro-level (large) perspective considers the big picture or high-level view and may incorporate micro-level issues. This approach is more philosophical and less specific in its use of data and statistics. For example, observing that morale or profits have diminished; and, • Micro-level (small) perspective focuses on a view of daily interactions and details. This approach is about single problems and the elements that contribute to them. For example, when a person's work performance is reviewed or training is offered. Level of analysis is also used in other fields such as social sciences and statistics. See also, *big picture*.

Phrase or Term	Definition

Leverage	*Leverage* is:
	• An investment strategy that uses borrowed money to increase the potential return of an investment. This may also refer to the amount of debt an organization uses to finance assets (i.e., the leverage ratio);
	• The ability to:
	  o Inflict loss or withhold gain; or,
	  o influence people or a situation to control the outcome.
	• The exertion of force by means of a lever, or an object used like a lever, to gain a mechanical advantage.
	During a negotiation, the party that is the most afraid of losing and has the most to lose has the least leverage and vice versa. One can acquire leverage by persuading the counterpart that they have something real to lose if the negotiation fails. Leverage is therefore an emotional concept that can be manufactured if it does not exist. Leverage is not necessarily power. For example, someone with has many bottles of water has leverage over someone who is powerful but lives in a desert and lacks water. See also, *leverage multiplier*, *mechanical advantage*, and *three types of leverage*.

Leverage Multiplier	A *leverage multiplier* is hidden but important information that can give a negotiator the advantage. Specifically, it is an unknown unknown, for the negotiator and possibly also for their counterpart. For example, the world did not ask Steve Jobs to create the iPad (first unknown). He uncovered and exploited its need without anyone else realizing it was needed (second unknown). This concept was developed by Chris Voss in the book *Never Split The Difference: Negotiating As If Your Life Depended On It* (with Tahl Raz). See also, *black swan*, *kinds of information*, and *leverage*.
Liaise	To *liaise* is to collaborate or cooperate on a matter of mutual concern. For example, the lead electrical and mechanical engineers must liaise to ensure that the project specifications are properly applied). See also, *coordinate* and *interface*.

License	A *license* is:
	• A document, plate, or tag that provides evidence a license is granted (e.g., a license plate on a vehicle); or,
	• Permission granted by an authority that allows one to:
	  o Act or engage in an activity that is otherwise unlawful (e.g., acquire a fishing license);
	  o Engage in a regulated business or occupation (e.g., incorporate a business); or,
	  o Own or use something (e.g., acquire a forklift operator's licence).
	See also, *certificate* and *permit*.

Phrase or Term	Definition
Life Cycle	A *life cycle* is a series of stages through which an asset (e.g., equipment or materials, a facility, or a product) passes during its usable existence. A life cycle covers the timeframe of *creation to disposal* or *start to finish*. A life cycle spans from the time an item is designed, constructed and installed, extends throughout its service life (i.e., including engineering, commissioning, operation, maintenance, inspection, repair, alteration, enhancement, integrity assessment, replacement, and decommissioning), and ends when the item is no longer a useful or working asset (as determined by a project or owner). See also, *A to Z*, *asset*, *asset integrity management*, *end-to-end*, and *life cycle costing*.
Life Cycle Costing	*Life cycle costing* is an accounting method where expenses are allocated over the life of an asset, facility, or product. For example, the life cycle costs are often lower for stainless steel products than for other alternatives despite a higher initial outlay, because stainless steel products generally last longer and require less maintenance. See also, *asset* and *life cycle*.
Life Hack	A *life hack* is novel method, shortcut, skill, or trick used to increase efficiency and productivity in everyday life. For example, a life hack could be a clever method of removing a dark stain from a white carpet. The phrase originates from the term hack (meaning to chop or cut with heavy or rough blows) and is used to describe an inelegant but effective solution to a specific problem. See also, *hack* and *self-help*.
Lifelong Learning	*Lifelong learning* is the ongoing, self-motivated, and voluntary pursuit of knowledge for personal reasons, professional reasons, or both, over a lifetime. It enhances active citizenship, competitiveness, employability, social inclusion, personal development, and self-sustainability. The concept recognizes that learning is not confined to childhood, youth, or the classroom; instead, it takes place in a range of situations and throughout life. Lifelong learning may be formal (e.g., taking a training course) or informal (e.g., reading a book or using a video tutorial). See also, *boot camp*, *continuing education*, *cross-train*, *lunch and learn*, *professional development unit*, and *workshop*.
Lifting Capacity	*Lifting capacity* refers to the maximum weight that a machine can safely lift. See also, *safe working load* and *safety factor*
Lifting Lug	A *lifting lug* is a mechanical device bolted, built in, or welded to equipment for the purpose of lifting it for installation, loading, removal, or unloading (e.g., usually a crane with cables). See also, *grounding lug*.
Light Tight Oil	*Light tight oil* is a phrase that describes a very low viscosity oil found in certain kinds of shale rock.
Like-for-like	*Like-for-like* is: • A comparison between similar business aspects (e.g., sales in 2019 compared to sales in 2020) or identical things from two time periods (e.g., status of a valve in 2019 compared to 2010); or, • Something of the same or similar quality or value that replaces a damaged or lost item (e.g., repairing vehicle damage caused by an accident). See also, *replacement in kind*.

Phrase or Term	Definition
Likert Scale, Likert-type Scale	A *Likert scale* or *Likert-type scale* is a scale commonly used in research that uses questionnaires to measure the intensity of people's attitudes, feelings, opinions, or perceptions. It is named after its inventor, psychologist Rensis Likert. He distinguished between a scale proper, which emerges from collective responses to a set of items (e.g., usually eight or more), and the format in which responses are scored along a range. A Likert scale refers only to the former. When responding to a Likert scale, respondents specify their level of agreement or disagreement for a series of statements on a symmetric agree-disagree basis. The Likert scale is used in business, education, marketing, psychology, social sciences, statistics, and other areas.
Limiting Belief	A *limiting belief* is something that constrains personnel in some way. Just by believing something is true, one is unable do, envision, say, or think whatever is inhibited. These may be beliefs about abilities, duties, permissions, or rights. Limiting beliefs may be related to one's self or self-identity, or to other personnel and the world in general. These beliefs may be about how: • I can/cannot; • I do/do not; • I am/am not; • Others are/are not; • Others will/will not; or, • The world works (e.g., motorcycles are dangerous but cute animals are not). The result of a limiting belief is that one is prevented from: • Achieving peak performance; • Making better or different choices; • Seeing better or different opportunities; • Seeing more than the negative aspects of life; or, • Seeing all of one's own true abilities or those of others.
Line Management	*Line management* refers to the personnel who manage a labour force that is directly involved in the delivery or production of products, services, or both (e.g., a restaurant manager or foreman). They report to higher-ranking management. See also, *labour force*, *lead hand*, *management*, *middle management*, and *senior management*.
Line of Fire	The phrase being in the *line of fire* describes being in harm's way. For example, if personnel were to stand if front of a pressure vessel when it exceeds its operating parameters, those personnel would be in the line of fire (by way of released energy). The three major categories of line of fire incidents are: • Caught-in or between equipment (e.g., being pinched or pinned between or in operating equipment); • Released energy (e.g., being burned by flame or steam from operating equipment); and, • Struck-by an object (e.g., being struck-by a falling object or moving vehicle). In the military industry, this phrase refers to: • The expected path of a bullet, missile, or projectile; or, • A position that is vulnerable to attack.

Phrase or Term	Definition
Line Pipe	*Line pipe* is steel pipe that is used for transporting materials through pipelines (e.g., natural gas, oil, petroleum, and water). Line pipe must meet certain regulations and specifications.
Linear Economy	A *linear economy* has a take-make-use-dispose model of production. This means that raw materials are mined and then processed to make a product, which is used and then discarded. See also, *circular economy*, *economy of repetition*, *economy of scale*, *gig economy*, *hydrogen economy*, *knowledge economy*, and *low-carbon economy*.
Linear Management	*Linear management* refers broadly to management practices that seek to impose structure on an organization – usually from above (i.e., by senior management). See also, *management*, *nonlinear management* and *senior management*.
Lining	A *lining* is a coating or layer of sheet material applied to the interior surface of a container to protect it from corrosion caused by the contents of the container, to protect the contents of a container from contamination caused by container material, or both. See also, *coating*.
Lip Service	*Lip service* is a declaration of advocacy, adherence, or allegiance expressed in words but not supported by actions. For example, to say: *I am in favour of recycling*, but then *not* recycling recyclables.
Lipstick on a Pig, Pearls Before Swine	To *put lipstick on a pig* is to make cosmetic or superficial changes to something – disguising it as more appealing, attractive, successful, or expensive than it really is. For example, *that proposal is a poor investment, and the pictures of happy people in the associated marketing brochures are only putting lipstick on a pig.* This phrase was coined by a San Francisco KNBR radio host in 1985 to describe a plan to refurbish Candlestick Park instead of constructing a new stadium for the San Francisco Giants. Similar concepts and cultural references to pigs have been used since ancient times. For example, one interpretation of the biblical metaphor pearls before swine is that you should not give that which is valuable to those who do not appreciate its value. There are several other interpretations of its meaning.
Liquefied Natural Gas	*Liquefied natural gas* is methane that is cooled to -161°C (258°F) to becomes a clear, colourless, odourless, non-corrosive, non-explosive, and non-toxic liquid. See also, *methane* and *natural gas*.
Liquid Labour Force	A *liquid labour force* is a team of personnel that can rapidly adapt and change per business needs. This is not a new concept; it is simply a new phrase. It comprises a mixture of traditional permanent employees, an adaptable mix of non-employees (e.g., consultants), and technologies (e.g., automation and robotics). Technology is not only a disrupter, but also an enabler: it transforms organizations and personnel to create a competitive advantage for their businesses. Thus, companies invest in technologies and tools to keep pace with constant change, and may evolve away from legacy work practices (e.g., personnel working 9 to 5 at an assigned office location). See also, *consultant*, *full-time equivalent*, *gig*, *independent contractor*, *labour force*, *side hustle*, *silo mentality*, and *virtual assistant*.
Liquid Organic Hydrogen Carrier	A *liquid organic hydrogen carrier* is an organic compound (e.g., dibenzyltoluene or toluene — an aromatic hydrocarbon) that can absorb and release hydrogen through chemical reactions. Thus, these compounds can be used as storage media for hydrogen. See also, *hydrogen*.

Phrase or Term	Definition
Liquid Penetrant Test, Dye Penetrant Test	A *liquid penetrant test* or *dye penetrant test* is a non-destructive method of examining non-porous materials (e.g., ceramics, metals, and plastics) for defects. This test uses capillary action, where a low surface tension fluid penetrates a surface-breaking discontinuity. The penetrant is a petroleum- or water-based liquid that is coloured by a dye. The dye is removed from the surface after a dwell time and then replaced by a white developer. The developer acts as a blotter, drawing out the penetrant trapped in voids. The penetrant stains the developer, indicating the presence and location of a flaw. Examination is performed under ultraviolet or white light, depending on the type of dye used – fluorescent or nonfluorescent (visible). See also, *discontinuity* and *non-destructive examination*.
Load Shifting	*Load shifting* refers to a short-term reduction in electricity consumption followed by an increase in production later, when the grid demand or power cost is lower. Dedicated generators or electricity storage facilities owned by a power consumer can be used to bridge high-load or high-price phases but play less of a role if production will eventually catch up again. See also, *base load* and *peak shaving*.
Load Test (No, Partial, or Full)	A *load test* is used to transmit *no*, *partial*, or *full*-rated power per purchase order requirements and at the rated input speed. A load is applied using a hydraulic or mechanical method (e.g., dynamometers and pony brakes) until equipment bearing and lube oil temperatures stabilize.
Load-out	*Load-out* is the transfer of materials or equipment onto a vessel or vehicle for transportation to the final destination or site.
Lobby	A *lobby* is an organized group of people who represent an area of business, interest, or society to achieve an outcome more favorable to its agenda or objectives. They conduct activities to influence or persuade politicians or public officials toward a desired action, decision, legislation, or policy (e.g., to oppose or promote a development or project). This includes parties from advocacy or interest groups (e.g., associations or the non-profit sector), corporations (e.g., for-profit or private sector) and other government officials. When politicians or public officials, who have a duty to serve their constituents' interests or the public good, can benefit by decision-making that serves the interests of parties instead of constituent interests, a conflict of interest exists. Therefore, lobbying is often regulated by laws to establish transparency by publicly registering lobbyists and lobby groups in an effort to prevent political corruption. A lobby can also be the main area, typically in the entrance of a structure, where people congregate. See also, *conflict of interest*, *for-profit*, *interest*, *non-profit*, and *rent-seeking behaviour*.
Localization	*Localization* is the practice of adjusting a product's characteristics and functional properties to accommodate the cultural, legal, linguistic, and political differences of a foreign country or market. For example, in English, writing is read from left to right. A graphic depicting a cleaning process would therefore logically show a dirty item on the left and a clean item on the right. In Arabic, however, writing is read from right to left. A graphic depicting a cleaning process in Arabic-speaking countries would therefore need to show a dirty item on the right and a clean item on the left. See also, *globalization* and *internationalization*.

<table>
<tr><th>Phrase or Term</th><th>Definition</th></tr>
<tr><td>Local Control Panel</td><td>A local control panel operates equipment or processes. For example, a dedicated local control panel operates a large underground refrigeration plant. See also, control panel and local control station.</td></tr>
<tr><td>Local Control Station</td><td>A local control station could be a complete process in a group of processes. For example, a large underground refrigeration plant in a deep level mine the local control station would be located on the surface. See also, local control panel and station.</td></tr>
<tr><td>Location Remoteness</td><td>Location remoteness is the distance between a large population centre and project-related infrastructure (e.g., human resources, logistical support, power, roads, and a supply chain). Remoteness may be defined for an onshore project as follows:<ul><li>Non-remote
Local with project-related infrastructure.</li><li>Semi-remote
Among the first projects in an area and project-related infrastructure is still under development; and,</li><li>Very remote
More than 200 km (124 mi.) away from a large population centre (i.e., greater than 50,000 people) and project-related infrastructure.</li></ul>An offshore project is remote if it operates in a new area or province relative to a company or an industry.</td></tr>
<tr><td>Lockout</td><td>Lockout is the isolation of energy from a system (e.g., equipment or process) by physically locking a system into a de-energized or safe mode. A lockout state prevents starting and operating the system. See also, car seal, lockout device, and tag out.</td></tr>
<tr><td>Lockout Device, Locking Device</td><td>A lockout device or locking device is a mechanical means of locking or securing (e.g., using an individually keyed lock) an energy-isolating device into a de-energized or safe mode that prevents the energization of a system. See also, car seal, energy-isolating device, and lockout.</td></tr>
<tr><td>Logical</td><td>Logical is:<ul><li>Capable of reasoning or of using reason in an orderly cogent fashion; and,</li><li>Of, being in accordance with, involving, or relating to logic; being formally true or valid.</li></ul>See also, anecdotal, empirical, logical fallacy, non sequitur, and word salad.</td></tr>
<tr><td>Logical Fallacy</td><td>A logical fallacy is an error in reasoning that undermines the logic of an argument. Faulty reasoning (e.g., an illegitimate, irrelevant, or invalid point) is often identified by the lack of evidence to support the claim. A logical fallacy may be committed intentionally to manipulate or persuade by deception, or unintentionally due to carelessness or ignorance. Fallacies usually involve arguments, although some involve only definitions, explanations, or other products of reasoning. The term fallacy is sometimes correctly used to broadly indicate a cause of a false belief.
It is beyond the scope of this document to list all logical fallacies. A List of Fallacies is published at: https://en.wikipedia.org/wiki/List_of_fallacies. See also, bait and switch, circular reasoning, cognitive bias, either/or, false equivalence, logical, red herring, righteousness fallacy, slippery slope, and straw man argument.</td></tr>
</table>

Phrase or Term	Definition
Logistics	*Logistics* is the part of supply chain management that plans and controls the efficient, delivery and storage of goods between the point of origin and the point of use, to meet project requirements (e.g., transport by land, sea, and air, including international and heavy-lifts). See also, *goods*.
Logistics Coordinator	A *logistics coordinator* performs a logistics role for a project. See also, *coordinate* and *logistics*.
Long Lead	*Long lead* is equipment or materials with long delivery schedules that must be purchased earlier than other products with shorter delivery schedules. Complexity, higher costs, limited production, and supplier location or transportation timelines (e.g., for world-wide shipment) negatively influence delivery schedules.
Long Pipe Fluid Highway®	The *Long Pipe Fluid Highway®* is a pipe manufactured from composite materials comprising a nylon or polyurethane inner lining, multiple resin impregnated glass reinforcing layers, and an outer protective casing. This pipe has many competitive advantages over alternatives (e.g., polyethylene or steel pipe), has a range of applications, and offers opportunities to address several limitations in existing pipe materials. It: • Accommodates extreme temperatures, and large-rapid temperature variations; • Has a high level of internal corrosion resistance; • Is extremely durable and strong, yet very lightweight; • Offers substantial savings for logistics and supply costs compared to polyethylene or steel pipe; and, • Produces only one third of the carbon emissions produced by polyethylene or steel pipe.
Loop	A *loop* is a section of pipeline designed and constructed in parallel to another pipeline, and also connected by construction and technology. Both pipelines transport the same product and share the same destination. A loop increases pipeline flow capacity without altering pipeline pressure or modifying an existing pipeline.
Loophole	A *loophole* is an ambiguity, inadequacy, or omission in a law, rule, or system, which can be used to legally avoid or evade its intent or purpose whether explicitly stated or implied. For example, *the contract had a loophole that made the contract unenforceable*. Originally, a loophole was an arrow slit or narrow vertical window in a wall through which an archer shot arrows. See also, *lacuna*.

Phrase or Term	Definition
Looks Good on Paper	Something that *looks good on paper* refers to an argument, contract, drawing, idea, or plan that seems acceptable, adequate, or plausible in theory, but needs to be modified for implementation or use, or is not adequate or appropriate in application, practice, or reality. For example, a new product design that appears to meet all criteria for aesthetics, durability, operability, and other factors, but would be too expensive to profitably produce (e.g., due to complexity or material cost). For instance, many patented and produced devices were never commercially successful. The pistol sword was developed between the 16th and early 20th centuries for civilian and military use. However, instead of being useful as two weapons, in practice it was never more than one substandard device – a heavy pistol and an off-balance sword. This can also occur with 3D modelling where something that appears acceptable in a computer model proves to be unacceptable or unworkable in practice. For example, the alignment of piping in a system appears acceptable but when constructed the piping clashes with a manual valve and prevents proper access for operation of the handwheel. This phrase also refers to someone with seemingly abundant resources (e.g., money), or excellent credentials (e.g., title) or qualifications (e.g., education, experience, and skills); however, has not acquired the assets or credentials claimed.
Lose-lose, Negative Sum Game	*Lose-lose or negative sum game* is situation where all participants lose, or the sum of winnings (positives) and losses (negatives) is negative. In this situation, each party realizes a loss (e.g., of money, property, rights, ownership, etc.). For example, a negotiation that produces no agreement but incurs legal fees is a lose-lose situation (to everyone involved, except the lawyer!). See also, *business wargame*, *game*, *finite mindset*, *infinite mindset*, *positive sum game*, *win-win*, *wargame*, and *zero-sum game*.

Loserthink	*Loserthink* is a concept that was developed in the book *Loserthink: How Untrained Brains Are Ruining America* by Scott Adams. He describes Loserthink as unproductive thinking that includes:
	• A hollow man argument;
	• A mischaracterization of what someone is trying to say and what their intent is, as a result of attempting to read their mind;
	• An inability to:
	  o Imagine alternative explanations; and,
	  o Keep ego out of decision-making.
	• A propensity to make too much of coincidences; and,
	• An inclination to word-think.
	See also, *cognitive dissonance*, *cognitive flexibility*, *Darwin Awards*, *diffusion of responsibility*, *groupthink*, *fundamental laws of stupidity*, *hollow man argument*, *lucky fool syndrome*, *mental prison*, *one-variable thinking*, *professional jealousy*, *reverse Darwinism*, *thinking mode*, and *word-think*.

Loss Aversion	*Loss aversion* refers to the tendency to avoid losses and instead acquire gains. For example, it is preferred to *not lose* \$5 rather than to find \$5. What distinguishes loss aversion from risk aversion is that the utility of a monetary payoff depends on what was previously experienced or was expected to happen. Risk aversion, for example, is choosing to put money into a bank account rather than buying a stock. See also, *risk* and *risk aversion*.

Phrase or Term	Definition
Lot	A *lot* is: 1. A quantity of product accumulated under conditions considered uniform for sampling purposes; or, 2. Items constituting a defined quantity of uniform product for purposes of proceeding collectively through a process. See also, *sample*.
Low Blow	A *low blow* is a slang phrase that describes an: • Illegal and unsportsmanlike blow that strikes an opponent below the waist (i.e., in the gonads); or, • Unfair, unprincipled, or unwarranted attack, comment, or criticism (e.g., to terminate personnel without cause one week before Christmas). See also, *termination without cause*.
Low-carbon	*Low-carbon is:* • Industry Power sources with minimal greenhouse gas emissions (e.g., carbon dioxide). • Metallurgy Steel (i.e., mild steel) with a low ratio of carbon to iron (i.e., 0.04 to 0.30% carbon); or, Stainless steel with a lower ratio of carbon compared to another grade. For example, 316L stainless steel has less carbon than 316 (the L stands for *low*). 316L is preferred for welding because it is less susceptible to weld decay (i.e., corrosion within the weld) than 316. See also, *carbon*, *grade*, *low-carbon economy*, and *net-carbon neutral*.
Low-carbon Economy, Low-fossil-fuel Economy, Decarbonised Economy	*Low-carbon economy*, *low-fossil-fuel economy*, or *decarbonised economy* is an economy based on low-carbon power sources that have a minimal output of greenhouse gas emissions (e.g., carbon dioxide) into the atmosphere. See also, *circular economy*, *economy of repetition*, *economy of scale*, *gig economy*, *hydrogen economy*, *knowledge economy*, *linear economy*, and *low-carbon*.
Low-hanging Fruit	*Low-hanging fruit* is a popular business expression that refers to actions, goals, or tasks that are most easily (or most quickly) achieved, and because of this, these tasks are usually performed first. The expression derives from the very literal task of picking fruit off a tree that is reachable by standing on the ground, first (before using a ladder). See also, *cherry pick*, *handpick*, *quick win*, and *Pareto Principle*.
Low Hydrogen	*Low hydrogen* is a phrase applied to consumables manufactured to deposit weld metal with a specified maximum limit to the diffusible hydrogen content or to a welding processes with an inherently low diffusible hydrogen level in weld metal (e.g., not more than the requirements of AWS A.3 or CSA W48). See also, *consumable*, *electrode*, *hydrogen embrittlement*, *low hydrogen electrode*, and *non-low hydrogen electrode*.

Phrase or Term	Definition
Low Hydrogen Electrode	A *low hydrogen electrode* produces a controlled or very low quantity of diffused hydrogen in the weld metal compared to other electrodes (i.e., a non-low hydrogen electrode). For example, the coating moisture level is less than 0.6% when tested at 980 C (1,800 F), not more than the requirements of AWS A.3 or CSA W48. Some codes and specifications refer to hydrogen control in terms of either requiring the use of low hydrogen electrodes or placing specific limits on diffusible hydrogen (e.g., <16 mL/100 g). Many codes and specifications use phrases like *low hydrogen electrode*, *low hydrogen weld deposit*. These phrases are defined or listed in CSA W59, but not in AWS A3.0. However, AWS A5.1 lists several electrode classifications with low hydrogen coatings. The purpose of controlling diffused hydrogen is to minimize the possibility of hydrogen-induced cracking. Low hydrogen electrodes are used to weld high pressure or high temperature service steels to avoid hydrogen entrapment in the weld metal. Hydrogen gas entrapped inside the weld metal will cause stress on the weld when it is exposed to higher pressures or temperatures. If the pressure exerted by the hydrogen inside the weld exceeds the tensile strength of the weld, cracks will occur, and the weld may ultimately fail. See also, *controlled hydrogen electrode*, *diffusible hydrogen designator*, *electrode*, *electrode oven*, *hydrogen embrittlement*, *low hydrogen*, *non-low hydrogen electrode*, and *weldability*.
Low-level	*Low-level*: • Means at or of a level below that which is average or normal (e.g., he is a low-level construction labourer); • Describes being of low importance, influence, or rank (e.g., a low-level official); • Involves or relates to people of minor or no authority or influence, or low administrative rank (e.g., a city bus driver is a low-level public servant); or, • Describes ideas or operations that are specific individual components of a systematic operation, wherein the focus is on the details of rudimentary micro functions. Example 1: the physics of gas and water molecules is a low-level description of the same system. Example 2: a project schedule that contains detailed the tasks needed to coordinate day-to-day work in specific areas. See also, *high-level*.
Lube (Lubrication) Oil Skid	A *lube (lubrication) oil skid* provides for the cooling, filtration, and storage of lubrication oil for large rotating equipment (e.g., compressors, pumps, and turbines). See also, *skid*.

Phrase or Term	Definition
Lucky Fool Syndrome, Self-Attribution Bias, Survival of the Most Idiotic	The *lucky fool syndrome* or *self-attribution bias* means one is inclined to take credit for all things that go well, but blames outside forces when things go wrong. In addition to this bias, individuals have difficulty differentiating luck and skill. *Survival of the most idiotic* is a satirical evolutionary theory proposed by Dr. Alan Grant in 2001. It describes the apparent luck which befalls people who make foolish decisions. The book *Fooled by Randomness* by Nassim N. Taleb explains the underestimation of the influence of happenstance on our lives. He said *lucky fools do not bear the slightest suspicion that they may be lucky fools — by definition, they do not know that they belong to such category*. This is also explained in the study *Self-Attribution Bias in Consumer Financial Decision-Making: How Investment Returns Affect Individuals' Belief in Skill*, by Arvid Hoffmann and Thomas Post. They identified how self-attribution bias can hurt investors and lead to repeated mistakes because *they attribute bad returns to factors beyond their control*. The study also identified that when one ignores the role of luck, one is also blind to bad investment behaviour like overtrading or under-diversification. See also, *bias*, *critical thinking*, *fundamental laws of stupidity*, *loserthink*, *reverse Darwinism*, and *thinking mode*.
Luddite	A *luddite* is someone who dislikes, or is unfamiliar with, new technology. Originating at the beginning of the Industrial Revolution, the Luddite Labour Movement protested how mechanized manufacturers used unskilled labourers to undermine the work of skilled craftsmen. The Luddites were textile workers and weavers who objected to the increased use of mechanized knitting frames and looms. Most were trained artisans who spent years learning their crafts and feared that unskilled labour would rob them of their livelihoods. Consequently, the Luddites began breaking into factories and destroying textile machines (the first major occurrence happened in 1811 in Nottingham, United Kingdom). The Luddites named themselves after Ned Ludd, who was rumoured to have destroyed a textile machine in 1779. There is no evidence, however, that Ludd was a real person. In the 20th century, luddite became a synonym for technophobe. See also, *Industrial Revolution* and *technophobe*.
Lunch and Learn	A *lunch and learn* is a seminar, which is typically catered, occurs at lunchtime, and is: • A formal or informal information or instructional training session for personnel; or, • A sales meeting, where a service provider or supplier presents products-services to existing or potential customers. See also, *boot camp*, *continuing education*, *cross-train*, *lifelong learning*, *on-the-job training*, *professional development unit*, and *workshop*.

Phrase or Term	Definition
M	1 A B C D E F G H I J K L **M** N O P Q R S T U V W X Y Z
Machine to Machine	*Machine to machine* refers to direct communication between devices using any communications channel, including wired and wireless connections. This communication is applicable to industrial instrumentation and can: • Enable a sensor or meter to communicate information it records (e.g., inventory level or temperature) to software that can use it. • Place an order and replenish inventory or adjust a process based on temperature. In the past, communication was more central and relied on a remote network of machines that relayed information to a central hub for analysis.
Machine, Mechanical Device	A *machine* or *mechanical device* is an apparatus that provides power using mechanical means (e.g., an engine).
Machining	*Machining* is any of various processes in which a piece of raw material is cut into a desired final shape and size by a controlled material-removal process. It is a part of the production of many metal products, but it can also be used on other materials (e.g., ceramic, composites, plastic, and wood).
Magnetic Particle Test, Mag Particle Test	A *magnetic particle test* or *mag particle test* is a non-destructive method of examining ferromagnetic materials (e.g., cobalt, iron, nickel, and some of their alloys – such as steel) for defects. This test uses an internal magnetic field that is generated in the test specimen. In locations where flaws (i.e., non-magnetic voids) exist, some of the field will leak into the air. Magnetic (i.e., iron oxide) particles are dusted over the magnetized area and attracted to external fields or magnetic flux leakage. The buildup of particles forms a flaw indication at any shallow subsurface or surface-breaking discontinuities. The particles may be applied either dry, or in petroleum- or water-based liquid. Before testing, a white contrast paint may be applied to the test specimen. Examination is performed under ultraviolet or white light, depending on the type of dye used – fluorescent or nonfluorescent (visible). See also, *discontinuity* and *non-destructive examination*.
Main Gang Crew	A *main gang crew* completes the bulk of the mainline welds along a right-of-way. Welds that would result in the inefficient use of their time are welded by other crews. See also, *crew*.
Mainline	A *mainline* is a pipeline used to carry product in large volume from one point to another for general or collective use. See also, *gathering line* and *service line*.
Mainstream Media, Mass Media	*Mainstream media* or *mass media* refers collectively to the various large mass news media (e.g., magazines, movies, newspapers, radio, and television) that influence people, and both represent generally accepted beliefs and opinions and shape prevailing thought. This media is designed to appeal to as large an audience as possible. See also, *alternative media* and *the press*.

Phrase or Term	Definition
Maintenance Repair and Operating Supply Items	*Maintenance repair and operating supply items* are items used in production processes but do not comprise the products themselves. These items may include: • Computers; • Consumables (e.g., cleaning, laboratory, and office supplies); • Equipment supplies (e.g., gaskets, lubricants, and repair tools); • Industrial equipment and parts (e.g., compressors, pumps, and valves); and, • Personal protective equipment (e.g., ear plugs, gloves, and safety glasses). These items are required keep supply chains running smoothly. Failure to routinely maintain and repair systems can cripple or bankrupt a business. Businesses that diligently maintain an inventory of these items and optimize supply chains can better supply products to customers in an economical, reliable, and timely manner.
Major Repair	A *major repair* is usually defined by a project specification rather than a code of construction. For example. ASME BPVC Section VIII Division 1 UG-78 and UW-38 or CSA Z662 7.12 do not use major/minor repair terminology. Typically, a major repair is one that will require additional inspection and/or prior approval (e.g., by the project), and may also require an approved repair procedure. A major repair may be defined as • Affecting a defined depth or percentage of wall thickness; • Occurring after post weld heat treatment; • Relating specifically to the materials and production of pressure equipment; • Requiring the removal and replacement of a major part of the pressure boundary, other than a nozzle, and is not an alteration; • Requiring the removal of a crack or crack-like indication; • Requiring subsequent repair operations to repair a repair (e.g., maximum two attempts without mandatory cut-out); See also, *alteration*, *cut-out*, *minor repair*, *repair*, and *repair procedure*.
Make-work Job, Make-Work Project	A *make-work job* or *make-work project* is an enterprise or task intended primarily to keep personnel busy with something that that serves little or no useful purpose. The benefits produced by performing the task are less than the cost required to support the effort. For example, digging holes and filling them up again. This may be done to: • Address periods of high unemployment (e.g., for economic relief); • Fulfill a government's social responsibility; • Give unemployed personnel something to do; • Help personnel learn skills required to obtain other productive employment; or, • Occupy personnel until there is a task to perform (e.g., sweep the floor until a truck arrives that needs to be unloaded, even though the floor was swept two hours ago). Make-work jobs occur in all sectors (e.g., business, government, and the military). See also, *busy work*, *fool's errand*, and *government job*.

Phrase or Term	Definition
Male Connector, Male Fitting	A *male connector* is hardware that connects to a cable, piece of hardware, or wire having one or more exposed, shielded electrical terminals, and constructed such that a female connector can be inserted snugly into it. *A male fitting is* fitting or pipe with threads on the outside (like a bolt) that screw into female threads located on the inside (like a nut). In electrical and mechanical manufacturing and trades, each half of a pair of mating connectors or fittings is conventionally designated male or female. The *female* connector is generally a receptacle that receives and holds the *male* connector. See also, *fitting*, *male connector*, and *male fitting*.
Management	*Management* is the: • Administration and organization of the activities and resources of a business or organization to achieve defined objectives; • Judicious use of means to accomplish an end; or, • Personnel who have the power and responsibility to coordinate efforts, make decisions, set strategies, and oversee or supervise a business or organization. See also, *administrative theory, functions of management*, *leadership*, *line management*, *managing the grey*, *managing up*, *middle management*, *principles of management*, *quality management*, *risk management*, *supplier quality management*, and *senior management*.
Management by Objectives, Management by Results	*Management by objectives* or *management by results* was first popularized by Peter Drucker in his book *The Practice of Management*. This process: 1. Defines specific objectives within an organization that management can convey to its personnel, who can then decide on how to achieve each objective in a sequence; 2. Allows managers to take work that needs to be done, one step at a time, to allow for a calm and productive work environment; and, 3. Helps personnel to see their accomplishments as they achieve each objective, which reinforces a positive work environment and a sense of achievement. An important part of this process is the measurement and comparison of a person's actual performance with the standards set. Ideally, when personnel themselves have been involved with the goal setting and choosing the course of action to be followed by them, they are more likely to fulfill their responsibilities. This can be described as a process whereby the superior and subordinate jointly identify common goals, define everyone's major areas of responsibility in terms of the results expected of them, and then use these measures as a guide for operations and to assess the contribution of each person. See also, *deliverology*, *goal*, *management*, *objective*, and *SMART*.
Management Information System	A *management information system* is a computerized information-processing system designed to support the activities of company or organizational management.
Management of Change	*Management of change* is a system that ensures safety, health, and environmental risks and hazards are properly controlled when an organization makes permanent or non-permanent changes to facilities, operations, procedures, documentation, or personnel. See also, *management* and *second-order effect*.

Phrase or Term	Definition
Managing in the Grey	*Managing the grey* is understanding a situation with clarity and applying appropriate strategies for informed decision-making. The concept was developed by Joseph L. Badaracco Jr. in the book *Managing in the Gray: Five Timeless Questions for Resolving Your Toughest Problems at Work*. His philosophy purports that resolution of grey area issues should: • Be approached as a manager; and, • Resolved as human beings. These activities are the core of leadership. Badaracco provides a way to resolve grey area issues by asking five questions in the context of managing difficult decision-making. This approach improves deliberation and judgment because the questions complement, correct, and strengthen each other. The questions are: 1. What are the net consequences? 2. What are my core obligations? 3. What will work in the world as it is? 4. Who are we? 5. What can I live with? See also, *management*, *leadership*, and *net net*.
Managing Up	*Managing up* means learning your manager's strengths and weaknesses and using that knowledge to improve the quality of your work and working relationships. Managing up is about deliberately and proactively recognizing a manager's thinking patterns. It is not about being manipulative, obsequious, or sneaky; instead, it is about being assertive, proactive, and strategic. Skillful attentiveness to interactions for yourself and your manager fosters a proactive relationship and avoids surprises. Managing up means doing excellent work and becoming a genuine source of help and value to your manager and organization. This is a strategic approach that allows professionals to take responsibility for their career needs, trajectories, and wants. Regardless of the type of manager you have, fostering a healthy relationship will reinforce your reputation as a valued asset. Personnel skilled at managing up: • Anticipate a manager's needs; • Know the correct way to approach to a manager about a problem; and, • Recognize what makes a manager angry and happy. See also, *leadership* and *management*.
Man-day	Not a preferred term – see *labour day*.
Manhole	A *manhole* is an access and egress point that allows personnel to travel within an underground or enclosed structure (e.g., a pressure vessel or tank), typically for maintenance and inspection purposes. See also, *hand hole*.
Man-hour	Not a preferred term – see *labour hour*.
Manpower	Not a preferred term – see *labour force*.

Phrase or Term	Definition
Mansplaining	*Mansplaining* is the explanation of something by a man, typically to a woman, in a manner regarded as condescending or patronizing, with the mistaken assumption that he knows more about it than her. Although Rebecca Solnit did not use this term in her essay *Men Explain Things to Me*, her later book (of the same title, co-authored with Ana Teresa Fernández) described exactly what it is. See also, *sexism* and *splaining*.
Manual	*Manual* is: • A device (e.g., valve) that is operated by hand; or, • A book of instructions, especially for operating equipment or learning a subject; a handbook (e.g., an installation operation and maintenance manual). See also, *manual valve*.
Manual Valve	A *manual valve* requires someone to adjust it by hand, using a hand wheel, lever, or geared mechanism. A manual valve cannot be operated automatically or remotely. See also, *actuated valve*.
Manufacturer	A *manufacturer* is an individual or company that assembles, designs, manufactures, or makes something. See also, *manufacturing*.
Manufacturer's Data Report	A *manufacturer's data report* contains the project equipment manufacturing and testing data that is certified by the supplier and a safety codes officer, or authorized inspector. Common forms include ABSA AB-25, AB-28, AB-83, AB-83F, API STD 650 Appendix L, and ASME U-1, U1A, U4.
Manufacturer's Standard	The *manufacturer's standard* is the manufacturer's method of doing something (e.g., using a standard operating procedure). This method may use internally-developed or proprietary procedures, industry or regulated procedures, or a combination of these. The method could involve delivering a service, making a product, managing a process, or supplying materials. It is often a lower cost alternative compared to using an owner's specification. See also, *specification*.
Manufacturing	*Manufacturing* includes all preparation, assembly, and testing, or other production activities performed at a supplier's facilities. The term manufacturing may be used where the products are typically mass produced (e.g., with a model number). See also, *fabrication*, *manufacturer*, and *lean manufacturing*.
Manufacturing Record Book	A *manufacturing record book* is a document that contains the final compilation of turnover, quality control, and other documentation per supplier data requirements (including as-built, certified, and record copies). The content and format of this document is typically specified in the purchase order. See also, *mud book* and *manufacturing record book index*.
Manufacturing Record Book Index	A *manufacturing record book index* is the table of contents from a manufacturing record book. The project usually reviews this index before required documentation is compiled and submitted. See also, *manufacturing record book*.

Phrase or Term	Definition
Mapping	*Mapping* the process of assigning information to a bonded or welded joint to enable easy identification of relevant information and records, such as: • Design (e.g., bonding procedure specification or welding procedure specification); • Production operations (e.g., date, post weld heat treatment and hydrostatic testing); • Production personnel (e.g., welder identification and qualification); • Quality (e.g., non-destructive examination and visual examination); and, • Traceability (e.g., base metal and consumable heat numbers).
Market Bottom	A *market bottom* is a trend reversal at the end of a financial market downturn where the beginning of an upward moving trend begins (i.e., a bull market). See also, *bear market*, *market index*, *market top* and *market trend*.
Market Economy	A *market economy* is an economic system in which economic decisions, including the pricing of goods and services, are made by businesses and consumers (the market). There is no government intervention in a purely free market or *laissez-faire* economy; however, no truly free market economy exists in the world. While a country may be capitalistic, the government may attempt to control or regulate economic activities or offer government programs. In this type of economy, there is a separation between the government and the market to keep the government's interests aligned with that of the market and to prevent the government from becoming too powerful. See also, *capitalism*, *centrally planned economy*, *economic system*, *goods*, *mixed economy*, and *traditional economy*.
Market Index	A *market index* is a hypothetical portfolio of investment holdings that represents a segment of the financial market with a calculation of its value made from the prices of the underlying holdings. Examples include the: • Dow Jones Industrial Average • Nasdaq Composite Index; and, • Standard & Poor's 500 Index. See also, *bubble*, *market bottom*, *market top*, and *market trend*.
Market Share	*Market share* is the percentage of total sales in an industry generated by a particular company. It is calculated by taking the company's sales over time and dividing it by the total sales of the industry over the same time. This metric is used to give a general idea of the size of a company in relation to its market and its competitors. It is also a metric of the popularity of a brand or product. The company with the largest market share in an industry is the market leader. For example, as of October 2020, Apple has a 24.99% worldwide market share of mobile vendor devices. See also, *advertising*, *brand*, *market share*, *marketing*, *mindshare*, and *positioning*.
Market Top, Market High	A *market top* or *market high* is a trend reversal at the end of a financial market upturn where the beginning of a downward moving trend begins (i.e., a bear market). See also, *bull market*, *market bottom*, *market index*, and *market trend*.

Phrase or Term	Definition
Market Trend	A *market trend* is a perceived tendency of financial markets to move in one a particular direction over time. The trends are classified as: • Secular for long time frames; • Primary for medium time frames; and, • Secondary for short time frames. In reality, prices in the future are unknown – a trend can only be determined in hindsight. See also, *bear market*, *bull market*, *market bottom market index*, and *market top*.
Marketing	*Marketing* is a conceptual framework that directs and guides the relationships of companies and clients-customers. It is attracting desirable consumers (e.g., those with interest and expendable incomes) and repelling undesirable consumers (e.g., those without interest and expendable incomes). Unlike advertising and social media, marketing comprises the systems used to enroll, reach, and retain clients-customers. See also, *attention interest desire and action principle*, *brand*, *advertising*, *copywriting*, *content marketing*, *market share*, and *mindshare*.
Marking and Labelling	*Marking and labelling* are the identifiers applied to packaging for product shipment, to communicate information that may include: • Compliance for export and import (e.g., customs and taxation); • Handling and storage requirements (e.g., dimensions, indoor or outdoor storage, and weight); • Hazardous goods and emergency responder instructions (e.g., flammable goods); • Identification of the shipper and receiver (e.g., names and addresses); • Product contents (e.g., identification and number of items or quantity); • Project name, purchase order number, and tag number, if any; • Shipping and receiving data (e.g., country of origin); or, • Specified languages (e.g., English and other languages). The information provided by marking and labelling may be defined by the original equipment manufacturer, project specifications, regulations, and other sources, or a combination. See also, *label*, *packaging*, *preservation*, and *shipping and handling*.

Markup	A *markup*:
	• Is a fee to cover an intermediary's profit or internal expense (e.g., accounting and administration), typically calculated as a percentage of the expense. For example, with:
	  o No markup, a meal expense for $23.45 is charged at $23.45; and,
	  o A 10% markup, a meal expense for $23.45 is charged at $25.80.
	• Are annotations in documents, diagrams and text that indicate changes and revisions, and provide comments, corrections, and explanations about document content. Use of markups to progress design and engineering work is conventional and may be provided by way of hard copy or digitally.
	See also, *flow-through* and *squad check*.

Phrase or Term	Definition
Mass Balance, Material Balance	*Mass balance* or *material balance* is the application of conservation of mass to the analysis of physical systems. By accounting for material entering and leaving a system, mass flows can be identified that might have been difficult to measure or unknown without this technique. The exact conservation law used in the analysis of the system depends on the context of the problem; however, all revolve around mass conservation (i.e., matter cannot disappear or spontaneously be created). This process is widely used in engineering and environmental analyses. See also, *heat balance*.
Massage	To *massage* means to: • Doctor or manipulate information (e.g., evidence, facts, figures, or statistics) to deceive, manufacture a more acceptable result, or support a position; or, • Knead, press, rub, squeeze, or tap the joints and muscles of the body using a masseuse's arms, hands, or instruments for patient relaxation or physiotherapy.
Master Inspection and Test Plan	*A master inspection and test plan*, also known as a summary, identifies all inspection and test plans that will be produced for a purchase order. Some equipment and materials require numerous inspection and test plans for the supplier and sub-suppliers, and many disciplines and or scopes of supply or work. These individual documents may not cross-reference each other and may be produced and managed separately. Therefore, a master or summary document is useful to identify the entire list and provide some information about each listed item. See also, *inspection and test plan* and *quality plan*.
Master of Business Administration, Master's in Business Administration, MBA	A *Master of Business Administration*, *Master's in Business Administration*, or *MBA* is a graduate degree that provides theoretical and practical business or investment management education. Graduates understand business management functions and may have a general, discipline-specific focus (e.g., accounting, finance, marketing, or relationship management), or industry-specific focus (e.g., entrepreneurship, information technology, or project management). See also, *degree* and *Master of Business Administration-ization*.
Master of Business Administration-ization, MBA-ization	*Master of Business Administration-ization* or *MBA-ization* is the criticism that business-school graduates care more about financial performance (e.g., numbers, PowerPoint presentations, and spreadsheets) than about delivering and improving products or services for their customers. An MBA education may focus on finance and following the numbers with mathematical models and ratios, instead of on business and producing innovators and leaders that are curious, humane, and moral—a major distinction. High-altitude postulating may be offered about how to manage profits and loses with few practical ideas about how to foster growth and innovation. A manager's effectiveness cannot be predicted by their academic achievement alone. In his article *Political Instability May Be a Contributor in the Coming Decade*, Russian-American scientist Peter Turchin predicted a period of growing instability in the United States and western Europe due, in part, to the overproduction of young graduates with advanced degrees. See also, *Master of Business Administration*.

Phrase or Term	Definition
Mastermind Group	A *mastermind group* is a peer-to-peer mentoring group with members who accelerate their success, create synergies, and solve problems by collaboratively sharing advice, contacts, experience, feedback, ideas, information, knowledge, perspective, and resources. This phrase was coined by author Napoleon Hill in his book *The Law of Success*. This concept was further refined in his subsequent book *Think and Grow Rich*. He said that those in tune with THE Mastermind (i.e., God, the Universal Power or your term for the all-powerful life force) have significantly more positive energy available – power that can be focused on their success. There are two types of mastermind groups: one is focused on individual success; the other, on group success. An optimal mastermind group meets regularly, with rotating leadership, and is limited to between 6 to 10 members. Mastermind groups have been used by some of the most successful people in history, including Alexander Graham Bell, Henry Ford, and Thomas Edison.
Match-marking	*Match-marking* is an activity that involves placing marks on adjacent separable parts of an assembly to aid in (re)assembly at site. This may be completed with a fit-up test, tagging, and marked-up drawings to further aid in the assembly of parts. Match-marking is especially useful for parts that are numerous, appear similar, are difficult to manipulate (e.g., must be lifted using a crane or forklift), or a combination of these factors. See also, *fit-up test*.
Material	*Material* is items used by a supplier for fabrication or manufacturing (e.g., bulks, components, consumables, elements, parts, pieces, and supplies). See also, *bulk items*.
Material Classification	The *material classification* is a category that is determined by a material's physical and chemical characteristics, and intended use. Material classification is applicable to processed and unprocessed (e.g., raw) materials. For steel, this is specified by grade and type. There are thousands of grades of steel (e.g., 316L or S31603, 4140 or 42CrMo4, and 359 Category II or X52 PL2) and four types of steel (e.g., alloy, carbon, stainless, and tool). Steel can be classified using these parameters: • Composition (e.g., alloy, carbon, or stainless); • De-oxidation process (e.g., oxygen removed during the steelmaking process for killed or semi-killed steel); • Finishing method (e.g., cold-drawn, cold-finished, cold-rolled, or hot-rolled); • Form or shape (e.g., bar, pipe, plate, rod, sheet, structural, or tube); • Heat treatment (annealed, normalized, or quenched and tempered); • Microstructure (e.g., austenitic, ferritic, martensitic, or pearlitic); • Physical strength (e.g., tensile and yield strength per ASME, ASTM, or CSA standards); • Production method (e.g., continuous cast or electric furnace); and, • Quality nomenclature (e.g., commercial, drawing, or pressure vessel quality). See also, *grade*, *material*, and *type*.
Material Request for Purchase	Not a preferred phrase – see *request for purchase*.
Material Request for Quotation	Not a preferred phrase – see *request for quotation*.

<table>
<tr><th>Phrase or Term</th><th>Definition</th></tr>
<tr><td>Material Requirements Planning</td><td>Material requirements planning is a production inventory control, planning, and scheduling system used to manage manufacturing processes. Most systems are software-based, but it is possible to manage these processes manually. A material requirement planning system is intended to simultaneously meet these three objectives:
• Ensure materials are available for production and products are available for delivery to customers;
• Maintain the lowest possible material and product levels in inventory; and,
• Plan delivery schedules, manufacturing activities, and purchasing activities.
See also, capacity requirements planning.</td></tr>
<tr><td>Material Requisition</td><td>Not a preferred phrase – see requisition, request for quotation, or request for purchase.</td></tr>
<tr><td>Material Test Report, Mill Test Report</td><td>A material test report or mill test report is a quality assurance document used in the metals industry to certify a material's chemical and physical properties, and compliance with specific material standards.</td></tr>
<tr><td>Materials Management</td><td>Materials management is a core supply chain function and scientific technique concerned with the control, organization, and planning of the flow of materials – from the initial purchase to the final destination. For example, managing the ongoing purchase of thousands of pipes, fittings, and flanges of many types from suppliers, and the shipment of these items to be free-issued for use at the final destination (i.e., to manufacturers or fabricators at their facilities, to constructors at site, or both). In materials management, the primary objectives are:
• Assure the supply of material;
• Ensure minimum deviation between actual and planned results; and,
• Optimize inventory levels.
Materials management is responsible to:
• Communicate material requirements to procurement and other project functions;
• Determine inventory levels and amount of material required at each location; and,
• Establish plans for material replenishment.
See also, five Rs of materials management, free-issued, management, and material manager.</td></tr>
<tr><td>Materials Manager</td><td>A material manager is responsible for the planning and procurement of materials, material and information flow, and the control and management of inventory to achieve business objectives. See also, material management.</td></tr>
</table>

Phrase or Term	Definition
Materials Responsibility Matrix	A *materials responsibility matrix* is an organizational document used to transpose a project's strategy into an itemized matrix of responsibilities for the control and supply of equipment and materials. The matrix assures the project does not overlap the supply of items and closes supply chain gaps between the constructors, engineering and procurement company, owner, project management office, and module yards. It typically evolves during project planning, with increasing levels of definition at each stage. The matrix serves as an important appendix to design-build and construction contracts, particularly when the owner or engineering and procurement company are responsible to deliver equipment and materials. For example, this avoids a scenario where the electrical building supplier's purchase order and the construction contractor's contract both require the same supply of stairs and platforms.
Matrix of Report Information	A *matrix of report information* summarizes report information at the beginning of a report distribution email. This matrix allows recipients to quickly identify relevant information, even when working on one or more projects. This matrix may include: 1. Project name; 2. Purchase order number; 3. Commodity; 4. Supplier and supplier location; 5. Assignment and discipline; and, 6. Quality surveillance representative name.
Matrix Organization	Personnel in a *matrix organization* report to more than one manager, with different accountability and reporting lines representing different organizational functions or projects. A matrix structure can boost employee engagement and cross-field approaches to problem solving, but it can also create ambiguity over personnel roles. See also, *accountability* and *reporting line*.
Mature Fine Tailings	*Mature fine tailings* are a gel-like material created from processing clay fines that are extracted from an oil sands mine. See also, *fine tailings*, *tailings*, and *oil sands*.
Mature Quality Management System	A *mature quality management system* is a system that has processes that are well-embedded into a business or organization's culture, with respect to product or service or quality management. The five maturity levels through which a business or organization progresses are: 1. Uncertainty; 2. Awakening; 3. Enlightenment; 4. Wisdom; and, 5. Certainty. See also, *management* and *quality management system*.
Maximum Allowable Working Pressure	*Maximum allowable working pressure* is the wall strength of a pressurized vessel (i.e., a pipeline or storage tank). The maximum allowable working pressure determines the amount of pressure vessel walls may safely accommodate during normal operation.
May	In this document, the word *may* is used where alternatives are equally acceptable.

Phrase or Term	Definition
McCarthyism	*McCarthyism* is the practice of smearing others with baseless accusations. This refers to the tactics of Senator Joseph McCarthy, who destroyed the careers of many prominent Americans by branding them Communists between 1950 and 1954.
Measure Twice and Cut Once, Measure Twice Cut Once	*Measure twice and cut once* or *measure twice cut once* means: 1. Figuratively: o Plan and prepare carefully before acting; or, o Think of what could go wrong or have a contingency plan before deciding. 2. Literally Double-check measurements for accuracy before cutting material (e.g., metal or wood). If material is cut too short, it must be discarded and a new piece must be used – which is a waste of material and time. The meaning of this phrase makes this common construction expression humorous: *I cut [the material] twice and [the material] is still too short*! See also, *contingency plan*, *do it right the first time*, and *double-check*.
Mechanical Advantage	*Mechanical advantage* is the: • Advantage gained using a machine system, mechanical device, or tool to transmit force (e.g., a lever); or, • Ratio of force produced by a machine to the force applied to it, used to assess the performance of a machine. See also, *leverage*.
Mechanical Integrity	*Mechanical integrity* is the management of critical process equipment to ensure it is: • Designed and installed correctly; and, • Operated and maintained properly. See also, *fixed equipment mechanical integrity*, *pressure equipment integrity management system*, *process equipment*, *reliability based mechanical integrity*, and *report by inclusion*.
Mechanical Joint	A *mechanical joint* is a connection made for leak resistance, strength, or both.
Mechanical Run Test	A *mechanical run test* runs equipment after bearing and lube oil temperature have stabilized for a predetermined period, and at a specified percentage of maximum flow rate, maximum flow rate or continuous speed. This test demonstrates compliance with purchase order requirements and may also be completed as an extended test, with additional specified procedures or requirements.
Mechatronic Technology	*Mechatronic technology* combines electronics and mechanical engineering. For example, an automated robotic welding machine that welds vehicle frames to vehicles' chassis on an assembly line. See also, *engineering*, *electro-mechanical equipment*, and *technology*.

Phrase or Term	Definition
Media Bias	*Media bias* is the prejudice or perceived bias that favours or disagrees with a group, person, or thing when compared with another by editors and journalists within the mass media. Media reports events and stories selectively, which is inherently unfair; it is the opposite of *fair and balanced* news. This phrase implies a pervasive or widespread bias that contravenes the standards of journalism, not the perspective of an individual article, editor, or journalist. See also, *brainwashing*, *bullshit*, *disinformation*, *fake news*, *false narrative*, mass media, *misinformation*, *propaganda*, *standards of journalism*, *the press*, *smells fishy*, *stretch the truth*, *truth*, *truthiness*, and *untruth*.
Medium-sized Business	A *medium-sized business* is defined by Industry Canada as a business with 101 to 500 employees. Other organizations may have different definitions. See also, *big business*, *medium-sized business*, and *small business*.
Megaproject Paradox	The *megaproject paradox* is that even with risky scenarios, megaprojects have never been more in demand, and the size and frequency megaprojects have never been larger. See also, *megaproject*, and *paradox*.
Megaproject, Gigaproject, Teraproject	A *megaproject*, *gigaproject*, or *teraproject* is a capital endeavour (e.g., large infrastructure construction project) that is characterized by: • A long timeline for development and construction (i.e., years); • Extreme or vast complexity (especially in organizational terms); • Involvement of multiple public stakeholders, private stakeholders, and large investment commitments (i.e., in the billions or trillions of dollars); and, • Potentially long-lasting impacts to the economy, environment, and society. According to Merriam-Webster, the first use of the term megaproject was in 1976. Mega is an adjective that means very large; but as a unit of measurement, it means a million. Giga means a billion. Tera means a trillion. Recent developments in the size of the very largest projects indicate that we are in the tera era of trillion-dollar projects. Projects with costs over CAD$ 1.34T (USD$ 1T) compare with the gross domestic product of the world's top 20 countries, and are similar in size to the national economies of Australia and Canada. See also, *gross domestic product*, *high-risk project*, *high-value project*, *infrastructure*, and *project*.
Member	A *member* is an individual angle, beam, plate or piece that becomes an integral part of an assembled frame or structure. See also, *beam* and *column*.
Mental Prison	A *mental prison* comprise the illusions and unproductive thinking that limit one's ability to clearly understand and act rationally. See also, *hollow man argument*, loserthink, *mental prison*, *one-variable thinking*, *professional jealousy*, *thinking mode*, and *word-thinking*.
Mentally Flexible, Open-minded	*Mentally flexible* or *open-minded* personnel are not afraid of errors or mistakes and are willing to embrace and try to understand other approaches, ideas, or perspectives. They know that any decision can result in new possibilities. They realize that wrong decisions are not *bad decisions*, because any decision is good if it is followed by another decision that is advantageous. See also, *mentally rigid*.

Phrase or Term	Definition
Mentally Rigid, Closed-minded	*Mentally rigid* or *close-minded* describes someone who does not understand or rejects all points of view except their own. Hence, these people often reject new, unfamiliar, or innovative approaches, ideas, or perspectives. Mental rigidity occurs when someone desires the elimination of uncertainty that arises from a situation but is only willing to explore conventional or known solutions. Mental rigidity inhibits creativity, exploration, imagination, and improvement. Mentally rigid people: • Assume that their perspective is the only correct perspective; • Believe there is only one way to do things correct. • Cling to the past and refuse to make progress or modernize; and, • Resist change and are xenophobic. See also, *mentally flexible*.
Mentoring	*Mentoring* is a personnel training system under which an experienced or senior personnel is assigned to act as an advisor, counselor, or guide to junior personnel or trainees. The mentor is responsible for providing constructive feedback and support to the person under their tutelage. See also, *coaching*, *cross-generational mutual mentoring*, *nextpert*, *peer mentoring*, and *reverse mentoring*.
Meritocracy	*Meritocracy* is organization, society, system, or government in which people are chosen and moved into positions of influence, power, and success based on examination or performance (e.g., ability, credentials, education, effort, and merit) rather than social class or wealth. While this concept is ancient, the term *meritocracy* was coined by sociologist Michael Dunlop Young and described in his book *The Rise of the Meritocracy*. See also, *government*, *kleptocracy*, *oligarchy*, *plutocracy*, and *populism*.
Messy	*Messy* is: • Extremely difficult or unpleasant (e.g., a messy project); • Lacking care, cleanliness, neatness, or precision (e.g., a messy presentation); or, • Marked by confusion or disorder (e.g., a messy desk). See also, *doodle*, *messy people*, *meticulous*, and *people are messy*.
Messy People	*Messy people* are individuals who appear to be chaotic, disorganized, sloppy, or untidy. Organizations and society value order and predictability; however, a chaotic workspace may not represent a chaotic mind. A messy person may: • Be brave and spontaneous (e.g., focus on the big picture instead of details); • Find inspiration in messiness (e.g., break free of tradition to produce fresh insights); or, • Not be concerned with the status quo (e.g., flout convention and propriety). See also, *big picture*, *doodle*, *messy*, *meticulous people*, and *people are messy*.

Phrase or Term	Definition
Metacognative Thinking, Metacognition	*Metacognative thinking* or *metacognition* is: • A critical awareness of one's own learning and thinking, and oneself as a learner and thinker; • Cognition about cognition, knowing about knowing, or thinking about thinking; • Knowledge about when and how to use a strategy for learning or problem-solving; and, • The processes used to assess, monitor, and plan one's own performance and understanding. Metacognitive thinking helps students and others (e.g., non-students and professionals) become aware of their strengths and weaknesses (e.g., as leaders, learners, readers, team members, test-takers, or writers). Recognizing the limit of one's ability or knowledge is key to understanding how to expand that knowledge or improve that ability. See also, *active learning*, *Bloom's Taxonomy*, *cone of learning*, *knowledge transfer*, *passive learning*, *thinking mode*, and *unlearning*.
Metadata	*Metadata* is data about data. For example, metadata about a photograph includes the camera that was used, and the date and location where it was taken. Another example, metadata about a drawing may include its description or title, issue date, and revision.
Metal	A *metal* is a ductile, fusible, opaque, and typically lustrous substance that: • Forms a positively-charged ion by a loss of electrons (i.e., one that would be attracted to the cathode in electrolysis); • Is a good conductor of electricity and heat; and, • Yields basic hydroxides and oxides. See also, *alloy*, *base metal*, *cathode*, *ferrous*, *mineral*, *non-ferrous*, *non-metallic*, and *ore*.
Metcalfe's Law	*Metcalfe's law* demonstrates the value of networks. It states that the effect of a telecommunications network is proportional to the square of the number of connected users of the system (n^2). It was first formulated in this form by George Gilder in 1993 and attributed to Robert Metcalfe regarding Ethernet. It was originally presented, c. 1980, not in terms of users, but rather of compatible communicating devices (e.g., fax machines and telephones). Only later, with the globalization of the Internet, did this law carry over to users and networks. Its original intent was to describe Ethernet purchases and connections.
Methane	*Methane* is a colourless, flammable, odourless gas which is the main constituent of natural gas. It occurs abundantly in nature and as a product of certain human activities (e.g., decomposition of landfill refuse and vegetable matter under water, which is referred to as marsh gas or swamp gas). It is among the most potent of the greenhouse gases. Its chemical formula is CH_4. See also, *greenhouse gas*, *hythane*, *liquified natural gas*, *natural gas*, *renewable natural gas*, and *synthetic methane*.

Phrase or Term	Definition
Meticulous	*Meticulous* is: • Marked by extreme or excessive care for the consideration or treatment of details (e.g., a meticulous plan); or, • Showing great attention to detail, neatness, or precision (e.g., a meticulous presentation). See also, *messy* and *meticulous people*.
Meticulous People	*Meticulous people* are individuals who perform tasks very carefully and with great attention to detail. Organizations and society value order and predictability; however, a meticulous person might focus on details that are irrelevant to the big picture. A meticulous person may: • Ask critical questions of themselves when reading, studying, or working; • Be a micromanager; • Be rigid with regard to grammatical details; • Excel at pattern recognition if given enough information; • Proofread their work multiple times; • Read others' emotions and intentions easily; or, • Retain random information or trivia with great detail. See also, *big picture*, *messy people*, *meticulous*, and *micromanagement*.
Micro Rebar™	*Micro Rebar*™ is a twisted steel rod 25-mm (1") long that is used in construction to strengthen concrete structures. It is mixed into concrete to create a superior building product. It has many advantages over conventional reinforcing bars, including the following: • Accommodates versatile applications for concrete structures (e.g., foundations, linings, slabs, and walls) and precast products (e.g., manholes, tanks, and vaults); • Allows construction schedules to be shortened with no installation time; • Eliminates the need for large storage areas and specialized lifting equipment; • Improves safety with the elimination of dirty, difficult to maneuver, infection-inducing, and sharp materials that pose tripping hazards; • Increases cracking resistance with structural durability against blasts, corrosion, earthquakes, and water penetration; • Is approved for use by building codes; • Reduces cost compared to reinforcing bar when labour and time savings, safety improvements, and customer satisfaction are all considered (as a con, when comparing direct material costs, reinforcing bar can be slightly less expensive than Micro Rebar™); • Reduces labour costs and eliminates rework caused by misplacement; and, • Resists corrosion three times longer than conventional uncoated steel reinforcing bars through use of a specialized coating. See also, *reinforcing bar*.

Phrase or Term	Definition
Microaggression	A *microaggression* is a subtle but offensive action or comment directed at a group or person based on a bias or stereotype. It may be a commonplace but unconscious or unintentional indignity that communicates a derogatory, hostile, or negative attitude. The person making the microaggression may be well-intentioned and unaware of the potential impact. Examples include: • Ascribe someone's intelligence based on their heritage, ethnicity, or race; • Assume someone is a foreigner or from a different country; • Looking at a watch or smartphone during a face-to-face discussion; • Refusing to acknowledge someone's presence; or, • Treat someone as a lesser being without equal rights. Microaggression was coined by American psychiatrist Chester M. Pierce in 1970 to describe the dismissals and insults from non-black Americans to African Americans, which he regularly witnessed. See also, *bias*, *phub*, *Black Indigenous and People of Colour*, and *racism*.
Micro-business	A *micro-business* is defined by Industry Canada as a business with fewer than five employees. Other organizations may have different definitions. See also, *big business*, *medium-sized business*, and *small business*.
Microdots	*Microdots* are invisible or coloured (e.g., yellow) dots that form an encoded design on a printed page (e.g., in a pattern that has meaning). These dots are typically barely visible to the naked eye. Depending on the microdot design, they may become more easily visible when magnified or photographed under a blue light-emitting diode light. Many colour printers add microdots to documents. Most people are unaware that microdots exist. These dots are encoded security information that may include the date and time that the pages were printed, and the printer's serial number. Microdots have existed for many years to prevent forgeries (e.g., bank notes) and identify unauthorized use (e.g., espionage). See also, *cryptography*, *encryption*, and *steganography*.
Micromanagement	*Micromanagement* is a management style where a manager closely controls or observes the work of the employees or subordinates. The micromanagement style is disliked by most because it makes employees feel that they cannot be trusted to manage their own (routine) tasks. See also, *delegation* and *management*.
Micrometre, micron	A *micrometre* or *micron* is a unit of length equal to one millionth of a metre (0.000001 m).
Microsoft	*Microsoft* is a worldwide leader in software, services and solutions. See more at the Microsoft website: https://www.microsoft.com/
Microstructure	*Microstructure* is the fine structure of a metal or other material, which can be made visible (e.g., by etching and polishing) and examined with a microscope.
Middle Management	*Middle management* is the intermediate personnel who are subordinate to senior management and responsible for managing lower levels of personnel. The main duty of middle management is to implement company strategy efficiently. See also, *line management*, *management*, and *senior management*.
Middle Office	The *middle office* refers to the functions of a business, typically in a financial services institution, to ensure that transactions are correctly executed and manage risk. See also, *back office* and *front office*.

Phrase or Term	Definition
Middle Ground	*Middle ground* is an area of compromise or standpoint midway between extreme or opposing objectives, options, or positions. For example, an agreement for business partners to share all expenses equally *is good middle ground* if profits are also equally shared. See also, *fifty/fifty*.
Middleware	*Middleware* is software that connects two separate systems (i.e., between an operating system and the applications running on it). It provides applications with common functionality and services. For example, application programming interface management, application services, authentication, data management, and messaging are services commonly managed using middleware. Middleware allows users to perform requests such as submitting forms on a web browser or allowing the web server to return dynamic web pages based on a user's profile. Middleware is a portmanteau of middle and software.
Midstream	The *midstream* sector of the oil and gas industry transports (by pipeline, rail, barge, oil tanker or truck), stores, and wholesale markets crude or refined petroleum products. See also, *upstream* and *downstream*.
Milestone	A *milestone* is a specific event or progress point in a project timeline. These mark the progress needed to successfully complete a project. See also, *milestone schedule*.
Milestone Schedule	A *milestone schedule* details the time relationships associated with project milestones. See also, *milestone*.
Mill	A *mill* is a factory that makes and processes materials (e.g., steel pipe or plate). See also, *mill test report,* and *rumor mill*.
Mill Test Report	Not a preferred phrase – see *material test report*. See also, *mill*.
Millennium Development Goals	The *Millennium Development Goals* are eight goals that comprise a 2007 United Nations plan with the goal of meeting the needs of the world's poorest people by 2015. Today, this work continues with an ambitious post-2015 development agenda comprising sustainable development goals. See also, *Sustainable Development Goals*.
Mind your Ps and Qs	*Mind your Ps and Qs* means *be on your best behaviour* or *watch your language or manners* and is typically told to children. There are several stories about the origin of this expression. One is that 17th century English bartenders would announce to customers *mind your Ps and Qs (pints and quarts of beer)* during a bar fight. This meant *I will not offer you another pint or quart of beer if it gets spilled during this fight.*
Mindfulness	*Mindfulness* is: • Being aware or conscious of something; or, • Focusing one's awareness on the present moment, especially as part of a meditative or therapeutic technique (e.g., deliberate acts of regulating attention through the non-judgmental observation of body states, emotions, and thoughts). The benefits of this practice include increased concentration and emotional intelligence, improved creativity and well-being, and reduced burn-out and stress. See also, *emotional intelligence* and *monkey mind*.

Phrase or Term	Definition
Mindshare, Mind Share	*Mindshare* or *mind share* describes the amount of consumer awareness or popularity surrounding a particular brand, company, idea, or product. It is one of the main objectives of advertising and promotion. It is also the consumer's perception of a particular brand or product compared to the competition as measured by the amount of discussion or mentions or generated by the media or public. For example, BAND-AID® is an adhesive bandage with such dominance in consumer minds that is often used as a catch-all phrase to describe any adhesive bandage – even if the adhesive strip is manufactured by a competing brand and uses a different brand name. See also, *advertising*, *BAND-AID®*, *brand*, *catch-all*, *market share*, *marketing*, and *positioning*.
Mineral	A *mineral* is a: • A man-made or synthetic substance having the chemical composition, crystalline form, and properties of a naturally occurring mineral (e.g., synthetic diamond or quartz); • An inorganic substance or ore (i.e., neither animal nor vegetable); or, • Any naturally occurring homogeneous substance usually obtained from the ground (e.g., coal, natural gas, petroleum, salt, sand, stone, sulphur, sand, and water). See also, *alloy*, *base metal*, *ferrous*, *metal*, *non-metallic*, and *ore*.
Mini Gang Crew	A *mini gang crew* is a poor-boy crew. See also, *crew*.
Minimum Backup Requirement	The *minimum backup requirement* is how much generating capacity a renewable energy system must have if it is to reliably produce the electricity needed when wind and solar power are unavailable. It is all the power needed on the hottest or coldest low wind nights. At night, solar power may not be generated. With no wind, wind power may not be generated. Nonetheless, the need for power is very high when it is very cold or very hot. This phrase was coined in the article *It Takes Big Energy to Back Up Little Wind and Solar* by David Wojick. See also, *renewable energy*.
Minimum Viable Product	*Minimum viable product* is the minimal or lean way to give consumers what they want without it necessarily being a fully realized idea or product. A minimum viable product recognizes that, in addition to efficiency and reliability, the product experience must also be desirable and lovable. See also, *lean*.
Minor Repair	A *minor repair* is a small repair that does not need approval by the buyer, engineer, or owner. See also, *major repair* and *repair*.
Minutes of Meeting	*Minutes of meeting* are notes that document topics and details of a meeting.
Mirroring, Isopraxism	*Mirroring* or *isopraxism* is when *some*one consciously or unconsciously scans and subtly imitates the non-verbal (e.g., body language) and verbal (e.g., tempo, tone, and vocabulary) behaviours of someone else to establish empathy and rapport. This occurs because people naturally attracted to what is familiar and similarly repulsed by what is different or foreign. As a negotiating technique, mirroring is focuses only on the counterpart's words and not their other behaviours. The last three words or the critical one to three words that someone has just said are repeated. This concept was developed by Chris Voss in the book *Never Split The Difference: Negotiating As If Your Life Depended On It* (with Tahl Raz). See also, *body language* and *paraverbal communication*.

Phrase or Term	Definition
Misinformation	*Misinformation* is false or inaccurate information that is inadvertently or mistakenly created or spread without the intent to deceive. See also, *brainwashing*, *bullshit*, *disinformation*, *double-dealing*, *doublespeak*, *fake news*, *false narrative*, *media bias*, *propaganda*, *smells fishy*, *stretch the truth*, *truth*, *truthiness*, *two-faced*, and *untruth*.
Mission	An organization's *mission* describes what business the organization is and is not involved in – presently and in the future. The aim is to provide focus for management and personnel. See also, *mission statement*, *purpose*, *values*, and *vision.* See also, *purpose*, *values*, *mission statement*, and *vision*.
Mission Critical	*Mission critical* is a factor in a system (e.g., component, equipment, personnel, process, procedure, or software) that is essential to a business operation, organization (e.g., project financing), or equipment (e.g., the oxygen system in a spacecraft). The disruption or failure of any mission critical factor has serious impacts (e.g., causes a catastrophe, complete failure, social turmoil, or explosion).
Mission Statement	A *mission statement* is a concise enunciation of the goals of an activity, project, or organization. A mission statement usually comprises a short paragraph. It is designed to provide direction and guidance for personnel and to inform clients. For example, Twitter's mission is: *The mission we serve as Twitter, Inc. is to give everyone the power to create and share ideas and information instantly without barriers. Our business and revenue will always follow that mission in ways that improve – and do not detract from – a free and global conversation*. See also, *mission* and *vision statement*.
Mixed Economy	A *mixed economy* is an economic system that combines aspects of capitalism and socialism to obtain the best of both. This protects private property and allows some economic freedom in the use of capital, but also allows governments to regulate economic activities to achieve social objectives. Almost all modern economic systems today are mixed economies. While the globalization of the world economy limits government intervention in free trade, governments retain mechanisms for exceptions to the free market economy (e.g., for social welfare). See also, *capitalism*, *centrally planned economy*, *economic system*, *market economy*, *socialism*, and *traditional economy*.
Mixed Reality	*Mixed reality* occurs when the real and virtual worlds are merged to produce a new environment in which digital ad physical objects can interact. See also, *augmented reality*, *digital reality*, and *virtual reality*.
Mnemonic	A *mnemonic* (pronounced *ni-mon-ik*) is a tool that helps you remember an idea or phrase using a pattern of associations, ideas, or letters. These devices include acronyms, images, outlines, poems, rhymes, songs, and other tools. For example, SMART is a mnemonic for remembering these criteria to set effective goals and objectives: Specific, Measurable, Achievable or Attainable, Relevant, and, Time-bound. *Mnemonic* is derived from the Greek phrase *mimnēskesthai* meaning *to remember*. See also, *acronym*, *abbreviation*, *initialism*, and *SMART*.
Mobilization (Mob)	*Mobilization* (*mob*) is an activity performed by a company or project to assemble, marshal, organize, and prepare personnel and resources for active project service.

<table>
<tr><th>Phrase or Term</th><th>Definition</th></tr>
<tr><td>Model</td><td>A model is:
• A graphical, mathematical (symbolic), physical, or verbal representation or simplified version of a concept, phenomenon, relationship, structure, system, or an aspect of the real world;
• In computer graphics, a three dimensional model is a mathematical representation of any surface of an object (either inanimate or living) that is developed in three dimensions using specialized software; or,
• Something that a copy can be based on because it is an extremely good example of its type, or an original item from which a copy or duplicate is made.
See also, model review.</td></tr>
<tr><td>Model Review</td><td>Performing a model review (i.e., three dimensional model) ensures that the facility design and layout conforms to project requirements and that these requirements are met:
• Budget;
• Constructability;
• Maintainability;
• Operability;
• Process; and,
• Safety.
A model review is critically important to the success of a project. All relevant disciplines need to participate in this review. It enhances or replaces the traditional squad checking and eliminates the need for interpretation of various drawings (e.g., detail, elevation, general arrangement, layout, location, and plan). It is the electronic equivalent of reduced-scale plastic model review. See also, model, model review report, and squad check.</td></tr>
<tr><td>Model Review Report</td><td>A model review report details the findings of a model review. See also, model review.</td></tr>
<tr><td>Modern Monetary Theory, MMT</td><td>Modern monetary theory or MMT is the concept that a government with a fiat currency system can-should create-print as much money as needed because it cannot become insolvent unless there is a political decision to do so. MMT is outside of mainstream or orthodox schools of economic thought. Traditional monetary theory suggests uncontrolled spending is fiscally irresponsible because debt balloons and inflation skyrockets. MMT does not think of government spending like a household budget for countries with complete control over their own fiat currency. Instead of thinking of government spending as expenses and taxes as income, MMT argues that fiscal policy is merely a representation of how much money the government is putting into or taking out of the economy. Only in this way does MMT consider fiscal policy similarly to how monetary policy is seen in the mainstream. See also, fiat currency, fiscal policy, monetary policy, and orthodox.</td></tr>
</table>

Phrase or Term	Definition
Modular Construction	*Modular construction* is a process in which a structure is constructed off-site at a fabrication or manufacturing facility, or modular (mod) yard. Construction is completed under controlled conditions using the project-approved materials and design. This construction method is faster than the stick-built construction method. Similar terms and phrases include construction integrated manufacturing, designed for manufacturing and assembly, offsite construction, panelized construction, and prefabrication. See also, *digital project delivery*, *module*, *modularity*, *prefabricated construction*, *prefabricated prefinished volumetric construction*, and *stick-built construction*.
Modularity	*Modularity* is: • In industrial design, a technique that builds larger assemblies or systems by combining smaller subassemblies or subsystems. For example, the use of a basic core unit onto which different components or modules can be fitted enables a variety of versions of the same unit to be produced. • In manufacturing, the degree to which something is produced with standardized dimensions, options, or parts for maximum flexibility and variety of use. For example, Volkswagen can use a single stamping die to produce floor pans for multiple Volkswagen vehicle types. This reduces vehicle weight by about 50 kg (110 lb) per vehicle by the number of required parts and standardized material use. This term has other meanings. See also, *modular construction* and *module*.
Module	A *module* for a project is usually a large steel-framed and skid-mounted structure that may contain piping, valves, electrical components, and instrumentation. A module is usually constructed off-site to facilitate the assembly of completed units on-site. A module may also contain other equipment (e.g., pumps and motors, tanks, and packaged equipment). See also, *modular construction* and *modularity*.
Module Factory Acceptance Test	A *module factory acceptance* test is factory acceptance test for a module. See also, *factory acceptance test* and *module*.
Mom and Pop Shop	A *mom and pop shop* is a family-owned, independent, or owner-operated small business. Typically, this type of business is not franchised and only operates at single location. See also, *small business*.
Monetary Policy	*Monetary policy* is actions by a county's central bank to promote sustainable economic growth and control the supply of money. See also, *fiscal policy* and *modern monetary theory*.

Phrase or Term	Definition
Money Laundering	*Money laundering* is the illegal process of making large amounts of money, especially cash that is generated by a criminal activity, appear to be produced from legitimate, legal sources. Monetary proceeds from criminal activity is considered *dirty* (informally); thus, this process launders dirty money to make it appear *clean* (even though it is still dirty). This may be accomplished by routing dirty funds through a deliberately-complex sequence of banking transfers or commercial transactions designed to confound and confuse auditors. It includes the: • Acquisition, possession or use of property, knowing that it is obtained from criminal activity; • Assistance or participation in the movement of funds to make illegitimate proceeds appear legitimate; • Concealment or disguising of the nature of how funds were obtained; and, • Conversion or transfer of property. See also, *white collar crime*.
Monkey Mind, Mind Monkey	*Monkey mind* (Chinese: *xinyuan;* Sino-Japanese: *shin'en*) or *mind monkey* is a Buddhist phrase with meanings that include confused, inconstant, indecisive, restless, uncontrollable, and unsettled. It describes a mind that jumps from thought to thought like a monkey jumps from tree to tree. This means that the mind is agitated and refuses to be focussed or quiet. A person has tens of thousands of separate thoughts each day, which are often related to the fears and pressures of life. Methods to tame a monkey mind include: • Be here now; • Binaural beat music; • Meditation; • Mindfulness; • Physical exercise; and, • Yoga. *Remember, Be Here Now* is a book about meditation, spirituality, and yoga written by the American Ram Dass (Richard Alpert). See also, *mindfulness*.

Phrase or Term	Definition
Monopoly	A *monopoly* is one company or consortium that controls all or nearly all of the market (e.g., assets, commodities, infrastructure, and supplies) for a product or service, and dominate that market sector (e.g., Tesla dominates and has a virtual monopoly in the electric vehicle market sector). Monopolies share characteristics that include: • Economy of scale A monopoly can produce products and services at lower costs because of their ability to purchase huge quantities of inventory at a volume discount (e.g., Walmart). Consequently, a monopoly can lower prices, engage in price wars, and undercut other market prices (e.g., sell product below the cost to manufacture it) to eliminate competition. • High barrier to entry By acquiring, or merging with, the competition, monopolies prevent competitors from entering the market by making the cost enter the market and compete too expensive. • Price discrimination A monopoly can change the price or quantity of a product by selling: o Higher quantities of product at a lower price (in an elastic market); and, o Lower quantities of product at a higher price (in a less elastic market). • Price maker A monopoly decides the price of a product or service and does not need to consider competition. Thus, it can raise prices at will. For example, when Turing Pharmaceuticals acquired the drug Daraprim in 2015, it immediately raised the price of one pill from USD$13.50 to USD$ 750.00. Consequently, the annual treatment cost for many patients became unaffordable (e.g., hundreds of thousands of dollars per year). • Profit maximizer A monopoly maximizes profits by exploiting lack of competition. • Single seller or supplier A monopoly is the only (or one of the only) seller or supplier in the market, so it becomes the same as the industry it serves. For example, from 1888 until the start of the 21st century, De Beers controlled up to 85% of rough diamond distribution. Competition has since dismantled the monopoly, so that today the De Beers Group only controls approximately 35% of the market. See also, *antitrust laws and regulations*, *cartel*, *consortium*, *cooperative*, *duopoly*, *economy of scale*, *joint venture*, *monopsony*, and *oligopoly*.
Monopsony	A *monopsony* occurs when there is only one buyer of a product or service. For example, some economists consider that Ernest and Julio Gallo, a conglomerate of wineries and wine producers, is a monopsony. This company is so large and has so much buying power over grape growers that grape wholesalers have no choice but to lower prices and agree to the company's terms. A monopoly may also have monopsony control of a sector of a market. See also, *cartel*, *consortium*, *cooperative*, *joint venture*, and *monopoly*.

Phrase or Term	Definition
Monte Carlo Simulation, Multiple Probability Simulation, Monte Carlo Technique	A *Monte Carlo simulation*, *multiple probability simulation*, or *Monte Carlo technique* is used to model the probability of different outcomes in a process that cannot easily be predicted due to the intervention of random variables. This technique is used to understand the impact of risk and uncertainty in forecasting and prediction models. See also, *forecasting*.
Monty Hall Problem	The *Monty Hall problem* is a statistical illusion that occurs because the brain's process for evaluating probabilities is based on a false assumption. Like optical illusions, the illusion can seem more real than the actual answer. In the television game show, *Let's Make a Deal*, one of three doors hides a prize; the other two doors have no prize. The host, Monty Hall, asks a participant to guess which closed door a prize is behind. The participant chooses a door, but the door is not opened right away. Instead, Monty opens one of the other two doors, and there is no prize behind it. At this moment, there are two closed doors, one of which was picked. The prize is behind one of the closed doors, but it is not known which one. Monty asks: *Do you want to switch doors*? Most people assume that both doors are equally likely to have the prize. Because there is no perceived reason to change, most stick with their initial choice. The correct answer, as proven by Marilyn vos Savant, is that the participant should switch their choice to have a 66% chance of winning. This is because there are only nine different combinations of choices and outcomes. Typically, we think of probabilities for independent, random events. Flipping a coin is a good example. The probability of a heads is 0.5 and we obtain that simply by dividing the specific outcome by the total number of outcomes. That's why it *feels* so right that the final two doors each have a probability of 0.5. For this method to produce the correct answer, the process you are studying must be random and have probabilities that do not change. The Monty Hall problem, however, does not satisfy either requirement. See also, *false assumption*.
Moonshot, Moon Shot	A *moonshot* or *moon shot* is: • A project or proposal that addresses a huge problem, proposes a radical solution, and uses breakthrough technology (e.g., Google's vision for driverless or self-driving automobiles) that is inspirational, credible, and imaginative; • A spacecraft mission to the moon; or, • An ambitious, exploratory, and ground-breaking project undertaken without any expectation of near-term benefit or profitability and possibly without a full benefit and risk assessment (e.g., Yahoo's vision to be on every smartphone and tablet, every day, for every Internet user). The term *moonshot* is derived from the American Apollo 11 project, which landed the first human on the moon in 1969. It also relates to a similar phrase (slang): *shoot for the moon*, which means to aim for a lofty target. See also, *future back process*, *pipe dream*, *radical*, *panacea*, and *unobtainium*.
Moore's Law	*Moore's Law* describes exponential growth in computing power. It was observed in 1965 by Gordon Moore, co-founder of Intel, that the number of transistors per square inch on integrated circuits had doubled every year since the integrated circuit was invented. Moore predicted that this trend would continue for the foreseeable future. See also, *Eroom's Law*.

Phrase or Term	Definition
Moral Equivalence	*Moral equivalence* is the claim that two opposed parties (e.g., countries, persons, or political parties) are doing the same thing and thus should be judged and treated the same way. This overlooks correctness; that is, which party is right and which party is wrong. For example, moral equivalence would ignore that, when two children fight, one is acting in self defence and the other is a bully. See also, *bully*.
More is More	*More is more* is the opposite of the saying less is more. It describes: • A place that boasts of excess and encourages consumerism on a wide scale (e.g., Beverly Hills, Dubai, or Seoul); or, • A kitchen sink approach. See also, *kitchen sink approach* and *less is more*.
MoSCoW Prioritization Method	The *MoSCoW prioritization method* allows a project manager to communicate with stakeholders about the importance of delivering specific requirements. The abbreviation indicates four categories of priority and importance for project requirements. Each requirement is prioritized as a: • Must have; • Should have; • Could have; or, • Won't have.
Most Likely Development	The *most likely development* principle applies to plan that is experience-based, reasonable, and realistic. It assumes that an activity or project will encounter risks that need to be identified and reduced or eliminated. See also, *everything goes according to plan* and *worst-case scenario*.
Mothball	*Mothball* has these meanings: 1. A small pellet of a pungent substance, formerly camphor but nowadays typically naphthalene or paradichlorobenzene, that is put among stored clothes to keep away moths;, 2. To dismiss from further consideration or reject for further use; and, 3. To place something into protective storage (e.g., put a boat in a boathouse): o To deactivate something and prevent deterioration (e.g., store a motorcycle in a heated garage over winter); or, o To withdraw from service or use and keep in reserve or put it aside (e.g., empty a generator of all its fluids, apply preservation oils and preservation requirements, and relocate to long-term storage for future use). See also, *mothballing*.
Mothballing	*Mothballing* is the long-term closure of a facility. Decommissioning and maintenance allow a mothballed facility to be brought back into service when required – although this would take longer than if the plant were idled. If a facility were simply shutdown and shuttered it may be rendered inoperable and unsuitable for future use. See also, *idling*, *mothball*, *preservation*, *shutdown*, and *shuttering*.

Phrase or Term	Definition
Mother Coil, Mother Pipe, Mother Plate	The original material that is used during fitting or pipe manufacturing is referred to as a: • *Mother coil* for the entirety of a hot rolled coil of steel processed from a single reheated slab which may be used to produce one or more pieces of pipe; • *Mother pipe* for the original raw pipe that is used to manufacture another finished product (e.g., a bend, fitting, or smaller pipe or tube); or, • *Mother plate* for the hot rolled plate of steel processed from a single reheated slab which may be used to produce one or more pieces of pipe. See also, *daughter coil*, *daughter pipe*, *daughter plate*, and *starting pipe*.
Motivation	*Motivation is:* • A reason or stimulus (e.g., internal or external factors, or both) that drives personnel to be continually committed to, and interested in, a job, role or subject – or comprises the reason to make the effort to attain a goal; or, • The desire to pursue personal or organizational goals and be productive.
Motto	A *motto* is a short phrase or sentence that expresses the beliefs or ideals of an individual, group, or organization, as an aim or guiding principle. See also, *tagline* and *slogan*.
Moving Parts	*Moving parts* describes components that move within a product or device (e.g., the fan blade in a fan). Generally, the more moving parts in a product: • The more complex the product or device; and, • The more difficult challenges are to resolve because of the interdependence and complexity of moving parts. An vehicle engine has numerous moving parts that rely on interdependent operation; hence, it is complex and not easy to fix if it breaks down. *A business with many moving parts* is a metaphor for an organization with numerous business units, departments, interfaces, and regions that operate a variety of equipment and employ diverse personnel. *The business has many moving parts* is a phrase that summarizes that the organization is complex – time is needed to explain details.

Phrase or Term	Definition
Moving the Goalposts	*Moving the goalposts,* a metaphor derived from goal-based sports (e.g., football, hockey, and lacrosse), means that the criterion or standard (i.e., the goal) of an argument, competition, or process has been changed while it is underway. The new standard or goal typically offers an advantage or disadvantage to one party (i.e., anything will meet the standard or nothing will meet the standard). This occurs when one party refuses to accept or concede that their opponent has met the standard. Moving the goalposts creates ambiguity in terms of if the original goal was met. Generally, moving the goalposts is considered unfair and often has deleterious consequences. For example, a sales team's targets, which were agreed-to at the start of the year, seemed ambitious but achievable. Then, six months later, management changed the business plan and the sales targets. Consequently, the sales team is expected to deliver more than originally expected for the same salary. This is moving the goalposts. In 2009, the goalkeeper for the Swedish football (soccer) team IFK Gothenburg, Kim Christensen, was caught trying to physically move the goalposts to make the goal smaller and gain an advantage. See also, *stretch target*.
Moving to the Left	*Moving to the left* is an expression or colloquialism that indicates the (project) schedule is being compressed and that activities need to be completed within a shorter timeframe than originally planned. This phrase also describes political beliefs that become more left-wing. See also, *float*, *left-wing*, and *moving to the right*.
Moving to the Right	*Moving to the right* is an expression or colloquialism that indicates the (project) schedule is being expanded or activities may be completed over a longer timeframe than originally planned. This phrase also describes political beliefs that become more right-wing. See also, *float*, *moving to the left*, and *right-wing*.
Mud Book	A *mud book* is a hard copy of a manufacturing record book and other specified supplier data shipped with the equipment. This is intended to be a copy of the final turnover document and is as complete as possible when delivered. However, if a shipment occurs immediately upon completion of fabrication or manufacturing, some documents may be incomplete or unavailable at that time. Thus, in reality, a mud book is subject to some change even though it is considered a *final* document. A mud book also provides site receiving and other personnel with supplier data necessary for receiving and construction, without requiring access to a document management system or printer. See also, *manufacturing record book*.
Mugshot	A *mugshot* is a photograph of someone after being arrested (taken from the waist up). It is usually prepared on a plain background with a side- and front-view. *Mug* is British slang term for *face* (18th century). A mugshot is also slang for a picture of any face. See also, *headshot*.
Multimeter	A *multimeter* or a *multitester*, also known as a *volt-ohm-milliammeter or VOM*, is a hand-held electronic measuring instrument that combines several measurement functions in one unit. A typical multimeter can measure voltage, current, and resistance. Analog multimeters use a dial display; digital multimeters use a numeric or graphical display. See also, *continuity test*.

Phrase or Term	Definition
Multiple Discovery, Parallel Invention, Simultaneous Invention	*Multiple discovery*, *parallel invention*, and *simultaneous invention* comprise the hypothesis that scientific discoveries and inventions may be made independently and simultaneously by more than one inventor or scientist. For example, the 17th century independent formulation of Calculus by Sir Isaac Newton, Gottfried Wilhelm Leibniz, and others. See also, *heroic theory of invention* and *discovery*.
Murphy's Laws, Murphy's Corollaries	*Murphy's Laws* are: 1. Nothing is as easy as it looks; 2. Everything takes longer than you think; 3. Anything that can go wrong will go wrong; 4. If there is a possibility of several things going wrong, the one that will cause the most damage will be the one to go wrong. Corollary: If there is a worse time for something to go wrong, it will happen then; 5. If anything simply cannot go wrong, it will anyway; 6. If you perceive that there are four possible ways in which a procedure can go wrong, and circumvent these, then a fifth way, unprepared for, will promptly develop; 7. Left to themselves, things tend to go from bad to worse; 8. If everything seems to be going well, you have obviously overlooked something; 9. Nature always sides with the hidden flaw; 10. Mother nature is a bitch; 11. It is impossible to make anything foolproof because fools are so ingenious; 12. Whenever you set out to do something, something else must be done first; and, 13. Every solution breeds new problems. *Murphy's Corollaries* are: 1. Left to themselves, things tend to go from bad to worse; and, 2. It is impossible to make anything foolproof because fools are so ingenious. These laws and corollaries are often cited by project management to encourage adequate project contingency planning. See also, *project management*.
Must	In this document, the word *must* is used where a provision is mandatory. See also, *shall* and *will*.
Myers-Briggs Type Indicator®	A *Myers-Briggs Type Indicator®* is an introspective self-reporting questionnaire that helps identify psychological preferences in people related to how they perceive their world and make decisions. It is based on the conceptual theory proposed by Carl Jung (1875 to 1961), which concludes that humans experience the world using four principal psychological functions – sensation, intuition, feeling, and thinking – and that one of these four functions is dominant most of the time. See also, *thinking mode*.

Phrase or Term	Definition
N	1 A B C D E F G H I J K L M **N** O P Q R S T U V W X Y Z
Nadir	The *nadir* is the lowest point. For example, the nadir of the professional boxer's career was getting knocked out in the first round by an amateur boxer. See also, *zenith*.
Name-dropping	*Name-dropping* is alluding to or mentioning the name(s) of a famous or important organization or person(s) one is associated with or knows or claims to know, to impress others. It is used to position oneself within a social hierarchy or create a sense of superiority by raising one's status. For example, *I was invited to a party at Bill Gates' home* or *I attended Harvard University*.
Nameplate Capacity, Installed Capacity, Nominal Capacity, Maximum Effect, Rated Capacity	*Nameplate capacity*, *installed capacity*, *nominal capacity*, *maximum effect*, or *rated capacity* is the intended full-load sustained output of a facility such as a chemical or fuel plant, electric generator, metal mine or refinery, or a power plant. For example, it is the number registered to classify the power output of a power station, usually expressed in megawatts. See also, *high capacity factor*, *nameplate*, and *power output*.
Nameplate Markings	*Nameplate markings* are the specific information provided by the original equipment manufacturer, as required by the code of construction, or both. These include: • Item numbers (e.g., registration, serial, model, and tag); • Manufacturing date; • Material of construction; • Maximum allowable working pressure and temperature; • Minimum design metal temperature; • Name of the manufacturer and/or logo; and, • Official code stamp (e.g., ASME or API). See also, *nameplate* and *nameplate rubbing*.
Nameplate Rubbing	*Nameplate rubbing* is an easy and rudimentary method of reproducing the markings etched or stamped on the surface of a nameplate. A nameplate rubbing is created by placing a piece of paper over the nameplate and rubbing the paper with charcoal or pencil such that the embossed or stamped portions of the nameplate transfer to the paper. With the advent of digital photography, projects now typically take pictures instead of creating a rubbing because this is faster and automatically digitizes results. Nameplate rubbings were originally used to avoid handwritten errors. See also, *nameplate* and *nameplate markings*.
Nameplate, Name Plate	A *nameplate* or *name plate* is an engraved lamacoid or stamped metal plate with data required by the purchase order, code of construction, or both. It is typically installed on equipment using rivets to help personnel identify equipment, and equipment certifications and design criteria. See also, *tag*, *nameplate capacity*, *nameplate marking*, *nameplate rubbing*, and *lamacoid*.
Naphtha	*Naphtha* comprises various volatile, often flammable, liquid hydrocarbon mixtures. Naphtha is a diluent (chemical) that reduces bitumen viscosity and allows bitumen to flow through a pipeline. See also, *diluent* and *condensate*.
National Oil Company	A *national oil company* is an oil company that is owned (wholly or holds a majority share) by a national government. See also, *international oil company*.

Phrase or Term	Definition
Narcissist, Narcissism	A *narcissist* is someone with a mental condition or personality disorder that has: • A deep need for excessive admiration and attention; • A lack of empathy for others; • An inflated sense of their own importance; and, • Troubled relationships. *Narcissism* originated in Greek mythology where a young man named Narcissus fell in love with his own image reflected in a pool of water. Pathological self-absorption was first identified as a disorder by Havelock Ellis in 1898. See also, *narcissistic defense mechanisms*.
Narcissistic Defense Mechanisms	*Narcissistic defense mechanisms* are the conscious or unconscious actions or reactions a narcissist uses to get what they want, including: • Denial of anything not considered to be in their best interest; • Distortion of emotions, facts, or realities (e.g., exaggeration or minimization) and lying; • Psychological projection to control someone else (e.g., using blame or guilt); or, • Repression of emotions, expressions, facts, and situations. See also, *defensive behaviour* and *narcissist*.
National Pipe Thread	*National pipe thread* or NPT is a US standard for tapered threads used on threaded pipes and fittings. In contrast to straight threads that are found on a bolt, a taper thread will pull tight and make a fluid-tight seal. See also, straight thread.
Native Advertising	*Native advertising* is advertising created and paid for by an advertiser (e.g., an Audi television commercial). See also, *advertising, commercial, sponsored content*, and *performance marketing*.
Native File Format, Filename Extension	*Native file format*, also known as the *filename extension*, is the file format used by an application to create or save files (e.g., Microsoft Excel uses .xltx, Microsoft Word uses .docx, and Adobe Acrobat uses .pdf). Many native file formats are proprietary. See also, *portable document format*.
Natural Gas	*Natural gas* is flammable, occurs naturally underground, and is often associated with hydrocarbons. It must be processed and converted into cleaner fuel before consumption, during which many useful by-products may be extracted (e.g., butane, carbon dioxide, ethane, propane, and nitrogen). Natural gas is the cleanest burning fossil fuel that emits about 40% less carbon dioxide and fewer air pollutants compared to coal. It can significantly reduce smog when used to generate electricity, especially when compared to burning coal. See also, *coal*, *fossil fuel*, *hydrocarbon*, *hythane*, *liquified natural gas*, *methane*, *petrochemical*, *petroleum*, *renewable natural gas*, and *synthetic methane*.
Natural Gas Liquid	Not a preferred phrase (but may be used interchangeably) – see *condensate*.
Natural Resource	A *natural resource* is a naturally occurring asset such as a material or substance (e.g., fertile land, forests, minerals, oil sands, and water). See also, *mineral*, *resource*, and *resource company*.

Phrase or Term	Definition
Naturally Occurring Radioactive Materials	Some *naturally occurring radioactive materials* comprise uranium, thorium, radon, and plutonium and reside in the Earth's crust in very low concentrations. These materials may be brought to the surface by human activities and natural processes (e.g., dissolution in ground water, earthquakes). These materials are found in many industrial processes including: • Coal mining and combustion; • Construction (e.g., excavation and building on granitic ground); • Fertilizer manufacturing (e.g., phosphates); • Metal mining and smelting; • Rare earth element production (e.g., titanium and zirconium); • Recycling; and, • Oil and gas exploration and production.
Nature	*Nature* is: • The basic or inherent features of something, especially when seen as characteristic of it; or, • The phenomena of the physical world collectively, including animals, landscape, plants, and other features and products of the earth, as opposed to human creations. See also, *nurture*.
Nature-based Solutions, Natural Climate Solutions	*Nature-based solutions* or *natural climate solutions* are sustainable management projects that protect, restore, or transform land such that the natural environment absorbs more carbon dioxide emissions from the atmosphere than it did previously. The solutions are cost-effective, inspired and supported by nature, and simultaneously: • Help build resilience to challenges such as climate change, disaster risk management, food security, human health, water pollution, and security; and, • Provide economic, environmental, and social benefits. These projects can lead to the marketing, sale, and trading of carbon credits. The projects improve degraded land and increase carbon sequestration with natural, managed, or newly-created forests, grasslands, or wetlands by: • Afforestation by planting trees where there were no trees (e.g., on arid land or adjacent to waterways); • Conservation to prevent the wasteful use of a resource (e.g., trees and water); • Improvement of water availability and quality (e.g., with constructed and natural wetlands); • Reduction of risks associated with water-related disasters (e.g., with floodplain restoration); • Reforestation to replant degraded forest reserves (e.g., by disease, fire, harvesting, or insect infestation); or, • Restoration of modified or natural ecosystems (e.g., return abandoned industrial land to its natural, pristine state). See also, *carbon credit*.

<table>
<tr><th>Phrase or Term</th><th>Definition</th></tr>
<tr><td>Negative Variance</td><td>Negative variance is the amount by which actual project performance does not follow planned project performance. Negative variances in budget or time show the project is more expensive or proceeding more slowly than planned. See also, positive variance.</td></tr>
<tr><td>Negotiation One Sheet, One Sheet</td><td>A negotiation one sheet is a document used to instill confidence and prepare a list of tools that may be used during a negotiation, which contains these sections:
1. The goal (i.e., best-case scenario);
2. A summary (i.e., facts leading up to the negotiation);
3. An accusation audit and labels (i.e., a list of the counterpart's arguments, with descriptions, intended to diffuse their arguments and extract information);
4. Calibrated questions (i.e., to identify and overcome deal-killers); and,
5. Non-cash offers (i.e., non-monetary but valuable items possessed by either party).
The entertainment industry uses a one sheet to summarize a product for publicity and sales, from which the phrase negotiation one sheet is derived. This concept was developed by Chris Voss in the book Never Split The Difference: Negotiating As If Your Life Depended On It (with Tahl Raz). See also, accusation audit, calibrated question, goal, and label.</td></tr>
<tr><td>Negotiating Style</td><td>A person's negotiating style is someone's approach to negotiation. There are basically three negotiating styles:
• Accommodator
Someone who builds rapport and values a continuous exchange of information. They are accommodating with the hope of reciprocity, optimistic, and sociable. They believe time is for relationships and love win-win solutions.
• Analyst
Someone who is diligent, methodical, and unhurried, who rarely shows emotion and wants to minimize mistakes. They are information aggregators and problems solvers who are skeptical and use silence to think. They believe time is for preparation and hate surprises.
• Assertive
Someone who is aggressive, candid, and direct who values action more than perfection. They want to be heard and respected. They believe time is money and love winning.
Many argue that the assertive style is the best. Each style can be effective; however, one may need elements from others to be most effective.
A common reason that negotiations fail is that the negotiators fail to understand and then adapt or respond to their counterpart's negotiating style. This concept was developed by Chris Voss in the book Never Split The Difference: Negotiating As If Your Life Depended On It (with Tahl Raz). See also, win-win.</td></tr>
<tr><td>Nemesis</td><td>A nemesis is:
• A formidable or long-standing and usually victorious opponent, rival, or enemy (i.e., an archenemy); or,
• One that inflicts retribution or vengeance (e.g., the inescapable agent of someone's or something's downfall).
Nemisis is the Greek goddess of retributive justice. See also, archrival, competitor, frenemy, and rival.</td></tr>
</table>

Phrase or Term	Definition
Net Net, Net-Net	*Net net* or *net-net* is • An investing technique developed by Benjamin Graham where a company is valued based solely on its net current assets (e.g., accounts receivable, cash, cash equivalents, and inventory); or, • The final and true result of something, determined after all allowances and subtractions have been made.
Net Positive Suction Head	*Net positive suction head* is the total absolute suction pressure determined at the suction nozzle and referred to the datum elevation, minus the vapour pressure of the liquid. This is expressed in metres or feet of head of the pumped liquid.
Net Worth	*Net worth* is the current value of all assets less the liabilities of a business or person. See also, *wealth*.
Net-carbon Neutral, Net-zero, Zero Carbon	*Net-carbon neutral*, *net-zero*, and *zero carbon* mean that an equivalent amount of carbon has been removed from the atmosphere than was put into it. This is when the amount of carbon emissions minus the carbon captured or offset equals zero. See also, *historical emissions*, *low-carbon*, and *scope 1, 2, and 3 emissions*.
Network Folder Structure	A *network folder structure* comprises folders and subfolders that reside in an electronic medium (e.g., hard drives, shared drives, and servers) that logically organize and store electronic data. For users, folders are typically displayed in a hierarchical tree structure – with main folders at the top and subfolders residing within the main folders to indicate hierarchy. This structure was developed for computer systems in the 1980s to mimic the manual paper folder systems of the time. Accessibility to folders may be granted (provided), restricted (partial access), or denied (no access) based upon an organization's confidentiality, business, personnel, and project needs. Common main folder names are: 1. Corporation (e.g., ABC Inc.) or project name (e.g., the KT Project); 2. Department (e.g., Engineering) or group name (e.g., SMEs); 3. Discipline name (e.g., Electrical); 4. Function descriptor (e.g., Squad Checkers); 5. Content type (e.g., Procedures); and, 6. Category (e.g., Phase 2 Work) or Topic (e.g., Topographical Surveys).
New Normal	*New normal* is a social state that emerges following a crisis (e.g., the effect the COVID-19 virus had on social behaviour in public areas in 2020), especially when it differs significantly from the previous prevailing social state. Other recent examples of new normals occurred after the: • 2001 terrorist attacks in New York City (e.g., increased security); • 2007-2008 global financial crisis (e.g., increased scrutiny of loans); and, • 2008-2012 global recession (e.g., lower economic and productivity growth, and high unemployment). A new normal crisis response incorporates new ways of thinking, applies new methods of working, learns from lessons learned, and seeks preservation from, and prolonged sustainment against, the deleterious consequences of a crisis. See also, *crisis* and *deleterious*.

Phrase or Term	Definition
Nextpert	A *nextpert* is the initial mentee of an original expert. See also, *mentoring* and *subject matter expert*.
Niche Market	A *niche market* is a subset of a larger market in which a specific product is focused. It has its own demographics, interests, needs, or preferences, which may be different from the larger market. For example, a vehicle manufacturer introduces a new, off-road or winter 4x4 vehicle for a small number of customers who desire it. See also, *bottom of the market* and *top of the market*.
Night Owl, Evening Person	A *night owl* or *evening person* is a someone who tends to stay up late at night. See also, *early bird*.
NIMBY	*NIMBY* is an initialism for the expression *not in my backyard*, which is characterized by the opposition to a proposed development in someone's local area. See also, *not in my back yard* and *obstructionist*.
Nipple	A *nipple* is a short length of pipe ≤12 in. that may or may not be threaded on one or both ends. See also, *pipe*.

Nitrogen Blanket	A *nitrogen blanket* is a layer or padding of inert nitrogen gas (i.e., at a small positive pressure to force out air, which contains oxygen) that is placed into equipment (e.g., exchanger, tank, tank car, or pressure vessel) during storage, transportation, or used to:
	• Prevent a chemical reaction of the product contained inside;
	• Protect the interior of the equipment from corrosion; or,
	• Reduce the level of dangerous (e.g., explosive) gases.
	These items shall be clearly marked with a warning to identify the presence of the nitrogen blanket to:
	• Prevent injury if the equipment is opened before the nitrogen gas pressure is bled off; and,
	• Advise the installers or users and others that:
	  o The nitrogen blanket shall be maintained at the required pressure and replenished in case of any pressure drop (e.g., due to leakage); and,
	  o An air purge may be required before the equipment is installed and commissioned.
	See also, *tank car*.

Phrase or Term	Definition

No	*No* means to:
	• Express:
	  o Denial, dissent, negation, or refusal (e.g., no, I will not);
	  o Doubt, incredulity, or surprise (e.g., no way!);
	  o The negative of an alternative or possibility (e.g., shall we proceed or no?); or,
	  o The opposite of a positive statement (e.g., no good);
	• Make a comparison (e.g., no better or worse);
	No can also mean something other than rejection and have an alternative or more real meaning. For example, I:
	• Am not ready to agree:
	• Do not understand; or,
	• Need more information, time, or both.
	See also, *yes*.

No-Orientated Question™	A *no-oriented question™* is asked to obtain a negative answer (i.e., *no*). It can leverage the natural human instinct to say *no* to get things done. Getting someone to say no is easy! For example, a no-oriented questions could be: • Do you want me to fail? • Is my idea bad? • Is this a bad time to talk? See also, *yes-oriented question*.
No Silo Rule	The *no silo rule* is a philosophy that encourages company leaders to see the big picture and collaborate for the good of the company. It is founded on principles of emotional intelligence, where emotions work to one's advantage, instead of to one's disadvantage. As a company grows, so too does the tendency for divisive behaviour – both consciously and unconsciously. For example, in silos, teams may compete against one another and work against one another's or the company's best interests. Elements of this philosophy include: • Critical thinking to use resources in one area to solve problems in another area; • Feedback loops to help personnel understand how their team's work affects other teams; • Personnel that are cross-trained in different departments; • Teams that are comprised of personnel from different departments; and, • Teams that are organized to work together within a cohesive and flexible company, with one bottom line, rather than as divisions with their own bottom lines. This philosophy was used successfully by Elon Musk at Tesla and Steve Jobs at Apple. See also, *big picture*, *cross-train*, *emotional intelligence*, *feedback loop*, *silo*, and *silo mentality*.
Nominal Pipe Size	*Nominal pipe size* or NPS is a North American set of standard sizes for pipe, which identifies pipe diameter and schedule. NPS is frequently and incorrectly referred to as *national pipe size*. See also, *dimensionless designator* and *schedule*.

Phrase or Term	Definition
Non Sequitur	A *non sequitur* is a conclusion or reply that does not follow logically from the previous statement. For example, fast cars burn more fuel; therefore, a rabbit is a better pet than a squirrel. Non sequiturs are often used for comedic effect. See also, *logic* and *word salad*.
Non-capital Project	A *non-capital project* is a small project. Typically, these projects involve the alteration, enhancement, or renovation of existing assets. If the project is below a specified threshold (e.g., dollar value, size, or scope), it may be managed as an operating expense and not a project. *See also*, *capital project*, and *project*.
Non-causal Relationship	A *non-causal relationship* is where a process or state, a cause, does not contribute to the production of another process or state, an effect. The cause is not responsible for the effect, and the effect is not dependent on the cause. For example, when the sun rises (cause), streets are wet (effect). The street is wet because it rained during the night, not because of the sun's movement. See also, *causal relationship*.
Non-conformance	A *non-conformance* is a characteristic that renders an item unacceptable for its intended use (e.g., an item that is blemished, defective, deficient, incomplete, or imperfect). A non-conformance is dispositioned using a non-conformance report or NCR. See also, *disposition*, *conformance*, *non-conformance report*, and *quarantine*.
Non-conformance Report	A *non-conformance report* describes the non-conforming issue and disposition with sign-off for acceptance. A non-conformance is a request to accept a non-conforming product with a specified disposition (e.g., repair, replace, or accept as-is). See also, *conformance*, *non-conformance*, and *non-conformance report coordinator*.
Non-conformance Report Coordinator (NCR Coordinator)	A *non-conformance report coordinator* or *NCR coordinator* ensures that personnel follow the correct non-conformance work processes, obtain required dispositions, dispositions are correct, close-out of non-conformances are timely, and all necessary follow-up actions are completed (e.g., through revised documentation). See also, *coordinate* and *non-conformance report*.
Non-conformance Report Log *(NCR log)*	A *non-conformance log* or *NCR log* is a project record of non-conformance reports that contains relevant details (e.g., description, disposition, originator, and status). A typical non-conformance log uses Microsoft Access, Excel or SharePoint software, which may also include counting, data sorting-filtering, and reporting features. A non-conformance log may also be a simple list of supplier non-conformances relevant to a purchase order. See also, *Excel* and *SharePoint*.
Non-conventional Non-destructive Examination	Not a preferred phrase – see *advanced non-destructive examination*.
Non-critical Dimension	A *non-critical dimension* has the largest tolerances, does not affect fit or function, or does not have regulatory or safety affects. These dimensions do not affect fit-up of assemblies or components, or impact the design intent or the overall performance of the item. See also, *critical dimension*.
Non-destructive Examination	*Non-destructive examination* is a test that examines an object (i.e., a structure, material, or component) without damaging or destroying the object – to ensure object (product) reliability and quality. See also, *advanced non-destructive examination*, *conventional non-destructive examination*, *destructive testing*, *non-destructive examination method* and *non-destructive examination technique*.

Phrase or Term	Definition
Non-destructive Examination Method	The *non-destructive examination method* refers to the type of scientific principle, penetrating medium, or equipment used during a non-destructive examination. See also, *non-destructive examination* and *non-destructive examination technique*.
Non-destructive Examination Technique	The *non-destructive examination technique* is a subset of the non-destructive examination method used (e.g., dry or wet for magnetic particle; water-washable or solvent-removable for liquid penetrant). See also, *non-destructive examination* and *non-destructive examination method*.
Non-destructive Testing	Not a preferred phrase – see *non-destructive examination*.
Non-expansible Fluid	*Non-expansible fluid* is liquid that does not expand. See also, *expansible fluid*.
Non-ferrous	A *non-ferrous* metal contains no iron (e.g., aluminium, brass, and copper). See also, *alloy*, *base metal*, *ferrous*, *metal*, *mineral*, *non-metallic*, and *ore*.
Non-government Organization, Non-governmental Organization, NGO, Civil Society	A *non-government organization*, *non-governmental organization*, *NGO*, or *civil society* is non-for-profit organization that functions independent of government. NGOs are categorized as international, local, national, and regional, and serve a political, social, or other cause. NGOs promote the interests of society or stakeholders, or may form a lobby for a group of companies or industry (e.g., for a business, environmental, or humanitarian purpose). For example, a well-known NGO is Doctors Without Borders or Médecins Sans Frontières (French). They provide medical assistance to people affected by conflict, disaster, epidemic, or exclusion from healthcare. This organization hasteams comprising tens of thousands of administrative, professional, and logistical personnel. See also, *government*, *lobby*, and *non-profit*.
Nonlinear Management	*Nonlinear management* refers broadly to management practices that emphasize adaptation to changing circumstances, flexibility, and self-organization. It runs counter to the concept of linear management. The defining characteristics of nonlinear management include encouragement of flexible working arrangements for personnel, outside the box thinking, and proactivity in responding to challenges. See also, *linear management*, *management*, and *outside the box*.
Non-low Hydrogen Electrode	A *non-low hydrogen electrode* has no diffusible hydrogen designator or does not produce an inherently low level of diffusible hydrogen (e.g., it will produce diffusible hydrogen levels in weld metal more than the requirements of AWS A.3 or CSA W48). See also, *diffusible hydrogen designator*, *electrode*, *low hydrogen*, and *low hydrogen electrode*.
Non-metallic	A *non-metallic* is a material that is not a metal (e.g., ceramic, fibreglass or fibre-reinforced plastic, graphite, glass, polyvinyl chloride or plastic and refractory). See also, *base metal*, *ferrous*, *metal*, *mineral*, *non-ferrous*, *refractory*, and *ore*.
Non-potable Water	*Non-potable water* is water that is not fit for drinking, cooking, or bathing. It is not treated to meet federal, state, provincial, or other standards but may be used for other purposes (e.g., for washing vehicles, irrigation, or fire fighting). It may come from an underground or surface source (e.g., a well or river), or other source (e.g., collected rainwater). See also, *potable water* and *wastewater*.

Phrase or Term	Definition
Non-profit, Not-for-profit	A *non-profit* or *not for profit* organization does not earn profit for its owners or shareholders. All the money donated to, or earned by, a non-profit organization is used to: • Pay operational expenses; and, • Pursue organization objectives. These organizations are usually tax-exempt charities because operations typically relate to religion, safety, science, or public service organization. Income sources may include funding from government contributions, membership dues, and private donations. The legal definition of a non-profit (e.g., a charitable organization or philanthropic foundation) and its tax treatment varies between and within countries. Although a non-profit organizations do not earn profit, annual budgets may be as large as for-profit organizations (e.g., millions or billions of dollars). See also, *charity*, *for-profit*, *philanthropy*, and *voluntary sector*.
Non-renewable Resource	A *non-renewable resource* cannot be replaced naturally at the rate it is consumed so it is, therefore, a finite or limited resource. For example, fossil fuels such as coal, natural gas, and oil are non-renewable resources. See also, *fossil fuel* and *renewable resource*.
Non-standard	Something that is *non-standard* is different from the usual type or version of that thing; it is not normal or is unique and unusual. It may be custom-built or modified to suit a specific design or purpose.
Non-witnessed	*Non-witnessed* describes an activity that is performed by the supplier without a witness and without notification to a quality surveillance representative. Quality control documentation may be required to document non-witnessed results, which may be reviewed later to verify that results are complete and acceptable. See also, *verification point*.
Non-zero-sum	*Non-zero-sum* is a situation where one's gain or loss does not necessarily result in the another's gain or loss. In this situation, gains and losses of all participants do not add up to zero; thus, everyone can potentially gain. For example, a country that produces apples trades with another country that produces bananas; both participants benefit from this arrangement. See also, *lose-lose*, *win-win*, and *zero-sum*.
Nootropic, Smart Drug	A *nootropic* or *smart drug* is a substance (i.e., a drug or supplement) that enhances cognitive functions (e.g., concentration, creativity, learning, memory, or motivation) and executive skills, or prevents the decline of these abilities. The use of these products is prevalent among many students. The term nootropics was coined by the Romanian Dr. Corneliu Giurgea in 1972. Today it is an umbrella term that may include everything from dietary supplements to prescription drugs. A product defined as a supplement may be legal if the manufacturer can demonstrate that it is safe. The manufacturer may not, however, make any claim that the product will prevent or treat any condition or disease. Supplements that advertise health claims or contain drugs may be illegal, depending upon jurisdiction. See also, *executive skills*.
Normative	*Normative* is information about the criteria or rules to conform to a standard, from which no deviation is permitted. These are prescriptive requirements that compel compliance and are denoted by the verbs *must* or *shall*. See also, *empirical* and *informative*.
North of	*North of* means more than (e.g., *north of a million dollars* means *more* than a million dollars). See also, *south of*.

<table>
<tr><th>Phrase or Term</th><th>Definition</th></tr>
<tr><td>Not in my back yard</td><td>Not in my back yard refers to the characterization of opposition by people to a proposed development in their local area (e.g., a proposal to construct a halfway house in your community). It carries the connotation that the opposition to the development is only because of its proximity; that they would support or tolerate it if it were built farther away. This opposition can be overcome through communication, impact mitigation, negotiation, and transparency.
This phrase may also be applied to those who advocate for a proposal but oppose implementation because it may adversely affect their lives or require personal sacrifices. People with these views are often called NIMBYs and their viewpoint is called NIMBYism. See also, build absolutely nothing anywhere near anyone, halfway house, NIMBY and obstructionist.</td></tr>
<tr><td>Nothing to Fear But Fear Itself</td><td>Nothing to fear but fear itself is a paraphrased citation from United States' President Franklin D. Roosevelt's inaugural address. His actual words were:<blockquote>... the only thing we have to fear is fear itself — nameless, unreasoning, unjustified terror which paralyzes needed efforts to convert retreat into advance.</blockquote>His speech was made in 1933 during the depths of the Great Depression, when panicked citizens began withdrawing money from banks en masse. President Roosevelt meant that people should avoid inappropriate fear that could unnecessarily cause panic or paralysis. What is feared may cause less damage than the solution to avoid it. So, if fear can be rejected and resisted, then a difficult situation can be overcome. Variations of this concept were expressed earlier by Michel de Montaigne, Francis Bacon, The Duke of Wellington, and Henry David Thoreau. See also, Great Depression and panic buying.</td></tr>
<tr><td>Notification</td><td>Notification is the act of making others aware of information through electronic (e.g., email), printed (e.g., a form), or other means. For example, the supplier shall provide advance notification for attendance by the quality surveillance representative and others at quality verification points that are witness and hold points (e.g., 5 or 10 days).</td></tr>
<tr><td>Nozzle</td><td>A nozzle is an appurtenance with a flanged connection that is a fitting. A nozzle controls the direction or characteristics of fluid flow into or out of an enclosed chamber (e.g., fitting, pipe, tank, or vessel). See also, reinforcing pad.</td></tr>
<tr><td>Nuclear Option</td><td>The nuclear option refers to an extreme maneuver or response to a situation with potentially serious consequences, such as:<ul><li>A drastic step or last resort (e.g., a wildcat strike);</li><li>A parliamentary procedure of the United States Senate (i.e., to override the three-fifths vote rule to close debate by a simple majority, rather than the two-thirds majority normally required);</li><li>An act (e.g., of retribution) that is unwarranted or overly severe in its consequences (e.g., to unfriend someone on social media); or,</li><li>To use a nuclear weapon in war.</li></ul>See also, simple majority and wildcat strike.</td></tr>
<tr><td>Nurture</td><td>Nurture is the process of caring for and encouraging the development growth of someone or something (e.g., a zookeeper caring for an elephant). See also, nature.</td></tr>
</table>

Phrase or Term	Definition
Nutpicking	*Nutpicking* occurs when personnel with the most extreme and alternative views are presented as an accurate representation or cross-section of the group. See also, *cherry pick* and *hollow man argument*.
Nuts and Bolts	*Nuts and bolts* are: • The essential elements of something; or, • Threaded fastening devices. See also, *bread and butter* and *fastener*.

Phrase or Term	Definition
O	1 A B C D E F G H I J K L M N **O** P Q R S T U V W X Y Z
Objective	An *objective* is: • A deliberate step needed to complete a larger goal; • A specific result that a person or system aims to achieve within a time frame and available resources; or, • A state of mind that is bias-free (i.e., not influenced by personal feelings or opinions), neutral, and bases judgement on verifiable evidence or facts (and *not* on attitude, belief, or opinion). The antonym for *objective* is *subjective*. In general, objectives are more specific and easier to measure than goals. Objectives are basic tools that underlie all planning and strategic activities. They serve as the basis for creating policy and evaluating performance. Examples of business objectives include minimizing expenses, expanding internationally, or earning profit. See also, *bias*, *business objectives*, *objective*, and *SMART*.
Objective Reflective Interpretive Decisional Questioning	*Objective reflective interpretive decisional questioning* is a structured conversation. It is led by a facilitator and moves towards a decision by asking four types of questions in sequence: 1. Objective Identify objective facts (what do we know about this?). 2. Reflective Identify subjective perceptions (how do we feel about this?). 3. Interpretive Identify perspective and potential impacts (what does this mean for you or us?). 4. Decisional Identify achievable, positive outcomes or the best course of action (what are we going to do?). See also, *calibrated question™*, *precision questioning*, *perspective-taking*, and *reflective questioning*.
Observation Point, Observed Test	An *observation point* or *observed test* is API's phrase for a *witness point*. See also, *witness point*.

<table>
<tr><th>Phrase or Term</th><th>Definition</th></tr>
<tr><td>Obstructionist</td><td>An obstructionist is someone who deliberately:
<ul>
<li>Delays, impedes, or prevents progress (e.g., by illegal, legal, non-violent, or violent means); or,</li>
<li>Delays or obstructs the business before a legislative body by parliamentary contrivances or legalistic maneuvers (e.g., a politician's filibuster).</li>
</ul>
For example, despite federal government and National Energy Board approvals of the Trans Mountain pipeline expansion project in Canada, opponents continually obstruct the project by all means:
<ul>
<li>Activists and protesters threatened civil disobedience such as:
<ul>
<li>Blockading rail shipment of construction materials;</li>
<li>Blocking access to pipeline construction sites and facilities;</li>
<li>Building structures on the pipeline right-of-way; and,</li>
<li>Occupying treetops along project routing.</li>
</ul></li>
<li>First Nations challenged the project in the Federal Court of Appeal with claims that consultations with Indigenous communities were insufficient; and,</li>
<li>The British Columbia provincial government pursued legal actions in the British Columbia Court of Appeal and the Supreme Court of Canada, attempting to regulate the movement of diluted bitumen from Alberta through its territory (e.g., using the Environmental Management Act to regulate an interprovincial project that falls under federal jurisdiction and imposing a permitting system that could impede shipments). This action was promised in the 2017 election campaign. These legal proceedings ended in 2020 when the case was dismissed.</li>
</ul>
See also, NIMBY, not in my back yard, and right-of-way.</td></tr>
<tr><td>Occam's Razor, Ocham's Razor, Ockham's Razor, The Law of Parsimony</td><td>Occam's Razor, Ocham's razor, Ockham's Razor, or the Law of Parsimony is the problem-solving principle that the simplest solution is most likely the correct one. This means that:
<ul>
<li>Other things being equal, simpler explanations are generally better than complex ones; or,</li>
<li>When problem-solving, select the solution with the fewest assumptions.</li>
</ul>
This idea is attributed to William of Ockham, who originally stated the idea as entities should not be multiplied without necessity. For example, if a vehicle will not start, check the fuel level before calling a tow truck to take it away for repair. See also, keep it simple stupid principle and The Sagan Standard.</td></tr>
<tr><td>Off Gas</td><td>Off gas is a gas that is produced as a by-product of a chemical or industrial process, or given off by a manufactured material or object. Off gas may or may not be harmful.</td></tr>
<tr><td>Off Specification, Off Spec, Out of Specification, Out of Spec, Off Quality</td><td>Off Specification, off spec, out of specification, out of spec, or off quality refers to a material or product that does not meet its specification or standard requirements. For example, refined oil that becomes contaminated with water and no longer meets the specified cleanliness specifications. See also, secondary steel.</td></tr>
</table>

<table>
<tr><th>Phrase or Term</th><th>Definition</th></tr>
<tr><td>Office Jerk, Work Jerk</td><td>An office jerk or work jerk is someone whose behaviour is so inappropriate and outrageous that it bullies or offends others, or distracts from the work-at-hand. The book The No Asshole Rule: Building a Civilized Workplace and Surviving One That Isn't by Robert I. Sutton lists dirty dozen common office jerk behaviours as:
1. Personal insults and status slaps;
2. Invading one's personal space;
3. Uninvited personal contact (e.g., physical and on social media);
4. Threats and intimidation;
5. Sarcastic jokes and mean-spirited teasing;
6. Email flaming;
7. Public humiliation and yelling;
8. Rude or frequent interruptions;
9. Two-faced attacks;
10. Dirty looks (e.g., eye rolls, glares, and stares);
11. Treating people like they are invisible; and,
12. Isolating people socially, physically, or informationally.
The office jerk engages in hurtful office politics, uses humour that is divisive or offensive, and may become an office pariah.
See also, bully, cyberbullying, flame war, flaming, office joker, office pariah, office politics, save face, toxic behaviour, and trolling.</td></tr>
<tr><td>Office Joker</td><td>An office joker is someone who works in an office and uses humor to:
• Boost morale;
• Build camaraderie;
• Bust boredom;
• Keep the company's history alive by retelling stories from the past;
• Make new employees and all personnel feel more included;
• Speak to management about uncomfortable truths; and,
• Strengthen the corporate cultural.
Personnel who tell jokes appear to be more confident and competent to their peers, provided the humor is positive, and not divisive or offensive. An office joker engages in positive office politics and uses humour to help us laugh at ourselves and find new ways of solving problems. See also, office jerk and office politics. See also, office jerk and office politics.</td></tr>
<tr><td>Office Pariah</td><td>An office pariah is someone who is socially awkward to the point of becoming an outcast, and is avoided by others in office and social settings. They may:
• Be avoided or ridiculed by others;
• Be disliked by all or most colleagues;
• Be unable to conform to group culture;
• Lack conversation etiquette or skills, or be ignorant of social or cultural norms; or,
• Lack meaningful connections with others.
See also, office jerk.</td></tr>
</table>

Phrase or Term	Definition
Office Politics, Workplace Politics	*Office politics* or *workplace politics* are the activities, attitudes, and behaviours that are used to get or keep advantage or authority within a business or organization. It is the informal way in which power is: 1. Administered or shared; and, 2. Affected by the personal relationships between the personnel who work there. Office politics also gives some the ability to use power beyond what is legitimate, or to gain personal advantage or to benefit. This can be either negative (e.g., seeking advantage through unfair means at the expense of others or the greater good) or positive (e.g., seeking to fairly promote oneself or the greater good). See also, *office jerk*, *office joker*, *politics*, and *save face*.
Offline	Not a preferred phrase – see *out of service*.
Offshore	To be *offshore* is to be situated away from land or a shore (i.e., in deep, open, or shallow water). Some businesses operate outside of the country in which they were founded (i.e., colloquially known as an offshore business) to take advantage of lower costs, less stringent regulations, or both. See also, *onshore*.
Offshoring	*Offshoring* is the practice of basing company or project processes or services overseas to leverage lower labour costs, equipment costs, material costs or combination. See also, *outsourcing*.
Off-the-shelf	*Off-the-shelf* describes a standard item that is produced with existing designs and production, which is typically available for immediate purchase and use (e.g., instruments, small motors, and valves). *Off-the-shelf* items are *not* custom-designed or made-to-order.
Oil Sands	*Oil sands* is bitumen-soaked sand. Oil sands may be found primarily in the Canadian province of Alberta within these geographic regions: Athabasca, Cold Lake, Peace River, and Wabasca. See also, *bitumen*.
OK, O.K., Ok, ok, okay	*OK*, *O.K.*, *Ok*, *ok*, or *okay* mean *all correct* and denote acceptance, acknowledgment, agreement, approval, assent, or indifference. The origin of this word is unclear with etymologies from Africa, Europe, and the United States. Many variations of this word are used in other languages, sometimes with a phonetic spelling to produce a similar pronunciation as in English. Hence, it may be the most frequently spoken-written word on the planet.
Old Boy's Club, Old Boy's Network, Old Boys' Society	An *old boy's club*, *old boy's network*, or *old boys' society* (club)is an informal club or group by which money and power are retained by wealthy white men through improperly exclusive or self-interested business relationships. These relationships are often informally established at functions in exclusive country clubs, executive skyboxes at sporting events, on high-priced golf courses, or through private-exclusive fraternities or social clubs. These clubs exclude social groups (deliberately or unintentionally), especially minorities and women. These social groups are traditionally excluded from, or not represented in the club and thus are excluded from *club* conversations, related social relationships, and the benefits of upward mobility gifted to club members. The phrase originated from British upper-class boy's attendance at certain public schools (e.g., at Eton or Oxford). Former students are *old boys*. See also, *crony capitalism* and *smoke-filled room*.

<table>
<tr><th>Phrase or Term</th><th>Definition</th></tr>
<tr><td>Old School</td><td>Being old school refers to:
• Adhering to conservative or traditional ideas, methods, policies, or practices;
• Anything old-fashioned or not modern; or,
• Characteristic or evocative of an earlier, established, or original form, manner, or style which is regarded highly or with respect.
For example, the old school approach of the engineering firm involved hard copy markups of piping and instrumentation diagrams.</td></tr>
<tr><td>Olet, O-let, or O'Let</td><td>An Olet, O-let, or O'Let is a branch connection fitting that is a joined to an outlet pipe by butt welding. Stress concentrations in the joint are minimized by integral reinforcement, eliminating the need for a reinforcing pad. Types of O-Lets include:
• Elbowlet;
• Flangeolet;
• Letrolet;
• Nipolet;
• Sockolet;
• Swepolet;
• Thredolet; and,
• Weldolet.
See also, branch connection, butt weld, fitting, integral reinforcement, and reinforcing pad.</td></tr>
<tr><td>Oligarchy</td><td>An oligarchy is a government or society that is controlled or ruled by a small number of elite members (e.g., as distinguished by business, political, military, or religious affiliations, education, nobility, or wealth, or a combination). Greek philosopher Aristotle pioneered the use of this term as meaning rule by the rich. See also, government, kleptocracy, meritocracy, plutocracy, and populism.</td></tr>
<tr><td>Oligopoly</td><td>An oligopoly is a market structure with a small number of companies, none of which can keep the others from having significant influence. There is no precise upper limit to the number of firms in an oligopoly, but the number must be low enough such that the actions of one firm significantly influences the others. For example, the leading vehicle manufacturers in the United States, Ford Motor Company, Fiat Chrysler Automobiles, and General Motors Company are powerful enough to influence the development, introduction, and price of new vehicle models. See also, duopoly and monopoly.</td></tr>
<tr><td>Omnivert</td><td>An omnivert is both an extrovert and introvert. An omnivert can be introverted in a situation they don't care about or when then do not want to be included. By contrast, they can also be extroverted when in a situation that they care about and wish to enjoy. They can be quiet and reserved, and loud and open. See also, extrovert and introvert.</td></tr>
</table>

Phrase or Term	Definition
On Hold	*On hold* is a phrase used to describe the editing convention that identifies a document or document section as *not approved for use* because it contains missing or incomplete information. A document or document section marked as *on hold* is typically scheduled to be updated at a later date and is typically subject to change. A project may identify a *on hold* document or document section using the cloud method or another editing convention. See also, *cloud method*.
On Line	Not a preferred phrase – see *in-service*.
On-the-job Training, OJT	*On-the-job training* or *OJT* is a hands on method of instruction used by personnel to learn needed competencies, knowledge, and skills to perform a job or task in the workplace. OJT: • May be facilitated by a colleague or external service provider; • May be needed because no available resources provide personnel with transferable skills or to provide company-specific training; • May occur as personnel perform their actual work or in a dedicated training room or workstation; and, • Uses existing workplace documents, equipment, instructors, knowledge, materials, and tools to teach personnel how to perform work effectively and safely. See also, *book smart*, *boot camp*, *continuing education*, *cross-train*, *hands on*, , *lifelong learning*, *lunch and learn*, *professional development*, *school of hard knocks*, *street smart*, *tool time*, and *workshop*.
On the Same Page	*On the same page* is a common expression that indicates personnel have the same amount of knowledge or understanding about a project. For example, *the project team seems to be on the same page so we should proceed with project execution*.
One Hole	*One hole* refers to a bolt hole configuration that does not conform to the industry standard, which requires all flange bolt holes to straddle the centreline. Using one hole is usually a mistake and only used on explicit request (e.g., to fit-up with an existing flange that cannot be, or is not, two hole). See also, *centreline*, *straddle*, and *two hole*.
One-dimensional	*One-dimensional* is: • Having one dimension and therefore showing only linear information such as height, length, or width; or, • Lacking depth or a having a single focus that is (likely too) narrow or superficial.
One-off	A *one-off* is something done or something that happens only once and is not part of a planned sequence (e.g., witnessed performance test of a pump).
One-variable Thinking	*One-variable thinking* is a shortcut to analyzing a complicated situation with multiple factors by concluding that only one factor was decisive. For example, the conclusion that the falling price of oil is because more people now prefer to use renewable energy. See also, *hollow man argument*, *loserthink*, *mental prison*, *one-variable thinking*, *professional jealousy*, *shortcut*, *renewable energy*, *thinking mode*, and *word-thinking*.
Onshore	To be *onshore* is to be situated on land. Also, businesses that operate onshore operate within the country in which they were founded. See also, *offshore*.
Open Data	*Open data* is data in the public domain that can be freely used by third parties. See also, *closed data*.

Phrase or Term	Definition
Open-ended Question	An *open-ended question* is a query that cannot be answered simply by the words *no* or *yes*, or with a static response. A more thoughtful and longer response is required, which can be compared to other information that can be obtained or is already known. This type of question creates a learning experience for the person being questioned, the questioner, or both. This is accomplished by asking reporter's questions or the Five Ws. See also, *closed-ended question* and *reporter's questions*.
Operated Valve	Not a preferred phrase – see *actuated valve*.
Operation Technology	*Operation technology* is the hardware and software (e.g., sensors) that cause or detect changes in the processes of physical devices (e.g., pumps and valves) through control, direct monitoring, or a combination. See also, *operational technology* and *technology*.
Operational Control	*Operational control* is the authority to manage day-to-day or normal business operations and processes. It does not include business priorities, strategies, or targets. See also, *operations management*.
Operational Excellence	*Operational excellence* streamlines business deliverables and processes and finds a balance between lowering costs and maintaining healthy profit margins. It also brings products to market cheaper and quicker, and more safely than conventional methods. Implementing best practice operating procedures to identify and mitigate potential hazards has been shown to reduce safety incidents by up to 50%. The seven elements required to build operational excellence are: 1. Continuous improvement; 2. Employee accountability; 3. Knowledge sharing; 4. Leadership; 5. Management of change; 6. Risk control; and, 7. Risk identification. Operational excellence and organizational excellence are closely related but have been developed separately by different industries and organizations. See also, *continuous improvement*, *excellence*, *organizational excellence*, and *operational excellence*.
Operational Excellence Management System	An *operational excellence management system* is a business strategy used to maintain awareness of how every area and process of operations can be improved. See also, *integrated management system*, *organizational excellence management system* and *management*.
Operational Technology	*Operational technology* is the organization that supports operation technologies such as sensors and a supervisory control and data acquisition system. See also, *engineering technology*, *information technology*, *operation technology*, *supervisory control and data acquisition*, and *technology*.
Operational Test	Not a preferred phrase – see *function test*.
Operations	*Operations* are the business activities in which a company engages to convert resources and materials into finished products or services, sell them to customers, and earn a profit.

Phrase or Term	Definition
Operations Management	*Operations management* is the design and control of business operations and production processes used to transform resources into products or services. It ensures that operations are consistent with established business objectives and plans, efficient, and effective. See also, *operational control*.
Operator	An *operator* is personnel who operate a facility system. Also, not a preferred term when the *operator* is equipment – see *actuator*.
Operator Qualification	An *operator qualification* allows personnel to operate equipment (e.g., for automated or mechanized bonding or welding, or non-destructive examination) or a system (e.g., a pipeline). Not unlike a driver's license, it verifies that an individual has been evaluated and approved to perform specific operations.
Opinion-based Decision-making	*Opinion-based decision-making* relies heavily on either the selective use of evidence (e.g., on a single study irrespective of quality) or untested views of individuals or groups, and is often inspired by ideological standpoints, prejudices, or speculative conjecture. See also, *decision-based evidence making*, *decision-making*, *evidence-based decision-making*, *evidence-based management*, and *value-based decision-making*.
Opportunity	In project management, an *opportunity* is a possibility that can contribute to project objectives, which is classified as a type of risk. This is because of the loss of potential gains from all alternative courses of action. See also, *opportunity cost* and *opportunity for improvement*.
Opportunity Cost	The *opportunity cost* of a course of action is the loss of potential gains from all alternative courses of action. See also, *opportunity*.
Opportunity for Improvement	An *opportunity for improvement* is a finding that is not a non-conformance, which has a potential for improving products and services or addressing future needs and expectations by correcting, preventing or reducing undesired effects (e.g., improving efficiency or reducing waste). See also, *corrective action*, *non-conformance*, *opportunity* and *preventative action*.
Opportunity Shaping	*Opportunity shaping* is a business-led process by which sponsors develop and gather information needed for key decisions, and then evaluate key attributes of a potential project. The value of the project is then allocated to various stakeholders: • Ensuring that each stakeholder retains enough project value to make the venture worthwhile; and, • To make the project environment sufficiently stable for successful execution. One of the keys to a successful opportunity shaping process is to examine reasons to end a project instead of examining reasons for continuing it.
Optimism Bias, Wishful Thinking	*Optimism bias* or *wishful thinking* are a cognitive bias that causes one to believe that they or their project are less likely to experience a negative event than others. It is also an overly-optimistic belief that a project will deliver greater benefits than what was planned, what was experienced by other projects, or can be proven. This bias is common and transcends age, ethnicity, gender, and nationality. Ironically, this bias typically grows stronger when negative or unforeseen events or circumstances occur. See also, *bias*, *cognitive bias*, *confirmation bias*, *self-serving bias*, and *thinking mode*.

Phrase or Term	Definition
Optimization	*Optimization* is the act, methodology, or process, of making something (e.g., design, decision, or system) as fully effective, functional, or streamlined as possible. See also, *debottleneck*.
Order Entry	*Order entry* is the process of taking a customer order, and scheduling the processes of dispatching and ensuring goods or services are delivered to the customer. After an order is entered, the supplier can view information about it and update it, as required. See also, *goods* and *positive order entry*.
Order of Magnitude	An *order of magnitude* is the approximate size of something, measured in powers of 10. This measurement is often used to compare the relative quality, quantity, or size of items.
Order of Precedence	*Order of precedence* is the sequential hierarchy or priority of the contract or procurement documents. If conflicts between documents exist, the order of precedence resolves the conflict by indicating which documents have more authority over other documents. The more complex an agreement and the more documents incorporated into the agreement, the more important it is to state the order of precedence. Order of precedence rules are determined by the project, the party issuing the contract or purchase order, and the legal or regulatory system. The following lists examples of document genres in a (fictitious) project's preferred order of preference: 1. Owner and general contractor contract; 2. Owner and general contractor contract provisions documents; 3. Specifications; 4. Drawings; 5. Amendments or omissions; and, 6. Supplier documents. Common project order of precedence rules: • Drawings with the same date that contain larger scale have priority over drawings with smaller scale; • Dimensions shown on drawings have priority over dimensions scaled from drawings; and, • Later-dated documents shall have priority over earlier documents of the same type. Common order of precedence for regulatory documents (Alberta): 1. Acts; 2. Regulations; 3. Referenced codes and standards; and, 4. Orders, directives, errata, interpretations, and STANDATA. The order of precedence may also be used to sequence attached or referenced documents listed in a contract or purchase order. See also, *contract*, *purchase order*, and *STANDATA*.

Phrase or Term	Definition
Ore	An *ore* is a: • Naturally occurring material containing a valuable constituent (i.e., a mineral or metal), which is mined and refined (e.g., bauxite, hematite, or uraninite); or, • Source from which valuable matter is extracted (e.g., a database). See also, *alloy*, *base metal*, *ferrous*, *mineral*, *metal*, and *non-metallic*.
Organization	An *organization* is a group of personnel under specific leadership to function as a single entity for a particular purpose. See also, *organizational excellence*.
Organization for Economic Cooperation and Development, OECD	The *Organization for Economic Cooperation and Development* or *OECD* is a group of countries with market-based economies, almost all of which are highly developed economically. Its 50 members include Canada, Japan, the United States, and most countries in Western Europe. The OECD designation is used to describe developed economies.
Organization of the Petroleum Exporting Countries Plus, OPEC+, Vienna Group	The *Organization of the Petroleum Exporting Countries Plus*, *OPEC+*, and *Vienna Group* is a cartel of 14 OPEC and 10 non-OPEC countries. Most notable of the non-OPEC members are Kazakhstan, Mexico, and Russia. They control 55% of global oil supplies and 90% of proven oil reserves. See also, *cartel* and *Organization of the Petroleum Exporting Countries*.
Organization of the Petroleum Exporting Countries, OPEC	The *Organization of the Petroleum Exporting Countries* or *OPEC* is a cartel of 14 countries. The first 5 members were Iran, Iraq, Kuwait, Saudi Arabia, and Venezuela. They control 35% of global oil supplies and 82% of proven reserves. See also, *cartel* and *Organization of the Petroleum Exporting Countries Plus*.
Organization of Mineral Exporting Countries, OMEC	The *Organization of Mineral Exporting Countries* or *OMEC* is proposed as a collaboration between countries with aggressive climate change targets and mining companies that have access to critical energy minerals. This phrase was coined by Trevor Hart and recognizes that geopolitical power could change from oil-dominated countries to those that produce the minerals used by renewable energy. These critical energy metals could replace fossil fuels and the Organization of Petroleum Exporting Countries as the dominant force in the energy market. Countries with major deposits of critical energy metals include Australia. China, Chile, Congo, and Indonesia. This concept was developed by KPMG International in the report *Resourcing the Energy Transition: Making the World Go Round*. See also, *critical energy metals*, *fossil fuel*, *mineral*, *Organization of Petroleum Exporting Countries*, *rare earth element*, and *renewable energy*. See also, *mineral*.

Phrase or Term	Definition
Organizational Chart, Org Chart, Organigram, Organogram	An *organizational chart*, *org chart*, *organigram*, or *organogram* illustrates an organization's relationships, with its role or reporting hierarchy and structure. It may be used as a: • Management tool; • Personnel directory; and, • Planning tool. These are common types of organizational charts: • Divisional Structure For instance, organized by resources and functions that support a geographic area or product line. • Flat Organizational or Horizontal Organizational For instance, with few or no levels of middle management between personnel and upper management. • Functional top-down or hierarchical For instance, each entity in the organization, except one, is subordinate to another single entity. • Matrix Organizational For instance, personnel may report to more than one leader or supervisor. An organizational chart typically does not define or describe individual responsibilities like a role and responsibility chart does. For example, only personnel names and positions may be depicted. This kind of chart was first created by Scottish-American engineer Daniel McCallum around 1854, was developed further in the early 19th century by the French engineer Henri Fayol. See also, *dotted line reporting*, *role and responsibility chart*, *reporting line*, and *solid reporting line*.
Organizational culture	*Organizational culture* encompasses behaviours and values that contribute to the unique psychological and social environment of a business. This affects the way people interact, the context within which knowledge is created, the resistance they have towards certain changes, and ultimately the way they share or do not share knowledge. This includes the organization's assumptions, beliefs, habits, environment, language, location, norms, symbols, systems, values, and vision. It represents the collective beliefs, principles, and values of organizational members. It may be influenced by factors such as history, management style, market, national culture, technology, type of employees, type of product, and strategy. See also, *corporate culture* and *values*.

Phrase or Term	Definition
Organizational Excellence	*Organizational excellence* are the efforts to establish processes and standards to engage and motivate employees to deliver products and services that fulfill customer requirements within business expectations. It is the achievement of consistent superior performance such as outputs that exceed expectations, needs, or objectives. Organizational excellence is designed for permanent change by focusing on these key elements: • Change management; • Knowledge management; • Process management; • Project management; and, • Resource management. Organizational excellence and operational excellence are closely related but have been developed separately by different industries and organizations. See also, *continuous improvement*, *excellence*, *management*, *operational excellence*, *organization*, and *organizational excellence management is system*
Organizational Excellence Management System	*An organizational excellence management system* is a comprehensive method of building organizational excellence culture and improving health, environment and safety performance to prevent high-consequence incidents and impacts. See also, *integrated management system, management, and organizational excellence management system*.
Organizational Memory	*Organizational memory* refers to the collective ability of personnel to store and retrieve knowledge and information. This includes formal records, and embedded and tacit knowledge retained by organizational culture or personnel, and processes. By contrast, traditional memory refers to the ability to acquire, retain, and retrieve knowledge. See also, *key knowledge*. *Lessons learned, and retention of critical knowledge*.
Original Equipment Manufacturer	An *original equipment manufacturer* is a company that produces parts and equipment that may be marketed by another manufacturer. This may or may not be a supplier to the project. See also, *tier supplier*.
Originator	An *originator* is the person who begins or initiates something (e.g., a non-conformance report process or requisition).
Orthodox	*Orthodox* is a: • A type of doctrine, interpretation, practice, or theory that is authorized or generally accepted by an organization (e.g., religious, political, or scientific). For example, instituting checks and balances in a government; or, • A type of activity, belief, or idea considered acceptable, normal, and traditional by most people. For example, Muslims participating in the *call to prayer*. See also, *checks and balances*, *dogma*, *outside the box*, and *unorthodox*.

Phrase or Term	Definition
Ostrich Effect	The *ostrich effect* is the attempt to avoid negative financial (or other) information using irrational means. For example, an investor who avoids negative financial information about their investments by deliberately not reading the newspaper or stock market reports. This phrase originates from the debunked legend that an ostrich will bury its head in the sand to avoid danger (thereby, irrationally avoiding danger). The term was coined by Dan Galai and Orly in their article *The "Ostrich Effect" and the Relationship between the Liquidity and the Yields of Financial Assets*. See also, *boiling frog syndrome* and *slippery slope*.
Other	Like the phrase *accept as-is*, *other* is a disposition that indicates a deviation from original requirements, with acceptance based on alternatives deemed acceptable to the project (e.g., extended warranty, supplied spare parts, or an enhanced service agreement). This disposition may require additional documentation for effective implementation.
Other Specified Tests	*Other specified tests* are performed for electrical equipment. These tests may include complete, routine, load-bearing, bearing-hosing, resonance, deceleration, enclosure integrity, full pressure, full load, full speed, optional, heat run, spare parts, sound level, speed, roll, tandem, or vibration-level.
Out of the Box, Right Out of the Box	*Out of the box* or *right out of the box* refers to the immediate functionality or usability of a newly purchased product, without the need for integration or upgrades. For example, removing an air freshener from its packaging and plugging it into the wall. This is sometimes known (especially in the technology industry) as *turnkey* or *plug and play*. See also, *plug and play* and *turnkey*.
Out of Tolerance	*Out of tolerance* indicates that: • A data point has drifted or shifted beyond the allowable tolerance; or, • The total allowable error or permissible variation within an item has been exceeded. No device, machine, or production operation (e.g., machining, testing, or welding) holds measurements precisely to the nominal value. Thus, there are acceptable degrees of variation. If a product is produced with a measurement that is out of tolerance, it is not a usable as-is, per the design intent. See also, *as-built*, *calibration*, and *tolerance*.
Outage	An *outage* is: • A failure or interruption in the functioning or use of something (e.g., a power plant); or, • The time when electrical power or another service (e.g., water) is not available. See also, *brownout* and *blackout*.

Phrase or Term	Definition
Outlier	An *outlier* is: • A person or thing that is atypical within a particular category, class, or group; • A pariah; • A statistical observation that is markedly different in value from the others of the sample; or, • Something that is situated away from or classed differently from a main or related body (e.g., a geological feature). See also, *pariah*.
Outside the Box	*Outside the box* is a metaphor that means *to think differently, unconventionally, or from a new perspective*. This phrase often refers to creative, novel or non-conventional thinking. See also, *dogma*, *orthodox*, *outside the warehouse*, *thinking mode*, and *unorthodox*.
Outside the Warehouse	*Outside the warehouse* is a metaphor that means *to think even more differently, unconventionally, or from a very new perspective, than thinking outside the box*. This phrase often refers to very creative, novel or unconventional thinking. See also, *outside the box*.
Outside-in Thinking	*Outside-in thinking* means to look at a business with an objective view, from the customer's perspective. To counter bias, focus must turn inward and toward the customer. With this type of thinking, decisions, policies, processes, products, systems, and tools are based on what best meets customer needs. See also, *customer relationship management*, *inside-out thinking*, *relational customer service*, *thinking mode*, and *transactional customer service*.
Outsourcing	*Outsourcing* occurs when a company purchases a scope of supply (e.g., equipment, materials, or services) that could not be provided internally, or from another domestic or foreign company. It often involves a business operation, specialty process or a non-core function (e.g., assembly, coating, engineering, fabrication, manufacturing, or testing). See also, *in-house* and *offshoring*.
Out-of-pocket	*Out-of-pocket* means: • Being without money or losing money. To be broke; • Costs paid using personal money. For example, an employee's business or work-related expenses that were purchased using a personal credit card (i.e., purchased out-of-pocket), which are later reimbursed by an employer. Or expenses not covered, shared, or reimbursed in part or whole by others (e.g., the portion of the expense that the insurance company does *not* pay for); • Of or pertaining to spending of cash instead of using credit; or, • Someone who is unavailable (e.g., to discuss anything or meet anyone).
Ovality, Out-of-roundness	*Ovality* or *out-of-roundness* is the cross-sectional distortion of a round hollow structural steel pipe or tube, where the deviation forms the shape of an oval. Ovality is measured by comparing the difference between major and minor inside, nominal, or outside diameters.
Over Pressure Protection by System Design	*Over pressure protection by system design* refers to the protection of pressure equipment without the use of a pressure relief device. See also, *overpressure* and *pressure relief device*.

Phrase or Term	Definition
Overqualified	*Overqualified* means that qualifications (i.e., education, experience, training, or a combination) exceed the requirements of a particular position or role. An organization may be reluctant to recruit personnel that are overqualified because they: 1. Will accept a *lesser* role (e.g., are seen as desperate or someone who may not meet expectations); 2. May be more senior than their colleagues or manager (e.g., apt to cause conflict with others, have an unwelcome or outdated approach, or be resistant to change or using new technology); 3. May leave if a better opportunity becomes available elsewhere (e.g., for career advancement or increased compensation); or, 4. Will expect, or negotiate for, higher compensation – even if the compensation was previously disclosed (or the compensation range was known). This term may be used as a euphemism when an employer does not want to reveal their true reason(s) for not hiring an applicant. See also, *ageism* and *euphemism*.
Over-commit and Under-deliver	To *over-commit and under-deliver* is when a service provider fails to follow through on a commitment to a customer or does not meet their expectations. See also, *under-commit and over-deliver*.
Overpressure	*Overpressure* is a condition where the maximum allowable design pressure of pressure equipment is exceeded. See also, *over pressure protection by system design* and *overpressure risk assessment*.
Overpressure Risk Assessment	An *overpressure risk assessment* determines the maximum upset pressure and sources of overpressure. It must use a documented, organized, and systematic approach that is conducted by qualified personnel on a multi-disciplinary team. See also, *overpressure*.
Overthink	To *overthink* means to analyze or think about something for too long or too much, such that the thinking becomes more harmful than helpful. For example, ruminating about the past (e.g., what should have been done) or worrying about the future (e.g., what should be done). Overthinking involves dwelling on: • A problem. By contrast, critical thinking or problem-solving involves thinking about a solution to the problem; or, • Negative feelings and emotions, and self-pity about lack of control. By contrast, self-reflection enables self-learning and acquisition of new perspectives. Time spent developing creative solutions or learning about oneself is productive; however, overthinking is not. For example, someone who overthinks a problem will spend hours trying to determine how to free a vehicle that is stuck in the mud instead of asking someone for a tow. See also, *critical thinking* and *paralysis by analysis*.
Owner, Sponsor	An *owner* or *sponsor* is an entity, usually a government or private company, that seeks to develop and benefit from a project. In most cases, the owner is also the monetary investor in the project. See also, *resource company*.
Owner's Inspector	An *owner's inspector* is the person responsible to the owner for ensuring the requirements for inspection, examination, testing, and certification of a pressure piping system are met. This terminology is used in ASME B31 *Code for Pressure Piping* documents. See also, *owner*.

Phrase or Term	Definition
Owner-User	An *owner-user* owns an integrity management system per *Pressure Equipment Safety Regulations* and has been issued a *Quality Management System Certificate of Authorization Permit*. An owner-user exercises control over operations, engineering, inspections, repairs, alterations, maintenance, pressure testing, and rerating of pressure equipment. See also, *owner*.

Phrase or Term	Definition
P	1 A B C D E F G H I J K L M N O **P** Q R S T U V W X Y Z

P3 Assurance, Projects Programs and Portfolios Assurance	*P3 assurance* or *projects programs and portfolios assurance* involves:
	• Ensuring that projects, programs, and portfolios are on course to:
	  o Fulfill objectives;
	  o Meet performance expectations; and,
	  o Meet requirements.
	• Satisfying sponsors and stakeholders.
	See also, *P3 management*.

P3 Management, Projects Programs and Portfolios Management	*P3 management* or *projects programs and portfolios management* refers collectively to the management of projects, programs, and portfolios. See also, *P3 assurance*.
Package	Not a preferred term – see *packaged equipment* and *procurement documents*.
Package Engineer	Not a preferred phrase – see *responsible engineer* or *discipline engineer*.
Package Team	Not a preferred phrase – see *requisition team*.
Packaged Equipment	*Packaged equipment* is complex, high dollar value, and multi-disciplined equipment (e.g., engineered, electro-mechanical, modular, and process systems) that may include free-issued, or sub-supplied assemblies or components. See also, *electro-mechanical equipment and integration*.
Packaging	*Packaging* is the material (e.g., cardboard, metal, paper, plastic, wood, or a combination) and methods used to contain, preserve, and protect a product during distribution, sale, storage, and transportation. Packaging may comprise three layers: 1. Primary packaging To contain and preserve the product, and provide brand recognition, instructions, and other information. For example, a metal can that contains silicone lubricant. 2. Secondary packaging To group primary packages together and facilitate handling and storage. For example, a carboard box containing cans of silicone lubricant. 3. Tertiary or shipping packaging To group secondary packages together for bulk shipment such as in a container or on a pallet. For example, a pallet loaded with boxes of silicone lubricant cans that are secured together with plastic-wrap. The materials and methods used for packaging may be defined by the original equipment manufacturer, project specifications, regulations, and other sources, or a combination. See also, *intermodal container*, *marking and labelling*, *preservation*, and *shipping and handling*.
Packing List	Not a preferred term – see *packing slip*.
Packing Slip	A *packing slip* is a document that describes the contents of a package or packages. This document permits the shipper to identify what has been sent and the receiver to verify receipt. A packing slip lists product descriptions and quantities, and may include additional useful information (e.g., dimensions, weights, and reference or stock keeping unit numbers).

Phrase or Term	Definition
Pacta Sunt Servanda	*Pacta sunt servanda* is a key concept in civil, canon, common, and international law that it shall not be easy to escape contractual liability by proving an event was unforeseeable. It means that the clauses of a contract are binding or enforceable, so nonfulfillment of an obligation is a breach of the contract. This Latin phrase means *agreements must be kept* or *treaties shall be complied with*. It is arguably the oldest principle of international law. See also, *force majeure*.
Paint	Not a preferred term – see *coating*.
Paint System	Not a preferred phrase – see *coating system*.
Pallet	A *pallet* is a plastic or wood platform on which cargo is stored to expedite loading and unloading of cargo.
Panacea	A *panacea* is a: • Cure-all or remedy for any illness or disease; or, • Solution for any difficulty or problem. For example, the vaccine was no panacea; it cured only 20% of infections. See also, *moonshot*, *pipe dream*, and *unobtainium*.
Panic Buying, Panic Purchasing	*Panic buying* or *panic purchasing* is the action of buying large quantities of a commodity or product after a disaster or because of fears of a forthcoming price increase or product shortage. Panic buying can lead to exacerbated shortages regardless of if the risk of a shortage is perceived or real. This is a type of herd behaviour is caused by: • Coping mechanisms to relieve anxiety and regain control; • Fear of the unknown caused by emotional pressure and uncertainty; • Perception of the threat of a crisis and scarcity of products; and, • Social psychological factors from the influence of one's social network. For example, during the onset of the COVID-19 pandemic in Canada in early 2020, there was an irrational public fear that supply chain disruptions would yield a shortage of toilet paper. This perceived falsehood by the public resulted in panic buying for many weeks and led to a shortage of toilet paper. In this way, the perceived fear became reality. It is in this way that panic buying can lead to a self-fulfilling prophecy. See also, *COVID-19*, *herd behaviour*, *nothing to fear but fear itself*, and *self-fulfilling prophecy*.
Paper Over	*Paper over* is an expression that means to: • Make a difference or disparity appear acceptable, especially to maintain an agreement or unity; or, • Conceal or hide an unpleasant situation (e.g., a disagreement or problem) with the purpose of persuading others that it does not exist or is not serious. See also, *paper up*.
Paper Tiger	*Paper tiger* is a literal English translation of the Chinese word *zhilaohu*, which refers to something or someone that claims or appears to be powerful, threatening, or both – but is ineffectual and unable to withstand a challenge (e.g., a politician with a symbolic post who has no real political capital). In North America, the phrase *all bark no bite* (which refers to a dog with a loud bark who is too afraid to bite) has a similar meaning. See also, *red herring*, *straw man argument*, and *weak man argument*.

Phrase or Term	Definition
Paper Up, Draw Up	*Paper up* or *draw up* is an expression that means to prepare a valid, binding, legal agreement (e.g., containing specific terms from an agreement in principle, or made verbally). For example, a contract with signatures. See also, *paper over.*
Paperhanging	*Paperhanging* is the: • Act of providing affirmation statements on an office wall at a company (i.e., to espouse a certain principle, purpose, mission, value, or vision) but not taking action to fulfill the statement. This is also known as having *words on a wall.* For example, using the words: *Always finish projects on time* – but, in practice, rarely or never finishing company projects on time; • Crime of counterfeiting fraud (i.e., writing a bad or dishonoured cheque to illegally acquire or borrow funds that do not or may not exist); or, • Trade of decorating walls with wallpaper.
Paradigm	A *paradigm* is: • A clear or typical example of something (e.g., a common single engine airplane); • An exemplar (e.g., a template); • A model that explains something or shows how it can be produced (e.g., blueprints); or, • The philosophical and theoretical framework of a discipline, within which the concepts, laws, methods, standards, and theories are produced (e.g., Buddhism). See also, *anecdotal.*
Paradox	A *paradox* is: • A statement that is seemingly contradictory or opposed to common sense and yet may be true. For example, every year 300 pieces of the Eiffel Tower (tower) are replaced. Because the tower contains 18,000 pieces, it is completely replaced every 60 years. In 2009, the tower was 120 years old, so it had been *replaced* twice. As a result, paradoxically, some consider the tower to be the same tower as the original, while others view the tower as different from the original or a completely new tower; or, • An action, person, or situation, having seemingly contradictory qualities or phases. For example, the theory that global warming will lead to the next Ice Age. See also, *Catch-22*, *panic buying*, and *common sense.*

Phrase or Term	Definition
Parallel Power	*Parallel power* is where an individual or group low in an organization's hierarchy are powerful enough to do whatever they want, even when it is contrary to organizational objectives. This power may be based on: • Privileged access to external or internal information, lack of enforcement, or preferential support, which enables them to bypass normal decision-making channels. For example, the nightshift labour force does not follow procedures because there is no mechanism or no one to enforce compliance; or, • Specialized expertise, such as when a subject matter expert reports to an administrator or manager without specialized expertise. For example, a project manager cannot authenticate a report. Hence, an engineer is requested to do so, but refuses because they have not reviewed and validated the work, and cannot accept professional responsibility for it. Parallel power may force senior management to: • Accept ideas from subordinates that are not necessarily in the organization's best interests. This may be because they have insufficient information to ask the right questions or that opposition seems illegitimate; or, • Make decisions without all the necessary information because subordinates will not provide it. If an organization's incentive and reward systems are designed such that autonomous decisions generally reflect organizational goals (i.e., organizational and personal objectives are aligned), there are fewer risks with parallel power, or it can be avoided altogether. This concept was developed by Ann Langley in the article *Between Paralysis by Analysis* and *Extinction by Instinct. (management decision-making approaches)*. See also, *authenticate* and *subject matter expert*.
Paralysis by Analysis, Analysis Paralysis	*Paralysis by analysis* or *analysis paralysis* is an expression that describes how a group or individual can over-analyze information or overthink such that they are unable to act or decide. This can occur because of the fear of better options, an inability to satisfice, information overload, perfectionism, or a combination. See also, *extinct by instinct*, *fear of better options*, *overthink*, *perfectionism*, and *satisfice*.
Paraverbal Communication	*Paraverbal communication* is the voice modulation of speech including the cadence (i.e., rate and rhythm), expression (i.e., use of drama or emotion), pitch, tone, and volume. By ensuring paraverbal communication is consistent with spoken words, message confusion and message with dual meaning may be avoided. For example, whispering *good night* to a partner at bed time is proper; yelling it is not. Paraverbal communication is *how* something is said more than *what* is said. This means that identical statements can have a completely opposite meanings, depending on the voice modulation of the speaker. For example, asking *what is wrong* could be stated in a caring and supportive way or in condescending and impatient way. Paraverbal communication also promotes the delivery of an intelligible message, effectively and efficiently, so that listeners do not become bored or lose focus. See also, *body language*, *mirroring*, *reflective questioning*, and *tone-deaf*.

Phrase or Term	Definition
Pareto Chart	The *Pareto chart* is a graphic representation of the Pareto Principle: 20% of input produces 80% of output using a vertical bar graph and a line graph. The bar graph represents the metrics of various business process components, ordered from the largest to the smallest. The line graph represents the cumulative total of these metrics. This tool visualizes what part of the process influences output the most. See also, *Pareto Principle*.
Pareto Principle, 80/20 Rule	The *Pareto Principle* or *80/20 rule* is also known as the law of the vital few, or the principle of factor sparsity. It states that, for many events, roughly 80% of the effects come from 20% of the causes. Management consultant Joseph M. Juran suggested the principle and named it after Italian economist Vilfredo Pareto, who noted in 1896, that about 80% of the land in Italy was owned by 20% of the population. It is an axiom in business management that 80% of sales come from 20% of clients. See also, *low-hanging fruit and Pareto chart*.
Paris Agreement	The *Paris agreement* is intended to strengthen the global response to the threat of climate change by keeping a global temperature rise this century below 2°C above pre-industrial levels and to pursue efforts to limit the temperature increase even further to 1.5°C. Additionally, it is intended to increase the ability of countries to deal with the impacts of climate change, and at make economies compliant with climate-resilient and low greenhouse gas goals. The Paris agreement is a landmark agreement to combat climate change and to accelerate and intensify the actions and investments needed for a sustainable low carbon future. Parties to the United Nations Framework Convention on Climate Change agreed to the Paris agreement at the Conference of the Parties 21st conference (COP 21) in Paris, France on December 12, 2015. See also, *climate change* and *greenhouse gas*.
Parking Lot	A *parking lot* is: • An area or space used for the parking or temporary storage of vehicles; and, • A figure of speech for when a topic arises that is not relevant to the current discussion or meeting agenda. The topic is *parked*, to be addressed later. This is also called *tabling the conversation*. When an idea is parked, it is recognized as important, but needs to be considered at another place or time. For example, at the budget meeting, new ideas for improving the company's recruiting processes were put in the parking lot.
Parkinson's Law	*Parkinson's law* is the adage that *work expands to fill the time available for its completion*. For example, if a report is requested as due on Friday, it will be produced by end of business day on Friday, regardless if can be produced sooner. Parkinson's law was first articulated by Cyril Northcote Parkinson in the book co-authored with Osbert Lancaster *Parkinson's Law: The Pursuit of Progress*. This law is also known as induced demand and may be applied to the growth of bureaucracy in an organization. It has implications for other areas of business including bandwidth, project management, resource allocation, and storage capacity. See also, *bureaucracy*, *Horstman's corollary*, and *Parkinson's law of triviality*.
Parkinson's Law of Triviality	*Parkinson's law of triviality* refers to the tendency to devote large amounts of time to unimportant details while simultaneously not giving essential tasks the time that they require. See also, *Horstman's corollary* and *Parkinson's law*.

Phrase or Term	Definition
Parlour Game	A *parlour game* is a form of indoor, group, structured play for adults (e.g., played in a parlour) such as boxed games or games with charades, drama, logic, or word-play. These games were extremely popular among the upper and middle classes in Great Britain and the United States between 1837 and 1901, and between 1920 and 1970. During these times, only the upper and middle classes had enough leisure time to play parlour games. A parlour game is also an expression used to accuse political opponents of using deliberately confusing or nebulous language when describing a position on an issue.
Part Number, Catalogue Number	A *part number* or *catalogue number* is a unique identifier used to define each standard part, including its form and function. See also, *serial number*.
Partial Penetration	*Partial penetration* is a weld where the fused joint is not completely solid, with remaining gaps between the base metals after welding. There are many ways to accomplish this (e.g., the weld joint is a groove that is not the full thickness of the base metal), but the final product is not solid metal at the joint. This may be specified as an incomplete joint penetration weld. Partial penetration is required for joints that will be not subjected to high stresses because this weld has lower strength than a full penetration weld. See also, *base metal*, complete joint penetration, *full penetration*, and *incomplete joint penetration*.
Participant	Not a preferred term – see *attendee*.
Partnership	A *partnership* is a business arrangement in which two or more parties cooperate to manage and operate a business. See also, *corporation*, *cooperative*, *joint venture* and *sole proprietorship*.
Party	A *party* is a corporation or person directly involved or interested in any act, affair, agreement, contract, transaction, or legal proceeding. Someone who enters into an agreement, contract, or other transaction is considered a party to the agreement. When a dispute results in litigation, the litigants are referred to as parties to the lawsuit. The phrases *first party*, *second party*, and *third party* define the function of a proponent (i.e., a corporation or person) that is involved in an activity (e.g., the inspection of a pump package). 1. First party or 1st party Services are provided by personnel employed by the supplier or ultimate owner or purchaser. 2. Second party or 2nd party Services are provided personnel employed the engineering and procurement company. 3. Third party or 3rd party Services are provided by personnel employed by an independent and impartial company. For example, progressive and final inspection is performed by a third-party inspection agency. The phrase *third party* is commonly used; *first party* and *second party* are uncommon. Payment for third-party services may be made by a first party or second party (per purchase order requirements). See also, *inspection agency*.

Phrase or Term	Definition
Passivation	*Passivation* is the treatment of stainless steel with chemicals (e.g., acid solution or paste) to remove contaminants (e.g., caused by grinding, machining, or mechanical damage), promote the formation of a passive film on the surface, and improve corrosion resistance. See also, *pickling*.
Passive Fire Protection	*Passive fire protection* is an integral component of structural fire protection and safety where structures are built with separate areas or compartments of manageable risk. These areas provide escape for occupants and protect firefighters by restricting the growth and spread of a fire and smoke using materials or methods such as coatings, enclosures, fire-resistant materials, and firestop sealants.
Passive Learning	*Passive learning* is an instruction method that involves engaging students with the course material through presentation (e.g., assigned reading or lectures). This learning method is teacher-centred. The student is responsible for absorbing the information, and accountable to make inquiries and pay attention. It promotes skills for defining, describing, listening, writing, and convergent thinking, where questions typically have only one correct answer. This is the traditional method of instruction in many modern schools. See also, *active learning*, *Bloom's Taxonomy*, *cone of learning*, *convergent thinking*, *knowledge transfer*, *metacognative thinking*, and *unlearning*.
Patent	A *patent* is a government authority or license conferring a right or title for a set period, especially the sole right to exclude others from making, using, or selling an invention. See also, *invention*.
Path Dependence, Path Dependency	*Path dependence* or *path dependency* explains how decisions for any given circumstance is limited by experienced events or past decisions, even if those circumstances may no longer be relevant. This can refer to outcomes that occur at a certain time or a long-term process. What has occurred in the past persists because of resistance to change – it is easier or more economical to continue on an established path rather than create an entirely new one.
Path-breaking	To be *path-breaking* is to accomplish a task in an innovative, new, pioneering, or trailblazing way. For example, in the past, the introduction of digital technology to decrease and eliminate the use of hard copy and paper documents was path-breaking. Today, implementation and improvement of some electronic data management systems is path-breaking (e.g., Amazon online, Netflix, Twitter, and Uber). See also, *electronic data management system*.
Pay It Forward	*Pay it forward* is an expression for describing the beneficiary of a good deed repaying the kindness to others instead of the original benefactor. The phrase may have been coined by Lily Hardy Hammond in 1916. See also, *Givers Gain®*, Golden Rule, *karma*, *law of reciprocity*, and *reciprocity*.
Pay the Price, Pay the Penalty	To *pay the price* or *pay the penalty* means to suffer the negative effects and consequences of one's own actions. For example, paying the price of eating too much food is indigestion and weight gain.
Peak Load Plant	A *peak load plant* is a power generation facility that provides a continuous supply of electricity, over a 24-hours and throughout the year. It supplies a minimum power generation requirement to meet fundamental consumer demand. Power is commonly produced using energy from sources such diesel fuel, jet fuel. Natural gas, solar, and wind. This type of facility can be turned on and off or its output quickly decreased or increased as needed. See also, *baseload plant*, *peak load*, and *power output*.

Phrase or Term	Definition
Peak Load, Peak Demand, Peak Load Contribution	*Peak load, peak demand*, or *peak load contribution* is the maximum electrical power demand over a specified time (e.g., on daily, monthly, seasonal, and yearly cycles). The actual point of peak demand is a single half-hour or hourly period, which represents the highest point of customer consumption of electricity. See also, *base load* and *peak load plant.*
Peak Oil	*Peak oil* is when the maximum rate of petroleum extraction is reached, after which the supply it is expected to enter terminal decline. This theory is based on the observed rise, peak, fall, and depletion of aggregate production rate in oil fields over time. It was first presented in 1956 by geologist M. King Hubbert, who predicted that United States' peak oil would occur in about 1970. Many experts now believe that global energy demand will plateau between 2030 and 2040, as industry shifts energy production to electricity (e.g., solar), energy efficiency, and more service-driven economic growth instead of oil production. See also, *depletion*.
Peak Shaving, Load Shedding	*Peak shaving* or *load shedding* refers to levelling out peaks in electricity use by commercial and industrial power consumers. Power consumption peaks are important in terms of grid stability, but they also affect the cost of power. In many countries electricity prices for large-scale consumers are set with reference to their maximum peak-load. The reason is simple: the grid load and the necessary amount of power production need to be designed to accommodate these peak loads. With peak shaving, a consumer reduces power consumption quickly and for a short period to avoid a spike in consumption and increased costs. This is possible by either activating an on-site power generation system, temporarily scaling down production, or relying on battery power. See also, *base load*, *load shifting*, and *peak load*.

Phrase or Term	Definition
Peak State	A *peak state* is a condition of excitement and positivity when one feels energized, focused, and ready to succeed. Peak state can be achieved by using a combination of: • Attitude Have a positive attitude; envision and expect favorable results. • Body language Use positive body language, such as good posture, direct eye contact and smiling to encourage energy, enthusiasm, and positivity. • Focus Focus on goals and results. • Language Use positive language (e.g., it is not a *problem*, it is a *challenge*). • Literature Educate and empower by reading informative, educational, or inspiring writing. • Music Listen to music that inspires and encourages. • Network Connect with knowledgeable, interesting, and inspiring people. • Nutrition Consume healthy and nutrient-rich foods. • Vision Have a positive vision; use past uplifting events to inspire future ones. See also, *body language* and *priming*.
Peer	A *peer* is someone who has equal standing others by belonging to the same social group based on age, background, grade, interest, profession, or status. For example, a peer of a student is another student. A teacher is not student's peer. See also, *peer group*, *peer mentoring*, *peer pressure*, and *peer review*.
Peer Group	A *peer group* is a social group with which someone associates. A peer group is likely to influence individual beliefs, behaviour, or both (e.g., associating with an entrepreneurial peer group is likely to make an individual more personally successful, professionally successful, or both). A peer group contains many distinct behaviour patterns and hierarchies (e.g., students experience a hierarchy of friendship groups based on popularity, and academic and social status). See also, *peer, peer mentoring, peer pressure*, and *peer review.*

Phrase or Term	Definition
Peer Mentoring	*Peer Mentoring* occurs when two or more colleagues transfer knowledge. This may be: • Collaborative When like-minded personnel dealing with similar challenges, experiences, and successes provide a source of education and personal development to each other; or, • Experience-based When a person who has experienced a challenge acts as a mentor to a person who is new to that experience, who is the mentee. The peer mentor may challenge the mentee with new ideas and encourage them to move out of their comfort zone. Most peer mentors are chosen for their confidence, sensibility, social skills, and reliability. See also, *coach*, *comfort zone*, *cross-generational mutual mentoring*, *mentor*, *peer*, *peer group*, *peer review*, and *reverse mentoring*.
Peer Pressure	*Peer pressure* is direct encouragement or influence from peers to change attitudes, behaviours, or values to conform. This may have negative effects, positive effects, or both. Peer pressure can affect those who are not members of the group. An organization (e.g., a corporation) is also susceptible to peer pressure from other organizations in the same industry, within organizational management, and from media or social media. See also, *cancel culture*, *peer*, *peer group*, *peer mentoring*, and *peer review*.
Peer Review	*A peer review* is the evaluation of academic, professional, or scientific work by others (peers) working in the same field. See also, *cold eyes review*, *discipline review*, *independent review*, *independent project review*, *peer*, *peer group*, *peer mentoring*, and *peer pressure*.
Penmanship, Handwriting	*Penmanship* or *handwriting* is the: • Art or skill of writing by hand (e.g., a professional calligraphist has immaculate penmanship); or, • Quality or style of one's handwriting (e.g., you have the penmanship of a child; I can barely read this chicken scratch. See also, *chicken scratch*, *writing*, and *writing style*.
People Are Messy, People Will People, People Will Be People	*People are messy*, *people will people*, or *people will be people* are expressions that describe how individuals can be cynical, difficult, egotistical, emotional, fallible, irrational, needy, petty, racist, unforgiving, unpredictable, and have other undesirable characteristics. See also, *messy*, *messy people*, and *playfulness acceptance curiosity empathy*.
People Person	A *people person* is someone who has strong communication skills, and enjoys or is skilled at interacting with others (especially socially or professionally). They enjoy being around and talking to people, and are liked by others. See also, *extrovert* and *introvert*.
Pep Talk	A *pep talk* is a brief, emotional, intense talk designed to encourage or influence others to expend more effort, feel more confident, or both.

Phrase or Term	Definition
Per Diem	*Per diem* is Latin for *by the day* or *for each day*. For a project, a *per diem* is an allowance or payment made to personnel for each working day (e.g., accommodation and meal expenses), typically to pay for travel-related expenses. See also, *inspection cost estimate*.
Percentage Point	A *percentage point* is a quality verification point that occurs at a specified frequency up to and including 100%. See also, *quality verification point*.
Perception is Reality	*Perception is reality* is an expression that describes subjective experience or understanding of reality, in contrast to an external and objective reality. Although perception may feel real, it is not necessarily factual or the same as the perception of others. See also, *everybody knows principle*.
Perfectionism	*Perfectionism* is a refusal to accept anything but perfect results. See also, *paralysis by analysis*.
Perfectly Confident	*Perfectly confident* is a concept developed in the book *Perfectly Confident: How to Calibrate Your Decisions Wisely* by Don Moore. Confidence is an attitude that considers facts and personal feelings, but one's confidence needs to be calibrated. The key to success is to avoid being both over- and under-confident, by being perfectly confident – the Goldilocks Zone for confidence. Too little confidence results in missing out on valuable opportunities; too much confidence leads to taking on excessive risk. See also, *Goldilocks Zone*.
Perfect Storm	A *perfect storm* is a: • Bad or critical situation, arising from several negative and unpredictable factors; • Rare combination of circumstances that drastically aggravate a situation; or, • Violent storm arising from a rare combination of adverse meteorological factors. The phrase was popularized by the Hollywood movie *The Perfect Storm* (2000). See also, *black swan theory*.
Perfluoroalkyl Substances, Polyfluoroalkyl Substances, Forever Chemicals	*Perfluoroalkyl substances*, *polyfluoroalkyl substances*, or *forever chemicals* are man-made chemicals that can have negative impacts on the environment and the health of animals, people, and plants. These chemicals comprise carbon-fluorine bonds, which are recognized as the strongest covalent bond in organic chemistry. They resist degradation in the environment by heat, oil, and water, and have a low volatility. Jurisdictions around the world are developing standards to regulate these chemicals. Because of their persistence in the environment, and their ubiquitous nature, these chemicals have been, or are being, phased out from use in many parts of the world. They may impact casualty or future property claims (e.g., a fire that was extinguished using aqueous firefighting foams where there was no access to a fire hydrant). See also, *casualty*.

Phrase or Term	Definition
Performance Marketing, Performance-based Advertising	*Performance marketing* or *performance-based advertising* is a comprehensive term for online advertising and marketing programs where advertisers pay only when a specific action occurs. These actions can include a click, generated lead, sale, and more. Performance marketing professionals—in agencies, with media companies, publishers, or otherwise—rely heavily on paid marketing channels including: • Affiliate marketing; • Native advertising; • Search engine marketing; • Social media advertising; and, • Sponsored advertising. This became more common with the use of the Internet, where it is now possible to measure user actions resulting from an advertisement. See also, *advertising* and *native advertising*.
Performance Test	A *performance test* demonstrates that equipment complies with design intent and specified purchase order requirements, and meets process guarantees.
Periodic	*Periodic* refers to something that occurs at regular or intermittent intervals.
Periodic Table, Periodic Table of Elements	The *periodic table* or *periodic table of elements* is the organized array of all the chemical elements in order of increasing atomic number (i.e., the total number of protons in the atomic nucleus), electron configuration, and recurring chemical properties. When the chemical elements are thus arranged, there is a recurring pattern called the *periodic law* in their properties, where elements in the same column or group have similar properties. Russian chemist Dmitry I. Mendeleyev first published a recognizable periodic table in 1869. It was not recognized until the 1920s that the order of the elements in the periodic system is in fact by their atomic numbers. Since 2016, the periodic table has 118 confirmed elements, from element 1 (i.e., hydrogen) to 118 (i.e., oganesson). The first 94 elements occur naturally while the remaining 24 are manmade and only occur when synthesized in a laboratory.
Permatemp	A *permatemp* is a temporary employee who works for an extended period for a single staffing agency. Permatemp a portmanteau of the words *permanent* and *temporary*. See also, *independent contractor*.
Permit	A *permit* is a document officially confirming that the formal approval of plans, equipment, or installations meets requirements of prescribed codes and regulations by the designated government agency or jurisdiction. A permit is an authorization to proceed with construction, operation, or reconfiguration of specific items at a particular site per approved drawings and standards. A permit is for a specific discipline (e.g., structural, electrical, or pressure equipment) and considers only code and regulatory requirements (not owner-specific requirements, such as specifications). See also, *certificate*, *certificate of authorization permit*, and *license*.
Perpetual Beta	*Perpetual beta* describes a program or software that is not intended to be released as a final version. Instead, the software will be analyzed, improved, and used as a working copy of what will likely become a final software product. See also, *beta test*, and *evergreen*.
Person Hours	Not a preferred phrase – see *labour hours*.

<table>
<tr><th>Phrase or Term</th><th>Definition</th></tr>
<tr><td>Personal Agenda</td><td>A personal agenda seeks an outcome that suits someone's needs; however, it often contradicts what others or the organization need. A personal agenda typically is self-interested manipulation to reach a goal, which creates a conflict of interest between personal and organization agendas. For example, when an executive rejects a potentially great idea because senior management will not personally benefit from its success. The conflict is that the executive gives personal goals higher priority than organization goals. See also, conflict of interest.</td></tr>
<tr><td>Personal Protective Equipment</td><td>Personnel wear personal protective equipment to minimize exposure to hazards that cause workplace injuries and illnesses (e.g., contact with chemical, physical, electrical, mechanical, or other workplace hazards). Common personal protective equipment includes coveralls, earplugs, earmuffs, gloves, hard hats, high-visibility reflective safety vests, safety glasses, safety shoes, and respirators.</td></tr>
<tr><td>Personal Values, Perspectives</td><td>Personal values or perspectives are human characteristics developed over time that influence human behaviour and thought processes, which can be categorized as:
1. Aesthetic;
2. Economic;
3. Political;
4. Religious;
5. Social; and,
6. Theoretical.
See also, value-based decision-making and values.</td></tr>
</table>

Phrase or Term	Definition
Personality Type Theory	*Personality Type Theory* is a hypothesis that categorizes people as: • Having one of two contrasting behaviours (i.e., Type A or B); • Having a combination of the above (i.e., Type A-B, not categorically Type A or B); or, • Having another, fourth type of behaviour (i.e., Type D). These are: • Type A personalities Type A personalities are often high-achieving workaholics that may experience job-related stress. They tend to be aggressive, ambitious, anxious, aware of time management, competitive, impatient, organized, proactive, and status-conscious. • Type B personalities Type B personalities who lack Type A behaviours that may experience less job-related stress. They tend to be more relaxed and less neurotic. • Type A-B personalities Type A-B personalities who are on a continuum where they are either more of a Type A or Type B personality. • Type D personalities Type D personalities who are distressed and tend to be negative (e.g., apprehensive, irritable, and gloomy) and socially inhibited (e.g., modest and lacking confidence). The Type A behaviour pattern does not always outperform those with Type B personalities; however, Type A students tend to earn higher grades than Type B students. The *Type A personality* was first described in the 1950s by cardiologists Meyer Friedman and Ray Rosenman. See also, *workaholic*.
Personalization	*Personalization* is a type of knowledge transfer that involves a *one-to-one* or *face-to-face* transfer of information. See also, *codification*, *knowledge transfer*, and *tribal knowledge*.
Perspective-taking	*Perspective-taking* is perceiving a situation or understanding a concept from an alternative point of view. See also, *precision questioning*.
PEST Analysis	*A PEST analysis* examines how political, economic, social, and technological factors might affect a project.
Peter Principle	The *Peter Principle* is a concept in management developed by Dr. Laurence J. Peter, which observes that people in a hierarchy tend to rise to their *level of incompetence*. This means that a person is promoted based on their success in previous jobs until they reach a level at which they are no longer competent. This is because skills in one job do not necessarily translate to another. The concept was explained in the book *Peter Principle: Why Things Always Go Wrong* by Dr. Laurence J. Peter and Raymond Hull.
Petrochemical	*Petrochemical* is, or relates to, a substance obtained by the refining and processing of natural gas or petroleum. See also, *coal liquefaction*, *natural gas,* and *petroleum*.
Petroleum	*Petroleum* is a liquid mixture of hydrocarbons present in geological formations that can be extracted and refined to produce fuels (e.g., diesel, gasoline, jet fuel, kerosene, and oil) See also, *coal*, *fossil fuel*, *hydrocarbon*, *natural gas*, and *petrochemical*.

Phrase or Term	Definition
Petrostate, Petro-state	A *petrostate* or *petro-state* is: • A country whose wealth stems from the sale of oil (e.g., Venezuela); or, • An oil-rich country with weak institutions; where power and wealth are controlled by a minority (e.g., Saudi Arabia).
Philanthropy	*Philanthropy* is the charitable giving of resources (e.g., money) to a cause on a large scale. Philanthropy is much more than just a charitable donation: it is an effort by an individual or organization based on an altruistic desire to improve human welfare; a private initiative for the public good focusing on quality of life. Philanthropy attempts to address the root cause of societal or other problems, such as environmental preservation. A wealthy individual may establish a non-profit organization or private foundation to fund philanthropic efforts. See also, *charity*, *non-profit*, *voluntary sector*, and *wealth*.
Photograph Opportunity, Photo-op	A *photograph opportunity* or *photo-op* is an accidental or arranged occasion for the someone (often the press) to photograph a celebrity, politician, or notable event. A planned, formal photo-op is intended to generate positive publicity with pictures for use in magazines, newspapers, online, on television, or combination. It has acquired a negative connotation, as a staged pseudo-event masqueraded as news. Bruce Whelihan, an aide to United States President Richard Nixon's Press Secretary Ron Ziegler, coined this term during that administration (1969 to 1974).
Phub	To *phub* is to snub someone by repeatedly and frequently interrupting them by ostensibly viewing data on a smart device (e.g., a smart phone) while having a conversation. A phub may completely ignore the other party or give them only partial (not full) attention. Phub is a portmanteau of telephone and snub. Phub was coined in 2012 by Adrian Mills at the McCann advertising agency. See also, *microaggression*.
Pickling	*Pickling* is the treatment of stainless steel with chemicals (e.g., acid solution or paste) to remove contaminants, high temperature oxides or scale, and low-chromium layers (e.g., caused by heat treating or welding), promote the formation of a passive film on the surface, and improve corrosion resistance. See also, *passivation*.
Piecemeal	*Piecemeal* is characterized by unsystematic partial measures taken gradually over time to complete work (e.g., building a brick wall by laying one brick once every day). Piecemeal work is *not* completed all at once. To perform work in a piecemeal fashion is to complete the work only partially, and then return to complete another portion of the work on another date. And so on.
Pig Catcher (Pipeline Instrument Gauge Catcher)	Not a preferred phrase – see *pig receiver*.
Pig Launcher (Pipeline Instrument Gauge Launcher)	*A pig launcher* or *piping instrumentation gauge launcher* is a section or station on a pipeline that physically launches a pig into a pipeline, which typically initiates the (non-tethered) inspection of a pipeline. See also, *pig receiver*, *pipeline instrument gauge (pig)*, and *station*.

Phrase or Term	Definition
Pig Passage Indicator, Pig Signaler, or Pig Sig	A *pig passage indicator*, *pig signaler*, or *pig sig* indicates, by way of signal, when a pig has left a launcher, passed a point in a pipeline, or arrived in a pig receiver. Generally, two types may be used: • The intrusive type uses a mechanical trigger indicator that intrudes into the pipe; and, • The non-intrusive type uses a magnetic indicator that does not intrude into the pipe. See also, *pipeline instrument gauge (pig)*.
Pig Receiver (Pipeline Instrument Gauge Receiver)	*A pig receiver* or *piping instrumentation gauge receiver* is a section or station on a pipeline that physically receives a pig, typically after the pig has inspected a pipeline. Subject matter experts and specialists analyze data gathered by a pig. See also, *pig launcher*, *pipeline instrument gauge*, and *station*.
Pig Trap (Pipeline Instrument Gauge Trap)	Not a preferred phrase – see pig *launcher* and *pig receiver*.
Pigeon Hole	A *pigeon hole* is a: • Category, which usually fails to reflect actual complexities (e.g., a stereotype); • Small recess in which a pigeon nests; or, • Small compartment, open at the front and forming part of a set, where mail or messages may be left (i.e., a cubby hole).
Piggyback	To *piggyback* means to: • Carry or haul something as an extra load or by railroad car (e.g., carry a cargo truck on a flat deck train car); • Carry someone on your back or shoulders; or, • Take advantage of, and use something, that already exists or operates as a support for other work. For example, to modify existing project instructions, plans, and procedures for use with a new and similar project. See also, *approved supplier list*.
Pile	A *pile* is a cast concrete, or driven or screwed steel member installed in the ground to support a structure.
Pilot Plant, Demonstration Plant	A *pilot plant* or *demonstration plant* is a small model facility used for testing processes under actual production conditions. See also, *proof of concept*, *prototype*, and *scale*.
Pink Collar	*Pink collar* constitutes or relates to the class of personnel whose duties generally relate to customer interaction, entertainment, sales, or other service-oriented work. See also, *blue collar* and *white collar*.

Phrase or Term	Definition
Pinocchio Effect	The *Pinocchio effect* describes how a liar will use more complex sentences with more third person pronouns (e.g., her, him, it, one, their, and they) than someone who is honest. A liar is more worried about being believed than an honest person; hence, the liar will use more words and work harder to be believable. This effect was discovered and named by Lyn M. Van Swol, Michael T. Braun, and Deepak Malhotra in the article *Evidence for the Pinocchio Effect*. Pinocchio is a fictional character created by Italian writer Carlo Collodi in the children's book *The Adventures of Pinocchio*. Pinocchio was a puppet made by a woodcarver named Geppetto, who had a dream that the puppet could become a real boy. The character Pinocchio frequently told lies, which would cause his nose to grow longer and longer. This is the most translated (300 languages) non-religious book in the world and one of the best-selling books ever published (80 million copies).
Pipe	A *pipe* is a cylindrical hollow structure used in architectural, pressure, and structural applications. Pipe is measured by the nominal inside diameter and wall thickness. Pipes and tubes may look similar, yet they have many differences and are rarely interchangeable. See also, *hollow structural steel*, *structural steel*, and *tube*.
Pipe Class	A *pipe class* is part of a piping specification that identifies the types of components (e.g., pipe schedule, material, flange ratings, valve types and materials, gaskets, fasteners, and all other component-specific requirements) approved for use with different fluids under different operating conditions (e.g., pressure, temperature and corrosive environments). See also, *fastener* and *schedule*.
Pipe Dream	A *pipe dream* is a desire, idea, or plan that is unlikely to be realized – an impossibility or wishful thinking (e.g., my pipe dream is to own a castle on every continent). The phrase is an allusion to unrealizable dreams experienced opium smokers. However, the origin of this phrase is American. It was first used in *The Fort Wayne Gazette* (September 1895) to describe events in Chicago. See also, *fool's gold*, *hopium*, *moonshot*, *panacea*, *unobtainium*, and *when pigs fly*.
Pipe Spool	A *pipe spool* is a prefabricated component of a piping system that includes pipes, flanges, and fittings.
Pipeline Instrument Gauge (Pig)	A *pipeline instrument gauge* or *pig* is a device that traverses the interior of a pipeline and performs cleaning, gauging, batching, and other pipeline-related maintenance operations. Some of the first cleaning *pigs* were made from straw and wrapped in wire or leather. Both types made a squealing noise while travelling through a pipe, giving the pig its name. In the oil and gas industry, pig is an initialism (abbreviation) derived from the initial letters of the term *pipeline instrument gauge*. See also, *smart pipeline instrument gauge*, *in-line inspection*, *pig receiver*, and *pig launcher*.
Piping and Instrumentation Diagram	A *piping and instrumentation diagram* or P&ID is a detailed drawing used by the process engineering discipline to illustrate piping and vessels in a process flow in relation to associated instrumentation and control devices.
Plagiarism	*Plagiarism* is the act of using or closely imitating the language, thoughts or ideas of an author without authorization and the representation of that author's work as one's own by not crediting the original author. See also, *copyright* and *public domain*.
Plain End	*Plain end* is a pipe end preparation with a square cut. See also, *bevelled end* and *threaded end*.

Phrase or Term	Definition

Plain English,
Plain Language,
Plain Writing
	Plain English, *plain language*, or *plain writing* is written information that is concise, clear, uses the fewest words possible, and uses layman's terms so an audience can easily understand the content. It is intended to:
	• Avoid complex and technical vocabulary;
	• Be appropriate to the audience's developmental or educational level, and familiarity with the topic; and,
	• Be free of abbreviations, clichés, and jargon.
	Writing in plain English:
	• Involves these methods:
	  o Active voice instead of a passive voice;
	  o Common, everyday words;
	  o Easy-to-follow design features (e.g., headers, lists, and tables);
	  o Reader-centred organization; and,
	  o Short paragraphs and sentences.
	• Avoids using:
	  o Complex or technical terminology;
	  o Legalese;
	  o Passive voice;
	  o Pronouns, such as *you*, *I*, and *we*; and,
	  o Run-on sentences.
	The phrase is commonly discussed when reviewing government or business documents because they typically *do not* use plain English. Other examples of documents that typically do not use plain English include contract, insurance, legal, warranty, and user agreement documents. See also, *abbreviation*, *buzzword*, *colloquial*, *dialect*, *Ebonics*, *filler words*, *gibberish*, *jargon*, *vernacular*, *word salad*, and *writing style*.

Plan	A *plan* is an approved formal document used to guide project execution and project control. A plan's primary uses are to document planning assumptions and decisions, facilitate communication among project stakeholders, and document approved baselines for cost, schedule, and scope. A plan may be summarized or detailed. See also, *corporate instruction*, *guideline*, *knowledge transfer*, *planning*, *program*, *plan*, *project charter*, and *project instruction*.
Plan B	Not a preferred term – see *contingency plan*.
Plan–Do–Check–Act, Plan–Do–Check–Adjust	*Plan–Do–Check–Act* or *Plan–Do–Check–Adjust* is a four-step project planning tool used for continuous improvement. It means: 1. Plan – Recognize an opportunity and plan a change; 2. Do – Implement the plan and test the change; 3. Check – Review the test, analyze the results, and identify what was learned; and, 4. Act/Adjust – Act or adjust based on what was learned to improve the process. This is also known as the Deming circle/cycle/wheel, the Shewhart cycle, and the control circle/cycle. See also, *continuous improvement*.

Phrase or Term	Definition
Plane	A *plane* is a: • Flat or level surface; or, • Surface in which, if any two points are chosen, a straight line joining them lies wholly in that surface. A plane may be described as: • Auxiliary, horizontal, inclined, and vertical; • East-west, north-south, and vertical; and, • X, y, and z axes. See also, *field fit weld*.
Planned Obsolescence, Built-in Obsolescence, Premature Obsolescence	*Planned obsolescence*, *built-in obsolescence*, or *premature obsolescence* is the economic or industrial design strategy of deliberately ensuring that the current version of a product becomes out-of-date or useless within a known period. This guarantees that consumers will seek replacements in the future, thus bolstering demand and profit. Obsolescence can be achieved through introducing a superior replacement model, or by intentionally designing a product with an artificially limited useful life. In either case, consumers will theoretically favour newer generational products over the older ones. For example: • Contrived durability with parts that fail so the consumer will buy its replacement when it fails. For example, tools manufactured using brittle metal or plastic that will break easily; • Perceived obsolescence by clever marketing of newer products with *better* or *more* features. For example, a newer model cell phone with a better camera; • Restricted ability to repair or replace parts (e.g., so entire replacement of the product is the most cost-effective option). For example, ensuring that it is necessary to replace the entire air conditioner when it fails and that it is not economical or possible to repair it; or, • Updated software that drives consumers to discard or discontinue using an old product and buy a new one. For example, in 2017 Apple deliberately slowed down older iPhone models through software updates that were timed to coincide with releases of a newer iPhone models. In 1924, the global lightbulb association, also known as the Phoebus Cartel, engineered a lightbulb with a short life, giving birth to planned obsolescence. This generates long-term sales volume by reducing the time between repeat purchases. This is also known as *shortening the replacement cycle*. Planned obsolescence is not illegal but could be considered unethical by certain environmental groups.
Planning	*Planning* is the establishment of actions, forecasts, policies, and procedures to achieve goals and objectives. It is the fundamental management function used to decide beforehand: 1. What is to be done; 2. When it is to be done; 3. How it is to be done; and, 4. Who is going to do it. See also, *forecasting*, *goal*, *objective*, *plan*, and *procedure*.

Phrase or Term	Definition
Planning Poker	*Planning poker* is a consensus-based estimation technique that attempts to avoid the anchoring effect. Project team members make estimates simultaneously and then discuss them until they reach a consensus agreement. See also, *anchoring effect*, *budget* and *Delphi technique*.
Plant	Not a preferred term – see *facility*.
Plant North	*Plant north* is a suitable reference for normalizing a facility orientation on drawings. This standard reference is required because facilities, roads, and structures are historically built out of alignment with true north; it is rare to find a facility or property that has its axis aligned with true north. A property will be displayed on a map that includes an arrow showing true north that will be at an angle (angle of declination) off true north. It is difficult design anything at an odd angle. So to simplify this, a new north is created for the facility (called the plant north), which is used for design, engineering, and construction. It is conventional to include the plant north and true north on the plot plan with the angle of declination. Other drawings (e.g., structural foundations or equipment and piping layout) may only show plant north as a reference point. See also, *plot plan* and *true north*.
Plate Heat Exchanger, Plate and Frame Heat Exchanger	A *plate heat exchanger* or *plate and frame heat exchanger* transfers heat using thin, slightly separated plates with very large surface areas and small fluid flow passages. See also, *heat exchanger*.
Plate Metal	*Plate metal* is any sheet metal, including steel, with a thickness of 6 mm (15/64") or more. See also, *sheet metal* and *structural steel*.
Platform as a Service, Application Platform as a Service, Platform-based Service	*Platform as a service*, *application platform as a service*, or *platform-based service* is a cloud computing service that provides a platform for customers to develop, manage, and run applications without the complexity of building and maintaining the platform infrastructure typically associated with developing and launching an application. For example, the Google App Engine. See also, *anything as a service*, *cloud computing*, *content as a service*, *infrastructure as a service*, *software as a service*, and *security as a service*.
Playfulness Acceptance Curiosity Empathy, PACE	*Playfulness acceptance curiosity empathy* or *PACE* is method used to support an recovery from developmental trauma. It is also useful managing emotional dysregulation. This method was developed by Dr. Dan Hughes and consists of performing: • Playfulness Act to diffuse conflict and make a connection (e.g., tell an appropriate joke). • Acceptance Accept the emotions and needs that drive behaviour, without judgment, but not necessarily the behaviour itself (e.g., say a reaction is understandable). • Curiosity Seek to understand why behaviour has occurred (e.g., ask questions). • Empathy Connect with feelings and show compassion (e.g., by active listening). See also, *active listening*, *developmental trauma*, *emotionally dysregulated*, and *people are messy*.

Phrase or Term	Definition
Plot plan	A *plot plan* is an architectural, engineering, or landscape diagram or drawing that identifies equipment layout, locations of structures, positions of roads, elevations, landscape features, and utility runs of existing or proposed facilities at a defined scale. See also, *plant north* and *true north*.
Plug and Play	*Plug and play* is a feature on a device or software that works as intended when first connected or used (i.e., plugged in), without the need for user intervention (e.g., software installation, driver installation, configuration, registration, login requirements, or other adjustments). For example, plugging a plug and play memory stick into a computer allows the device to be used immediately. See also, *out of the box* and *turnkey*.
Plutocracy	*Plutocracy* is a government or society that is controlled or ruled by people of great income or wealth. See also, *government*, *kleptocracy*, *meritocracy*, *oligarchy*, and *populism*.
P-Number	Assigning a *P-Number* reduces the number of required welding and brazing procedures for base metals. This number is assigned by the ASME *Boiler and Pressure Vessel Code*. See also, *A-Number* and *base metal*.
Poetic Injustice	*Poetic injustice* is an outcome in which vice is rewarded or virtue is punished, usually in a manner that is ironically or peculiarly inappropriate. For example, an employee loses their job and the next day loses their vehicle in a fire. See also, *karma* and *poetic justice*.
Poetic Justice	*Poetic justice* is an ideal outcome in which vice is punished or virtue is rewarded, usually in a manner that is ironically or peculiarly appropriate. For example, an employee loses their job but wins the lottery the next day. The phrase was coined by Thomas Rymer in *The Tragedies of the Last Age Consider'd*. See also, *Givers Gain®*, *karma*, *law of reciprocity*, *poetic injustice*, and *reciprocity*.
Poetic License	*Poetic license* is when someone departs from normal rules or the truth to create a certain effect with language (e.g., alliteration) or embellish a story (e.g., hyperbole). For example, lyricist Bernie Taupin wrote the song *Daniel* about a Vietnam veteran who wants to get away. He made Daniel's destination *Spain* instead of America because *Spain* rhymes with *plane*.
Point of Sale System	A *point of sale system* allows the processing and recording of transactions between a company and their consumers, at the time in which goods and/or services are purchased. See also, *goods* and *point of sale*.
Point of Sale, Point of Purchase	The *point of sale* or *point of purchase* is when products or services are sold (e.g., at a cashier terminal). It may also be a point of customer order or return (e.g., a place where a customer can return a product at a brick-and-mortar store). The seller calculates the amount owed by the customer and then the customer makes a payment for the product or service. After receiving payment, the seller often issues a receipt for the transaction as proof of payment. See also, *brick-and-mortar business* and *point of sale system*.

<table>
<tr><th>Phrase or Term</th><th>Definition</th></tr>
<tr><td>Pointing-and-calling, Shisa Kanko</td><td>Pointing-and-calling or shisa kanko (Japanese) works on the principle of associating one's tasks with physical movements and vocalizations to prevent errors by raising the consciousness levels of workers — according to the Japanese National Institute of Occupational Safety and Health. Rather than rely on a worker's eyes or habits alone, each step in a task is reinforced audibly and physically in real time to ensure the step is complete and accurate.</td></tr>
<tr><td>Polarize</td><td>Polarize means to:<ul><li>Cause people to divide into two very distinct factions or groups (e.g., against or in favour of a proposed development or project, or to view something as only black and white);</li><li>Persuade someone to favour or oppose; or,</li><li>Restrict vibrations, partially or wholly, to one direction (e.g., of a transverse wave, especially light), or a definite pattern.</li></ul>See also, black and white, politicization, and radicalization.</td></tr>
<tr><td>Policy</td><td>A policy is:<ul><li>A set of ideas or plans, selected from among alternatives, that describe acceptable procedures and general goals that are used for decision-making, especially in business, economics, or politics;</li><li>A statement of intent implemented as a procedure or protocol;</li><li>An insurance contract against a specified damage, injury, or loss;</li><li>Expedient, prudent, or wise conduct or management; or,</li><li>The actions or attitudes regarding an issue.</li></ul>See also, policy layering.</td></tr>
<tr><td>Policy Layering, Policy Drift, Pancaking</td><td>Policy layering, policy drift, or pancaking is when new goals and measures are added to existing goals and measures. It involves the introduction of new policies, regulations, requirements, or standards without accommodating or dismantling the existing ones. This can result in unintended consequences (e.g., the introduction of requirements that conflict, are duplicates or overlap, are more complex or costly than existing policies, or are inefficient and overly-prescriptive), often mandating the need for elaborate systems to manage certification, compliance, validation, and verification. This concept is described by Robert Lyman in article The Clean Fuel Standard: The Latest in Climate Pancaking.

As an example, Lyman explains that under the Pan Canadian Framework on Clean Growth and Climate Change the Canadian federal, provincial, and territorial governments have established a framework of 237 policies and measures to reduce greenhouse gas emissions. Few Canadians examine how this plethora of measures interact, overlap, and affect others (i.e., people and industries). The Clean Fuel Standard is a set of regulations (circa 2020) that covers all fossil fuels used in Canada, with the objective to reduce Canada's greenhouse gas emissions by 30 million tonnes per year by 2030. This adds another policy layer to an already crowded set of climate policies that affect energy pricing in Canada. This complexity makes it very difficult to assess the policy's incremental cost and effect. See also, Clean Fuel Standard and policy.</td></tr>
</table>

Phrase or Term	Definition
Political	*Political* is: • Concerned with, or involving acts against, a government or political system; • Related to the government or public affairs of a country or region (e.g., municipality, province, or state); • Related to the ideas or strategies of a particular group or party in politics or public affairs; • Related to the making of governmental policy (as distinguished from the administration of policy); or, • To be active or interested in politics. See also, *agnostic*, *apolitical*, *government*, *non-governmental organization*, *politics*, *political suicide,* and *politicization*.
Politics	*Politics* is the: • Academic study of government (i.e., political science); or, • Activities associated with the governance of a country or region (e.g., municipality, province, or state), especially the conflict or debate among individuals or parties that have or seek to achieve power. Politics is also associated with the governance of non-governmental organizations like a for-profit company, joint venture, non-profit charity, partnership, or project. For example, a project may be comprised of stakeholders including more than one owner and more than one engineering and procurement company that may work together collaboratively or competitively to advise their common or separate interests. See also, *for-profit*, *government*, *joint venture*, *non-governmental organization*, *joint venture*, *non-profit*, *office politics*, *partnership*, *political*, *political suicide, politicization*, and *project alliance*.
Political Suicide, Politicide	*Political suicide* or *politicide* is an action or vote that is so unpopular with constituents that it will likely cause a politician to lose the next election or resign from public office. For example, a strong or unexpected deviation from a party's core values can be unacceptable to its constituents or supporters, resulting in a loss of confidence. For example, in some electoral districts, it is political suicide for a politician to support the constriction of a pipeline. American author William Safire observed, however, that *actions unpopular on their face can be taken as evidence of courage*. For example, On September 7, 1977 United States President Jimmy Carter signed the Panama Canal Treaty, which ceded the United States' control of the Panama Canal to the Panamanians and guaranteed the permanent neutrality of the canal beginning in 2000. Many opposed giving up a symbol of the United States' power and technological prowess, but the colonial-type administration of the strategic waterway irritated Latin Americans and especially Panamanians. The worst-case scenarios envisioned by the opponents (e.g., exorbitant tolls and seizure of the canal by the Soviets) never occurred. Thereafter, the Panama Canal was no longer run as a public utility. Instead, it became one of the most profitable transportation enterprises on the planet benefitting its shareholder – the Republic of Panama. Politicide is a portmanteau of *political* and *suicide*. See also, *political*, *politics*, *status quo* and *values*.

Phrase or Term	Definition
Politicization	*Politicization* is the act or process of causing an activity or event to become political in character or tone. This refers to how concepts, entities, or facts, that are considered as objective, outside of politics, or both, become topics for interpretation and subjective discussion. For example, leaders or proponents may manipulate data about an issue such as climate change to push an ideological or political agenda. See also, *political*, *politics*, *radicalization*, and *weaponization*.
Pollyanna Principle, Pollyannaism, Positivity Bias	The *Pollyanna principle*, *Pollyannaism*, or *positivity bias* is the tendency to remember pleasant things more accurately than unpleasant things. Research indicates that the subconscious mind tends to focus on the optimistic, while the conscious mind tends to focus on the negative. The name is from the novel *Pollyanna* by Eleanor H. Porter, which describes a girl named Pollyanna who attempts to find something to be optimistic in every situation. The phrase *Pollyanna principle* was coined by Margaret Matlin and David Stang to describe how the brain processes information that is pleasing and agreeable, and in a more exact and precise manner when compared to unpleasant information. See also, *bias* and *thinking mode*.
Poor-boy Crew	A *poor-boy crew* or mini gang crew completes the special short sections of a pipeline at steep hills and rail or road crossings, and prefabricated fittings, and any other welds that the main gang did not complete. See also, *crew*.
Populism	*Populism* is activities and ideas that are intended to appeal to ordinary people, who feel that their concerns are disregarded by established elite groups, by giving them what they want (e.g., tax cuts and higher wages). Populist policies may focus on boosting the popularity of a politician in power and on securing votes, instead of on achieving the best possible outcomes for the citizens – especially in the long term. Populism is both a left-wing and right-wing phenomenon. See also, *kleptocracy*, *left-wing*, *meritocracy*, *oligarchy*, *plutocracy*, and *right-wing*.
Pork Barrel	*Pork barrel* is when a ruling political party or politicians channel public money (pork) to particular constituencies based on political considerations, often for unnecessary and wasteful projects, at the expense of the public interest. This is done to please legislators or voters and win votes. The phrase may have originated in the United States with a pre-Civil War practice of giving barrel of salted pork to slaves. This was a common food product in 19th century households and could be used as a measure of a family's financial well-being. See also, *political*.
Pornography	*Pornography* is graphic material (e.g., pictures, videos or writing) that describes or depicts erotic behaviour and is intended to cause sexual excitement. This is considered by some people to be harmful, immoral, or unpleasant, and is illegal in some areas and countries. See also, *climate porn*.
Portable Document Format	*Portable document format* (PDF) is a file format developed by Adobe in the 1990s to present documents, including text formatting and images, in a manner independent of application software, hardware, and operating systems. See also, *native file format*.
Portfolio	A *portfolio* is a collectively managed set of programs and projects. See also, *portfolio balancing*.
Portfolio Balancing	*Portfolio balancing* is an organizational project management technique that involves selecting and tailoring a portfolio's components, so they can be managed in line with organizational objectives. See also, *portfolio*.

Phrase or Term	Definition
Portland Cement	*Portland cement* is the most common type of cement. It is a fine powder that is produced by heating limestone and clay minerals in a kiln, and adding 2 to 3% gypsum. See also, *cement*, *composite*, and *mineral*.
Portmanteau word	A *portmanteau word* blends two words into one. For example, the portmanteau word: • *Brunch* is derived from *breakfast* and *lunch*; • Motel is derived from *motor* and *hotel*; • Podcast is derived from *iPod* and *broadcast*; and, • Weblog is derived from *web* and *log*.
Positioner	A *positioner* is: • An analog, digital (smart), electrical, electro-pneumatic, or pneumatic device that adjusts a valve actuator's position based on a control signal; or, • A mechanical device for placing or holding an item in position during an operation (e.g., for drilling). It may also be used to rotate the item (e.g., for welding). See also, *actuator*.
Positioning	*Positioning* is the mental category that a brand occupies in the minds its customers and how it is distinguished from competitor products and services. To position their brand, a business may emphasize a product or service's distinguishing features (e.g., for products: what it is, what it does, and how it works; for services: how it works and benefits,) or create a suitable image (e.g., inexpensive or premium, utilitarian or luxurious, entry-level or high-end). Positioning enables the brand to be recognized correctly from the perspective of the business. For example, the vehicle brand Bugatti is positioned in most customers' minds as extreme expense, luxury, and quality. See also, *advertising, brand*, *customer centric*, *market share*, and *mindshare*.
Positive Change	*Positive change* is any form of change, redesign, or planning that begins with a comprehensive analysis, dialogue, and inquiry about an organization's positive core values. It involves the knowledge of multiple stakeholders by linking this knowledge to an organization's strategic change agenda and priorities. An organization's positive core values may be difficult to quantify unless they are documented. See also, *appreciative inquiry* and *values*.
Positive Material Identification	*Positive material identification* is the analysis of a metallic alloy to establish its composition and identification by reading the quantities of its constituent elements (by percentage) using non-destructive examination methods. See also, *non-destructive examination*.
Positive Order Entry	*Positive order entry* is paying attention to details and ensuring that a customer order is complete and correct. It involves actively engaging customers and obtaining feedback and using customer management tactics. This level of service builds trust with the customer and promotes repeated business. The supplier must manage their supply chain to deliver satisfactory products and services. See also, *customer relationship management*, *order entry*, *relational customer service* and *transactional service*.
Positive Variance	*Positive variance* is the amount by which actual project performance exceeds planned project performance. Positive variances in budget or time show the project is less expensive or is proceeding faster than planned. See also, *negative variance*.

Phrase or Term	Definition
Post	*Post* is a prefix (originally Latin) that has these meanings: • After (e.g., postdate); • Behind, following, or posterior (e.g., postlude); and, • Later than or subsequent to (e.g., post-shipment or post weld heat treatment).
Posttraumatic Stress Disorder, PTSD	*Posttraumatic stress disorder* or *PTSD* is a mental health condition that occurs in people who have experienced or witnessed a terrifying or traumatic event (e.g., a serious accident, a terrorist act, murder, violence, war, or the threat of death or serious injury). Symptoms may include anxiety, flashbacks, nightmares, or uncontrollable thoughts about the event. Most people who experience a traumatic event have temporary difficulty adjusting and coping but get better with good self-care and time. Someone whose symptoms get worse, interfere with their day-to-day functioning, or last for months or years may have PTSD.
Post Bend Heat Treatment	*Post bend heat treatment* is any heat treatment after bending pipe used to improve the properties of the bend.
Post Weld Heat Treatment	*Post weld heat treatment* is any heat treatment used after welding to improve the properties of the weldment, including the heat affected zone. See also, *heat affected zone*.
Postmortem	A *postmortem* is the analysis or discussion of: • A project after completion; or, • An event after it has occurred. See also, *hotwash* and *premortem*.
Potable Water, Drinking Water	*Potable water* or *drinking water* is water fit for drinking, cooking, and bathing. It is treated to meet federal, state, provincial, or other standards (e.g., for the removal of bacteria, debris, fecal matter, microorganisms, toxic chemicals, and viruses) and chemicals (e.g., chlorine or fluoride) to improve public health. It does not contain sufficient saline material to be regarded as mineral water. It may come from an underground or surface source (e.g., a well or river). See also, *mineral*, *non-potable water,* and *wastewater*.
Powder Coating	*Powder coating* is a type of coating that is applied as a free-flowing, dry powder. The main difference between a conventional liquid paint and a powder coating is that powder coating does not require a solvent and is cured under heat. It is usually used to create a hard finish, mainly for metals, that is tougher than conventional paint. See also, *coating*.

Phrase or Term	Definition
Power	*Power* is how fast energy is used or transmitted – the amount of energy divided by the time it took to use the energy. It is a physical concept that has different meanings, depending on context and available information. It can be: • A measurement of how much work someone or something performs over time; • A calculation as a function of velocity (i.e., how quickly mass is moved); or, • The product of current and voltage (i.e., electrical power). If the context and measurements are known, the appropriate formula can be selected to calculate power output. *Power* is also: • An official or legal authority, capacity, or right; • The ability to do something or act in a particular way, especially as a faculty, quality, or strength; or, • The ability or capacity to control, direct, or influence others or the course of events. See also, *energy* and *power output*.
Power Density	*Power density* is a measure of power output per unit volume. It is not as commonly used a measurement as energy density is; however, it is useful for energy system conversations. If a system has a high-power density, it can output large amounts of energy based on its volume. See also, *energy density* and *power output*.
Power Law	The *power law* is a functional relationship between two quantities, where a relative change in one quantity results in a proportional relative change in the other quantity. It is independent of the initial size of those quantities so that one quantity varies as a power of another. For example, considering the area of a square in terms of the length of its side – if the length is doubled, the area is multiplied by a factor of four. More than a hundred power-law distributions have been identified in physics (e.g., sandpile avalanches), biology (e.g., species extinction and body mass), the social sciences (e.g., city sizes and income), and in other fields.
Power Output	*Power output* is the is a measure of the rate at which work is done or similarly, at which energy is transferred. See also, *baseload*, *peak load*, *energy density*, *nameplate capacity*, and *power density*.
Power User	A *power user* uses advanced features of computer hardware, software, firmware, operating systems, programs, or websites (*not used* by the average person). They may not have extensive technical knowledge of the systems they use; however, they may be capable of computer programming or system administration. A power user has the competence (e.g., training), desire, or both to make the most intensive use of computer programs or systems. This may be a formal or informal role assigned to a specialist. Such personnel often retain their normal job role and also function as a power user to support business or project.

Phrase or Term	Definition
Power-to-gas Technology	*Power-to-gas technology* uses electrical power to produce a fuel-gas. When using surplus power from wind generation, the concept is sometimes called *windgas*. These systems use electrolysis to produce hydrogen, which can be used directly or converted into liquid petroleum gas, methane, or syngas. These gasses may be burned to produce heat, converted back into electricity using conventional generators or gas turbines, or used as chemical feedstock. Power-to-gas technology allows energy from electricity to be stored and transported in the form of natural gas that can be compressed, transported, and stored, often using existing infrastructure. See also, *feedstock*, *hydrogen*, *natural gas*, *power-to-X*, and *syngas*.
Power-to-X	*Power-to-X* describes electricity conversion, energy storage, and reconversion technologies that use surplus electric power (e.g., to convert electrical energy into gaseous or liquid chemical energy). This allows for the decoupling of power from the electricity sector for use in other sectors (e.g., chemicals or transportation). The X in *Power-to-X* is a variable that may refer to: • Ammonia; • Chemicals; • Food; • Fuel; • Gas; • Heat; • Hydrogen; • Liquid; • Methane; • Mobility; • Power; or, • Syngas. See also, *hydrogen* and *power-to-gas*.
Precast Concrete	*Precast concrete* is a construction product produced by casting concrete in a reusable form or mold. It is cured in a controlled environment and then transported to the construction site to be installed. In contrast, standard concrete is poured into site-specific forms and cured on site. See also, *composite* and *form*.
Precision Questioning	*Precision questioning* is a method of asking questions to acquire specific answers and assist with critical thinking and problem solving. It uses a highly structured, one-question/one-answer discussion format to help: • Conduct deep analysis; • Make difficult decisions; or, • Solve complex problems. This process can also be used to help others clarify possibly misunderstood arguments. See also, *calibrated question™*, *objective reflective interpretive decisional questioning*, *perspective-taking*, and *reflective questioning*.

Phrase or Term	Definition
Predicament	A *predicament* is a chronic, difficult, embarrassing, perplexing, trying, or unpleasant situation that cannot be permanently resolved, yet requires a response. The response to the predicament cannot eliminate the predicament because circumstances can only be managed. For example, a hurricane is a predicament that needs a response such as preparation, clean-up, or both; preventing the hurricane (the underlying predicament) is not possible. See also, *challenge*, *crisis,* and *problem*.
Prefabricated Construction	*Prefabricated* or *prefab construction* involves building components (e.g., pipe spools) off-site, at a fabrication facility, or a manufacturing facility. Prefab construction is completed under controlled conditions, using project-approved materials and designs – but in a shorter time frame than stick-built construction. See also, *modular construction*, *prefabricated prefinished volumetric construction*, and *stick-built construction*.
Prefabricated Prefinished Volumetric Construction	*Prefabricated prefinished volumetric construction* is a modular construction process where buildings are designed and planned to allow for the use of standardization of modules. This allows for efficient off-site manufacturing of partially to fully fitted out non-structural modules that can be rapidly assembled on-site. For example, this may be used to supply all washrooms as fully competed units for the construction of an apartment, condominium, or hotel. See also, *modular construction* and *prefabricated construction*.
Prefabrication Meeting	A *prefabrication meeting* is held before start of production by the supplier quality surveillance coordinator and includes the quality surveillance representative and supplier's quality personnel. Other personnel (e.g., from expediting and engineering) may also attend this meeting. The purpose of this meeting is to review technical, quality, and inspection requirements, including communications and responsibilities, with the supplier to facilitate successful execution. See also, *kick-off meeting*, *supplier* and *supplier quality surveillance coordinator*.
Preheat	*Preheat* is the application of heat to a base metal before performing a cutting, drilling, forming, or welding process.
Pre-inspection Meeting	Not a preferred phrase – see *prefabrication meeting*.
Premortem	*A premortem* is the analysis of a planned course of action that, for the purposes of discussion, is assumed to have failed. This is used to identify potential causes of failure or weaknesses in a plan, in a non-confrontational environment. See also, *postmortem*.

Phrase or Term	Definition
Present Forward Process	The *present forward process* is a view that tomorrow and the future will be much like today (i.e., extrapolating from the past to the present and into the future). It uses unchecked assumptions that the existing business model can simply be duplicated and extended into the future. In stable times, this approach helps organizations to optimize resource allocation. In turbulent times this approach can hinder organizations when critical market changes or disruptions are overlooked. For example, Blockbuster Video (Blockbuster) once dominated the video rental market in the United States. After opening in 1985 with one store, it franchised: • Over 400 stores nationwide after just three years; and, • Over 9,000 stores at its peak in the late 1990s. Blockbuster was so focused on a present forward process that they overlooked the lucrative, emerging, digital video-on-demand business upon the advent of the Internet. By 2001 Blockbuster still did not offer video streaming services to customers. Within a few years, Blockbuster's profits started declining because Netflix and other competitors reimagined video streaming (over the Internet) and undercut Blockbuster's brick-and-mortar video rental model using disruptive innovation. The last Blockbuster movie was rented on November 9, 2013. Blockbuster is now bankrupt. See also, *brick-and-mortar business, disruptive innovation, forecast planning, future back process, and scenario planning*.

Phrase or Term	Definition

Preservation	*Preservation* is the preventive maintenance and protection of equipment and materials after production, during shipment and storage, and before being used. The actions, materials, and methods for preservation may include:
	• Apply a:
	  o Continuous (e.g., fog or mist) or periodic lubrication;
	  o Corrosion inhibitor or rust-preventive; or,
	  o Desiccant or a nitrogen blanket.
	• Charge batteries (e.g., monthly) to prevent discharge (i.e., going flat);
	• Coat-exposed machined surfaces with corrosion inhibiting grease (e.g., a shaft);
	• Cover with plastic wrap (e.g., to be waterproof);
	• Drain condensation periodically (e.g., monthly);
	• Dry-out with compressed air;
	• Fill with oil containing rust-preventive;
	• Install a cap on or remove and plug grease fittings (i.e., on greased bearings);
	• Place product in a secure location indoors (i.e., heated or unheated), or outdoors;
	• Position valves in a closed, half-open, or open position (depending upon type);
	• Record preventive maintenance by item or tag number in a maintenance log;
	• Rotate the shaft of rotating equipment (e.g., bi-weekly);
	• Seal openings (e.g., install blind flanges and plug vents);
	• Store product off of the ground (e.g., off the earth, floor, or pavement on cribbing or pallets); and,
	• Supply heat to protect against condensation or freezing (e.g., electric heater).
	Visually examine for damage to preservation or product (e.g., monthly). The materials and methods used for preservation may be defined by the original equipment manufacturer, project specifications, other sources, or a combination. This may be specified as:
	• Domestic. This includes sources inside North America (e.g., shipped from a local supplier with one carrier such as by a truck;
	• International. This includes sources outside North America (e.g., shipped from an offshore supplier with more than one carrier such as by truck, then ship, and then truck;
	• Long-term. For example, the product will be shipped and stored longer than six months; or,
	• Short-term. For example, the product will be shipped and used within six months.
	Typically, domestic and short-term preservation requires less durable preventive maintenance and protection than international and long-term preservation. For example, waterproofing is usually required for the latter but not the former. Due to a long lead time or remote location, a project may purchase products well in advance of when they are needed, maintain a large inventory, or a combination. Spare parts, which may not be used for months or years, usually require preservation for long-term storage. See also, *dry-out*, *long lead*, *marking and labelling*, *mothballing*, *nitrogen blanket*, *packaging*, *rotating equipment*, *shipping and handling*, and *winterization*.

Phrase or Term	Definition
Pressure Equipment	*Pressure equipment* contains expansible fluid under pressure (e.g., fittings, boilers, pressure vessels, pressure piping systems, and thermal liquid heating systems). See also, *expansible fluid* and *pressure vessel*.
Pressure Equipment (Safety) Regulations	The *Pressure Equipment (Safety) Regulations* is the implementation of the *Pressure Equipment Directive* in the United Kingdom. See also, *Pressure Equipment Directive*
Pressure Equipment Directive	The *Pressure Equipment Directive* 2014/68/EU (formerly 97/23/EC) of the European Union sets out the standards for the design and fabrication of pressure equipment (i.e., generally over one litre in volume and having a maximum pressure more than 0.5 bar gauge). It also sets the administrative requirements for the conformity assessment of pressure equipment, without local legislative barriers. It has been mandatory throughout the European Union since 2002. A 2014 revision became effective in 2016. These standards and regulations are like the *American Society of Mechanical Engineers* codes for pressure equipment. See also, *Pressure Equipment Regulations*.
Pressure Equipment Integrity Management System	A *pressure equipment integrity management system* ensures that pressure equipment is designed, constructed, installed, operated, maintained, and decommissioned per the *Safety Codes Act*. See also, *fixed equipment mechanical integrity*, *integrated management system*, *integrity management program*, *mechanical integrity*, and *Safety Codes Act*.
Pressure Safety Device	A *pressure safety device* is a fail-safe fitting used to provide a means of relieving excess pressure or vacuum (i.e., by opening), which could implode or rupture containment or pressure equipment. It is the last line of defense for safety. The most common types are: • Pressure/vacuum relief valve, typically used to protect an atmospheric storage tank; • Relief valve, typically used for liquid service; • Rupture disk, a non-reclosing device used to protect equipment from an over pressurization or vacuum condition. Also known as a pressure safety disc, burst disc, bursting disc, or burst diaphragm; • Safety relief valve, includes the operating characteristics of both a safety valve and a relief valve, and may be used in either application; • Safety valve, typically used for steam or vapour service; and, • Temperature and pressure safety relief valve, typically used on potable water heaters. In addition to its pressure-relief function, it also has a temperature-sensing element which causes the device to open at a predetermined temperature regardless of pressure 99°C (210°F). See also, *fail safe*, *fitting*, and *potable water*.
Pre-startup Safety Review	A *pre-startup safety review* is conducted before commissioning (i.e., startup) of a modified or new facility to: • Assess the status of assets, procedures, and training; • Ensure that installations meet the original design or operating intent; • Ensure the facility is ready for *start-up* status; and, • Identify and re-assess all potential hazards introduced from changes made during the detailed engineering or construction project phases.

Phrase or Term	Definition
Pressure Test	A *pressure test* is a test of pressure equipment or systems to ensure it is capable of safely withstanding its rated pressure and is leak-tight. The testing media may be hydrostatic (e.g., water, glycol, or methanol), pneumatic (e.g., compressed air, halogen, helium, or nitrogen), or a gaseous-fluid combination.
Pressure Vessel	A *pressure vessel* is a container designed to hold gases or liquids at a pressure substantially different than ambient pressure (e.g., internal or external). The definition of a pressure vessel varies by code of construction, jurisdiction and country. See also, *atmospheric pressure*, *code of construction*, *external pressure*, *internal pressure*, *tank*, *thick wall pressure vessel*, and *thin wall pressure vessel*.
Pressure Welder	A *pressure welder* holds a pressure welder certificate of competency per *Alberta Pressure Welder Regulation*. This permits the welder to engage in pressure welding, subject to the limitations prescribed in the regulation or described on their performance qualification card. See also, *welder*.
Pre-Test	A *pre-test* is a duplicate test performed by the supplier, before a witness or hold point, to ensure successful test results. A pre-test is advantageous for a project and supplier because it ensures that all components and systems are functional, and all manufacturing or fabrication and quality requirements are completed before the specified test occurs. However, if a purchase order does not specify a pre-test, it may not be performed because of additional costs and time required for pre-testing.
Preventative Action	A *preventative action* is a solution meant to mitigate or eliminate a potential problem. For example, assessing nonconformance- or quality-related trends, or providing training to improve personnel competency and knowledge and therefore avoid costs associated with mismanagement of assets. See also, *corrective action* and *opportunity for improvement*.
Preventative Action Recommendation or Request	A *preventative action recommendation or preventative action recommendation request* is a tool (typically a document) that recommends or requests improvements to an organization's processes. The purpose of this tool is to eliminate or prevent causes of non-conformities or other related undesirable situations. See also, *Corrective Action Recommendation or Request*.
Price Discovery, Price Discovery Mechanism	*Price discovery* or *price discovery mechanism* is the process of determining a price in the marketplace through interactions of buyers and sellers. It is used to set the proper or spot rate of an asset, commodity, currency, or security. The process examines intangible and tangible factors such as economic and geopolitical environment, investor risk attitudes, and supply and demand. See also, *spot rate*.

Phrase or Term	Definition
Principles of Management	The *principles of management* are: • Authority to give orders and take responsibility. A manager's respect is earned by their intelligence, experience, rank, and values; • Centralization to balance upper management decision-making with labour force decisions (i.e., a place for everyone and everyone in their place); • Chain of command to provide personnel with guidance regarding their positions, the organization's hierarchy, and who to speak to (this was suggested as the now-familiar organizational chart that clearly depicts an organization's structure); • Collective or team interests to have priority over individual or personnel interests, including managers; • Discipline to have everyone follow the rules, with organization-labour force agreement to provide clarity; • Division of work with personal each assigned a task for which to become proficient, Thus, as personnel become more assured, efficient, and skilled, productivity will increase; • Equity to provide fair treatment of all personnel by management, with kindness and integrity, so teams perform their duties with dedication and loyalty; • Esprit de Corps to promote morale, team spirit, and unity; • Initiative to encourage personnel to develop and implement plans for improvement; • Order in the labour force to have the right person in the right place and the right materials in the right place (this pre-empted the just-in-time strategy); • Remuneration to obtain personnel satisfaction with fair financial and non-financial rewards; • Stability of tenure to maximize efficiency with minimal personnel turnover and role changes – personnel who are good at their job and secure are happier and more productive; • Unity of command to ensure that personnel receive instructions from only one supervisor, otherwise authority, discipline, order, and stability are threatened; and, • Unity of direction to ensure that teams with the same objective work under the direction of one manager, using the same plan, which is essential to the coordination of strength, focus of effort, and unity of action. The principles of management were developed in the early 19th century by the French engineer Henri Fayol. His focus on administrative skills rather than technical skills provided early acknowledgement of management as a profession. See also, *administrative theory*, *functions of management*, *just-in-time*, *management*, and *organizational chart*.

Phrase or Term	Definition
Primary Recovery	*Primary recovery* is a process by which usable petroleum is drawn out from beneath the earth's surface. During this stage, there is sufficient underground pressure to force the oil to the surface. Reservoir drive comes from several natural mechanisms that include: • Expansion of the associated gas initially dissolved in the crude oil; • Expansion of the associated petroleum gas at the top of the reservoir; • Gravity drainage resulting from the movement of oil within the reservoir from the upper to the lower parts where the wells are located; and, • Natural water displacing oil downward into the well. See also, *enhanced oil recovery* and *secondary recovery*.
Primavera	*Primavera* is an enterprise project portfolio management software that includes collaboration and control capabilities, product management, and project management. It was launched in 1983 by Primavera Systems Inc., which was acquired by Oracle Corporation in 2008.
Prime	Something *prime* (noun) is: • At an active or early stage (e.g., spring or youth); • Optimum (e.g., prime rib or in physically prime shape); • The best or chief individual or part (e.g., prime real estate); or, • The lowest available interest rate offered commercially by a bank (e.g., prime rate). To *prime* (verb) is to: • Apply the first coating or prepare to paint; • Coach, instruct beforehand, or stimulate; • Prepare a motor to start (e.g., prime the engine); and, • Prepare for work by charging or filling with something (e.g., prime a pump). See also, *priming*.
Prime Contractor	The *prime contractor* is a contractor with a purchase order from a project. See also, *contractor*.
Prime Mover	A *prime mover* is a source of motive power (e.g., an engine, motor, or turbine).
Prime Supplier	The *prime supplier* is a supplier with a purchase order from a project. See also, *supplier*.
Primer Coat, Prime Coat	A *primer coat* or *prime coat* is a material (e.g., paint) that is applied on an uncoated surface. This coating is specifically formulated to adhere to and protect the surface, and provide a suitable surface for subsequent coats. See also, *coating system* and *topcoat*.
Priming	Priming is the act of taking time to adjust emotions and thoughts, and to live life in a peak state. This act is most powerful when completed in the morning and sets a powerful and productive mindset for the day. It can cultivate positive emotions and drastically improve quality of life. This concept was explained by Tony Robbins in the article *How Should I Start Each Day? What is Priming?*. See also, *peak state.*

Phrase or Term	Definition
Prisoner's Dilemma	The *Prisoner's Dilemma* is a paradox in decision-making when two parties prioritize their own self-interest so an optimal outcome cannot be produced for everyone. Both parties choose to protect their interests at the expense of the other, so that the outcome is worse than if they had cooperated with each other. For example, the risk of nuclear war is increased by countries that own a nuclear arsenal. If all countries disarm, the risk is eliminated. If one country disarms, but the others do not, that country will place itself at a disadvantage. The Prisoner's Dilemma gets its name from a scenario such as the following. Two criminals are arrested and put into separate interrogation rooms but refuse to admit anything. A police officer then tells one prisoner: *Your partner in the other room is confessing, so they will get a prison sentence of only one year. Because you are remaining silent, your sentence will be five years in prison. If you also confess, you and your partner will only be sentenced to three years.* The police officer then repeats this to the other prisoner. What the police officer does not say to either prisoner, is that the evidence against them is weak, so if neither prisoner confesses, their sentence would be a maximum of two years. The Prisoner's Dilemma concept was originally described by Merrill Flood and Melvin Dresher, and later formalized and named by Albert W. Tucker. This concept is of interest in economics, politics, and sociology. See also, *free rider problem* and *game theory*.
Private Sector	The *private sector* is the part of the economy that is operated or owned by individuals and privately-owned for-profit businesses (*not* by governments). The antonym for *private sector* is *public sector*. See also, *for-profit*, *public sector* and *voluntary sector*.
Privy	*Privy* (adjective) is something: • Belonging or relating to personal capacity rather than a corporate or official capacity (e.g., only the mother is privy to that familial information); • Private or secret (e.g., a decision); or, • Shared as a secret (e.g., privy to the decision). A *privy* (noun) is: • A legal interest (e.g., in a contract, legal action, property, right, or transaction); • A toilet; or, • Knowledge of a private matter. Privy, when used as a compound noun phrase, can be: • A *privy council*, which is a body that advises a head of state (e.g., monarch, president, or prime minister); • A *privy mark*, which is a small mark in the design of a coin; and, • A *privy purse*, which is a British sovereign's private income.
Pro Bono, Pro Bono Publico	*Pro bono* or *pro bono publico* (Latin for *public good*) describes professional work donated or undertaken voluntarily without payment. This provides professional skills or services at no cost to organizations or people who cannot afford them. See also, *donation* and *volunteering*.

Phrase or Term	Definition
Proactive	A: • *Proactive* business strategy anticipates possible future challenges. For example, a project procures material from two or more suppliers instead of one as a contingency for one supplier failing to supply all the necessary materials; and, • *Proactive* person, policy, or action creates or controls a situation by causing something to happen instead of responding to it after it has happened. For example, personnel who hold toolbox talks. See also, *toolbox talk*, *reactive*, and *retroactive*.
Probabilistic Schedule Assessment	*Probabilistic schedule assessment* involves examining the elements on or near the critical path for a project and testing how varying the duration of critical elements changes the overall schedule duration. It asks the question: *how likely is it that everything will go according to plan*? See also, *critical path*.
Probability	*Probability* is a numerical description of how likely an event is to occur or how likely it is that a proposition is true. For example, the tossing of a coin has only two outcomes, heads or tails, so both are equally probable. The probability of either heads or tails is the same: ½, 0.5 or 50%. See also, *frequency*.
Probability and Impact Matrix	A *probability and impact matrix* is a visual framework for categorizing risks based on their probability of occurrence and impact. See also, *qualitative risk analysis*.
Probability of Failure	*Probability of failure* is the likelihood that equipment will fail in the future. It comprises half of the risk equation used to determine risk as part of the risk-based inspection methodology. See also, *consequence of failure*, *risk*, *risk equation*, and *risk-based inspection*.
Problem	A *problem* is: • A difficulty accepting or understanding; • A matter or question raised for consideration, inquiry, or solution; • A proposition in mathematics or physics stating something to be done; • A source of difficulty, distress, doubt, perplexity, uncertainty, or vexation; or, • An intricate unsettled question. A problem requires a solution to fix or solve it and enable a return to the original condition. For example, locking keys in a vehicle is a problem because it can be fixed by unlocking the vehicle. The word *problem*, which has negative connotations, is often replaced in business with *challenge* because challenge has positive connotations. See also, *challenge*, *crisis,* and *predicament*.
Problem Action Log	Not a preferred phrase – see *action item log* and *rolling item action log*.
Procedure	A *procedure* is an established sequence of actions or instructions to be followed when solving a problem or accomplishing a task. See also, *corporate instruction* and *project instruction*.
Procedure Qualification Record	A *procedure qualification record* is a formal written document that describes the testing and qualification of a welding procedure specification (e.g., details essential, supplementary, and non-essential variables).

Phrase or Term	Definition
Process Control System	A *process control system* is a combination of control and chemical engineering disciplines that uses industrial control systems to achieve a production level of consistency, economy and safety which could not be achieved by human manual control. See also, *integrated control and safety system*, *robotic industrial process automation*, and *safety instrumented system*.
Process Equipment	*Process equipment* houses and controls chemical reactions (e.g., for manufacturing plastics or refining oil), and controls and routes flow to and from storage (e.g., in a tank or pressure vessel). It must resist degradation because of its contact with the process fluid, which may be corrosive or cause erosion. Two major types of process equipment are: fixed equipment (e.g., a tank or pressure vessel) and rotating equipment (e.g., a compressor or pump). See also, *fixed equipment*, *pressure equipment*, *process fluid*, and *rotating equipment*.
Process Fluid	*Process fluid* is: • A substance (e.g., a gas or liquid) contained by process equipment (e.g., for manufacturing plastics or refining oil). For example, gas contained in a vessel; or, • Any fluid or solution that may be biologically, chemically, or otherwise contaminated or polluted in a form or concentration to constitute a health, pollution, or system hazard if introduced into a public or potable water system. See also, *potable water* and *process equipment*.
Process Industry	The *process industry* processes resources to produce feedstock or other products (e.g., chemicals or petrochemicals). See also, *feedstock* and *petrochemical*.
Process Intensification	*Process intensification* is a chemical engineering development that leads to more energy efficient technology, especially those that are substantially cleaner, safer, or smaller. The improvement is expected to be substantial (e.g., tens to hundreds of percent) and the benefits are expected to be substantial. For example, the use of a static mixer instead of mechanical agitation has the benefits of lower energy costs and an uncomplicated design with no moving parts. See also, *static mixer*.
Process Safety Management	*Process safety management* is a program designed to manage the integrity of processes and systems used to handle hazardous substances (i.e., highly hazardous chemicals) that are a safety risk to personnel (i.e., employees and the public). The purpose is to reduce the frequency and severity of incidents by preventing the unplanned release of explosive, flammable, reactive, or toxic gases or liquids during their handling, manufacturing, moving, storage, and use. See also, *highly hazardous chemicals* and *process industry*.

<table>
<tr><th>Phrase or Term</th><th>Definition</th></tr>
<tr><td>Procurement</td><td>Procurement is the act of obtaining or buying products and services. The process includes preparation and processing of a demand, and the end receipt and approval of payment. Procurement is often performed by the procurement discipline and involves:
• Disposals and other related functions;
• Financing and cost control;
• Inventory control and stores;
• Making the purchase;
• Price negotiation;
• Purchase planning;
• Specifications development;
• Standards determination;
• Supplier research and selection;
• Supply contract or purchase order administration; and,
• Value analysis.
The process of procurement is often part of a company's strategy because the ability to purchase certain materials will determine if operations will continue. A business will not be able to survive if it's price of procurement is more than the profit it makes selling actual product.</td></tr>
<tr><td>Procurement Documents</td><td>Procurement documents are all the documents attached to, and referenced in, a purchase order or requisition (e.g., the attachments are Sections 1 to 4 with data sheets, drawings, and specifications, and the references are codes of construction and industry standards). See also, purchase order, requisition, and three bids and a buy.</td></tr>
<tr><td>Procure-to-pay,
Purchase-to-pay</td><td>Procure-to-pay or purchase to pay is a single system that enables the integration of part of the procurement process, specifically purchasing with accounts payable. This may involve:
1. Supply management;
2. A cart or requisition;
3. A purchase order;
4. Receiving;
5. Invoice reconciliation; and,
6. Accounts payable.
Maintaining multiple systems within procurement and accounts payable functions creates inefficient and risky processes. Multiple systems make it difficult for management to see across the business and understand finances. Personnel are forced to re-enter data with the potential for errors. Suppliers may exist in more than one database, perhaps under slightly different names, which increases the risk of double payments.
This system is simple in theory, but the process is complex and touches many departments. It does not manage sourcing. See also, electronic commerce and source-to-pay.</td></tr>
</table>

<table>
<tr><th>Phrase or Term</th><th>Definition</th></tr>
<tr><td>Product</td><td>A product is:<ul><li>An article or substance that is fabricated, manufactured, or refined for sale; or,</li><li>A thing that is the result of an action or process.</li></ul>See also, cargo, fabrication, goods, and manufacturing.</td></tr>
<tr><td>Product Analysis</td><td>Product analysis is the chemical analysis of a sample taken from a finished product or material that is representative of the finished product. This may or may not be representative of (i.e., the same as) the heat analysis because further processing may occur. See also, heat analysis.</td></tr>
<tr><td>Product Cost Rollup</td><td>Product cost rollup is an accounting and management procedure that helps to determine the cost of a product. It is like a bill of material that is prepared as a nested table to include all the constituent parts, subassemblies, and the final assembly. The cost of these items can be summarized with labour, machine, and overhead costs to determine the actual cost for the entire assembly. See also, assembly and bill of material.</td></tr>
<tr><td>Product Realization</td><td>Product realization is the activities and processes necessary to bring a product into existence. Simply stated, it is realizing the product. It is a clear idea of what the product will be, usually expressed in drawings, specifications, statements, or other equivalents. Based on this clear idea, quality management can then objectively determine if the product was realized and how efficient the process was.</td></tr>
<tr><td>Production</td><td>Production is the action or process of making (e.g., assembling, constructing, fabricating, or manufacturing) from components or raw materials.</td></tr>
<tr><td>Professional</td><td>Professional means:<ul><li>A person engaged or qualified in a profession (e.g., engineering or technologist);</li><li>Connected with or relating to a profession (e.g., engineering or technology); or,</li><li>Engaged in a specified activity as one's main paid occupation (i.e., not as a pastime).</li></ul>See also, professional development unit. Professional engineer, and professional technologist.</td></tr>
<tr><td>Professional Development Unit</td><td>A professional development unit is a continuing education requirement, for personnel to maintain their professional certification or credentials. This may include:<ul><li>Attend technical conferences or meetings;</li><li>Coach or mentor peers;</li><li>Create content for learning resources;</li><li>Engage in scheduled informal discussions;</li><li>Give a presentation;</li><li>Participate in training or self-directed e-learning;</li><li>Study a technical article, book, or whitepaper;</li><li>Volunteer for a technical society; or,</li><li>Write a technical article or whitepaper.</li></ul>See also, boot camp, continuing education, cross-train, lifelong learning, lunch and learn, on-the-job training, professional, and workshop.</td></tr>
</table>

<table>
<tr><th>Phrase or Term</th><th>Definition</th></tr>
<tr><td>Professional Engineer, P.Eng</td><td>A professional engineer or P.Eng is someone licensed to practice engineering a jurisdiction (e.g., country, province, state, or territory) who:
• Adheres to a strict code of ethics;
• Continually upgrades their knowledge;
• Is ethically and legally responsible for their work and holds public safety as paramount; and,
• Maintains the highest levels of competence, as judged by their peers.
Prerequisites include successful:
1. Graduation from an accredited engineering program (e.g., with a Bachelor of Science degree);
2. Completion of work experience and an internship with on-the-job training under the supervision of a professional engineer;
3. Registration with a jurisdictional licensing body;
4. Completion of professional practice examinations; and,
5. Recertification, reregistration, or both (e.g., annually).
See also, degree, engineer, engineering, ethics, ethical, iron ring, and technologist.</td></tr>
<tr><td>Professional Jealousy</td><td>Professional jealousy is:
• The distress felt when someone else gets what one wants, or the resentment felt when someone else appears to be more successful than oneself. People who focus on another's success can lose focus on their own performance. For example, being jealous about someone else's accolades, attention, happiness, and promotions can distract from achieving professional goals; and,
• When personnel criticize or disrespect the work of former or other personnel because they are not present or no longer work at the company and are unable to defend themselves. For example claiming that:
o An old design is flawed or unimaginative while promoting ones' own design as brilliant and innovative, or,
o Product received from a sub-supplier, which meets code, quality, and other requirements, is substandard and would have been better made if produced by the supplier's personnel.
See also, hollow man argument, loserthink, mental prison, one-variable thinking, professional jealousy, and word-thinking.</td></tr>
</table>

<table>
<tr><th>Phrase or Term</th><th>Definition</th></tr>
<tr><td>Professional Technologist, P.Tech.</td><td>A professional Technologist or P.Tech. is someone licensed to practice engineering technology in a jurisdiction (e.g., country, province, state, or territory) who:
• Adheres to a strict code of ethics;
• Continually upgrades their knowledge;
• Is ethically and legally responsible for their work and holds public safety as paramount; and,
• Maintains the highest levels of competence, as judged by their peers.
Prerequisites include successful:
1. Graduation from an accredited engineering technology program (e.g., with an applied degree or diploma);
2. Completion of work experience and an internship with on-the-job training under the supervision of a professional engineer or geologist;
3. Registration with a jurisdictional licensing body;
4. Completion of professional practice examinations; and,
5. Recertification, reregistration, or both (e.g., annually).
See also, degree, diploma, engineer, engineering, ethics, ethical, professional engineer, and technologist.</td></tr>
<tr><td>Professional Work Product</td><td>A professional work product is a technical output that requires authentication and validation (e.g., professional documents, such as project drawings, plans, reports, and specifications). It is any output of professional services that contains technical information that is relied upon by others when making decisions or taking action. The product can be:
• Electronic:
    o For example, a model, simulation, software, or any other computer application that cannot be reproduced in a physical or electronic format; and,
    o For example, an electronic document or image.
• Physical. For example, a hardcopy photograph, physical document, plastic film, or slide.
See also, authentication.</td></tr>
<tr><td>Profit and Loss Statement</td><td>A profit and loss statement is a financial document that summarizes costs, expenses, and revenues incurred during a specified period, usually a fiscal quarter or year. These records provide information about a company or project's ability-inability to generate profit by increasing revenue, reducing costs, or both. This may also be referred to as an earnings, expenses, or income statements, or a statements of income, financial results, or operations financials. See also, quarter.</td></tr>
</table>

Phrase or Term	Definition
Program	A *program* is a: • Behaviour, process, or trait determined by, or as if by, the genetic material of an organism; • Curriculum, prospectus, or syllabus (e.g., an academic program); • Printed outline of the sequence to be followed, features to be presented, and persons participating in a performance or presentation (e.g., a talent show program); • Series of coded software instructions to control the operation of a computer or machine (e.g., a computer program); or, • Set of related activities or measures (e.g., a plan or system) which may be used to achieve a goal (e.g., a safety program that includes several Emergency Services components). See also, *goal* and *plan*.
Programmable Logic Controller	A *programmable logic controller* or programmable controller is an industrial digital computer, which has been ruggedized and adapted for the control of industrial processes, that requires high reliability control, ease of programming, and process fault diagnosis. See also, *controller*.
Progress Report	A *progress report* is an assessment that provides details about project health, including: • If the project is expected to be completed on time and budget; • What problems have been encountered; • What resources have been expended; and, • Which sub-goals have been accomplished. This report documents project or process progress, recurrently (e.g., monthly). Management uses these reports to determine if changes are necessary. The report may contain these sections: • Activities and progress; • Appendices; • Concerns and issues; • Dashboard or traffic lights; • Executive summary; • Financial results; • Quality status; and, • Risk management. See also, *status report*.
Progressive Inspection, Progressive Sign-off	*Progressive inspection* or *progressive sign-off* are activities that develop gradually or in stages. These proceed step-by-step before final inspection and during production. See also, *inspection* and *sign-off*.

Phrase or Term	Definition
Project	In the context of this document, a *project* is an oil and gas, oil sands, or pipeline project, or a project in another industrial sector (e.g., industrial, infrastructure, mining, or power). The owner-user client typically retains a consulting company for the engineering and procurement, and possibly other functions. The *project* may refer to all the stakeholders collectively or to specific functions that may act on behalf of the project (e.g., the buyer, engineer, manager, or owner). See also, *capital project*, *service provider*, *infrastructure*, *megaproject*, and *non-capital project*.
Project Alliance, Alliance Contracting	A *project alliance* or *alliance contracting* is a model for major capital project delivery, where a public sector agency (i.e., the owner) works collaboratively with private sector parties (i.e., non-owner participants such as engineering and procurement companies and contractors). All participants are required to act with integrity, make best-for-project decisions, and work together in good faith. Working as an integrated, collaborative team, the parties make unanimous decisions on all key project delivery issues. The project alliance structure capitalizes on the relationships between the participants, encourages effective integration with the owner, and removes organizational barriers. The most significant difference between traditional contracting methods and alliance contracting is that in alliancing, all project risk and reward outcomes are shared collectively by participants. Collaboration and collective risk sharing are intended to promote improved decision-making and innovation. If significant or substantial risk is allocated to an individual participant, then it may not be a project alliance and those characteristics (i.e., collective risk sharing) may not be appropriate or required. See also, *best-for-project*, *good faith*, *joint venture*, and *project alliance agreement*.
Project Alliance Agreement	A *project alliance agreement* is the contractual agreement between project alliance participants. See also, *project alliance*.
Project Baseline	A *project baseline* comprises the budget and schedule allocations set during the initial and planning phases of a project. Assuming the scope of the project remains unchanged, it may be used to determine variance from the planned budget or schedule. See also, *project*, *negative variance*, and *positive variance*.
Project Cancers	*Project cancers* are misunderstanding, omission, oversight, delay, error, and redundancy that yield cost overruns, frustration, rework, schedule delays, and waste, which can cause a project to fail. See also, *failure*.
Project Charter, Project Definition	A *project charter* or *project definition* is a document that defines the purpose and the scope of a project. It works as the legal authorization and plan for project business processes. It is an agreement between the project manager, sponsor, and stakeholders. It may include basic information such as the project's assumptions, business needs, constraints, major deliverables, objectives, organization, preliminary cost, resources, schedule, and scope. See also, *charter*.

Phrase or Term	Definition
Project Control, Project Controls	*Project control* or *project controls* is: 1. The work process used to create reports, develop plans, and measure performance for project cost, schedule, and resource outcomes. This is done by gathering and analyzing data, comparing actual with planned performance, and communicating with project teams to support effective and efficient decision making. Project control can include the development of corrective actions, and management of change. 2. A discipline that works closely with project management that manages and regulates project budgets. See also, *project management*.
Project Coordination	Not a preferred term – see *project management*.
Project Cost Management	*Project cost management* is an information system used to control, estimate, and measure costs through the project life cycle. It aims to complete a project within budget. See also, *life cycle* and *project management*.
Project Definition Rating Index	*Project definition rating index* is a methodology used by capital projects to measure the degree of scope definition, identify gaps, and take appropriate actions to reduce risk during front-end planning. It is an easy-to-use and powerful tool that identifies and precisely describes each critical element in a scope definition package. It also enables project teams to quickly identify risk factors related to desired outcomes for cost, operating performance, and schedule. See also, *project management*, *risk*, *risk assessment*, *risk management*, and *scope creep*.
Project Engineering	*Project engineering* is typically the management of engineering and technical projects – which may include management of budget, personnel, and scheduling. It often involves a hands on approach to directly manage front-line workers, liaise with project management and teams, provide on-site training, and react quickly to challenges or innovate solutions.
Project Execution Plan	A *project execution plan* is the governing document that describes how a project will be managed, organised, and executed. It is a project-specific plan that provides the main method of communicating the objectives and how they will be accomplished to all stakeholders. The project execution plan is an active (i.e., evergreen or live) document that should be updated with current and future project plans and procedures.

Phrase or Term	Definition
Project Financing	*Project financing* enables the project sponsor to recover the investment, and operating and maintenance expenses for a project. For example, a private company (i.e., sponsor) receives a concession from the private or public sector to finance, design, construct, own, and operate an asset stated in the concession contract. This enables the project sponsor to recover the investment, and operating and maintenance expenses for the project. Sponsor investment recovery usually occurs during the concession period. The rate of increase on return is often tied to a combination of internal and external variables, allowing the sponsor to reach a satisfactory internal rate of return for the investment. The forms of project finance are: • Build–lease–transfer; • Build–operate–transfer; • Build–own–operate; • Build–own–operate–transfer; • Design–build–finance–operate; • Design–build–operate–transfer; and, • Design–construct–manage–finance. The first project financed as a build–operate–transfer project was the China Hotel. It was built in 1984 and became China's first 5-star hotel.
Project Instruction	A *project instruction* is a documented, systematic method that allows for personnel to perform project job tasks correctly, safely, and consistently. This allows the job motions to be performed and enables an understanding of why it is important do something a specific way. A project instruction is tailored to a specific client and facility. See also, *charter*, *corporate instruction*, *knowledge transfer*, *plan*, and *project charter*.
Project Management	*Project management* is the work process of initiating, planning, executing, controlling, and closing the work of a team to achieve specific objectives. This is accomplished by applying knowledge, resources, skills, systems, and tools to meet the project requirements. Unlike a standard business process, project management is temporary and comprises final deliverables that must be produced following a pre-defined finite budget and timeline. See also, *digital project delivery*, *extreme project management*, *management*, *project control*, *project definition rating index*, *project engineering*, *project management office*, and *project manager*.
Project Management Body of Knowledge, PMBOK®	The *project management body of knowledge* or *PMBOK®* is a set of guidelines and standard terminology for project management that is generally recognized as a good practice. The guidelines recognize 49 processes categorized into 5 basic process groups and 10 knowledge areas that are typical of most projects, most of the time. Ultimately, it is a document created from work overseen by the Project Management Institute. See also, *body of knowledge* and *good practice*.

Phrase or Term	Definition
Project Management Diamond	The *project management diamond* is a visual metaphor with four corners that illustrates relationships between cost, quality, scope, and schedule. It expresses the idea that none of the four aspects can be amended without affecting the others. For example, a project can be completed faster by increasing budget or cutting scope. Similarly, increasing scope may require equivalent increases in budget and schedule. Cutting budget without adjusting schedule or scope will lead to lower quality. In practice, however, trading between constraints is not always possible. For example, adding additional money or people, or both, to a fully staffed project can slow it down. Moreover, on poorly run project, it is often not possible to improve budget, scope, or schedule without adversely affecting quality. See also, *management*, *project management triangle*, and *scope triangle*.
Project Management Office	A *project management office* is an external or internal group within a business that defines and maintains project management procedures and standards. It develops metrics, economies of repetition, and standardization to improve and maintain productivity. It supports project excellence by providing necessary decision-making information, documentation, guidance, strategies, and structure. The different types of project management offices are: • Centralized, which is business-wide; • Controlling, which requires compliance; • Decentralized, which focuses on a specific program or project; • Directive, which controls and manages projects; and, • Supportive, which consults. See also, *economies of repetition* and *project management*.
Project Management Triangle	The *project management triangle* is a visual metaphor with three corners that illustrates relationships between cost, scope, and schedule. It expresses the idea that none of the three aspects can be amended without affecting the others. Quality is a central theme. Projects must: • Be delivered on time; • Be delivered on budget; • Meet customer quality requirements; and, • Meet the agreed scope – no more, no less. The project management triangle is also referred to as the triple constraint, iron triangle, or project triangle. See also, *management*, *project management diamond*, and *scope triangle*.
Project Manager	A *project manager* is responsible to plan, procure, and execute a project that has a defined scope, start, and finish. They assess risk, ensure the project runs on budget and on time, and manage conflicts and issues.
Project Performance Indicators	*Project performance indicators* are measures used to assess project performance, usually with reference to project or performance baselines. These typically include cost, scope, and schedule, statuses. See also, *key performance indicators*.
Project Status Report	Not a preferred phrase – see *progress report* and *status report*.

Phrase or Term	Definition
Project Team	A *project team* is a group of engineering and procurement personnel, and subject matter experts. This may be an integrated team that includes client engineering and procurement, completions and commissioning, operations, and maintenance personnel. See also, *cadre*, *construction management team*, *remote team*, *requisition team*, *supplier team*, and *virtual team*.
Project-supplier Meeting	A *project-supplier meeting* is a meeting held between the project and supplier teams. These include: • Ad-hoc; • Bid clarification; • Drawing review; • Kick-off; • Pre-factory acceptance test; • Prefabrication; and, • Rolling action item log. See also, *project team* and *supplier team*.
Proof of Concept	A *proof of concept* is evidence, typically derived from an experiment or pilot project, that demonstrates an idea (e.g., business proposal or design) is feasible. See also, *pilot plant*.
Proof Test	A *proof test* is a pressure test used to determine the actual yield or failure (burst) pressure of equipment, which is then used to calculate the maximum allowable working pressure. See also, *maximum allowable working pressure*.
Propaganda	*Propaganda* is biased, misleading, or selective communication used to promote, publicize, or persuade others towards a point of view or cause. Propaganda is rooted in allegations, arguments, selective facts, one-sided arguments, half-truths, distorted ideas, lies, rhetoric, and rumors designed to further or harm a cause. For example, Alexander the Great intimidated opposing armies by leaving very large sized armour and helmets upon retreat to make his soldiers appear to be giants. Historically, governments use propaganda to influence, manipulate, and persuade national populations to take actions, and adopt attitudes, and conform to behaviours and beliefs en masse. See also, *brainwashing, bullshit, disinformation, fake news, false narrative*, *media bias*, *misinformation*, *rhetoric*, *smells fishy*, *social engineering*, *stretch the thought control, thought crime*, *truth*, *truth*, *truthiness*, and *untruth*.
Proprietary	*Proprietary* is something that is made, offered, or sold under the exclusive rights of property ownership as governed by copyright, patent, and trade secret laws by an agent, proprietor, or seller. Proprietary items usually have distinctive characteristics or features, and are often incompatible with competing items. See also, *intellectual property*, *proprietor*, *proprietary*, and *seller*.
Proprietor	A *proprietor* is a sole owner, or one of the owners of something. See also, *proprietary*.
Pros and Cons	*Pros and cons* are the arguments for and against something. Pros are good or beneficial points. Cons are bad or negative points. For example, one pro of purchasing an vehicle is increased mobility; however, one con is expense. Pro and con are Latin words that mean *for* and *against*, respectively.

Phrase or Term	Definition
Prospect	A *prospect* is a potential client or customer that is qualified based on their buying authority, financial capacity, and willingness to purchase a product or service (e.g., an executive is a good prospect to purchase a luxury home). *See also, referral* and *sales lead*.
Prospective Cost	A *prospective cost* is a future cost that may be avoided or a past cost that can be recovered if action is taken. Depending on the scenario, this can lead to unintended loss and negative or positive business results. For example, to: • Miss a progress payment milestone, thereby delaying receipt of the payment, which results in increased interest costs because a loan cannot be repaid; or, • Replace or retrofit lower-efficiency equipment with higher-efficiency equipment, to improve performance and reduce the cost of energy consumption. See also, *sunk cost*.
Protective Coating	A *protective coating* is applied to a substrate to protect the substrate from corrosion, erosion, or both. See also, *coating system*.
Prototype	A *prototype* is the first: • Example or model of something, from which later forms are developed (e.g., a scale model of a new airplane design or a software development model); or, • Full-scale and often functional form of a new design or type of something (e.g., the first flying drone). See also, *proof of concept*, *pilot plant*, and *scale*.
Pseudoscience	*Pseudoscience* is a system of assumptions, beliefs, methods, practices, or theories erroneously regarded as scientific. These are claimed to be both factual and scientific but are incompatible with the scientific method. The term is considered pejorative. See also, *junk science* and *scientific method*.
Psychological Flexibility	*Psychological flexibility* is how well someone can accommodate changing situational demands. It is: • How well one can balance different life domains, desires, and goals; and, • The ability to see things from different perspectives and then choose the most accurate or effective point of view. For example, the astronauts on Apollo 13 had such great psychological flexibility because they managed to avert a tragedy in space when faced with serious technical failure. This also means being able to adapt to the realities in life and avoiding habits and ruts. This concept was developed in the article *Two superpowers for 2021: psychological flexibility and sense of direction* by Dr Justyna Kucharska. See also, *goal*, *sense of direction*, and *rut*.
Psychomotor	*Psychomotor* relates to the origin of movement from conscious mental activity.
Public Domain	Something that is in the *public domain* is not under copyright protection and may be used freely (e.g., certain products, music, art, writing, etc.), without the permission of the former copyright owner and at no cost. Public domain is not a place; instead, it is a classification of legal use status. See also, *infringement*, *intellectual property*, and *copyright*.

Phrase or Term	Definition
Public Occupancy	*Public occupancy* is any building where members of the general public are likely to be present; the use or intended use of a building, or part thereof, for the shelter or support of animals, persons, or property.
Public Sector	The *public sector* is the part of the economy that is operated or owned by governments (not by individuals and privately-owned for-profit businesses). *Private sector* is the antonym for *public sector*. See also, *for-profit*, *private sector* and *voluntary sector*.
Public Speaking, Oration	*Public speaking* or *oration* is the: • Act or process of making or performing a formal or informal speech to a live audience; or, • Art of effective oral communication with an audience. Public speaking may be used to encourage, entertain, inform, motivate, or persuade. It plays a large role in the professional world and is believed that 70% of all jobs involve some form of public speaking. According to most studies, for most people, the fear of public speaking is greater than the fear of death. Comedian Jerry Seinfeld said that this means if the average person has to go to a funeral, they would rather be in the casket than give the eulogy. See also, *Lasswell's communication model* and *stress*.
Public-private Partnership	A *public-private partnership* is partnership between the public sector (e.g., a government) and the private sector (e.g., a community) to build public infrastructure (e.g., a hospital, road, or school) or deliver a service (e.g., operate a water treatment plant). See also, *private sector* and *public sector*.
Punch Item, Punch List Item	A *punch item* or *punch list item* is a deficient or inadequate item on a punch list. See also, *action item* and *punch list*.
Punch List	A *punch list* is a list of deficient items, inadequate items, or both, that may or may not be included in the supplier's scope of supply (e.g., changes, clarifications, and recommendations). Punch lists may be produced during inspections, factory acceptance tests, walk downs, commissioning activities and other tests. Suppliers and construction personnel commonly use punch lists and procedures. See also, *punch item*.
Pundit	A *pundit* is an expert in a field or subject (e.g., politics, technology, social sciences, or sports) who provides authoritative analysis, commentary, or opinion to the mainstream or mass media. The term is derived from a Hindi word *pandit* and the Sanskrit word *pandita*, meaning *learned one*. See also, *mainstream media*.
Pup	A *pup* is: • An 8.5-m (28-ft) trailer; • An off-cut from a length of line pipe; or, • The shortened form of puppy (a baby dog).

Phrase or Term	Definition
Purchase Order	A *purchase order* is a commercial document and the first official offer issued by a buyer to a supplier. It identifies equipment and material types and quantities, and agreed-upon deliveries, including prices for products, services, or both. A purchase order controls the purchasing of products and services from an external supplier to assemble, fabricate, or manufacture at the supplier's facility. By contrast, a construction contract specifies how and what a contractor will alter, construct, or repair at the owner's specified location. An engineering and procurement company or project purchase order typically comprises these sections: • Section 1, Scope of Supply (SOS) and Purchase Order Notes; • Section 2, Technical Documents List (TDL) and Technical Notes; • Section 3, Supplier Data Requirements (SDR); and, • Section 4, Inspection Requirements Document (IRD). See also, *award*, *contract*, *blanket purchase order*, *dummy purchase order*, *change order*, *holdback*, *order of precedence*, *procurement documents*, *supplier of choice agreement*, *terms and conditions*, and *unpriced purchase order*.
Purchase Order Award, Purchase Order Cut	*Purchase order award* or *purchase order cut* is the process of the buyer issuing a commercial document to a supplier for the purchase of products, services, or both. This process is often stated as *awarding* or *cutting* a purchase order.
Purchase Order Revision	Not a preferred phrase – see *master service agreement* and *change order*.
Purpose	*Purpose* is the reason why something created or done, or for which something exists. For a business, purpose explains what the organization is about, what it stands for, and why it exists. The purpose may include a: 1. Reason for being, which answers questions such as: • What is the point? • Why do we do this? • Why do we exist? • Why do we make this product? • Why do we provide this service? 2. Social ambition, which answers questions such as: • How we contribute to the greater good or provide positive benefit? • Is there a condition or injustice being addressed by our service? • Is there a wrong we are trying to make right? • What human need are we trying to meet? • Why is the world a better place because of us? 3. Profit motive, which answers questions such as: • How do we create value for our stakeholders and grow? • How do we maximize profits and shareholder value? See also, *mission*, *purpose*, *purpose statement values*, and *why*.
Purpose Statement	A *purpose statement* inspires personnel to do good work by expressing an organization's positive impact on the lives of whomever they serve (e.g., clients, customers, patients, or students). See also, *purpose*.

Phrase or Term	Definition
Pushing a Rope Uphill	*Pushing a rope uphill* is an idiom that denotes: • A futile exercise that cannot succeed because a rope is intended to be pulled. Pushing a rope uphill is not more plausible than pushing a rope at all, as nothing can be gained; or, • Influence that is more effective at moving things (e.g., objects or opinions) in one direction compared to another (i.e., you can pull, but not push). See also, *Catch-22* and *herding cats*.

Pushover	A *pushover* is:
	• Someone who is:
	  o Easily defeated or influenced (e.g., compared to the champion, everyone was a pushover);
	  o Incapable of effective resistance (e.g., citizens living in low income areas are pushovers in the eyes of municipal lawmakers); or,
	  o Unable to resist an appeal or attraction (e.g., when customers smell fresh-baked bread, they become a pushover).
	• Something accomplished easily and without difficulty (e.g., it was a pushover to convince Roy to try the new ice cream flavour).

Phrase or Term	Definition
Q	1 A B C D E F G H I J K L M N O P **Q** R S T U V W X Y Z
Quad Random	Not a preferred phrase – see *random length*.
Qualitative	*Qualitative* relates to measuring characteristics or the quality of something (e.g., the quality of steel could be *high grade*). Qualitative data cannot be expressed as a number This contrasts with *quantitative*, which measures something by a numeric value (e.g., assigning a number to a risk probability). See also, *quantitative*.
Qualitative Risk Analysis	*Qualitative risk analysis* is a project management technique that subjectively analyzes risk, probability, and impact. The risks are categorized on a probability and impact matrix, and those deemed significant may undergo a quantitative risk analysis. See also, *qualitative* and *quantitative risk analysis*.
Quality	In the context of this document, *quality* is the standard by which characteristics of a product or service are measured. For example, acceptable quality is achieved when a product or service satisfies the requirements stated in a purchase order. Whereas low quality or unacceptable quality is typically ascribed to a product or service that does not fulfill requirements in a purchase order, or is considered deficient, defective, or incomplete.
Quality Assurance	*Quality assurance* is the planned and systematic activities implemented within a quality system that provides confidence that a product or service will fulfill requirements for quality. Organizations use quality assurance to investigate an issue, determine root cause, and implement process improvements. Quality assurance focuses on defect prevention. See also, *quality control*.
Quality Assurance/Quality Control	*Quality assurance/quality control* has many interpretations because of the multiple definitions for the words *assurance* and *control*. This usually refers to actions performed that ensure the quality of a product, service, or process. See also, *quality assurance* and *quality control*.
Quality Control	*Quality control* is the operational techniques and activities used to fulfill requirements for quality. Quality control is used to test all products or representative samples to ensure they meet specified requirements. Quality control focuses on defect detection. See also, *quality assurance*.
Quality Control Check List	A *quality control check list* is a document that helps ensure all important steps or actions have been taken. This document outlines quality standards and product requirements that a supplier is shall meet and provides objective criteria for inspection and acceptance. It may be less detailed than an *inspection and test plan*. See also, *quality control traveler*.
Quality Control Traveler	A *quality control traveler* is a check list. This term is associated with fabrication and manufacturing industries, and exists per ASME codes of construction. This term refers to how a checklist *travels* with an item (e.g., equipment) as requirements for its production are competed and verified. See also, *quality control check list*.
Quality Management	*Quality management* ensures that an organization, product or service is consistent using four main components: quality planning, quality assurance, quality control, and quality improvement. See also, *management*, *risk management*, and *supplier quality surveillance*.

Phrase or Term	Definition
Quality Management Plan	A *quality management plan* is a document for application to the *Safety Codes Council* for accreditation to administer safety codes under the *Safety Codes Act* (e.g., for the electrical discipline per Alberta Municipal Affairs). See also, *management* and *quality management system*.
Quality Management System	A *quality management system* directs and controls an organization's quality requirements. A quality system comprises policies, processes, practices, and tools used by management to direct the resources of an organization. See also, *integrated management system*, *management*, *quality*, *quality management plan*, *Six Sigma*, and *total quality management*.
Quality Manager	A *quality manager* (also a quality control or quality assurance manager) is responsible for ensuring that the quality management system is efficient and fully implemented. They also ensure that the product or service provided is fit for purpose, consistent, and meets external and internal requirements. This includes customer expectations and legal compliance. A quality manager is responsible for delegating or performing quality-related tasks and working closely with other departments (e.g., engineering, production, and repair). See also, *quality management system*.
Quality Plan	A *quality plan* is a document that specifies the processes, procedures, and associated resources that will be applied by whom and when, to meet the requirements of a purchase order or scope of work (e.g., ISO 10005, *Quality Management Systems — Guidelines for Quality Plans*). See also, *inspection and test plan*.
Quality Surveillance	*Quality surveillance* is the short form of the phrase *supplier quality surveillance*. See also, *supplier quality surveillance*.
Quality Surveillance Coordinator	Not a preferred phrase – see *supplier quality surveillance coordinator*.
Quality Surveillance Level	The *quality surveillance level* provides the supplier with information for planned inspection activities including scope and frequency of inspections.
Quality Surveillance Plan	Not a preferred phrase – see *assignment instruction*.
Quality Surveillance Point	Not a preferred phrase – see *verification point*.
Quality Surveillance Representative	A *quality surveillance representative* is responsible for performing quality surveillance or inspection activities. This representative's services are typically contracted through an agency; however, these services may also be performed by others (e.g., a subject matter expert, or an employee from an engineering and procurement company or client).
Quality Trilogy	The *quality trilogy* is an approach to managing quality using these three aspects: 1. *Quality planning* to develop the products and processes required to meet customer needs; 2. *Quality control* to meet product and process goals; and, 3. *Quality improvement* to achieve unprecedented levels of performance.
Quality Verification Point	A *quality verification point* is identified in the inspection requirements document or an inspection and test plan as: witness, hold, verification, or other activities (e.g., audit, document review, in-process, observed, or non-witnessed). For the supplier, these are quality control activities. For the engineering and procurement company, these are inspection activities. See also, *inspection and test plan*.

<table>
<tr><th>Phrase or Term</th><th>Definition</th></tr>
<tr><td>Quantitative</td><td>Quantitative relates to measuring the quantity of something (e.g., number of anchor bolts). Quantitative data must be expressed as a number. This contrasts with qualitative, which may not be measured by a number (e.g., the quality of anchor bolts). Only quantitative data can be analyzed statistically. See also, qualitative.</td></tr>
<tr><td>Quantitative Risk Analysis</td><td>Quantitative risk analysis is the mathematical analysis of risk probability and impact. In project management, it is not a substitute for qualitative risk analysis. Instead, quantitative analysis is conducted after qualitative analysis and assesses risks that qualitative analysis has identified as significant. See also, qualitative and qualitative risk analysis.</td></tr>
<tr><td>Quarantine</td><td>To quarantine is to isolate:<ul><li>Animals, people, or plants that have arrived from elsewhere or (may) have been exposed to a contagious or infectious disease. This includes before a sickness is diagnosed or if no symptoms are present and after a sickness is diagnosed or symptoms are present; or,</li><li>Equipment or materials that are non-conforming and need to be dispositioned for quality control purposes. This may be done by placing the item(s) in a secured area, sealed area, through tagging, or by other means, or for identification as not acceptable for use.</li></ul>See also, disposition, non-conformance, and quality control.</td></tr>
<tr><td>Quarter</td><td>A quarter (Q) is:<ul><li>A three-month period on a company's financial calendar that acts as a basis for periodic financial reports and payment of dividends. It refers to one-fourth of a year and is often accompanied by a year, (e.g., Q1 2019 or Q1/19). The standard calendar quarters are:<ul><li>Q1 (January, February, and March);</li><li>Q2 (April, May, and June);</li><li>Q3 (July, August, and September); and,</li><li>Q4 (October, November, and December).</li></ul></li><li>Each of four corresponding or equal parts into which something is or can be divided (e.g., a pie chart with four pieces).</li></ul></td></tr>
<tr><td>Quereinsteiger,
Quereinsteigerin,
Lateral Entrant</td><td>Quereinsteiger (masculine) and Quereinsteigerin (feminine) are German for personnel who change their careers. Loosely translated, it means lateral entrant – but there is no equivalent word in English. It is a catch-all description for someone who works in an industry or job with no previous experience.
The first part of the word – quer – can refer to being a contrarian or unconventional, so the term has a stigma. This concept is described by Barbara Woolsey in the article The German 'Quereinsteiger' ending career-changing stigma. See also, catch-all and contrarian.</td></tr>
<tr><td>Quick Fix</td><td>A quick fix is a fast problem-solving technique that is usually temporary (e.g., using a portable air pump to continually fill a leaking vehicle tire instead of repairing or replacing the tire). See also, band-aid.</td></tr>
</table>

Phrase or Term	Definition
Quick Opening Closure, Rapid Opening Closure	A *quick opening closure* or *rapid opening closure* ensures fast (e.g., 30-second), reliable, and safe access to equipment or a pipeline system (e.g., to a pressure vessel or pig launcher and receiver). Integral safety features may also prevent unsafe operation. Common brands for these closures include GD Bandlock™, PECO®, Sentry™, TOOL-LESS®, TWINLOCK, and Yale™.
Quick Response Code, QR Code	A *quick response code* or *QR Code* is barcode that contains information about the item to which it is attached. The barcode comprises black squares arranged in a square grid on a white background. Applications include document management, item identification, marketing, product tracking, and time tracking. This barcode was originally designed in 1994 for use in the Japanese the automotive industry. It quickly became popular outside of the industry because it has faster readability and greater storage capacity than the conventional universal product code. See also, *barcode* and *Universal Product Code*.
Quick Win	A *quick win* is an improvement that can be delivered quickly, has immediate benefit, and is visible. A quick win does not need to be profound or have a long-term impact on an organization; however, it needs to be something that stakeholders agree is a good thing. For example, selling products in groups of four or six, instead of individually, to increase overall revenue. See also, *low-hanging fruit*.

Phrase or Term	Definition
R	1 A B C D E F G H I J K L M N O P Q **R** S T U V W X Y Z
Racism	*Racism* is antagonism, discrimination, prejudice, or stereotyping based on ethnic group or race (e.g., the belief that one ethnicity or race is intrinsically inferior or superior to another). See also, *ageism*, *Black Indigenous and People of Colour*, *discrimination*, *intersectionality*, *racist*, and *sexism*.
Racist	*Racist* is characterized by or showing antagonism, discrimination, prejudice, or stereotyping based on ethnicity or race. See also, *racism*.
Radical	*A radical*: • Advocates extreme measures to restore or retain a political system; • Is an organization (e.g., political party or part of a political party) with political policies, practices, and views that support extreme change; • Is concerned with, or tends to concentrate on, the fundamental aspects of something (e.g., radical re-examination); • Favours extreme or fundamental changes in existing conditions or institutions, or has extreme viewpoints (e.g., economic, political, or social); or, • Is something or a way of doing something that is very different than what is conventional, traditional, or usual. See also, *left-wing*, *politics*, *radicalization*, and *reactionary*.
Radicalization	*Radicalization* is the act or process of causing an individual or group to adopt more extreme positions or views in opposition to the status quo (e.g., political, religious, or social). This can originate from a broad consensus either against or for progressive change in society, and can result in both nonviolent and violent actions. In a democratic system of government, political polarization embodies the tension of binary political ideologies and partisan identities. See also, *polarize*, *politicization*, *politics*, *progressive*, *radical*, and *weaponization*.
Radiograph	A *radiograph* is digital or photographic image of the internal composition of something (e.g., a casting or weldment) produced by gamma rays or X-rays passed through it and being absorbed to different degrees by different materials. See also, *gamma ray*, *weldment*, and *X-ray*.
Radiographic Test, Industrial Radiography	A *radiographic test* or *industrial radiography* is a non-destructive test method used to examine the internal structure and integrity of a test specimen using penetrating radiation. This method is based upon either gamma rays from a radioactive source or X-rays from an electromagnetic source. The denser or thicker a material, the more radiation will be absorbed. The test specimen is placed between a source of radiation and a digital plate (i.e., a digital image capture device) or sheet of radiographic film. A discontinuity in the specimen will absorb less radiation than the specimen itself, so its presence and location will be indicated on the digital plate or film by an area of higher or darker exposure. See also, *discontinuity*, gamma ray, *non-destructive examination*, *radiograph*, and *X-ray*.
Radius	The *radius* of a circle or sphere is any of the line segments from its centre to its circumference or perimeter, or their length.

Phrase or Term	Definition
Random Inspection	A *random inspection* is an arbitrary or periodic verification of compliance (e.g., first off or in-process).
Random Length	*Random length* is pipe that is produced and delivered in various lengths. This means that the pipe is not cut to a specific length and varies within a specified range. In the API SPEC 5L 44th Edition, six plain-end random lengths are listed in metres (with feet in parentheses). It lists lengths of 6 (20), 9 (30), 12 (40), 15 (50), 18 (60), and 24 (80). Other random lengths may be specified by agreement. The terms double, single, quad, and triple random lengths are no longer mentioned in the standard; however, industry continues to use this terminology. These phrases were defined as follows: • The length of double random length pipe is usually 8 to 14 m (30 to 40 ft). A pipe length specified as *longer than double random pipe* exceeds 14 m (40 ft); • The length of quad random pipe is usually 24 m (80 ft); • The length of single random or mill length pipe is usually 4 to 7 m (13 to 23 ft); and, • The length of triple random pipe is usually 18 m (60 ft). See also, *tolerance* and *plain end*.
Random Walk Theory	The *random walk theory* suggests that movements in any currency are random and unrelated to their past movements or trends. Currency risk can be complicated and difficult to evaluate and understand but the basic concept can be simple when explained in plain English. See also, *Big Mac Index*, *plain English* and *theories of exchange rates*.
Rapoport's Rules, Rappaport's Rules, Dennett's Rules	*Rapoport's Rules*, *Rappaport's Rules*, or *Dennett's Rules* are named after game theorist Anatol Rapoport and philosopher Daniel Dennett. This set of rules is intended to encourage critical and productive discourse. The rules seek to avoid straw man arguments like the steelman argument. This will avoid the backfire effect that criticism often provokes. The rules to compose a successful critical commentary are to: 1. Attempt to re-express your opponent's position so clearly, fairly, and vividly, that they may say, thanks, I wish I had thought of stating it that way; 2. List any points of agreement, especially if these are not matters of general or widespread agreement; 3. Mention anything you have learned from your opponent; and, 4. Offer any criticism or rebuttal and not before steps 1 to 3 have been taken. See also, *backfire effect*, *steelman argument*, *straw man argument*, and *weak man argument*.

Phrase or Term	Definition
Rare Earth Element, Rare-Earth Mineral	A *rare earth element* or *rare-earth mineral* is one in a group of elements referred to as the lanthanide series in the periodic table of elements. These elements are used in a variety of applications for the aerospace, automotive, clean energy, electronics, and defence industries. Magnets represent the single largest and most important end use for rare earth elements, accounting for 21% of total consumption. These consist of: • Heavy rare earth elements Dysprosium, erbium, holmium, lutetium, terbium, thulium, ytterbium, and yttrium. These are produced mainly in China and are in limited supply; and, • Light rare earth elements Cerium, europium, gadolinium, neodymium, praseodymium, promethium, samarium, and scandium. These are produced in global abundance and are in surplus supply. Scandium and yttrium, while not true rare earth elements, are also included in this categorization because they exhibit similar properties to the lanthanides. See also, *critical energy metals*, *mineral*, *Organization of Mineral Exporting Countries*, and *periodic table of elements*.
Rat Hole	Not a preferred phrase – see *weld access hole*.
Rate Sheet	A *rate sheet* is a document containing details about charges and rates of services or products being sold. For example, straight and overtime labour, vehicle allowance, other applicable expenses, currency conversion, and if expenses are flow-through as 100% cost or marked-up. See also, *flow-through*.
Reactionary	A *reactionary* is a person, or a set of views, that is opposed to political or social liberalization or reform (e.g., ultraconservative) or points of view and policies meant to restore a past status quo. For example, Britain, where the idea of free trade was born, is withdrawing from the European Union, the world's largest free market, because of fears that its national identity and sovereignty are under threat. This could be considered reactionary by some. The usage arose because of the French Revolution (1789 to 1799). See also, *conservative*, *political*, *radical*, *right-wing*, and *status quo*.
Reactive	*Reactive* can be described as: • A behaviour that is not internally motivated but manifests in response to the actions of others or a situation; • A business strategy that responds to unanticipated events only after they occur; or, • An unstable chemical that reacts with its surroundings (i.e., air, water, light, heat, pressure) or certain other chemicals. Reactive chemicals start uncontrolled or unwanted reactions with the formation of corrosive, noxious, or toxic by-products or spoilage, an increase in temperature, or a release of energy. See also, *proactive* and *retroactive*.
Ready for Operations Team	A *ready for operations team* comprises a portion of the Commissioning or Operations group at site. See also, *construction management team*.

Phrase or Term	Definition
Relative Deprivation	*Relative deprivation* is the way a group or individual experiences dissatisfaction from not having the same resources as others around them (e.g., activities, amenities, diet, or lifestyle). This sense of discontent may be emphasized when someone is deprived of something that is believed to be an entitlement and describes feelings or measures of economic, political, or social deprivation that are relative rather than absolute. This occurs where such comparisons are unfavourable. For example, when someone does not have the resources to take annual vacations, but members their peer group take annual vacations regularly. See also, *peer*, *peer group*, and *relative gratification*.
Relative Gratification	*Relative gratification* is the way a group or individual experiences satisfaction from having more resources than others around them (e.g., activities, amenities, diet, or lifestyle). This may be emphasized when someone has something that is believed to be an entitlement and describes feelings or measures of economic, political, or social deprivation that are relative rather than absolute. This occurs where such comparisons are favourable. For example, when someone has the resources to take annual vacations but members of their peer group lack the resources to do the same. See also, *peer*, *peer group*, and *relative deprivation*.
Rebound Effect, Take-back Effect	The *rebound effect* or *take-back effect* is the reduction in expected gains from new technologies that increase efficiency of resource use because of behavioural or other systemic responses. These responses usually tend to offset the beneficial effects of the new technology or other measures taken. For example, driving a more fuel-efficient vehicle results in fuel costs, which can influence driving behaviour: the vehicle may be used more often, be driven longer distances, and a bicycle, mass transit, or walking may be used less. Daniel Khazzoom developed the idea that there is a less than a one-to-one correlation between gains in energy efficiency and reductions in energy use because of a change in the *price content* of energy in the provision of the final consumer product. This concept was originally observed by William Stanley Jevons and suggests that, to avoid the rebound effect – cost savings from any efficiency gains should be taxed to keep the cost of use the same. See also, *Jevons effect* and *Khazzoom-Brookes postulate*.
Receivership	*Receivership* is a process in which a legally appointed receiver acts as custodian of a company's assets or business operations, as with a bankruptcy. A bankruptcy court, creditor, or governing body may appoint a receiver. In most cases, a receiver is given ultimate decision-making powers and has full discretion in deciding how to manage received assets. See also, *bankruptcy*.

Phrase or Term	Definition
Receiving Inspection	*Receiving inspection* is an activity to verify the acceptable condition, description, quality, and quantity of equipment, and material received from suppliers. These activities may include the following: • Categorize items as complete and suitable for use or if items are damaged or there is a shortage of items; • Count and measure items including number of packages and units, and items shipped loose; • Open or partially open packaging for examination or verification and photography; • Prepare non-conformance, material receiving, shortages and damages, or other reports; • Review shipping and purchase order documents; and, • Verify material markings, labels, and tagging. See also, *non-conformance report*, *shipping documents*, and *purchase order*.
Recency Illusion	*Recency illusion* is the belief or impression that a language usage or a word is of recent origin when, in fact, it was established long ago. This is not restricted to linguistic phenomena; it is also the belief that things *you* have noticed recently have in fact recently occurred. The illusion is caused by selective attention. See also, *bias, Baader-Meinhof phenomenon, selection bias, and selective attention*.
Recession	A *recession* is a period of temporary economic decline during which employment, industrial production, trade, and wholesale-retail sales are reduced, generally identified by a fall in gross domestic product in two successive quarters. See also, *depression*, *Great Depression*, *gross domestic product*, *quarter*, and *recession*.

Reciprocity	*Reciprocity* is a:
	• Mutual exchange of something (e.g., privileges) between individuals or groups when each allows or gives something to the other (e.g., a trade agreement);
	• Quality or state of being reciprocal (e.g., mutual action, dependence, or influence);
	• Social norm of someone responding to an action with another similar action that is:
	  o Negative
	  When one's action has a negative effect on someone else a similar action that has an approximate equally negative effect is returned. For example, an insult that is returned.
	  o Positive
	  When one's action has a positive effect on someone else a similar action with an approximate equally positive effect is returned. For example, a compliment that is returned.
	See also, *Givers Gain®*, *Golden Rule*, *law of reciprocity*, *poetic justice*, and *relational customer service*.

Recognized and Generally Accepted Good Engineering Practices, RAGAGEP	*Recognized and generally accepted good engineering practices* or *RAGAGEP* are the basis for reliable and safe engineering, inspection, maintenance, and operational activities. These are based on industry codes, recommended practices, standards, technical articles or reports, or similar documents.

<table>
<tr><th>Phrase or Term</th><th>Definition</th></tr>
<tr><td>Recommended Practice</td><td>A recommended practice is a documented industry-wide agreement for methods, processes, or techniques that are accepted as superior to other alternatives. It produces results that are better than those achieved by other means, and mitigates challenges because of its standardized, logical approach. A recommended practice may be optional for use or required by the code of construction or procurement documents. See also, code of construction, procurement documents, and specification.</td></tr>
<tr><td>Recondition, Refurbish, Remanufacture, Retrofit, Revamp</td><td>Recondition, refurbish, remanufacture, retrofit and revamp are similar terms that mean to:<ul><li>Add an accessory or component to something that it did not have when it was made;</li><li>Repair or improve a broken, damaged, or outdated structure;</li><li>Manufacture something into a new product, especially using existing items or materials;</li><li>Overhaul or repair equipment;</li><li>Rebuild, repair, and restore equipment to meet or exceed the original equipment manufacturer's specifications; or,</li><li>Renovate or retrofit something, especially a building.</li></ul></td></tr>
<tr><td>Record</td><td>A record is documented information created, received, and maintained as evidence and used as an asset by a project to fulfill business and legal obligations. This includes hard copy and electronic documents produced by the owner, engineering and procurement company, and supplier (e.g., calculations, certificates, catalogs, drawings, reports, and specifications). Records may be issued as approved, reviewed, or final (as-built copies). See also, as-built drawing.</td></tr>
<tr><td>Red Flag</td><td>A red flag is a:<ul><li>Metaphor for the sign of a problem requiring attention or consideration. For example, the interviewee's answers or the interviewer's questions; or,</li><li>Physical flag that signals danger or warnings. For example, a flag identifying a high wildfire danger or a large pothole in a street.</li></ul>See also, false flag, red flag mechanism, and white flag.</td></tr>
<tr><td>Red Flag Mechanism</td><td>A red flag mechanism is a metaphor for a practical and powerful tool to turn information into information that cannot be ignored. It creates a climate where the hard truth is heard, and groupthink can be avoided. For example, each participant (e.g., in a classroom or meeting) can stop the discussion once, without penalty or restriction. They can then:<ul><li>Offer a disagreement;</li><li>Present an analysis;</li><li>Respond to another participant;</li><li>Share a personal experience;</li><li>Suggest something;</li><li>Voice an observation; or,</li><li>Whatever.</li></ul>This mechanism offers feedback in real time that cannot be ignored. See also, groupthink and red flag.</td></tr>
</table>

<table>
<tr><th>Phrase or Term</th><th>Definition</th></tr>
<tr><td>Red Herring</td><td>A red herring is a logical fallacy that intentionally misleads or distracts from an important or relevant theme and leads an audience to a false conclusion.
Consider this example:
There is a lot of commotion regarding saving the environment. We cannot make this world an Eden. What will happen if it does become Eden? Adam and Eve got bored there!
The idea of Adam and Eve getting bored in Eden is a red herring. It distracts readers from the real theme, which is about environmental damage.
A red herring may be used:
• As part of a rhetorical strategy (e.g., in politics);
• Inadvertently (e.g., evidence that later proves to be irrelevant); or,
• Intentionally (e.g., as in fiction).
In 1807, English polemicist William Cobbett popularized this phrase by telling a story about using a kipper (a strong-smelling smoked fish) to distract dogs from chasing a rabbit. See also, bait and switch, circular reasoning, either/or, false equivalence, logical, logical fallacy, paper tiger, rhetoric, righteousness fallacy, slippery slope, and straw man argument.</td></tr>
<tr><td>Red Ocean Strategy</td><td>Red ocean strategy is the conventional competition-based approach to business. That is, to beat the competition by offering better choices for consumers in terms of product or service differentiation and lower cost.
Red ocean is an analogy that describes the limited potential of the known market space and all the industries that exist. In red oceans, industry boundaries are defined and accepted, and the competitive rules of the game are known. Companies must work to outperform competition and take greater shares of product or service market demand. As the market space crowds, prospects for growth and profits reduce. And thus this cutthroat competition bloodies the ocean. Hence, red ocean.
See also, blue ocean strategy, disruptive innovation, and value innovation.</td></tr>
<tr><td>Red Tape</td><td>Red tape is an idiom that refers to excessive regulation or rigid conformity to formal rules that are considered bureaucratic, redundant, repetitive, and hinder or prevent action or decisions. It is usually used in reference to the machinations of corporations, governments, and large organizations. Red tape includes filling out forms, having multiple people or committees approve a decision, obtaining licenses or permits, and generally, low-level rules that make business more difficult, slower, or both. Cutting red tape refers to a reduction of bureaucratic obstacles.
Etymologically, scholars believe the phrase originated in the early 16th century, when red tape was used to bind the most important administrative documents. This provided a method of separating important documents from other ordinary administrative documents (which were bound with string).
See also, bureaucracy.</td></tr>
<tr><td>Red-line Drawing</td><td>A red-line drawing is an intermediate or final drawing that shows revisions made to the previous revision of the drawing. The term red-line refers to an (older) method of revising a hard copy drawing by hand using a pen with red ink. See also, as-built drawing.</td></tr>
</table>

Phrase or Term	Definition
Reengineering	*Reengineering* involves the extensive redesign or rethinking of core processes to achieve major performance improvements. It focuses on optimizing key performance areas such as efficiency and quality. It often involves restructuring organizations so that multi-functional teams can manage processes from start to end.
Reference-class Forecasting	*Reference-class forecasting* is a process that addresses confirmation bias by forcing decision-makers to consider cases that do not necessarily justify the preferred course of action. It is useful to compare a project under consideration to other similar projects that have already been completed. For example, if a city wants to build a 10-km public transit line with four stations, it should look at other cities that have completed similar lines to understand the actual costs and timeline dynamics. See also, *bias*, *confirmation bias*, and *optimism bias*.
Referral	A *referral* is an opportunity to present a value proposition and sell a product or service to a prospect that needs or wants it. This is because a referral source (e.g., an associate, colleague, or friend) has told the provider about the prospect's business and possible interest in learning more. The prospect is therefore expecting to be contacted, and already knows about the provider and what is being offered. There are five referrals levels: • Level 1 to 3 referrals are made by an advocate or promoter; and, • Level 4 and 5 referrals are made by a facilitator. These five referral levels each include the elements of the lower levels where the referral source offers: 1. The name and contact information of the referral to the provider; 2. Supplementary material (e.g., marketing literature or a website) to the prospect; 3. A personal experience to the prospect (e.g., a verbal or written testimonial); 4. An introductory call or meeting arrangement, or both for the prospect and provider; and, 5. An in-person meeting to introduce the prospect and provider. A referral can also be defined as one of these three types: 1. An *advocate* or *reactor* is a referral source that hears a prospect tell them about a need or want and asks if they know of a product or service provider that can offer a solution. For example, a prospect says *we need to find a caterer for our banquet—can you recommend a professional?* 2. A *promoter* is a referral source that hears of a prospect's need or want (without being asked) and offers the details and prospect's contact information to the provider (or vice versa) as a solution. For example, a referral source says *you will need a reliable caterer for your banquet—can I recommend a professional?* 3. A *creator* is a referral source that listens to a prospect's need or want and offers a solution that they have not asked for or described as a need or want. They discover the opportunity and offer a solution, For example, *your banquet will require a lot of planning in addition to the catering. Can I also recommend the services of a professional event planner?* See also, *cold call*, *prospect*, *sales lead*, *value proposition*, and *warm call*.

Phrase or Term	Definition
Reflective Questions	*Reflective questions* are used to debrief students about their learning process and to encourage improvement and self-assessment. This provides them with an opportunity to explore their experiences, ideas, and knowledge. Reflective questions elicit personal and thought-provoking responses, and do not have right or wrong answers. Reflective questions also provide important feedback to peers and teachers. Examples of reflective questions include: • What did you enjoy the most or the least about this project? • What new knowledge or skills did you acquire during this project? • What would you do differently if you did this project again? See also, *reflective questioning*.
Reflective Questioning	*Reflective questioning* is used to put another person's statements in one's own words and then verify with them to see if what they mean has been understood. Repeating or reflecting their message as a question gives them an opportunity to clarify and explain their message. This is also a powerful way to let another person know that they are cared for by listening carefully. See also, *active listening*, *body language*, *calibrated question™*, *objective reflective interpretive decisional questioning*, *paraverbal communication*, *precision questioning*, and *reflective question*.
Refractory	*Refractory* is a mineral or other substance that retains its strength and form at high temperatures, and is resistant to decomposition by heat, pressure, or chemical attack. Refractory is a non-metallic material used in environments above 538°C (1,000°F). Refractory may be cast, formed, moulded, or pre-cast). See also, *mineral*, and *non-metallic*.
Regression to the Mean, Regression Toward the Mean	In statistics, *regression to the mean* or *regression toward the mean* is the phenomenon that arises if a random variable is extreme on its: 1. First measurement but was closer to the mean or average on its second measurement; or, 2. Second measurement but was closer to the mean or average on its first measurement. For example, if a business has a highly profitable quarter, despite the underlying reasons for its performance being unchanged, it is likely to do less well the next quarter. See also, *quarter*.
Regulatory Requirement	A *regulatory requirement* is a compulsion, demand, obligation, or prohibition through a regulation that is placed on a regulated party, its activities or operations. These are rules that are established in a law, policy, regulation, specification, standard, or statute.
Reinforcing Bar, Rebar	*Reinforcing bar* or *rebar* is a common steel product used in construction to strengthen concrete structures. It is produced in a variety of grades and two types: weldable and non-weldable. See also, *composite*, *grade*, *Micro Rebar™*, and *weldability*.
Reinforcing Pad, Repad	A *reinforcing pad* or *repad* is a plate formed and welded around a nozzle to the shape of the pipe, tank or vessel, for extra strength. See also, *integral reinforcement* and *nozzle*.

Phrase or Term	Definition
Reject	*Reject* is a disposition when repair or rework is deemed too costly or extensive, may have unnecessary risk, or may compromise project schedules. Reject typically involves the replacement in-kind of equipment, materials, or both, with new items. Reject may also be referred to as scrap.
Relational Customer Service	*Relational customer service* involves building long-term relationships with customers, emphasizes fulfilment of customer experience or needs, and fosters loyalty (e.g., a realtor's service when buying or selling a home). Often, a service provider or supplier is differentiated from competitors by customer satisfaction and this may be the difference between a purchase order award and no award. For example, it is likely you would contact the same realtor for future real estate transactions if you were satisfied with your realtor's past performance. See also, *customer centric*, *Givers Gain®*, *positive order entry*, *outside-in*, *reciprocity*, and *transactional customer service*.
Reliability	*Reliability* is the: • Ability of an apparatus, machine, or system to consistently perform its intended or required function or mission, on demand and without degradation or failure, with results characterized by maximized profitability, safety, and stability; • Probability that a product, service, or system will perform its intended function adequately for a specified time or will operate in a defined environment without failure; • Overall consistency of a measure (i.e., similar results are produced under consistent conditions); or, • Quality of being accurate, honest, or trustworthy or of performing consistently well. See also, *reliability-centred maintenance* and *reliability based mechanical integrity*.
Reliability Based Mechanical Integrity	*Reliability based mechanical integrity* is a risk assessment and management process used to assess inspection and maintenance decisions and evaluate the risk of failure for fixed equipment and other assets (e.g., rotating equipment and instrumentation). See also, *risk-based inspection*, *risk of failure*, *mechanical integrity*, *reliability*, and *reliability-centred maintenance*.

Phrase or Term	Definition
Reliability-centred Maintenance	*Reliability-centred maintenance* ensures that preventive or predictive maintenance tasks are performed in a cost-effective, efficient, reliable, and safe manner. It is a process used to determine what must be done to ensure that physical assets continue operating correctly within the present operating context. Ultimately, by using this process, organizations seek to develop a specialized maintenance schedule for each critical asset. The four basic principles of a reliability-centred maintenance program are to: 1. Address failure modes by importance; 2. Define applicable maintenance tasks and select the most effective ones in the case of important failure modes; 3. Identify failure modes, which are the ways in which something might fail (i.e., errors or defects – especially ones that affect the customer – either actual or potential); and, 4. Preserve system functions. See also, *asset performance management*, *reliability*, and *reliability based mechanical integrity*.
Religion	A *religion* is: • A cause, principle, or system of beliefs held to with ardour and faith (e.g., climate change); or, • An institutional or personal system of attitudes, beliefs, and practices about the service and worship of God (e.g., Buddhism, Christianity, Hinduism, Islam, or Judaism) or the supernatural. See also, *climate change*.
Remediate	To *remediate* is to: • Provide a remedy to redress something or correct an error; or, • Restore by reversing or stopping environmental damage.
Remote Team	A *remote team* has personnel that collaborate, usually electronically, from different geographic locations. This is not necessarily the same as a virtual team. See also, *virtual team*.
Renewable Energy, Clean Energy	*Renewable energy* or *clean energy* is collected from renewable resources (e.g., geothermal heat, rain, sunlight, tides, waves, and wind), which are naturally replenished. For example, energy obtained from solar panels. See also, *green energy*, *hybrid energy system*, *renewable energy certificate*, and *renewable resource*.
Renewable Energy Certificate	A *renewable energy certificate* is a market-based instrument (i.e., a tradable, non-tangible energy commodity) that certifies the bearer owns one megawatt-hour of electricity generated from renewable energy. Once the power provider has fed the energy into the grid, the certificate can then be sold as an energy commodity. For example, it may be sold to an organization as a carbon credit to offset their emissions of carbon dioxide or other greenhouse gases. These are also known as green tags, tradable renewable certificates, or renewable energy credits. See also, *greenhouse gas* and *renewable energy*.

Phrase or Term	Definition
Renewable Hydrogen, Renewable H_2, Low-carbon Hydrogen, Low-carbon H_2	*Renewable hydrogen*, *renewable H_2*, *low-carbon hydrogen*, and *low-carbon H_2* is hydrogen that is produced using electricity from a renewable energy source. It has a lower carbon intensity compared to other types of production (e.g., blue, brown, or grey hydrogen). See also, *carbon intensity*, *clean hydrogen*, *colours of hydrogen*, *fossil fuel hydrogen*, *hydrogen*, and *renewable energy*.
Renewable Natural Gas, Sustainable Natural Gas, Biomethane, Biogas	*Renewable natural gas*, *sustainable natural gas*, *biomethane*, or *biogas* it is airborne form of methane released by the decomposition of organic waste material. It is produced from biomass such as animal- and plant-based material, food waste, landfill materials, and sewage, and used as a replacement for fossil fuel natural gas. See also, *24/7*, *methane*, *natural gas*, and *synthetic methane*.
Renewable Resource	A *renewable resource* can be used repeatedly and does not run out because it is naturally replaced; thus, essentially, it has an endless supply. For example, geothermal pressure, and solar and wind energy. Other resources may be considered renewable even though some time or effort must go into their renewal (e.g., fish, oxygen, and wood). Most metals are also renewable. Although these are not naturally replaced, they are not destroyed during their extraction and use, and can be recycled. See also, *green energy*, *non-renewable resource*, and *variable renewable energy*.
Rent-seeking Behaviour	*Rent-seeking behaviour* is when a business or person uses their position or resources to get some additional benefit from a government. The most common occurrence is when a company or industry lobbies the government to receive special grants, subsidies, and tariff protection. The term rent in economics means receiving a payment that is over the costs involved in the production of the item or keeping the item in service. These actions do not produce any benefit for the community-at-large and only redistribute taxpayer's resources. See also, *crony capitalism* and *lobby*.
Repair	*Repair* is a disposition that identifies work necessary to restore an item to a safe and satisfactory condition if there is no deviation from the original registered design or alterations. An approved repair procedure may be required. See also, *major repair*, *minor repair*, and *repair procedure*.
Repair Crew	A *repair crew* performs repair welding whenever non-destructive examination (e.g., visual, ultrasonic, or radiographic testing) identifies welds with defects that do not meet the acceptance criteria of the code(s) of construction (e.g., *CSA* Z662, *Oil and Gas Pipeline Systems*) or owner specifications. See also, *crew*.
Repair In-kind	A *repair in-kind* is a repair that uses the same or similar materials in the same or a similar location, which complies with the current code of construction or specification.
Repair Procedure	A *repair procedure* is a document used to define the methods, scope of work, and strategies for making complex repairs. A repair procedure may require approval by others (e.g., project, safety codes officer or both) See also, *procedure*, *major repair*, and *repair*.
Repeatable	*Repeatable* describes a sequence of activities that may be easily and efficiently replicated. These processes are economical because they have established operations and typically avoid negative variances. See also, *negative variance* and *positive variance*.

Phrase or Term	Definition
Replacement In-kind	A *replacement in-kind* is the substitute of an item (e.g., chemical, equipment, material or procedure) that satisfies the design specification for whatever item is being replaced. This can be an identical item or an alternative replacement item (if the alternative item does not adversely affect the use of the item or associated items). For example, replacing a valve with the same type of valve that is made by a different manufacturer. See also, *like-for-like*.
Report by Exclusion, Manage by Exception	To *report by exclusion* or *manage by exception* means that if nothing is found unacceptable during an inspection, then nothing is reported. Thus, only unacceptable results are reported or managed. See also, *report by inclusion*.
Report by Inclusion	To *report by inclusion* means that an inspection report contains the following minimum information (e.g., per API RP 572): • A description of the damage mechanisms and the extent of damage, and for damage mechanisms with data that can be trended – the corrosion or damage rate; • A description of the extent of inspection (e.g., the percentage of all areas examined) including all areas that were not examined and reason(s) why they were not examined; • A description of the inspection technique used including all parameters needed to reproduce the inspection (e.g., calibration parameters for non-destructive examination); • A repair plan, if needed, to describe what shall be repaired and when, and what acceptance criteria shall be used to deem the repair fit for continued service; • The acceptance and rejection criteria used; • The equipment original design conditions and any safe operating limits from the current operating procedure; and, • The inspector's or technician's company, credentials, name, and the date of inspection. The report-by-inclusion approach is required by various mechanical integrity programs and provides these benefits: • A better record for other subject matter experts to make recommendations; • A better understanding of: o The inspection results (i.e., what was discovered); and, o What was examined and the classification of inspection effectiveness (e.g., as defined in API RP 581 Section 3.1.37). • A more accurate inspection history for the owner-user that facilitates better decisions about: o If a damage mechanism is present; o Inspection deferral; o Repair or replacement; and, o Shutdown inspection planning. See also, *mechanical integrity*, *owner-user*, *report by inclusion*, and *subject matter expert*.

Phrase or Term	Definition
Reporter's Questions, Five Ws, Five Ws and How	*Reporter's questions* or the *Five Ws* are basic inquiries used for problem solving and research that begin with or include these words: • What; • When; • Where; • Who; and, • Why. Open-ended questions require factual answers for a report to be considered complete. Some include a sixth question in this list, *how*, which is then referred to as the *Five Ws and How*. See also, *closed-ended question* and *open-ended question*.
Reporting Line, Line of Reporting	A *reporting line or line of reporting* is a professional chain of command between someone who is responsible for oversight of personnel (e.g., a leader, supervisor, manager) and the personnel themselves. An organizational chart illustrates line of reporting. See also, *dotted line reporting*, *organizational chart*, and *solid line reporting*.
Reporting on Reporting	*Reporting on reporting* is when project management or stakeholders require source data to be transcribed from the original software to another software program or file format for reporting purposes. The collection of these data may be required from several sources that do not export or output data that correlates with the required reporting format or function, or both. Thus, a third software (e.g., another spreadsheet) may be used to filter, prepare, and sort the data for input as the reporting data. This work process may be time-consuming and at high risk for counting, data input/output, or other errors.
Request for Information	A *request for information* is a supplier's request to the project for clarification before proceeding with requested work. Typically, a supplier will make this request when a product may not meet purchase order requirements, or when requirements are unclear or conflicting (e.g., if a drawing and specification provides vague, dissimilar or ambiguous information about the same topic). See also, *request for information coordinator*.
Request for Information Coordinator (RFI Coordinator)	A *request for information coordinator* or *RFI coordinator* ensures that personnel follow the correct request for information work processes, obtain required dispositions, dispositions are correct, the close-out of request for information is timely, and that all necessary follow-up actions are completed (e.g., design change notices). See also, *coordinate* and *request for information*.
Request for Proposal	Not a preferred phrase – see *request for quotation*.
Request for Purchase	A *request for purchase* is a document that is used to award a supplier a purchase order to supply specific products or services for a project. Engineering issues this document to Procurement, who issues the purchase order award to a supplier.
Request for Quotation	A *request for quotation* is a document that invites a supplier to bid to the supply specific products or services for a project. Engineering issues this document to Procurement, who then invites suppliers to bid and compete for the purchase order award. See also, *bid* and *bidder*.
Required at Site Date	The *required at site date* is the date materials or equipment are required to be received at site, after shipment from the supplier's facility.

Phrase or Term	Definition
Required on Site Date	Not a preferred phrase – see *required at site date*.
Requisition	A *requisition* is a document issued by Engineering to Procurement that is used to solicit quotations from bidders, or purchase products or services from a supplier. It usually contains information organized into these sections: • Section 1, *Scope of Supply*; • Section 2, *Technical Documents List*; • Section 3, *Supplier Document Requirements*; and, • Section 4, *Inspection Requirements Document*. See also, *procurement documents*, *request for quotation*, and *request for purchase*.
Requisition Team	A *requisition team* are personnel responsible for preparing, issuing, and managing requisitions and purchase orders. This team comprises a buyer, responsible engineer, supplier quality surveillance coordinator, expediter, discipline engineers, subject matter experts, and personnel with other roles (depending on project size, composition and complexity). See also, *buyer*, *discipline engineer*, *expediter*, *project team*, *responsible engineer*, *subject matter expert*, and *supplier quality surveillance coordinator*.
Research and Development	*Research and development* are business activities directed toward the innovation, introduction, and improvement of products and processes. Typically, the goal is to commercialize new products and services and introduce these to market, with increased growth and profitability. In Europe, this is known as research and technical (or technological) development.
Residual Risk	*Residual risk* is a risk that has not or cannot be addressed by risk avoidance, or risk mitigation or procedures.
Resiliency	*Resiliency* is: • An ability to adjust easily or to recover from change or misfortune (e.g., a difficult career or life event). For example, even though the horse faltered at the crest of the hill, it was resilient enough to complete the journey; or, • The capability of a strained body to recover its original size and shape after deformation (i.e., being bent, compressed, or stretched). For example, the balloon retained its resiliency after it was deflated.
Resource	A *resource* is: • A stock or supply of facilities, materials, money, personnel, and other assets that can be drawn on by a management or an organization function effectively (e.g., direction from a director); or, • An action or strategy that may be adopted in adverse circumstances (e.g., implementing a skeleton crew during difficult economic times). See also, *natural resource* and *talent*.
Resource Company	A *resource company* develops, explores for, and/or produces naturally occurring assets or raw materials (e.g., mineral, natural gas, or oil) for economic gain. See also, *mineral*, *natural resource*, *owner*, and *resource holder*.
Resource Holder	*A resource holder* is the owner of natural resources that will be developed by a project. A resource holder is typically a government. See also, *resource company*.

Phrase or Term	Definition
Resource Levelling	*Resource levelling* is a technique that involves amending the project schedule to keep resource use below a set limit. Projects use this strategy when it is important to impose limits on resource use. Resource levelling can affect a project's critical path. See also, *critical path*, *resource*, and *resource smoothing*.
Resource Optimization Techniques	*Resource optimization techniques* seek to reconcile demands and supplies for resources. Depending on whether project duration or limiting resource use is prioritized, these can be used to amend activity start and finish dates in ways that do or do not affect a project's critical path. See also, *critical path, resource leveling, resources, and resource smoothing*.
Resource Smoothing	*Resource smoothing* is a technique that makes use of a float when allocating resources to avoid affecting total project duration when project time constraints are important. Resource levelling does not affect a project's critical path. See also, *critical path*, *float*, *resource*, and *resource smoothing*.
Responsibility	*Responsibility* is the state of having care or control over someone, important duties, independent decision-making. It may be an obligation to do something, as part of one's job role. Responsibility is typically assigned before a task is undertaken. The degree of responsibility is determined by whomever has accountability. See also, *accountability* and *delegation*.
Responsible Care	*Responsible care* is a program of the Chemical Manufacturers Association that promises to maintain the highest environmental standards regardless of facility location (i.e., in any country).
Responsible Engineer	A *responsible engineer* is the engineer who initiates a requisition or is responsible for the technical direction of a purchase order. See also, *discipline engineer and engineer*.
Resume, Résumé	A *resume* or *résumé* is a summary of one's professional background and career that typically includes professional accomplishments, certifications, education, experience, objectives, qualifications, skills, and training, and often a personal profile. There are two main kinds of resumes. • Functional. Functional resumes are organized by job function or area of skill; and, • Reverse-chronological. Reverse-chronological resumes list work experience from most-recent to least-recent experience. A resume is used: • By a company to inform clients of the skills a job candidate possesses. A corporate resume focuses on the company's business. A corporate profile or resume, like a website, is designed to showcase a company. It is a summary of accomplishments, details, clients, history, mission(s), offerings (i.e., products and services), personnel, and teams; or, • As a marketing tool by a job applicant to secure new employment. A personal resume may be adapted to suit each individual industry or job application. A resume is often accompanied by a cover letter. In South Asian countries such as Bangladesh, India, and Pakistan, the term *biodata* is often used instead of *resume*. Europe and the United Kingdom use *Curriculum Vitae* or *CV* (Latin for *course of life*) instead of *resume*. See also, *cover letter* and *task action results formula*.

<table>
<tr><th>Phrase or Term</th><th>Definition</th></tr>
<tr><td>Retention of Critical Knowledge</td><td>Retention of critical knowledge involves capturing key knowledge in an organization so it can be used again later. It can be applied in these three domains:<ul><li>Assets
To capture non-routine and unique characteristics of sites and assets, thereby ensuring effective transfers, safe operations, and strengthened performance.</li><li>People
To prevent increased costs and performance issues related to effective personnel and resources.</li><li>Projects
Improved project execution through capture and re-use of best practices and lessons learned from earlier projects. Ensuring that past mistakes are not repeated.</li></ul>See also, key knowledge, lessons learned, and organizational memory.</td></tr>
<tr><td>Retroactive</td><td>Retroactive is something that:<ul><li>Extends in application, effect, influence, or scope to conditions that existed or originated in the past or to a prior time. For example, a retroactive pay raise; or,</li><li>Takes effect from a date in the past (e.g., date prior to enactment, imposition, or promulgation). For example, a retroactive tax.</li></ul>See also, proactive and reactive.</td></tr>
<tr><td>Return Merchandise Authorization</td><td>A return merchandise authorization, return authorization or return goods authorization is a document used during the warranty period to return a product for a refund, replacement, or repair.</td></tr>
<tr><td>Return on Capital Employed</td><td>Return on capital employed is a financial ratio that measures a company's profitability and the efficiency with which its capital is used. It measures how well (or how poorly) a company generates profits from its financial assets. See also, capital, financial assets, and return on investment.</td></tr>
<tr><td>Return on Investment</td><td>Return on investment, often expressed as ROI, is the expected financial gain of a project expressed as a percentage of total project investment. It is used to assess the overall profitability of a project. See also, capital. Energy return on the amount of energy expended, and return on capital employed.</td></tr>
<tr><td>Reverse Darwinism, Regressive Evolution</td><td>Reverse Darwinism or regressive evolution occurs when a less evolved species is better equipped to survive than a more evolved species. For example, several species of cave dwelling animals, including crabs, fish, and salamanders, have evolved to become blind because sight is not necessary in dark caves. See also, autistic savant, brain fart, brain freeze, Darwin Awards, Darwinian theory, creative error, fundamental laws of stupidity, hiding hand principle, inverted Darwinism, loserthink, lucky fool syndrome, and savant.</td></tr>
</table>

Phrase or Term	Definition
Reverse Engineer, Reverse Engineering, Back Engineering	To *reverse engineer* or *reverse engineering* or *back engineering* is the: • Process by which a man-made object is deconstructed to reveal its architecture, designs, or to extract knowledge from the object; or, • Reproduction of another manufacturer's product following detailed examination of its construction or composition. It is applicable to the fields of chemical, electronic mechanical, and software engineering and has its origins in the analysis of hardware for commercial or military advantage.
Reverse Mentoring	*Reverse mentoring* is mentoring provided by lower-level personnel to senior management personnel to expose management to their perspectives. Personnel are paired by those who differ in background, ethnicity, gender, race, or other aspects of diversity. See also, *cross-generational mutual mentoring*, *mentoring*, *peer mentoring*, and *shadow board*.
Review Cycle	The *review cycle* is process through which the engineering and procurement company reviews a document submitted by the supplier. A review cycle is measured by the time it takes to complete the review of one or a set of documents. A review code is assigned and then the document is returned to the supplier through the Document Control department. Resubmittal (i.e., revision) may be required to incorporate comments or make changes, or as a certified or final record. Different reviews have different requirements. See also, *review code*.
Review Point	A *review point* is a quality verification point for the review of documentation or another element, which verifies completion or fulfillment of purchase order requirements.
Rework	*Rework* is a disposition that identifies additional work necessary to correct a deficiency and restore an item to safe and satisfactory condition, using the originally planned work processes. See also, *hidden work* and *scope creep*.
Rhetoric	*Rhetoric* is: • Language designed to have an motivative or persuasive effect on its audience, which may be regarded as lacking in meaningful content or sincerity. For example, the substance of a politician's speech during an election year; or, • The art of speaking or writing to effectively communication or persuade. For example, a lawyer writing persuasive arguments for a case. From ancient to modern times, rhetoric has played a myriad role in education, and for counsellors, historians, lawyers, orators, poets, politicians, *sloganeering*, and statespersons.
Ridgeline Plot, Joyplot	A *Ridgeline plot* or *Joyplot* shows the distribution of a numeric value for several groups. Distribution can be represented using histograms or density plots, all aligned to the same horizontal scale and presented with a slight overlap. See also, *horizon chart*.
Right of Entry	*Right of entry* refers to one's right to take or resume possession of land, or the right of a person to go onto another's property without trespassing. See also, *trespass*.

Phrase or Term	Definition
Righteousness Fallacy, Self-righteousness Fallacy, Fallacy of Good Intentions	The *righteousness fallacy*, *self-righteousness fallacy*, or *fallacy of good intentions* is the idea that one is benevolent or correct because their intentions are good or pure, and they have the facts or truth on their side. This happens when the intent of an action is considered more important than the results. For example, California has one of the most regulated housing markets in the United States, yet it has the highest homeless population in the country and, per capita, ranks 49th in housing supply in the United States. Nobel-Prize-winning economist Milton Friedman observed this as uncanny proclivity of politicians *to judge policies and programs by their intentions rather than their results*. Having a desire to help and helping, however, are two separate things. For example, in: • Business, the goal of maximizing short-term shareholder value that in practice undermines long-term shareholder value. For example, cease to provide support for old, or even relatively new, products, or use gimmicks such as non-GAAP accounting measures to briefly boost stock value; or, • Medicine, the widespread erroneous use of the lobotomy procedure (i.e., inserting a surgical device into the prefrontal cortex to destroy it) during the 1940s and 1950s to *improve* mental health. See also, *bait and switch*, *circular reasoning*, *either/or*, *false equivalence*, *generally accepted accounting principles*, *logical*, *logical fallacy*, *red herring*, *slippery slope*, and *straw man argument*.
Right-of-Way	A *right-of-way* is a tract of land on which an organization has the right to construct, manage or both. A right-of-way typically connects to another tract of land. It is a type of legal easement granted or reserved over land for construction of transportation and other assets (e.g., for electrical transmission lines, pipelines, canals, highways, and railways).
Right-side Up	*Right-side up* means with the top side facing up. For example, place the box right-side up. See also, *upside down*.
Right-wing	The *right-wing* is the conservative or reactionary section of a political party or system. This labelling system originated from the seating pattern of the French National Assembly, which put liberals on the left, moderates in the middle, and conservatives on the right. Someone who is right-wing usually supports tradition and the status quo. See also, *conservative*, *moving to the right*, *left-wing*, *political*, and *reactionary*.
Risk	*Risk* is the product of the probability of the failure of an item or process and the expected consequences should that failure occur. It is an educated guess about what the future holds, the range of things that might happen, and how probable each occurrence will be. Taking a risk exposes the possibility of loss or injury in the hopes of achieving a gain or reward. See also, *calculated risk*, *consequence of failure*, *de-risk*, *dumb risk*, *hazard*, *high-risk project*, *high-value project*, *probability of failure*, *project definition rating index*, *risk acceptance*, *risk aversion*, *risk assessment*, *risk compensation*, *risk equation*, *risk management*, *risk measurement*, *risk mitigation*, *risk normalization*, *risk register*, *risk tolerance*, *smart risk*, and *Swiss cheese model*.

Phrase or Term	Definition
Risk Acceptance	*Risk acceptance* involves acknowledging a risk and *not* taking pre-emptive action to avoid it. For example, inspection is not required for a product before shipping because the cost would be excessive; hence, receiving inspection may be performed instead. See also, *risk*.
Risk Appetite	*Risk appetite* is the amount and type of risk an organization is willing to accept in anticipation of gains. It is not the same as risk tolerance, which is the amount of variation in performance measures that an organization is willing to accept. See also, *risk* and *risk tolerance*.
Risk Assessment	A *risk assessment* is a systematic process of evaluating the potential risks involved in a projected activity or undertaking. See also, *calculated risk*, *project definition rating index*, *risk*, and *Swiss cheese model*.
Risk Assessment, Root Cause Analysis, Installation, Maintenance and Monitoring, and Removal Confirmation Date Procedure	A *risk assessment*, *root cause analysis*, *installation*, *maintenance and monitoring*, and *removal confirmation date procedure* is used with an engineered pressure enclosure. This product provides a temporary means to safely contain certain pressure equipment leaks until equipment can be properly repaired. See also, *engineered pressure enclosure*, *risk assessment*, and *root cause analysis*.
Risk Aversion	*Risk aversion* refers to the tendency to, when exposed to uncertainty, attempt to lower that uncertainty. It is the hesitation to agree to a situation with an unknown payoff, rather than another situation with a more predictable payoff – but with possibly a lower expected payoff. For example, choosing to put their money into a bank account with a low but guaranteed interest rate, rather than buying a stock that may have a higher return but has a greater risk of losing value. Loss aversion, for example, is preferring to not lose $5 rather than to finding $5. See also, *loss aversion* and *risk*.
Risk Compensation	*Risk compensation* is a theory which suggests that people typically adjust their behaviour in response to the perceived level of risk, becoming more careful where they sense greater risk and less careful if they feel more protected. For example, in one study, motorists drove closer to the vehicles ahead of them when vehicles were fitted with anti-lock brakes. By contrast, shared space is a highway design method that consciously aims to increase the level of perceived risk and uncertainty, thereby slowing traffic and reducing the number of and seriousness of injuries. The reduction of predicted benefit from regulations that intend to increase safety is referred to as the Peltzman effect, in recognition of Sam Peltzman. He controversially suggested in 1975 that offsets due to risk compensation were virtually complete, so that regulations had not decreased highway deaths. See also, *risk*.
Risk Cube	A *risk cube* is a more complex method of representing the risk matrix and risk evolution or horizon chart. Using a three dimensional graphic, a third axis is added to track the adjacent or parallel risks that have lateral effects. This type of risk graphic requires dexterity and expertise to effectively capture the risks and their adjacent complications. It is a powerful tool to help personnel think laterally when developing mitigations, plans, and strategies for project execution and other purposes. See also, *lateral effect*, *risk matrix*, and *risk management*.

Phrase or Term	Definition
Risk Equation	The *risk equation* is used to determine risk as part of the risk-based inspection methodology. The equation is: Risk = consequence of failure + probability of failure See also, *consequence of failure*, *probability of failure*, and *risk*.
Risk Graph, Profit Graph	A *risk graph* or *profit graph* is a two-dimensional graphical representation that displays the range of profit or loss possibilities for an options trade. The horizontal axis shows the price of an underlying security at its expiration date, while the vertical axis shows potential profit or loss. See also, *risk management*.
Risk Management	*Risk management* is the evaluation, identification, and prioritization of risk exposures – followed by coordinated and economical application of resources to control, minimize, and monitor the impact or probability of unforeseen events. Risk for: • A project is that equipment and materials will be delivered incomplete, incorrect, or late. This results in costs, schedule overruns, or both. Risk mitigation is provided by supplier quality surveillance programs with methods such as the use of a criticality ratings, quality surveillance levels, and quality verification points (e.g., review, witness, and hold points); or, • An operating system is that damage will occur to an asset, the environment, personnel, or product, or a shutdown of operations. This results in unplanned costs, lost revenue, or both. Integrity management programs provide risk mitigation using corrosion studies, fitness for service assessments, and in-line inspection. Risk management may also maximize the realization of opportunities; however, this is not typically the main purpose of a project. See also, *business continuity risk*, *diversification*, *quality management*, *management*, *project definition rating index*, *risk*, *risk cube*, *risk graph*, *risk matrix*, *risk register*, *supplier quality surveillance*, and *Swiss cheese model*.
Risk Matrix	A *risk matrix* is a two-dimensional graphical representation used to assess, easily illustrate, and evaluate risk. It is used during a risk assessment to define the level of risk by considering the category of probability or likelihood against the category of consequence severity. Examples of the various degrees of probability include frequent, likely, occasional, seldom, or unlikely. Examples of severity can include catastrophic, critical, marginal, or negligible. Activities that are considered to be both frequent and catastrophic would be considered extremely high risk, while activities that are both unlikely and negligible would be considered low risk. This is a simple mechanism to increase visibility of risks and assist decision making. It is used with a risk register and other risk tools. See also, *risk cube*, *risk management*, and *risk register*.
Risk Measurement	*Risk measurement* is an: • Assessment of an organization's exposure to risk; • Estimation of a range of possible outcomes using data; or, • Evaluation of the possibility and magnitude of a risk. See also, *risk*.

Phrase or Term	Definition
Risk Mitigation	*Risk mitigation* involves decreasing the probability of a negative risk occurring, and protecting project objectives from impacts of negative risk. See also, *risk* and *risk management*.
Risk Normalization, Normalization of Deviance	*Risk normalization* or *normalization of deviance* is the gradual process through which unacceptable practice or standards become acceptable. As deviant behaviour is repeated without catastrophic results, it becomes the norm for the organization. Almost every action requires some risk. The key is to make conscious decisions about what risks are acceptable and prevent poor habits for forming that create unnecessary risk. Risk normalization can creep into and affect seemingly mundane aspects of business. For example, the loss of NASA's space shuttles Challenger in 1986 and Colombia in 2003 were determined to be the result of risk normalization. See also, *risk* and *shortcut*.
Risk Register	A *risk register* is a record of identified risks. Risks are logged in the register including actions that have been completed to mitigate risk. Additional information about each risk may also be included (e.g., category, impact, owner, probability, and reference). Addressing a risk may be: • Recommended, but not undertaken; or, • Undertaken, but not recommended. See also, *risk*, *risk cube*, *risk management*, and *risk matrix*.
Risk Tolerance	*Risk tolerance* is the level of variation in performance measures that an organization is willing to accept. It is not the same as risk appetite, which is the level and type of risk an organization is prepared to accept in anticipation of gains. See also, *risk appetite*.
Risk-based Inspection	*Risk-based inspection* is a risk assessment and management process focused on the loss of containment from pressure or process equipment, or piping systems. It uses a qualitative or quantitative assessment of the consequence of failure and probability of failure to categorize and prioritize inspection activities. It is used to identify and understand risk, risk drivers, and to establish the lifecyle stage of equipment. See also, *condition-based inspection*, *condition-based monitoring*, *consequence of failure*, *corrosion study*, *damage mechanism review*, *probability of failure*, *process equipment*, *qualitative*, *quantitative*, and *reliability based mechanical integrity*.
Risk-free	*Risk-free* is an activity, asset, investment, or transaction whose future benefit is known and certain. It is something that does not involve any risk and is obtained with total certainty. For example, staying home and watching a Netflix movie is risk-free compared to travelling to see a movie at a theater. See also, *risk*.

Phrase or Term	Definition
Risks Assumptions Issues and Dependencies Log	A *risks assumptions issues and dependencies log* is a project management tool that records developments in these four aspects of project work for the stakeholders' benefit and for an end-of-project review: 1. Risks Events that can have an adverse impact if they occur. 2. Assumptions Resources assumed to be available to contribute to project success. 3. Issues Matters that need to be addressed and considered. 4. Dependencies Things or triggers that are a beneficiary of project outcomes, or that on which the project depends.
Rival	A *rival* is a group, person, or thing competing with another for the same or similar objective, or for superiority in the same field of activity. See also, *competitor*.
Road Ban	A *road ban* is a road restriction from an accident, construction, dimensional constraints, maintenance, or seasonal conditions. In Alberta for example, seasonal restrictions for heavy haul or service rigs are set per this schedule: • Spring: Weather dependent, subject to thaw depth readings of at least 25 cm and begins at same time as Period 1 weights for service rigs • Post-thaw: June 16 (may begin one week earlier or one week later) • Summer: July 1 • Fall: September 1 • Winter: Weather dependent, subject to frost depth readings of 75 cm or more. In Alberta, frost and thaw depths are monitored from a network of electronic thermistors throughout the province, which are used to determine the start of Spring/Period 1 seasons, Winter/Period 4 seasons and placement of thaw or frost line. Periods 2 and 3 do not experience frost.
Robbing Peter to Pay Paul	*Robbing Peter to pay Paul* means to take money from one person or thing and give it to another, especially when it results in the elimination of one debt by incurring another. For example, *borrowing money to pay a credit card debt*. Today, Rudyard Kipling's adaptation of the phrase *robbing selected Peter to pay for collective Paul* (1919) is used to criticize the concepts of collectivism and income redistribution.
Robot	A *robot* is a programmable machine that carries out a series of actions autonomously or semi-autonomously. Most interact with the physical world using actuators and sensors. A robot is more flexible than a single-function machine (e.g., an alarm clock) because it is reprogrammable. See also, *robotics*.

Phrase or Term	Definition
Robotic Industrial Process Automation	*Robotic industrial process automation* is technology that combines traditional physics-based and process control methods with the transformative benefits of machine learning to enable autonomous control and enhance operations. This increases throughput production and energy efficiency, and improves safety through continuous and seamless operation. Much like many airplanes, which are built with a standard autopilot that assists (but does not replace) the pilot, this technology assists an operator in a local control station. It controls physical processes using control systems and physical machines to automate industrial process with real-time optimization that are otherwise checklist-, procedure-, and rule-driven. See also, *local control station*, *process control system*, *robot*, *robotic process automation*, and *robotics*
Robotic Process Automation	*Robotic process automation* is technology that allows computer software or a robot to be configured to mimic human actions within digital systems to execute business processes that are manual, repetitive, and rule-based. These robots capture data and manipulate applications like humans (e.g., communicate with other systems, interpret data, and trigger responses to perform repetitive tasks), but perform non-stop and with no errors. For example, robots can copy and paste data, extract data, fill in forms, log into applications, move files and folders, and scrape browsers. See also, *robotic industrial process automation* and *robotics*.
Robotics	*Robotics* is a discipline dedicated to building, designing, programming, and using robotic equipment and related machines. Robotics involves robots. Robotics may or may not use automation. See also, *automation*, *robot*, *robotic industrial process automation*, and *robotic process automation*.

Phrase or Term	Definition
Rockefeller Habits	The *Rockefeller Habits* Rockefeller Habits is a business administration theory that was formulated by Verne Harnish. In his book *Scaling Up: How a Few Companies Make It … and Why the Rest Don't*, he provides practical tools for entrepreneurs related to company growth. He formulated his theory after reading the biography of the incredibly wealthy John D. Rockefeller, who used these principles as the foundation for successful management in the late 19th and early 20th centuries. Harnish explains how an organization can grow stronger by eliminating weaker elements and focusing on its strengths. The basics of the Rockefeller habits or the Rockefeller theory are Priority, Information, and Rhythm. These three points are important for a company's growth. According to Harnish, company growth does not follow a straight line; it has ups and downs. Successful growth primarily has to do with timing. For instance, responding to a growing market needs to occur at exactly the right time. The theory focuses on John D. Rockefeller's 10 business habits: 1. Inspiring management team; 2. Joint deadlines; 3. Clear information; 4. Managing goals; 5. Open communication; 6. Customer reviews; 7. Core ideology; 8. Lear market proposition; 9. Key performance indicators; and, 10. Weekly management team meetings.
Role and Responsibility Chart, Responsibility Assignment Matrix, Linear Responsibility Chart	A *role and responsibility chart* (also known as RACI matrix), *responsibility assignment matrix,* or *linear responsibility chart* describes participation by role when completing deliverables or tasks for a project or business process. RACI is an abbreviation derived from these four key responsibilities: responsible, accountable, consulted, and informed. Projects use this chart for clarifying and defining roles and responsibilities in cross-functional or departmental projects and processes. There are many alternatives to, and variations of, the RACI model. See also, *organizational chart*.
Rolling Action Item Log	A *rolling action item log* is a spreadsheet used to record action items, responsibilities, and resolutions for project-supplier procurement meetings. New items are progressively added and completed items are hidden or struck-out. This log provides a record that can be used for each subsequent meeting instead of minutes that need to be copied or reproduced. See also, *minutes of meeting* and *rolling action item log meeting*.
Rolling Action Item Log Meeting	A *rolling action item log meeting* is project-supplier procurement meeting held on a regular basis (e.g., weekly) to review items that require action or resolution. See also, *rolling action item log*.

Phrase or Term	Definition
Rolling Blackout, Rotational Load Shedding, Feeder Rotation	A *rolling blackout*, *rotational load shedding*, or *feeder rotation* is an intentionally engineered electrical power shutdown where electricity delivery is stopped for non-overlapping periods of time in different parts of the distribution region. This is a last-resort measure used by an electric utility company to avoid a total blackout of the power system when the demand for electricity exceeds the power supply capability of the network. See also, *blackout*.
Rolodex, e-Rolodex	A *rolodex* is a desktop card index used to record contact information (i.e., names, addresses, and telephone numbers). Rolodex cards are removable from the rotating spindle on which they are attached. The spindle allows quick and easy access to contact information. The rolodex is dated; it has largely been replaced with electronic applications. Today, contact information that was stored in a rolodex is now stored in an *e-rolodex* (e.g., in applications, websites, and on electronic devices where it is conveniently accessible and electronically searchable).
Room Temperature	*Room temperature* is a temperature between 10 and 38°C (50 and 100°F).
Root Cause	The *root cause* is the primary reason an event occurs. See also, *root cause analysis*.
Root Cause Analysis	The *root cause analysis* is a methodology aimed at discovering the primary cause of a challenge, issue, problem, or quality concern. This analysis is completed in three stages: 1. Open Phase – Brainstorming ideas about what might be causing the concern; 2. Narrow Phase – Reducing the ideas to as few as possible by proving others to be incorrect; and, 3. Closed Phase – Validating the final short-list of root causes. See also, *brainstorm* and *root cause*.
Rotating Equipment	*Rotating equipment* describes non-stationary equipment that uses kinetic energy to move fluids, gases, and other process materials (e.g., compressors, engines, gearboxes, generators, pumps, and turbines). See also, *electro-mechanical equipment*, *fixed equipment*, and *process equipment*
Round Table, Roundtable	A *round table* or *roundtable* is a: • Conference for discussion or deliberation with several participants, which may be at a physical table, remotely (e.g., via Skype), or both; • Discussion or meeting where each participant has equal status and equal time to present their views; • Group gathered together for a conference or discussion; or, • Method to end a business meeting that involves asking each attendee for final comments, concerns, and questions.

Phrase or Term	Definition
Rounding Error, Roundoff Error	A *rounding error* or *round-off error* is a mathematical miscalculation or quantization error caused by altering a real number to an integer (e.g., change 7.428 weeks to 52 days) or one with fewer decimals (i.e., change 7.4285714 weeks to 7.4 weeks). This is the difference between the calculated approximation of a number and its exact mathematical value. Rounding errors occur by inexactness in the representation of numbers and the arithmetic operations used with them. The significance of a rounding error depends on the circumstances. For example, the value of the number pi (i.e., π) which is 3.141592653589793238 at 18 decimal places, is often rounded to 3.14 or three decimal places for practicality. Precise measurements of large distances may be acquired from the increased accuracy of using more decimal places such as with a Global Positioning System. Google map's standard is seven decimal places, which is the practical limit of commercial surveying. Rounding may be inconsequential enough to be ignored in most cases; however rounding errors can have a cumulative effects (e.g., in computerized engineering and financial calculations). In these cases, rounding may need to be avoided or rectified. A rounding error can be especially problematic when a rounded input is used in a series of calculations, causing an error to compound or propagate, so that the final answer is erroneous. For example, dividing 1 by 3 on a calculator yields 0.333333333333. Rounding to the nearest two decimal places yields 0.33. Multiplying 0.33 by 3 yields 0.99, instead of yielding 1 had rounding not been used. The phrase *rounding error* may be used as an expression to indicate an expense that is immaterial and relatively negligible. For example, a USD$ 90.00 business meal may be referred to as a rounding error for an organization that earns billions of dollars annually. See also, *algorithm*, *error propagation*, and *Global Positioning System*.
Routings	*Routings* are the specific steps that must be executed in a production environment to create an assembly, part, or subassembly. It includes a sequence of production operations, identifies the work centre that will complete each operation, and identifies the expected time to perform the operation.
Royal Canadian Air Force Exercise Plans, 5BX, XBX	The *Royal Canadian Air Force Exercise Plans* are also known as: • *5BX* or five basic exercises (an 11-minute plan for men); and, • *XBX* or ten basic exercises (a 12-minute plan for women). These exercise plans were developed by Dr. Bill Orban in the late 1950s and then published as *The 5bx Plan: For Physical Fitness* and *The XBX Plan: For Physical Fitness*. These exercises are intended to maintain physical fitness without requiring equipment or a large area. At that time, this challenged the conventional wisdom that taught it was necessary to use equipment and have a gruelling workouts to maintain physical fitness. These exercises became popular with the public and the plans were translated into thirteen languages. Prince Philip, Duke of Edinburgh was a well-known practitioner who was remarkably fit and mobile throughout his life. He could walk without a cane and had good posture until his passing at the age of 99 years.

Phrase or Term	Definition
Rubber Chicken Circuit	The *rubber chicken circuit* is the gruelling series of speaking events at public dinners and luncheons that politicians must attend to gain support and raise funds. The food often includes chicken, which is a less expensive meet than beef or seafood. It is also a meat that appeals to most religious preferences (e.g., for Christians, Hindus, and Muslims) and thus is more likely to be eaten by most circuit attendees. The meal may be cooked hours earlier and then reheated, giving the meat a rubbery texture and an etymology for this phrase.
Rubric	A *rubric* is: • An established custom, rule, or tradition (e.g., celebrating a national holiday); • An explanatory or introductory commentary (e.g., a sidebar); or, • A statement of function or purpose (e.g., the purpose of the project is to build a bridge). *Rubric* also has other meanings. *Rubric* originates from *ruber* (Latin for *red*). Ancient writers (circa 1300) inserted rubrics, which are special instructions or explanations, into manuscripts using red ink to set them off from black ink used in the main text. See also, *key knowledge*.
Rule of Three	The *rule of three* is a negotiation strategy used to get someone to agree to the same thing three times in the same conversation. This is effective because of human nature: it is difficult to repeatedly fake conviction or lie. It is useful to avoid a counterfeit yes and obtain a true commitment and triples the effect of the agreement. To avoid being too assertive or repetitive, use this three-step approach: 1. Ask for an agreement (e.g., make a statement to which they can respond with *that is true* such as *would you agree that* ...); 2. Summarize and repeat their statement (e.g., make a statement to which they can respond with *that* is right such as, *you mean that* ...); and, 3. Ask a calibrated question (e.g., ask how *that* can be accomplished such as *how can we get from A to B?*). Alternatively, the same question may be asked three different ways. For example: 1. Do you agree it is true that our project cannot be completed by Monday unless someone with high-level expertise such as you works on Saturday? 2. Could you work a few hours on Saturday to complete the final copy? 3. Would you be agreeable to taking Monday off in lieu of Saturday so you can prepare the final copy? This technique will also uncover a problem before it occurs. For example, if someone answers yes to an initial inquiry, but it is a *counterfeit yes* (which means *no*). This can be revealed by subsequent inquiries, which may reveal that there is not yet a true agreement. This concept was developed by Chris Voss in the book *Never Split The Difference: Negotiating As If Your Life Depended On It* (with Tahl Raz). See also, *A to B*, *calibrated question*, *win-win*, and *yes*.
Rule of Thumb	A *rule of thumb* is a broadly accurate guide or principle that is not intended to be strictly accurate or reliable for every situation. It refers to an easily learned and applied procedure or standard that is practical but not exact or sophisticated. It is based on practical experience rather than a calculation or theory. For example, a good rule of thumb is to check the vehicle for fuel before driving it.

Phrase or Term	Definition
Rumour Mill	A *rumour mill* is a group who start and spread rumours. See also, *mill*.
Run and Gun	A *run and gun* project is one that is managed with few procedures or guidelines. Run and gun also refers to a(n): • Fast and freewheeling style of basketball; • Offensive system in American football, also known as the *run and shoot offense*; and, • Video game genre.
Run it Up the Flagpole	*Run it up the flagpole* means to suggest or propose an idea to learn how others react to it (e.g., favourably, indifferently, or unfavourably). *Let's run it up the flagpole and see if anyone salutes it* is an expression that was popular in the United States during the late 1950s and early 1960s that has since become cliché. See also, *plain English* and *trail balloon*.
Run Test	Not a preferred phrase – see *mechanical run test*.
Rupture	To *rupture* is to break or burst suddenly (e.g., a pipe or vessel). This occurs when the design and safety parameters for a material or pressurized system are exceeded. See also, *leak before break*.
Rust	*Rust* is a corrosion product comprising various hydrated iron oxides and iron oxides. This term properly applies only to iron and ferrous alloys. See also, *corrosion* and *flash corrosion*.
Rut	A *rut* is a: • Deep furrow or track worn by regular passage (e.g., from animals, humans, vehicles, the wind, or another source); • Fixed or usual practice. Especially a fixed way of thinking or a monotonous routine (e.g., working from home has put me in a rut; I never do anything or go anywhere exciting); • Periodic and often annually recurring state when most male animals exhibit breeding behaviour (e.g., it is rutting season for deer); or, • Situation that is boring and difficult to change (e.g., my 9 to 5 rut is exhausting but I need the money for groceries and rent). See also, *9 to 5*.

Phrase or Term	Definition
S	1 A B C D E F G H I J K L M N O P Q R **S** T U V W X Y Z
Sacrificial Anode	Not a preferred phrase – see *galvanic anode*.
Saddle	A *saddle* is a support structure for a horizontal pressure vessel. See also, *shell* and *skirt*.
Safe Work Procedure	A *safe work procedure* is a step-by-step process that must be followed when an operation deviation causes an incident or loss. It is a risk management tool that describes the: • Associated hazards; • Most efficient and safest way to perform a task; • Risk score associated with the associated hazards (e.g., using a risk matrix); • Steps involved to complete the activity without incident; and, • What personal protective equipment is required. See also, *standard operating procedure*.
Safe Working Load, Normal Working Load	*Safe working load* or *normal working load* is the maximum safe force that lifting equipment, a lifting accessory or a device can exert to lift, suspend, or lower a given mass without breaking. This value is usually marked on the equipment by the manufacturer. It is a calculation of the minimum breaking load or minimum breaking strength, divided by its safety factor, For lifting equipment this is usually ten to one (1/10 or 10:1), although it is application-dependant. Other safety factors are ¼, 1/5 and 1/6. See also, *lifting capacity* and *safety factor*.
Safety Codes Act	The Alberta *Safety Codes Act* is safety legislation for pressure equipment (i.e., boilers and pressure vessels) and these disciplines: electrical, plumbing, gas, fire, and building. Jurisdictions outside of Alberta have similar legislative requirements.
Safety Codes Council	The *Safety Codes Council* was established by the Government of Alberta in 1993. It administers portions of the Alberta safety system established in the *Safety Codes Act*, including: • Accrediting municipalities, corporations, and agencies that sell permits and inspect the work carried out under these permits; • Administering the Alberta Master Electrician Program; • Certifying and training safety codes officers who perform inspections; and, • Working with industry through sub-councils to make recommendations for codes and standards. The Council derives its authority from the *Safety Codes Act* and the Alberta Minister of Municipal Affairs. See also, *Safety Codes Act*.
Safety Codes Officer	A *safety codes officer* is the authorized inspector for a jurisdiction. For example, in Alberta this is someone employed with the *Alberta Boilers Safety Association*.

Phrase or Term	Definition
Safety Factor, Factor of Safety, Safety Margin	A *safety factor*, *factor of safety* or *safety margin* expresses how much stronger equipment, materials, or structures are than they need to be for their intended use. (e.g., a member, part, structure, or system). The terminology may be confusing because reference books, standards, and other sources use the terms differently. Careful designers or engineers refer to a calculated value as a realized factor of safety and a required value as a required factor of safety. For example, the Brooklyn Bridge in New York, which was constructed in 1883, now bears the load of about 145,000 vehicles per day. During the time of construction, it was common for bridges to collapse from inadequate aerodynamic stability, load capacity, or material failure. Out of caution, it was designed to have a factor of safety of 6. The cables are estimated to have a safety factor of 4 due to faulty wire. Thus, the bridge can sustain today's (unplanned) loading and it continues to be durable almost 140 years later! Many things are intentionally built stronger than needed to accommodate normal stress, normal use, natural degradation, emergency situations, misuse, and unexpected loads. This may be: • A measure of the reliability of a design as a calculated value; or, • A constant required value that is imposed (e.g., by a contract, regulation, specification, or standard) to which something must conform or exceed. The use of a safety factor does not imply that a design, item, or structure is *safe*. Many end-use, engineering design, installation, production, and quality assurance factors influence if something is safe for a particular use. The safety factor should always meet or exceed the required design factor; if it does not, the design is inadequate. A design safety factor is measured as a ratio of the maximum strength to the intended load. See also, *design factor*.
Safety Instrumented System	A *safety instrumented system* consists of an engineered set of hardware and software controls which are especially used on critical process systems. See also, *integrated control and safety system* and *safety instrumented system.*
Safety Integration Test	A *safety integration test* is a software test that involves individual equipment or hardware items that are combined and tested as a group. The purpose of this test is to expose faults in the interaction between integrated units.
Safety Moment, Safety Minute, Safety Talk, Safety Chat, Safety Brief	A *safety moment*, *safety minute*, *safety talk*, *safety chat*, or *safety brief* is a concise discussion or mini-meeting about a specific workplace health or safety topic (e.g., how to avoid slips, trips, and falls). Many companies delegate this task to managers, who typically hold these 2- to 10-minute meetings regularly (i.e., at the start of a meeting or shift). See also, *toolbox talk*.

Phrase or Term	Definition
Sagan Standard, Extraordinary Claims Require Extraordinary Evidence	*The Sagan Standard* or *extraordinary claims require extraordinary evidence* is an adage coined by astronomer Carl Sagan that means that the more unlikely a certain claim is – the greater the standard of proof that is expected of it. So, if a claim is considered extraordinary (e.g., it is highly unlikely because it contradicts the scientific consensus), the claim it is held to a higher standard of proof. For example, if someone claims they saw a unicorn, then stronger evidence would be required than if they claimed they saw a horse. It is well-known that horses exist, but no meaningful or substantial evidence supports the existence of unicorns – which makes that claim extraordinary. This concept can help discredit false information and make better decisions by understanding when a higher standard of proof should be expected. It is also important to understand how to avoid overapplying this concept, which can lead to the incorrect dismissal of a valuable claim. This concept is not entirely originally Sagan's; it was previously described by Swiss psychologist Théodore Flournoy and French scientist and philosopher Pierre-Simon Laplace in the 19th century and by earlier scientific thinkers. See also, *Occam's razor*.
Salary Compression	*Salary compression* is when there is little or no difference in salary between personnel but large differences in qualifications, responsibilities, or skills. This occurs for many reasons. For example, when it is difficult to recruit or retain personnel for certain positions, or when salary structures are misaligned with market conditions. See also, *salary inversion*.
Salary Inversion	*Salary inversion* is when an employer pays a higher compensation to inexperienced personnel or new recruits when compared to the salaries of existing and experienced personnel. This occurs for many reasons. For example, when the demand for suitably qualified personnel exceeds the supply. See also, *salary compression*.
Sales Lead	A *sales lead* is an opportunity to sell a product or service to a prospect that needs or wants it, because someone (e.g., an associate, colleague, or friend) has provided this information. When a sales lead is contacted, they may not be expecting contact, know who is calling, or what is being offered. This type of prospect is not a cold call, because a mutual connection who is aware of their needs or wants has provided information about them (i.e., a reason to believe they are interested in purchasing a product or service). See also, *cold call*, *prospect*, *referral*, *value proposition*, and *warm call*.
Sales Order Number	A *sales order number*, like a job number, is a unique number or code used to identify a scope of supply and material quantities. This number generally indicates that no additional production effort will be applied to the product (e.g., assembly). Rather than using the customer's purchase order number, an internal sales order number allows the internal audit control of completeness to be monitored. See also, *job number*.
Sample	A *sample* is one or more units of product or a quantity of material that is taken from a lot for the purposes of inspection and to determine if the lot will be accepted (based on the sample). See also, *lot* and *sample size*.
Sample Size	The *sample size* is the number of units in a sample. See also, *sample*.
Sanction	*Sanction* is an approval or official permission or for action (e.g., for a project to proceed).

Phrase or Term	Definition
Sandbagging	*Sandbagging* is hiding the difficulty, skill, or strength of something, or someone early in an engagement. It may refer to: • In games (e.g., billiards or poker), deliberately playing below one's actual ability to fool opponents into accepting higher stakes bets in future games, or to lower one's competitive rating to play in a future event with a higher handicap and consequently have a better chance winning; or, • Suing for a breach of a contractual representation or warranty, despite knowing at the time of the contract that it was untrue. *See also, business wargame*, *game*, *games against nature*, *gamesmanship*, *gamification*, *lose-lose*, *win-win*, *wargame*, and *zero-sum game*.
Sandblasting	Not a preferred term – see and *abrasive blasting*.
Sandbox	A *sandbox* is: • A box that contains sand for children to play in; • A metaphor for the type of collaboration that can take place in communal or open office spaces, face-to-face and virtual meetings, team building activities, and other shared or common places where personnel interact and work together. This can eliminate organizational silos and foster synergy; and, • A virtual space in which new or untested software can be securely created and tested (e.g., software stored on a network folder). See also, *silo mentality* and *synergy*.
SAP® Business Suite	*SAP® Business Suite* is a bundle of business software applications that provide integration of information and processes, collaboration, industry-specific functionality and scalability. It is based on the NetWeaver technology platform.
Satisfice	To *satisfice* means satisfying and sufficing. It is when the first reasonable option is chosen, instead of comparing alternatives and choosing the best option. This strategy used by personnel in high-stakes decision-making where changing conditions, limited information, time-pressure, and vague goals (e.g., chess masters, fire fighters, nuclear power plant operators, and pilots) limit the ability to consider various alternative plans. It entails establishing one reasonable plan and then doing a quick mental test for possible problems. If there are none, this becomes the plan of action. This strategy is also used in other situations. For example, when searching on the Internet or a website, users click links that they think will lead them to the information they are seeking. *Satisfice* is a portmanteau of *satisfy* and *suffice*. See also, *fear of better option* and *fear of missing out*.
Savant	A *savant* is someone who: • Has detailed knowledge in a specialized field (e.g., literature or science); or, • Is affected with a developmental disorder (e.g., autism or an intellectual disability) who exhibits exceptional brilliance or skill in some limited field (e.g., mathematics or music). See also, *autistic savant* and *reverse Darwinism*.

Phrase or Term	Definition
Save Face	To *save face* is to: • Afford or give someone an opportunity to avoid embarrassment, humiliation, or shame; • Preserve dignity or retain respect; or, • Try to regain favourable standing after something embarrassing has happened. This concept is a core social value in Asian and other cultures and has been part of the English language since the 19th century. Embarrassment can be caused inadvertently or willfully. Disagreeing indirectly instead of directly with someone is one method of saving face. For example, saying that *driving fast is a good way to quickly get to a destination, but adding ... have you considered speed limits?* Another example, during a hiring process, telling a candidate: *your experience and skills are impressive but we decided to hire a more junior candidate who has less experience* (instead of directly saying *we are not going to hire you)*. See also, *bully*, *office jerk*, and *office politics*.
Saving Grace	*Saving grace* is: • A redeeming characteristic or quality that makes it possible to accept someone or something when it has one or more negative characteristics. For example, the old vehicle's (*old* is a negative characteristic in most vehicles) redeeming grace is that it is inexpensive (one redeeming grace) and well-maintained (another redeeming grace); or, • The redeeming grace of God. In Christian theology, it is a state of sanctification by God.
Scaffolding, Scaffold	*Scaffolding* (plural) or a *scaffold* is a temporary construction structure used to provide workers access to a structure above ground.
Scalable	*Scalable* describes something can be changed in size or scale (i.e., scale up means to increase; scale down means to decrease). For example, John scaled up his mechanic shop to a national franchise within the first five years. See also, *scale*.
Scale (of a Facility)	The *scale of a facility* describes its purpose and size. It may be a: 1. Laboratory, lab, or bench scale for testing of chemical processes, materials, or methods on a small scale, such as on a workbench; 2. Demonstration or pilot scale for proof of concept with an increased size as a percentage of a total actual production for the application. Usually this is a facility installed on-site at the location where a full-scale facility will be constructed and operated; and, 3. Full scale for commercial operation as the final designed and constructed system, with the outcomes from the laboratory and demonstration scale incorporated to optimize production. See also, *proof of concept*, *pilot plant*, and *prototype*.

Phrase or Term	Definition
Scarcity Mindset	A *scarcity mindset* is the belief that there will never be enough for everyone (e.g., health, intelligence, judgment, knowledge, money, relationships, time, and willpower), resulting in feelings of anxiety, fear, and stress. Scarcity affects one's feeing and thinking by orienting the mind automatically and towards unfulfilled needs, often overlooking other goals in the process. However, scarcity can help prioritize choices and result in more effective decisions. Seven signs of a scarcity mindset are believing: 1. Big things are impossible (e.g., situations in life are fixed or permanent); 2. There are limited resources in the world (e.g., it is impossible to be generous to others and oneself); 3. In pessimism and being resentful for others' success and not appreciating the good things in one's own life; 4. Change is an intimidating part of life that does not lead to positive outcomes; 5. Waiting for things to happen and then reacting; 6. One already *knows everything*, which severely limits the potential for growth and learning; and, 7. Negative beliefs, emotions, and thoughts, and focus is on *what is not working*. This mindset was described by Stephen Covey as seeing life as having only so much. The act of taking means others will have less. See also, *abundance mindset*, *goal*, and *thinking mode*.
Scenario Planning, Scenario Thinking, Scenario Analysis	*Scenario planning*, *scenario thinking*, or *scenario analysis* is a strategic planning method organizations use to make flexible long-term plans. First, several challenges, opportunities, or scenarios are planned. Then forecasts are adjusted to consider how it would be necessary adapt to actual results. For example, how can the supply chain be adjusted to accommodate an expected adverse weather event (e.g., flood or hurricane) or lack of resources (e.g., labour disruption or pandemic). The flexibility of this method makes it popular for farmers and military organizations. See also, *forecast*, *forecast planning*, *future-back process*, and *planning*.
Schedule	A *schedule* is: 1. A list, table, or inventory of items; 2. A non-dimensional number that designates a pipe wall thickness; or, 3. An arrangement or plan for activities or events to take place at a particular time and date. See also, *dimensionless designator* and *nominal pipe size*.
Schedule-driven	A *schedule-driven* business or project operates to deliver by an established, inflexible, final delivery date such that this delivery date becomes the single main overriding consideration when making project management decisions. On this type of project, the schedule is given highest priority over all other project elements, including cost, quality, resources, and risk. Consequently, schedule-drive projects are at higher risk of cost overruns, safety incidents, and other negative outcomes than cost-driven and value-driven projects. See also, *cost-driven* and *value-driven*.

Phrase or Term	Definition
School of Hard Knocks, University of Hard Knocks	The *school of hard knocks* or *university of hard knocks* means that someone obtained an education through hands on experience, on-the-job training, and learning by failures and successes (trial and error), and without any or minimal academic or formal education. This is known as acquiring an education *the hard way* (e.g., hard knocks) and may also be referred to as the *school of life* or *university of life*. See also, *book smart*, *easy way*, *hands on*, *hard way*, *on-the-job training*, and *street smart*.
Science of Risk	The *science of risk* is how to: • Define risk; • Measure risk; • Identify the type of risk; and, • Manage risk. Se also, *risk*.
Scientific Method	The *scientific method* is an empirical method of acquiring knowledge using deductive logic that has characterized the development of science since the 17th century. It involves: 1. Careful observation and application of rigorous skepticism about what is observed, given that cognitive assumptions can distort an interpretation; 2. Formulating hypotheses, by induction, based on such observations; 3. Experimental- and measurement-based testing of deductions made from the hypotheses; and, 4. Elimination or refinement of the hypotheses based on the findings. The purpose of an experiment or measurement is to determine if the observations agree with or conflict with the predictions derived from a hypothesis. See also, *abductive logic*, *deductive*, *design thinking*, *inductive logic*, *junk science*, *logic*, *logic*, and *pseudoscience*.
Scope	*Scope* is what a project will produce (product scope) and the work needed to produce it (project scope). It is the sum of the products, services, and results produced in a project, and is often documented using a scope statement and a work breakdown structure, which are approved by the project owner. See also, *project definition rating index*, *scope creep*, and *work breakdown structure*.

Phrase or Term	Definition
Scope 1, 2, and 3 Emissions	*Scope 1, 2, and 3 emissions* are three categories of carbon emissions identified by the *Greenhouse Gas Protocol*. They comprise: 1. Scope 1 Direct emissions from an organization's owned or controlled sources. For example, those emitted onsite by company equipment. 2. Scope 2 Indirect emissions from purchased energy. For example, fuel used for equipment and electricity that is used onsite. 3. Scope 3 All other indirect emissions in the organization's value chain that are not owned or controlled. These are usually the greatest share of the carbon footprint, such as business travel, procurement, waste and water, and emissions from the use of organization products. See also, *Greenhouse Gas Protocol*, *historical emissions*, and *net-carbon neutral*.
Scope Creep	*Scope creep* refers to unanticipated changes, or continuous or uncontrolled growth in a project's scope after the project begins. It adds features and functions to the project scope without addressing the effects on time, costs, resources, or schedule, and without customer approval. This typically occurs when the project scope is improperly defined, documented, or controlled. It is generally considered harmful – but project change is inevitable; thus, the possibility of scope creep is inevitable. For example, the client requests *gold plating* with additional components or testing that results in cost escalations, schedule delays, or both. See also, *bells and whistles*, *gold plating*, *hidden work*, and *scope*.
Scope of Supply	A *scope of supply* specifies the project materials and equipment, project services to be purchased, or both, and pricing. This information is typically contained in Section 1 of the purchase order, with each item identified on a separate line. Lines may identify tag numbers and offer very detailed descriptions. See also, *purchase order* and *scope of work*.
Scope of Work	A *scope of work* specifies the project materials and equipment, project services to be contracted, or both. See also, *contract* and *scope of supply*.

<table>
<tr><th>Phrase or Term</th><th>Definition</th></tr>
<tr><td>Scope Triangle, Quality Triangle, Project Management Triangle</td><td>The scope triangle, quality triangle, or project management triangle is a model of the three primary constraints of project management:
1. Time
Time is the available time to deliver the project.
2. Budget and costs
Budget (money and resources) available to pay required costs.
3. Quality
Quality represents the fit-to-purpose products that the project must render to be successful. With regards to quality:
a. The quality of work is constrained by the project's budget, deadlines, and scope (i.e., the features);
b. The project manager can make intelligent compromises that accommodate or flex to constraints; and,
c. Changes in one constraint necessitates changes in others to compensate or quality will suffer.
This is also referred to as triple constraint, iron triangle, and project triangle. See also, management, project management diamond, and project management triangle.</td></tr>
<tr><td>Scorecard, Performance Scorecard</td><td>A scorecard or performance scorecard is a graphical representation or statistical record of an entity's progress or performance over time (e.g., business unit, project, personnel, or supplier) as measured against specified goals, key performance indicators, objectives, or a combination. See also, dashboard, key performance indicator and traffic lights.</td></tr>
<tr><td>Scrappy Design Thinking</td><td>Scrappy design thinking uses the questions: Why? Who? What? How? ... to solve a problem. Like Scrappy Project Management, Scrappy design thinking focuses on the customer in terms of desired outcomes rather than being hampered by self-limiting beliefs about what is possible or stuck in a routine. It seeks to find a solution by answering these questions:
1. Why?
Why is this problem important? Understand the problem before you attempt to solve it.
2. Who?
o Who are the stakeholders (i.e., which stakeholders are impacted or involved)?
o Who will judge the success of the outcomes?
3. What?
What is the goal? What outcomes would exceed stakeholder expectations?
4. How?
How outcomes can be achieved (e.g., with a design or prototype) is determined only after establishing why, who, and what. For projects, this is typically a project schedule.
This above can be repeated, as necessary, to refine outcomes. See also, design thinking, Scrappy Project Management, and scrappy thinking.</td></tr>
</table>

Phrase or Term	Definition
Scrappy Project Management	*Scrappy Project Management* balances leadership with process (i.e., checklists and clear goals) to achieve, not only project success and what seems impossible, but what is in fact merely difficult. For example, pilots, even those with thousands of hours of flight time, use checklists with every flight to ensure that nothing is completed out-of-sequence or missed. This concept was developed by Kimberly Wiefling in the book *Scrappy Project Management: The 12 Predictable and Avoidable Pitfalls Every Project Faces*. See also, *scrappy design thinking* and *scrappy thinking*.
Scrappy Thinking	*Scrappy thinking* is a mindset that makes the most available resources, even if there are very few. Being *too busy* (e.g., inadequate time to be creative or procrastination) and *excuses* (e.g., finding reasons to give up) are the nemesis of scrappy thinking. A scrappy thinker uses the strategies of an underdog to stand out and succeed by: • Finding opportunities where others find barriers; and, • Using ideas to think differently instead of spending money. The three key elements of scrappy thinking are to: 1. Reframe a negative by turning it into a positive; 2. Take small risks; and, 3. Use constraints to be more creative. Scrappy thinking is outthinking the competition, especially when the competition cannot be outspent. For example, developing an effective and powerful viral campaign on social media (for a few hundred dollars) instead of costly, conventional television advertisements. See also, *contingency plan*, *scrappy design thinking*, *Scrappy Project Management*, *scrum*, and *underdog*.
Scrubber	A *scrubber* is equipment (e.g., a tank or pressure vessel) that uses a solution or water for purifying gases or vapours. See also, *tank* and *pressure vessel*.
Scrum	*Scrum* is a methodology in the information technology industry that allows a team to self-organize and make changes quickly, per agile principles. See also, *agile*, *scrappy*, and *scrum master*.
Scrum Master	The *scrum master* manages the process for how information is exchanged within the scrum team. They have the role of coaching, facilitating, mentoring, and teaching. See also, *scrum*.
S-curve	An S-*curve* is a mathematical graph that depicts relevant cumulative data for a project (e.g., cost or labour hours) plotted against time. This graph typically forms a curve shaped like a loose, shallow *S*; hence, its name.
Seal Weld	A *seal weld* is any weld designed primarily to provide a specific degree of leak-tightness by sealing joints. The purpose of a seal weld is to contain a gaseous or liquid fluid. See also, *leak-tight* and *strength weld*.

Phrase or Term	Definition
Sea-lawyer, Sea Lawyer	A *sea-lawyer* or *sea lawyer* is: • A captious or querulous sailor who tends to criticize orders and regulations rather than obey them, or always argues; • A sailor who is (considered by I or others) a legal expert when at sea, but only because few can dispute their expertise (e.g., only those on the boat can dispute the sailor's expertise). However, when on land, this sailor is no longer an expert because their expertise can be challenged or questioned by those more educated, experienced, and knowledgeable; or, • Someone who knows more about regulations and rules than they do about the technical aspects of a job. Sea-lawyer is generally a derogatory term that is synonymous with barracks lawyer, breakroom lawyer, Internet lawyer, and jailhouse lawyer. See also, *sea-lawyering*.
Sea-lawyering	*Sea-lawyering* is using a legal technicality to assign blame to others or evade responsibility. For example, using parsing of words, semantics, wordplay, and similar tricks to mislead. See also, *sea-lawyer*, *technicality*, *weasel out*, and *weasel words*.
Seam	A *seam* is a welded joint. See also, *joint* and *seamless*.
Seamless	*Seamless* fittings and pipe are manufactured by forging or forming (not by welding). Larger diameter fittings and pipe manufactured from plate are welded and not seamless. See also, *forging* and *wrought*.
Search Engine, Internet Search Engine	A *search engine* or *Internet search engine* is a computer program (i.e., algorithm) that searches for and identifies items in a database that correspond to characters or keywords specified by the user to find information and websites on the Internet. See also, *algorithm* and *search engine*.
Search Engine Optimization	*Search engine optimization* is the process of increasing the quality and quantity of website traffic by increasing the visibility of a website or a web page to users of a web search engine. It refers to the improvement of unpaid results (i.e., natural or organic results) and excludes direct traffic/visitors and the purchase of paid placement. Search engine optimization may target different kinds of searches (e.g., academic, industry-specific vertical, image, news, or video search engines). As an Internet marketing strategy, search engine optimization considers how search engines work. Optimizing a website may involve editing its content, adding content, and modifying hypertext markup language and associated coding to both increase its relevance to specific keywords and remove barriers to the indexing activities of search engines. See also, *algorithm* and *search engine*.
Seasonal Affective Disorder, Seasonal Depression	*Seasonal affective disorder* or *seasonal depression* is a mood disorder characterized primarily by depression during winter seasons. Common symptoms include having low energy, overeating, and sleeping too much. This condition is typically associated with winter depression, but springtime lethargy or other seasonal mood patterns are also common. Seasonal mood variations are believed to be related to light (or absence thereof). Seasonal depression is common in Nordic and northern countries.

Phrase or Term	Definition
Second Guess	To *second guess* is to: • Anticipate or predict (i.e., by guesswork); • Be indecisive; • Criticize or judge someone with hindsight (i.e., after the results of the actions or decisions are known); or, • Make a decision, and then later question the decision (e.g., buyer's remorse). See also, *armchair critic, backseat driver, buyer's remorse*, and *hindsight is 20/20*.
Secondary Recovery	*Secondary recovery* is a process by which usable petroleum is drawn out from beneath the earth's surface. During the lifetime of a well, the well's pressure falls such that there is insufficient underground pressure to force the oil to the surface. External energy is therefore applied to the reservoir by injecting fluids to increase reservoir pressure; hence, increasing or replacing the natural reservoir drive with an artificial drive (e.g., gas lift, gas reinjection and water injection). See also, *primary recovery* and *enhanced oil recovery*.
Secondary Steel, Off Spec Steel	*Secondary steel* or *off spec steel* is steel that does not meet original customer or design specifications because of defective chemical or mechanical properties, improper dimensions, or inadequate quality. The manufacturer then downgrades the material and sells it to a customer who will accept the defects, typically for a discounted price. See also, *off specification*.
Second-hand Stress	*Second-hand stress* is the affect that one's attitudes, behaviours, and emotions, which are highly contagious, can have on others. For example, most teams adopt the personality of their leader. An anxious leader's team will be reluctant to change its beliefs. An intellectually curious leader's team will be flexible and less risk adverse. A trusting and warm leader's team will be cohesive. In his blog Humor at Work, Michael Kerr stated that *yes, those annoying, fun-sucking fun-suckers really can infect an entire team with their fun-sucking attitudes*! This is the same concept for how second-hand tobacco smoke can cause lung cancer in a non-smoker. See also, *hand* and *second-hand*.
Second-hand, Second-hand News	*Second-hand* is: • An experience or information that is accepted on another's authority or as hearsay and not from personal experience or original investigation. For example, I heard that *my brother caught the biggest fish in the tournament, second-hand*; • Something that has had a previous owner or is not brand-new. For example, a second-hand vehicle; or, • Something that is a derivative. For example, *that is not a new idea; it is second-hand*. *Second-hand news* is information that is not obtained directly from the original source (e.g., when the New York Times newspaper reprints a story from the Associated Press). See also, *brand-new, first-hand, hand, second-hand stress*, and *third-hand*.

Phrase or Term	Definition
Second-order Effect	Every action has a consequence, and each consequence has another consequence. This is the *second-order effect*. A change made to a process or system may have second-order, and even third-order, effects that may affect the functionality of a process or system. These effects may be undesirable or unintended. For example, using Google maps and almost never getting lost, but then not gaining understanding of the area. The second-order effect is lacking the knowledge to navigate in the same environment if the phone battery dies or if access to the phone is removed. See also, *1-way and 2-way door decision-making*, *Cobra effect*, *lateral effect*, *management of change*, *superforecasting*, *spillover effect*, and *unanticipated consequences*.
Secured Document	A *secured document* is protected by security features, such as a password to open the document, or does not allow copying, editing or printing document contents.
Security as a Service	*Security as a service* is cloud computing service that integrates security services into a corporate infrastructure over the Internet on a subscription basis. This can be more cost effective than for corporations or individuals to provide their own services (when the total cost of ownership is considered). These services often include authentication, anti-malware, anti-spyware, anti-virus, intrusion detection, and others. See also, *anything as a service*, *cloud computing*, *content as a service*, *infrastructure as a service*, *platform as a service*, *software as a service*, and *total cost of ownership*.
Selection Bias, Selection Effect	*Selection bias* or *selection effect* is the selection of data, individuals, or groups for analysis that does not guarantee proper randomization, which is required. This taints the sample group because it is not representative of the population intended to be analyzed. This tainting distorts the results of the statistical analysis. If selection bias is not considered, then an unknown number of conclusions may be inaccurate and unreliable. See also, *bias*, *confirmation bias, and recency illusion*.
Selective Attention	*Selective attention* involves focusing on a specific aspect of a scenario while simultaneously ignoring other aspects. It allows one to relegate unimportant details to the background and focus on what matters. It is also the tendency not to notice, and more quickly forget, stimuli that cause emotional discomfort, contradict previous beliefs, is irrelevant, or a combination. For example, listening carefully to one conversation while simultaneously ignoring other conversations. See also, *recency illusion* and *selective retention*.
Selective Retention	*Selective retention* is the process whereby personnel more accurately remember messages that are closer to their beliefs, interests, and values, than messages that contrast with the same. Selective retention narrows mental information flow; memory retention becomes more selective. See also, *selective attention*.
Self-enforcing Agreement	A *self-enforcing agreement* is an agreement that is enforced only by the parties to it; no external party can or needs to enforce or interfere with the agreement (e.g., a lawsuit in a court of law). It may be an informal or formal agreement (i.e., a contract) that will stand as long as the parties believe it is mutually beneficial and it is not breached by any party. See also, *contract*.

Phrase or Term	Definition
Self-framing Building, Rigid Frame Building	A *self-framing building* or *rigid frame building* is a pre-engineered building that uses roll formed steel (i.e., sheet metal) roof and wall panels as significant parts of the structural system. These buildings are supplied with all necessary components to construct a complete building system (e.g., with doors and windows). Self-framed building components may be shipped for erection at site, or smaller buildings may be fully assembled at a facility and then transported to site.
Self-help	*Self-help* is self-guided improvement to better oneself or overcome emotional or personal challenges without professional support. Many resources are available to support self-help, including books, Internet sources, and support groups. Self-help topics are numerous and include business, education, psychology, and psychotherapy. See also, *life hack*, *self-taught*, and *support group*.
Self-motivated, Self-starter	*Self-motivated* or *self starter* describes someone who is motivated to achieve or do something because of their own desire, enthusiasm, or interest – and without needing influence or support from others. Personnel with self motivation can find a reason and strength to complete something, even when challenged. This may be a simple task or something larger, such as starting a business, career, or pursuing a university education.
Self-serving Bias	*Self-serving bias* is how we collect information that proves us *correct* and allows us to ignore information that proves us *self incorrect*. It attributes positive events to us and negative events to external factors. Not everyone can perform above average — but we can harbour this illusion. However, by doing so, we overlook crucial facts. For example, some ignore the intelligence of others to make themselves feel smarter than is actually the case. Thus, self-serving bias allows us to rationalize our own mistakes. See also, *bias*, and *optimism bias*.
Self-sustaining, Self-sufficient	To be *self-sustaining* or *self-sufficient* is to continue in a healthy state (i.e., to maintain oneself or itself) by independent effort (i.e., with minimal or no outside assistance). Sustainable living is where what is consumed by an individual is also produced by the individual (i.e., by living off-the-grid or subsistence agriculture). For example, food is raised and prepared by the same individual who consumes it. See also, *sustainability*.
Self-taught	To be *self-taught* is to acquire knowledge and skills by teaching oneself, instead of by formal instruction or training. For example, a pianist who learns to play the piano without learning how to read music or receiving lessons from an instructor. Master guitarists Eric Clapton, Jimi Hendrix, and Prince all claim to be self-taught. See also, *academic*, *life hack,* and *self-help*.
Seller	Not a preferred term – see *supplier*.
Seller Drawing and Data Requirements	Not a preferred phrase – see *supplier data requirements*.

Phrase or Term	Definition
Senior Management, Executive Management, Top Management, Upper Management	*Senior management*, *executive management*, *top management*, or *upper management* is generally the management team comprised of individuals at the highest professional level of an organization (e.g., executives). They are responsible for the day-to-day management of an organization (e.g., a business unit or company) and may include a: • Chief executive officer; • Chief financial officer; • Chief information officer; • Chief operating officer; and, • Chief risk officer (or chief risk management officer or chief risk management officer). Other corporate roles may also be designated using chief in the title, a convention that is becoming more common (e.g., chief business officer, chief commercial officer, chief marketing officer, and chief people officer). See also, *C-suite*, *line management*, *management*, and *middle management*.
Sense of Direction	A *sense of direction* is the ability to: • Know, without explicit guidance: o Where something is located (e.g., the island is on the lake); or, o Where you are physically located (e.g., 200 m south of the lake); o Which direction you should be moving (e.g., to reach the lake, we need to canoe north up the river for about 200 m). • Understand how to achieve one's goals and what matters in the big picture or long term. For example, before I summit the mountain I must acquire mountaineering equipment, train with knowledgeable mountaineers, and reach an adequate level of physical fitness. This concept was developed in the article *Two superpowers for 2021: psychological flexibility and sense of direction* by Dr Justyna Kucharska. See also, *big picture*, *goal*, and *psychological flexibility*.
Sequence	A *sequence* is: • A consequence, result, or subsequent development; • A continuity of progression; • A continuous or connected activity or series; or, • An order of executing or succession.
Serial Number	A *serial number* is a unique number in a series of numbers used for identification (e.g., on a document, manufactured item or valve). For a manufactured item, the serial number is typically included on the *name plate*. See also, *name plate*, *part number*, and *tag number*.
Service Line	A *service line* carries product from the mainline to the customer. Se also, *gathering line* and *mainline*.
Service Provider	A *service provider* is an individual or organization that provides advice or a service to a client including, but not limited to, agents, architects, brokers constructors, engineers, managers, planners, surveyors, and technologists. See also, *client*.
Set Up	*Set up* means that an item or package is assembled in its complete state. See also, *knocked down*.

Phrase or Term	Definition
Seven Principles for Effective Verbal Intervention	The *seven principles for effective verbal intervention* provides the most effective and safest method to intervene with an individual who is out-of-control (e.g., belligerent, hostile, or noncompliant). An effective response to this defensive behaviour is often key to avoiding a physical confrontation. The seven principles are to: 1. Remain calm; 2. Isolate the individual; 3. Heed your (and their) body language; 4. Keep it simple; 5. Use reflective questioning; 6. Use silence; and, 7. Heed your (and their) paraverbal communication. While there is no guarantee that an out-of-control person will not act out physically, these principles help reduce the anxiety and defensiveness that often precede dangerous behaviour. This concept was developed by the Crisis Prevention Institute in the resource *Seven Principles for Effective Verbal Intervention*. See also, *body language*, *defensive behaviour*, *reflective questioning*, and *paraverbal communication*.
Seven Wastes, Seven Wastes of Lean, 7 Wastes	The *seven wastes*, *seven wastes of lean*, or *7 wastes* are at the root of all unprofitable activity within and organization. The wastes are: 1. Defects; 2. Overproduction; 3. Transportation; 4. Waiting; 5. Inventory; 6. Motion; and, 7. Processing. The worst of all the 7 wastes is overproduction because it includes all other types of waste.
Sexism	*Sexism* is antagonism, discrimination, prejudice, or stereotyping based on a person's gender or sex (e.g., the belief that one gender or sex is intrinsically inferior or superior to another). See also, *ageism*, *Black Indigenous and People of Colour*, *discrimination*, *intersectionality*, *racism*, and *sexist*.
Sexist	*Sexist* is characterized by or showing antagonism, discrimination, prejudice, or stereotyping based on a person's gender or sex. See also, *sexism*.
Shadow Ban, Stealth Banning, Ghost Banning, Comment Ghosting	*Shadow banning*, *stealth banning*, *ghost banning*, or *comment ghosting* is the act of blocking or partially blocking a user or their content from an online community so that it is not readily apparent to the user that they have been banned. See also, *blacklist* and *deplatforming*, *doxing*, and *cancel culture*.

Phrase or Term	Definition
Shadow Board	A *shadow board* is a: • Device for organizing tools (e.g., workstation wall with hooks to hang tools) that defines where each tool should be placed when it is not in use. It has outlines or shadows of the tools marked on it so personnel can quickly identify which tools are in use or missing. These are used in the workplace to improve a facility's lean six sigma capabilities (e.g., reduce loss and time spent looking for tools); or, • Group of young professionals directed by an actual company board to perform business challenges and tasks. The shadow board, also known as a mirror board, reports its results to the actual board. The purpose of a shadow board is to educate and provide experience to the next generation of company management personnel. See also, *board*, *cross-generational mutual mentoring*, *reverse mentoring*, and *six sigma*.
Shall	In this document, the word *shall* is used where a provision is mandatory. See also, *will* and *must*.
SharePoint	*SharePoint* is Microsoft software that was launched in 2001. It is a web-based collaborative platform that integrates with Microsoft Office suite of software. It is a highly-configurable electronic network document management and storage system. Corporate Internet and intranet websites are often built using Microsoft SharePoint software. See also, *wiki*.
Sharp End, Pointy End, Business End	*Sharp end*, *pointy end*, and *business end* are casual expressions that refer to the part of an object that carries out a procedure or does something important (e.g., the tip of a spear). This expression infers involvement of high levels of competition, danger, or difficulty. See also, *smart end*.
Sheet Metal	*Sheet metal* is any metal that is formed into flat pieces, including steel. The thickness or gauge of sheet metal is based on its weight in pounds per square foot, per inch of thickness. The larger the gauge number, the thinner the metal. For example, gauge 10 steel is 3.43 mm (0.135") thick and weighs 5.625 psf. Sheet metal thinner than 0.2 mm (13/640 in.) is called foil. Sheet metal 6 mm (15/64") or thicker is termed as plate metal or structural steel. See also, *plate metal* and *structural steel*.
Shell	A *shell* is the cylindrical structure of made of rolled plate(s) or pipe that is used to produce a pressure vessel or tank. A pressure vessel shell may vary in thickness. See also, *head*, *pressure vessel*, *tank*, *thick wall pressure vessel*, and *thin wall pressure vessel*.
Shell and Tube Heat Exchanger	A *shell and tube heat exchanger* consists of a shell with a tube bundle inside it. One fluid runs through the tubes, and another fluid flows over the tubes (through the shell) to facilitate heat transfer between the two fluids. See also, *heat exchanger*, *tube bundle*, and *shell*.

Phrase or Term	Definition
Shipping and Handling, Shipping Storage and Handling	*Shipping and handling* or *shipping storage and handling* are the methods used to care for and transport products from the original equipment manufacturer to the final destination. This phrase may also refer to the costs for delivery, labour, and storage that are incurred to supply an order. Information provided may include caution, safety, or warning messages or instructions about how to: • Create a cargo unit for shipping (e.g., bundle, nest, or strap); • Lift or move items (e.g., lift only from the bottom); • Load and unload, including repeated handling; • Package and unpackage; • Prevent contamination, damage, or deterioration (e.g., protect against abrasion, breakage, corrosion, moisture, radiation, temperature variation, shock, vermin, or vibration); • Provide additional covering (e.g., a tarp) for transport on an open deck truck or trailer; • Secure (inside or outside) and during transit (e.g., in a ship, train, or vehicle); • Stack items (e.g., do not stack, and upright or vertical orientation only); or, • Use temporary bracing (e.g., painted orange to identify removal before use). The methods used for shipping and handling may be defined by the original equipment manufacturer, project specifications, and other sources, or a combination. This may include requirements for marking and labelling, packaging, and preservation. These may be provided as separate documents. See also, *final destination*, *marking and labelling*, *packaging*, and *preservation*.

Shipping Documents, Documents with Shipment	*Shipping documents* or *documents with shipment* are the completed forms that accompany a shipment. Typically, shipping documents include these details:
	• Customer name;
	• Date shipped;
	• Method of shipment; and,
	• Quantities and specifications of goods shipped.
	Shipping documents may include:
	• A bill of lading;
	• A packing slip;
	• A waybill;
	• An invoice;
	• Insurance documents; and,
	• Supplier documents requested with the shipment in the purchase order (i.e., per supplier document requirements). This is also known as the mud book and may contain:
	  o Assembly or erection drawings;
	  o Operating instructions; and,
	  o Quality documents (e.g., manufacturer's data report or material test report).
	See also, *goods*, *invoice*, *mud book*, and *packing slip*.

Phrase or Term	Definition
Shipping Plan	A *shipping plan* synchronizes the delivery of assemblies, materials, and subassemblies to a construction or other site. It avoids sending materials before they are needed and ensures there is enough space to store them. The site labour resource plan must be synchronized with the shipping plan to ensure maximum efficiency and meet project milestone dates.
Shipping Release	A *shipping release* authorizes shipment and provides detailed shipping instructions. It is issued by an expediter to a supplier or others responsible for shipment. A shipping release is always issued after an inspection release. A release may be for the complete or partial supply of a purchase order. See also, *inspection release*.
Shop	Not a preferred term – see *facility*.
Shop Equipment	*Shop equipment* is equipment that is owned and used by the supplier for in-house testing. It is not purchased for the project for installation and operation. Shop equipment may be calibrated for testing purposes.
Shop Traveler	Not a preferred phrase – see *quality control check list* and *quality control traveler*.
Shop-fabricated Tank	A *shop-fabricated tank* is a tank that is built in a shop. Tank sizes range between 13,638 L (3,000 gal.) to 227,305 L (50,000 gal.). These tanks are built and assembled at the supplier's facility and then transported to site where they are installed. Size limitations are based on the size of equipment and facility capacity, and size and weight restrictions applicable to transportation. See also, *field-erected tank* and *tank*.

Shopping Cart Theory	The *shopping cart theory* claims that what one does with a shopping cart in the grocery store parking lot reveals their true character. The origin of this theory is unknown. It attaches a moral component to the Platonic notion of self-governance, which is a community or person's ability to regulate their own behaviour so that they can co-exist within a functioning society. Philosopher John Locke suggested that *the ability to self-govern is the source of true freedom*, but this can also reveal if one is a bad or good citizen and person. For example, with a shopping cart, do you:
	• Leave it:
	  o Where you parked your vehicle and not return it to its proper place (e.g., the cart corral)?
	  o With another misplaced cart so you don't have to return it to the its proper place but still receive your coin back (many stores require a coin to be deposited to take a cart, which is returned when the user connects it to another cart or returns it to its proper place)?
	• Return it:
	  o To its proper place to protect parked cars from damage and save grocery store personnel the time it takes to return it themselves?
	  o From a willingness to do what is right, even if no one is forcing you and it is only to get your coin back?
	• Was there a dire emergency that prevented you from returning the cart?
	One's disposition to break or follow other social norms similarly reveals one's true character. For example, littering vs. picking up garbage (including another's garbage), or keeping a wallet with money in it vs. returning it to its owner.

Phrase or Term	Definition
Short Change	To *short change* means to: • Cheat someone by giving them insufficient money as change; or, • Treat someone unfairly by withholding something of value.
Short List, Shortlist	*Short list* or *shortlist* is a list of selected candidates, items, or options for final consideration (e.g., award, job, position, or prize).
Shortcut	A *shortcut* is a method of performing something more directly and quickly than by using a conventional method. A shortcut is typically less thorough than a conventional method and may not conform to the associated procedure. See also, *one-variable thinking*, *risk normalization*, and *workaround*.
Shorthand	*Shorthand* is a: • Method of rapid writing using abbreviations and symbols, especially for taking dictation; in legal settings; or, • Short and simple method of expressing or referring to something.
Should	In this document, the word *should* is used where a provision is preferred.
Shovel-ready Project	A *shovel-ready project* is a project where engineering, permitting, and planning are advanced enough that – with sufficient funding – construction can begin immediately or shortly. For example, to build a new asset such as a highway interchange. See also, *shovel-worthy project*.
Shovel-worthy Project	A *shovel-worthy project* is a project where limited resources will be used strategically for construction that is prioritized to have greatest community impact in the short term. For example, to build flood protection (e.g., a dam or diversion canal), repair an aging asset (e.g., a bridge), or upgrade energy-efficiency (e.g., improve power output at a power plant by using the same amount of energy). See also, *shovel-ready project*.
Shutdown	A *shutdown* is a planned interruption or outage of a section of a facility, or an entire facility, usually initiated to perform major maintenance and repair operations, and to inspect, test, and replace materials and equipment. See also, *idling*, *shuttering*, and *mothballing*.
Shuttering	*Shuttering* is the permanent closure of a facility. Typically, the only way for such a facility to be returned to operation is through the sale of the facility to a new owner who puts it back in service. A shuttered facility may or may not be mothballed. See also, *idling*, *mothball*, *mothballing*, and *shutdown*.
Side Hustle	A *side hustle* is any type of employment or a second job undertaken in addition to one's full-time job. A side hustle is generally freelance or piecework in nature, providing a supplemental income. Side hustles are often things a person is passionate about, rather day a typical day job worked to make ends meet. See also, *consultant*, *entrepreneur*, *full-time equivalent*, *gig*, *independent contractor*, and *solopreneur*.

Phrase or Term	Definition
Signal-to-Noise Ratio	*Signal-to-Noise ratio* is a measure, used in engineering and science, that compares the level of a desired signal to the level of background noise. It is defined as the ratio of signal power to the noise power, often expressed in decibels. A ratio higher than 1:1 (greater than 0 dB) indicates more signal than noise. While it is commonly quoted for electrical signals, it can be applied to any form of signal such as financial trading signals. Signal-to-noise ratio is sometimes used metaphorically to refer to the ratio of useful information to false or irrelevant data in a conversation or exchange. For example, in online discussion forums and other online communities, off-topic posts and spam are regarded as *noise* that interferes with the *signal* (theme) of appropriate discussion.
Sign-off	*Sign-off* means to acknowledge or approve something by, or as if by, a signature on a document. See also, *final sign-off*.
Silent Majority	The *silent majority* is the largest part of a population comprising people who are not actively involved in politics and do not express their political opinions publicly, but together have great power. This phrase was popularized by United States President Richard Nixon, who claimed that Vietnam War protesters comprised a minority, while a silent majority supported it. The phrase was used in the 19th century as a euphemism referring to all the people who have died. Today, this phrase is used around the world.
Silo	A *silo* is: • A tall metal bin or pit used on a farm to store animal feed (e.g., a grain silo); • An independent category, department, or group that deliberately functions apart from others, especially in a way that limits-hinders communication, cooperation, and sharing of information (e.g., accounting and business development departments that refuse to share key financial information); or, • An underground structure used to store and launch a missile (e.g., a missile silo). See also, *no silo rule*, *silo mentality*, and *siloed work*.

Phrase or Term	Definition
Silo Mentality, Stovepipe Mentality	A *silo mentality* or *stovepipe mentality* is a mindset when personnel or teams are inward-looking and do not easily share information with others, even those who work within the same organization. This thinking restricts the flow of information within an organization between management and personnel, and inhibits or prevents cross-organizational communication. A silo structure can benefit an organization by keeping personnel focused on their areas of expertise, providing direction and focus, and preventing distraction or involvement in other unrelated areas. It creates a sense of belonging and pride that a large organization, due to its sheer size, might not be able to offer, and provides for the effective division and organization of the labour force. A silo structure can hurt an organization by causing disconnected and disjointed decision-making that is based on local context or personnel requirements focused on internal deliverables or outputs instead of external deliverables and customer needs. It does not consider information outside of the silo and is incapable of dealing with frequent cooperation across silo boundaries and non-standard problems or providing holistic resolutions. For example, *empire building* and *turf war* are expressions that describe the phenomenon of departmental-focused organizations that fight for budget dollars, or control over direction or resources, with the intent to win regardless of the overall effect on the organization. Consequently, information shared between silos becomes scarce when it becomes a source of power. See also, *labour force*, *no silo rule*, *silo*, and *siloed work*.
Siloed Work	*Siloed work* is work completed by departments or groups within an organization who deliberately do not share information, knowledge, or resources for competitive, contractual, security, or other reasons. Siloed work encourages personnel who are on the same team to compete with, or work against, one another. See also, *no silo rule*, *silo*, and *silo mentality*.
Silver Bullet	In folklore, a *silver bullet* is one of few weapons that are effective against a werewolf, witch, vampire, or other monster. In military slang, a *silver bullet* is a thermometer, especially one used for measuring rectal temperatures. In most contexts, a silver bullet is a solution that provides an effective, instant, and simple resolution to a complicated or long-standing problem. The phrase is almost always used to state that such a solution does not exist. For example, *the project is understaffed but there is no silver bullet to recruit all the personnel that are needed*. See also, *panacea*.
Simple Majority	A *simple majority* is a voting requirement to obtain a higher number of votes cast for any one candidate, issue, or item compared to the second-highest number. This may not constitute an absolute majority and may refer to a requirement to obtain: • More than half of all ballots (e.g., a majority such as 50% + 1); or, • The most ballots (i.e., more than any other option with more than two options or plurality). See also, *first-past-the-post*, *absolute majority*, and *supermajority*.

Phrase or Term	Definition
Single Point of Contact	A *single point of contact* is a person or a department serving as the coordinator or focal point of information concerning an activity, program, or project. A single point of contact is used when accuracy is important and information is time-sensitive. For a supplier, the buyer is the project's single point of contact before purchase order award and the expediter is the project's single point of contact after purchase order award. See also, *buyer*, *expediter*, and *supplier*.
Single Point of Failure	A *Single point of failure* is a component whose failure would cause the collapse of an entire system. Example 1: if an application is hosted on one server, the entire application collapses (shuts down) if the server is compromised. In this example, the single point of failure is the server. Example 2: if the Sun goes supernova, the collapse of the Earth's solar system is guaranteed. In this example, the single point of failure is the Sun. See also, *business continuity risk*.
Single Random	Not a preferred phrase – see *random length*.
Single Source	*Single source* is a procurement method without competitive bidding where only one supplier is chosen to provide needed products or services. This typically occurs after a thorough analysis of possible suppliers. The selection of suppliers is based on the: • Preference for brand or supplier, typically based on past experience with a supplier; • Price, especially if no other supplier can beat or match it; or, • Quality of products, especially if products and services offered are superior to others. See also, *bid*, *competitive bid*, and *sole source*.
Sink or Swim	To *sink or swim* means to fail (sink) or succeed (swim) entirely by their own efforts. For example, newly hired personnel are offered a short orientation before being assigned to work independently with their clients. After that, personnel sink or swim.
Sinn Paradox, Green Paradox	The *Sinn paradox* or *green paradox* is the concept that, as environmental policies become greener or stricter over time – this induces owners of fossil fuel resources to accelerate extraction. This enables their profits to be converted into investments that offer higher yields from decreased profits in fossil fuels, which also contributes to accelerated global warming. *The Green Paradox* is a book authored by German economist Hans-Werner Sinn. See also, *fossil fuel* and *global warming*.
Sit Down	*Sit down* is an informal expression that means *to meet with others*, especially to discuss and resolve challenges or conflicts. For example, let's have a sit down and discuss details of the contract. See also, *stand down*, *stand up* and *where you stand determines where you sit*.

<table>
<tr><th>Phrase or Term</th><th>Definition</th></tr>
<tr><td>Sit on the Fence</td><td>To sit on the fence is to:
• Avoid or delay choosing or deciding, or defer siding with one of two possibilities (e.g., delaying purchasing a new vehicle because of indecision);
• Be undecided about something (e.g., the client is still sitting on the fence regarding approval of the proposed project);
• Not commit or take a side in a dispute (e.g., a judge must sit on a fence while the trial is underway; after which, he can get off the fence and make a judgement based on the evidence); or,
• Remain neutral when a decision or choice is required (e.g., after last night's shutout, few fans were on the fence about their hockey team's Stanley Cup prospects);
Common reasons for sitting on the fence:
• A desire to remain neutral or on good terms with opposing sides;
• A lack of courage or indecisiveness;
• An apathic or uncommitted attitude about the situation; or,
• Unwillingness to acquiesce with a decision with which one disagrees.
See also, walk the line.</td></tr>
<tr><td>Site</td><td>Site is the location of the owner's facility that is being constructed. See also, project.</td></tr>
<tr><td>Site Acceptance Test</td><td>A site acceptance test analyzes the functionalities of interfaces at site and can only be performed at site. This test is typically performed immediately after a system is permanently installed at site.</td></tr>
<tr><td>Situation Objective Action Results Aftermath Method, SOARA Method</td><td>The situation objective action results aftermath method or SOARA method is a job interview technique developed by Hagymas Laszlo and Dr. Alexander Botos. It uses the following structure to progressively obtain information relating to a recent challenge:
• Situation – what was a recent challenge and situation did you found yourself in?
• Objective – what did you need to achieve?
• Action – what did you do, why did you do it, and what were the alternatives?
• Results – what was the outcome of your actions and did you meet your objectives?
• Aftermath – what did you learn from this experience and have you used this learning again?
See also, situation task action result method and task action result formula.</td></tr>
</table>

Phrase or Term	Definition
Situation Task Action Result Method, STAR Method, Situation Task Action Result Reflection Method, STARR Method	The *situation task action result method* or *STAR method* is a job interview technique used to gather information about a specific capability that the job requires. It uses the following structure to progressively obtain information relating to a recent challenge: • Situation Describe a recent challenging professional situation. • Task What did you need to achieve? Some methods use *target* rather than *task*. This is because people who describe a target they set themselves instead of an externally imposed task demonstrate their motivation to grow and perform. • Action What did you do? Why did you do it? What were the alternatives? • Results What was the outcome of your actions? Did you meet your objectives? What did you learn from this experience? Have you used this learning again? This is also known as the *situation task action result reflection method* or *STARR method*. The interviewee can then surmise what the candidate would do in a future situation that is better, different, or the same. This aims to gather insight into the interviewee's ability to iterate and learn. See also, *situation objective action results aftermath method* and *task action result formula*.
Situational Awareness, Situation Awareness	*Situational awareness* or *situation awareness* is the perception of environmental elements and events with respect to space or time, the comprehension of their meaning, and the projection of their future status. This awareness is a critical, yet elusive, foundation for successful decision-making across a broad range of situations, many of which involve the protection of assets and human life (e.g., air traffic control, aviation, ship navigation, emergency response, health care, and operations that involve military, nuclear power plants, and offshore oil rig operations). Assessing dangerous situations, being aware of one's surroundings, and identifying potential threats is more of a mindset than a hard skill; it can be exercised to a certain degree by anyone with adequate discipline and will. Inadequate situational awareness is a primary driver for human error.
Six Sigma	*Six Sigma* is a set of techniques and tools used to improve business processes. Each Six Sigma strategy follows a defined sequence of steps and has specific value targets (e.g., reduce costs, process cycle time, or pollution, or increase customer satisfaction or profits). These strategies seek to improve the quality of the output by identifying and removing the causes of defects and minimizing variability. The quality-management tools are also used outside Six Sigma. See also, *doing the right things right the first and every time*, *quality*, *Six Sigma quality*, *Six Sigma tools*, and *total quality management*.
Six Sigma Quality	*Six Sigma quality* generally describes a well-controlled process. See also, *Six Sigma*.
Six Sigma Tools	*Six Sigma tools* are the problem-solving tools used to support Six Sigma and other process improvement efforts. For example, business process mapping, the five whys, and the seven wastes. See also, *business process mapping*, *five whys*, *seven wastes*, and *Six Sigma*.

Phrase or Term	Definition
Skid	A *skid* is a portable steel frame with mounted equipment that facilitates handling and transportation with cranes, flatbed trucks, or trailers. A baseplate or metal building may be mounted to a skid. See also, *baseplate*.
Skill set	A *skill set* is a: • Category of abilities, experience, and knowledge necessary to perform a job or task; and, • Person's range of abilities or skills that can be applied to a creative or professional endeavour. Categories of a skill set includes: • Hard skills, which are quantifiable and teachable. They include specialized abilities and knowledge; • Hybrid skills, which include a combination of non-technical and technical skills; • Job-specific skills, which are necessary for employment; • Soft skills, which are emotional intelligence (i.e., human relations, interpersonal, and people skills); and, • Transferable skills, which are skills that can be used in different career fields.
Skilled Trade	A *skilled trade* is a skilled labour occupation that requires work experience, on-the-job training, and often an apprenticeship or formal vocational education, but typically not a bachelor's degree. For example, a boilermaker, electrician, insulation installer, iron worker, machinist, millwright, non-destructive testing technician, parts person, pipefitter, painter, sheet metal worker and welder. See also, *apprenticeship*, *trades person*, and *unskilled labour*.
Skirt	A *skirt* is a support structure for a vertical pressure vessel. See also, *saddle* and *shell*.
Slippage	*Slippage* is the negative variance between planned and actual activity completion dates. It may also refer to the general tendency of a project to be delayed beyond planned completion dates. See also, *negative variance*.
Slippery Slope	A *slippery scope* is a logical fallacy wherein a conclusion is based on the premise that if A and B happen, then eventually – through progressive small steps – X, Y, and Z will also occur. Thereby, equating A and Z. Hence, if Z is undesirable, then A must be prohibited. For example, banning off-road vehicles from environmentally-sensitive areas (A) means off-road vehicle use is environmentally unsound (B), so off-road vehicles use will be banned in all areas (Z). See also, *bait and switch*, *boiling frog syndrome*, *circular reasoning*, *either/or*, *false equivalence*, *logical*, *logical fallacy*, *ostrich effect*, *red herring*, *righteousness fallacy*, and *straw man argument*.
Slogan	A *slogan* is a: • Motto associated with a group such as a movement or political party (e.g., environmental conservatists: there is no Planet B): • Phrase or word used to express a characteristic goal, position, or stand to be achieved (e.g., M&Ms: *melts in your mouth, not in your hands*); or, • Short attention-getting and memorable phrase used in advertising or marketing (e.g., MacDonald's: *I'm lovin' it*). See also, *motto* and *tag line*.

<table>
<tr><th>Phrase or Term</th><th>Definition</th></tr>
<tr><td>Sloganeering</td><td>Sloganeering is to:<ul><li>Attempt to persuade others by repeating slogans instead of explaining ideas (e.g., Red Bull: it gives you wings); or,</li><li>Coin or use slogans to sway opinions (e.g., America: Uncle Sam wants you).</li></ul>Sloganeering is employed in advertising and marketing, and diplomacy, government, and politics. See also, rhetoric and slogan.</td></tr>
<tr><td>Slurry</td><td>A slurry is a semiliquid mixture (e.g., fine particles of clay, coal, ore, or sand suspended in water or other liquid). See also, tailings and ore.</td></tr>
<tr><td>Small Business</td><td>A small business is defined differently by different organizations. See next examples:<ul><li>Canada Small Business Financing Program
A small business is a business with CAD$ 5 million or less in annual revenue.</li><li>Canadian Bankers Association
A small business is a business with a credit limit of CAD$ 500,000 or less.</li><li>Export Development Corporation
A small business is a business with less than CAD$ 1 million in annual export sales.</li><li>Industry Canada
A small business for:
Suppliers is a business that has between 5 and 100 employees for suppliers
Service providers is a business that has between 5 and 50 employees.</li></ul>Other organizations will have a different definitions. See also, big business, medium-sized business, micro-business, and mom and pop shop.</td></tr>
<tr><td>Small Steps Method</td><td>The small steps method is a strategy to advance towards a goal in small, incremental advancements by creating the conditions favourable to success. This requires focus on completing small tasks consistently, which gradually yield the ultimate objective. For example, to run a marathon, which is 42.195 km (26.2 mi.), a runner will train by running short, medium, and then long distances to build the stamina required to run this distance. Thus, by completing the small steps consistently (small, medium, and long distances), the ultimate goal of completing a marathon can be completed. See also, continual improvement.</td></tr>
</table>

Phrase or Term	Definition
Small Talk, Chit Chat	*Small talk* or *chit chat* is casual or informal and polite or positive conversation about uncontroversial or unimportant matters, especially preceding professional and social events. The ability to conduct small talk is a social skill and a type of social communication that, on the surface, may seem to fulfill little purpose – but is a bonding ritual and a strategy for defining or managing interpersonal relationships between colleagues, friends, and new acquaintances. It is often used before engaging in more functional or serious topics of conversation. For example, *before the weekly meeting starts everyone engages in small talk about what they did on the weekend or the weather*. It may work as: 1. An introduction and set the mood, especially when the speakers do not know each other. For example, *hi, I joined this company last week – what is your position here*?; 2. A means of filling prolonged uncomfortable or awkward conversational silence. For example, *did you watch the football game on the weekend*?; and, 3. A means to end a conversation or social interaction with a friendly or positive gesture, so as to not reject the other person or save face. For example, *it was great to meet you, but I need to make a phone call now*. See also, *save face*.
Smart	To be *smart* is to possess above average acumen or intelligence. A *smart* device is electronic and programmable (e.g., a cell phone). See also, *SMART*, *smart pipeline instrument gauge*, *smart practice*, and *smart project*.
SMART	*SMART* is an abbreviation and mnemonic used to identify the criteria to set effective goals and objectives (e.g., for employee-performance management, personal development, and project management) that stands for: • **S**pecific; • **M**easurable; • **A**chievable or **A**ttainable; • **R**elevant; and, • **T**ime-bound (i.e., with a deadline or defined end point). The principal advantage of SMART is that the objectives are easier to understand, and it is clear when they have been completed. For example, if a goal is not SMART, it may be arbitrary (e.g., something that sounds good but is not related to a specific need – for example, to write four articles every month) or not truly achievable. Or, if the goal is difficult to gauge or measure, it will be difficult to gauge progress. This will lead to disappointment and frustration when the goal is ultimately not achieved, which is exacerbated by a lack of data that means the reasons for failure will not be known. This abbreviation was first attributed to George T. Doran in 1981 and is commonly associated with Peter Drucker's *management by objectives* concept. See also, *abbreviation*, *deliverology*, *goal*, *management by objectives*, *mnemonic*, *objective*, and *smart*.

Phrase or Term	Definition
Smart Charging, Intelligent Charging, Dumb Charging	*Smart charging* or *intelligent charging* is a system where an electric vehicle and a charging device share a data connection, and the charging device shares a data connection with a charging operator. This allows the charging device owner to remotely manage, monitor, and restrict the use of their devices through a cloud to optimize energy consumption. *Dumb charging* is a when a traditional device is not connected to a cloud during charging. See also, *cloud* and *vehicle-to-grid*.
Smart End, Dumb End	When two people use a tape measure to acquire a measurement, the *smart end* is the end of the tape measure where the measurement is read. The *dumb end* is the end of the tape measure opposite the smart end. See also, *sharp end*.
Smart Pipeline Instrument Gauge	Not a preferred phrase – see *pipeline instrument gauge* and *in-line inspection*.
Smart Practice	A *smart practice* is the ability or the means of achieving a goal or objective in a safely and cost-effective manner. See also, *best practice*, *good practice*, and *smart project*.
Smart Project	A *smart project* is intelligent and forward-thinking. See also, *forward-thinking*, *thinking mode*, and *smart practice*.
Smart Risk	A *smart risk* is one where much can be gained from success, but even major failure would not be catastrophic. It is wise because it is clear, calculated, planned and rational, and based on sound analysis and facts. A smart risk is taken to push ourselves and our organizations out of a comfort zone to achieve increased growth, innovation, results, and satisfaction. For example, to identify and plan a new business opportunity based on market insights and trends instead of focusing only growing current existing opportunities. See also, *comfort zone*, *dumb risk*, and *risk*.
Smart Technology	*Smart technology* is equipment or hardware that incorporates digital technologies. For example, a pump or motor with digital capabilities (i.e., operation technology). See also, *digital*, *dumb technology*, *operation technology*, and *technology*.

Phrase or Term	Definition
Smells Fishy	Something that *smells fishy* means that: • Someone is lying. For example, an old vehicle that has low kilometres (miles). This smells fishy because old vehicles generally have travelled very far (hundreds of thousands of kilometres-miles). This vehicle's odometer may have been tampered-with, which is illegal; or, • Something is doubtful, dubious, improper, odd, questionable, suspect, suspicious, or unlikely. For example, the owner of the vehicle will only show the vehicle at night; never during the day. This smells fishy because the vehicle owner may be attempting to hide body damage, rust, or another defect that is visible only in light. Another example: something smells fishy about buying an expensive new watch from a vehicle in a parking lot – these watches are likely fake or stolen because it is not possible or sustainable to make a profit selling them if they were real or legally-acquired. This expression originated in the early 17th century, when consumers would test the freshness of a fish by it odour. A freshly-caught fish has little odour. A fish with a strong odour is not fresh. Thus, fishmongers who misrepresented their products by claiming they were selling fresh fish, but really were selling older fish, are the originators of this popular expression. See also, *bullshit*, *bullshit detector*, *ethical*, *ethics*, *false narrative*, *misinformation*, *propaganda*, *stretch the truth*, *truthiness*, *white lie*, *untruth*, and *wear and tear*.
Smoke and Mirrors	*Smoke and mirrors* is an expression that describes a: • Disguise that draws attention away from an embarrassing or unpleasant matter. For example, a government's new proposal for tax reform offers no substantive changes, but takes attention away from previous policy failures; • False or misleading impression of reality, which when examined closely, proves to be a deception. For example, a government introduces a new tax but offers a rebate to citizens that is promised to return more money than will be paid, resulting in a net benefit; or, • Technique to create a (seemingly magical) illusion (e.g., a magician who makes an object appear to hover in mid-air). Smoke and mirrors is also used to describe the machinations of big business or political obfuscation. During the 19th century German charlatan Johann Georg Schröpfer created illusions with a hidden projector, mirrors, and smoke to create what he claimed were apparitions. See also, *big business* and *wag the dog*.
Smoke-filled Room, Smoke-filled Back Room	A *smoke-filled room* or *smoke-filled back room* is a round-table-style decision-making process or secret political gathering. The phrase suggests a gathering of cigar-smoking men who are well-connected power brokers that meet privately to make decisions or nominate a dark horse political candidate without regard for the interests of others outside the group (e.g., society). See also, *crony capitalism*, *dark horse*, *old boy's club*, *politics*, and *stakeholder*.

Phrase or Term	Definition
Snellen Chart	A *Snellen Chart* measures visual acuity by presenting letters that are progressively smaller in size. Typically, a patient will attempt to read letters on the chart with one eye covered. The patient's visual acuity is determined by which letters the patient can read at 6 m (20 ft). The Snellen Chart is named after the Dutch ophthalmologist Herman Snellen, who developed it in 1862. Many ophthalmologists, optometrists, and others use an improved version of this chart known as the *LogMAR Chart* or *Bailey-Lovie Chart*. It was developed in 1976 by the National Vision Research Institute of Australia and yields a more accurate estimate of visual acuity than a Snellen Chart. See also, *20/20*.
Snoop Test	A *snoop test* is a sensitive pneumatic leak test using Snoop® or another similar solution (e.g., detergent-based) that is sprayed over the equipment or system being tested. Bubbles will then form wherever a leak occurs, which are formed by escaping gas.
Social Engineering	*Social engineering* is: • The use of deception to manipulate individuals into providing access (e.g., to an account, building, information, or network), confidential or personal information (e.g., passwords), or valuables (e.g., money). It is any act that influences a person to take an action that may or may not be in their best interest. The techniques used are based on specific cognitive biases, which are exploited in various combinations for an attack; or, • In Social sciences, the use of centralized planning to manage social change and regulate future behaviour and development of a society. It is the manipulation of a group of people through laws and principles, or other means, which tells them what is and is not acceptable. For example, fake news that appears to be real but is actually government-sponsored propaganda. Another example: applying taxes on tobacco products, permitting smoking in designated areas only, and requiring specific warning labels on tobacco packaging to reduce lung cancer caused by smoking and second-hand smoke. See also, *cognitive bias, cybercrime, fake news*, and *propaganda*.
Social Justice	*Social justice* is the concept of fair and just relations between the individual and society, as measured by the distribution of wealth, opportunities for personal activity, and social privileges. In Western and older Asian cultures, the concept of social justice often refers to the process of ensuring that individuals fulfill their societal roles and receive their benefits from society. In the current global context, emphasis is on breaking barriers to social mobility, the creation of safety nets, and economic justice. See also, *social justice warrior*.
Social Justice Warrior	*Social justice warrior* is a pejorative phrase that describes an individual who promotes socially progressive views, including civil rights, feminism, and multiculturalism. It is pejorative because the inference is that a social justice warrior pursues personal validation in the guise of a faux deep-seated social conviction and are disingenuous. The phrase originated in the late 20th century and was a neutral or positive phrase to describe someone involved in social justice activism. See also, *activist*, *hacktivist*, and *social justice*.

Phrase or Term	Definition
Social Licence, Social Licence to Operate	*Social licence* or *social licence to operate* can be determined as contractual grounds for the legitimacy of activities and projects for which a company is involved. It refers to the level of support and approval of a company's activities by its stakeholders. See also, *corporate social responsibility*.
Socialism	*Socialism* is an economic system in which a collective or the government owns most or all capital and capital goods. The production of goods and services is based on central planning – known as a command or planned economy. China and the Soviet Union employ central planning. See also, *capital*, *capitalism*, *economic system*, and *goods*.
Socket Weld	A *socket weld* is a pipe attachment detail in which a pipe is inserted into a recessed area of a fitting, flange, or valve for welding. Socket welded fittings are mainly used for small pipe diameters (i.e., small bore) and, generally, for NPS 2 piping or smaller. See also, *butt weld*, *Gap-A-Let®*, and *lap weld*.
Software Acceptance Test	A *software acceptance test* is a type of factory acceptance test that involves testing the computer code. See, also *hardware acceptance test*.
Software as a Service	*Software as a service* is cloud computing service that uses a software distribution and licensing on a subscription basis to host and make applications available to Internet customers. Sometimes referred to as *on-demand software*, Microsoft referred to this as *software plus services*. See also, *anything as a service*, *cloud computing*, *content as a service*, *infrastructure as a service*, *platform as a service*, and *security as a service*.
Softwood	*Softwood* is wood produced from a conifer tree such as fir, pine, or spruce. See also, *hardwood*.
Solder	*Solder* is a fusible metal alloy used to join base metals. See also, *alloy* and *soldering*.
Soldering	*Soldering* is a process in which two or more base metals are joined together by melting solder (filler metal) into a joint. Soldering does melt the base metals. Soldering occurs at temperatures between 90 and 450°C (190 and 840°F). See also, *brazing*, *filler metal*, *solder*, and *welding*.
Sole Proprietor	A *sole proprietor* is the single owner of a business, who alone is responsible for all liabilities – debts owed to third-party creditors – and is unincorporated. See also, *sole proprietorship* and *unincorporated*.
Sole Proprietorship	A *sole proprietorship* is a business with a single owner who alone is responsible for all liabilities, including all debts a company owes to third-party creditors. See also, *corporation*, *independent contractor*, *joint venture*, *partnership*, *sole proprietor*, and *unincorporated*.
Sole Source	*Sole source* is a procurement method without competitive bidding where only one supplier provides needed products, services, or both. The sole source has either established a monopoly or is the only provider within a geographic region. See also, *bid*, *competitive bid*, *monopoly*, and *single source*.
Solenoid	A *solenoid* is a cylindrical coil of wire that acts as a magnet when carries an electrical current.
Solid Line Reporting	*Solid line reporting* refers to personnel that have direct accountability to others (e.g., the discipline engineers report to their lead engineer). See also, *accountability*, *dotted line reporting*, *organizational chart*, and *reporting line*.

Phrase or Term	Definition
Solopreneur	A *Solopreneur* is a professional who establishes and operates a business, and prefers to work alone. Most work with others, as needed, but do not hire employees or work with partners. Instead, they hire contractors and pay them by the hour or project. Solopreneur is a portmanteau of *solo* and *entrepreneur*. See also, *entrepreneur*.
Solution in Search of a Problem	The phrase *a solution in search of a problem* has both negative and positive connotations. 1. Negative A solution is created when there is no problem to be solved. Thus, it is useless. 2. Positive There is a problem, but people do not know it is a problem. For example, a Proctor & Gamble cross-functional team observed stay-at-home mothers mopping their floors. They observed that the mothers were spending more time cleaning their mops than cleaning the floor. This observation led to the invention of the Swiffer Mop. This was a situation where a solution was created to solve an unarticulated problem. This is clearly different from creating a solution when it will not solve a problem. A similar phrase is *a hammer in search of a nail*. If you are the hammer, everything looks like a nail!
Sour Crude Oil	*Sour crude oil* is an oil product that contains a high percentage of impurity sulphur (i.e., more than 0.5%). See also, *sour service*.
Sour Natural Gas	*Sour natural gas* is a gas product that contains a significant amount of hydrogen sulphide (H_2S). That is, more than 5.7 mg/m^2 or 4 ppm by volume under a standard temperature and pressure. See also, *sour service*, and *standard temperature and pressure*.
Sour Service	*Sour service* describes equipment (e.g., a pressure vessel or tank) or a piping system that processes or transports sour crude oil or sour natural gas (product). For service to be considered sour service, the product H_2S level must be above a certain threshold – which may vary by agency, application, country, province, or state. For example, NACE MR0175/ISO 15156 may or may not be specified. Sour service subjects materials that come in contact with product to corrosion. See also, *sour crude oil*, *sour natural gas*, and *sweet service*.
Source-to-Pay	A *source-to-pay* is the entire end-to-end procurement process – from spend management, strategic sourcing, and supplier management – to purchasing, performance management, and accounts payable. See also, *electronic commerce* and *procure-to-pay*.
South of	*South of* means less than (e.g., *south of a million dollars* means *less* than a million dollars). See also, *north of*.
Spanner	Not a preferred term – see *wrench*.

Phrase or Term	Definition
Speak Truth to Power	*Speak truth to power* is a non-violent political tactic used by dissidents to resist authoritarianism, ideocracy, or oppression. It means confronting an individual or an institution in a position of power (e.g., academia, business, government, society or group, the military, or a religious organization). It demands a moral response to a problem, rather than an easy, expedient, or selfish response. The speaker uses criticism instead of flattery, frankness instead of persuasion, and truth instead of falsehood or silence. It is not about moral superiority but changing someone's actions or attitude. For example, the watershed event in which an unnamed Chinese man blocked a military-led convoy of tanks with his body during the Chinese 1989 Tiananmen Square protests in China. This is a symbolic act because the man was widely viewed as speaking to his truth to defend himself and fellow citizens, even risking deleterious consequences. Speaking truth to power is therefore an act of bravery because of the speaker often risks harm or life for acts that are largely symbolic. This phrase was first used in 1955 in *Speak Truth to Power: A Quaker Search for an Alternative to Violence*.
Specialization, Division of Labour	*Specialization* is how an individual or organization focuses on a particular production method or skill to efficiently produce a product or service. It is an agreement within a group or organization where each of the members best suited for a specific task assumes responsibility for its successful execution. Specialization is also referred to as the *division of labour* because it is a process through which economic parties develop and use their strengths, and trade with others to efficiently complete a task. This means that individuals and organizations rely on others (specialists) to produce product or service that cannot be efficiently produced themselves. For example, a customer purchases a meal prepared at a restaurant instead of purchasing the necessary groceries and preparing a meal at home. Individuals and organizations do what they do best to earn money to pay others for the things they cannot or prefer not to do or produce. See also, *self-sustaining*.
Specification	A *specification* is a technical standard comprising a set of documented requirements to be satisfied by a material, design, product, or service. The purpose of a specification is to define: • Requirements (e.g., design or product); • Design or products (e.g., designed or final produced solution); and, • In-service or maintenance parameters (e.g., conditions after operation and wear). See also, *data sheet*, *design specification*, *pipe class*, *recommended practice*, *specifications*, and *standard*.
Specification Break, Spec Break	A *specification break* or *spec break* is the location where the specification of a piping system functionality changes. This is usually designed to occur at a flange, to transition from one design criteria to another, or vice versa: • Code of construction (e.g., ASME B31 to CSA Z662); • Corrosive to non-corrosive service; • Higher to lower pressure; and/or, • Higher to lower temperature.

Phrase or Term	Definition

Specification Creep	*Specification creep* is when a specification is:
	• Continuously amended to address new technologies or when obsolete, outdated, or unnecessary information is removed. Text that is deleted, however, may:
	  o Become needed for reference or use later; or,
	  o Have related text that is not deleted within the same document (e.g., cross-references). The meaning of this text then often becomes unclear.
	• No longer readily available for use because it is inaccessible after one or more merger and acquisitions, or one or more transfers of business data to other servers.
	See also, *code creep*.

Specification Excerpt, Specification Reference	A *specification excerpt* or *specification reference* is a verbatim citation of a portion of a specification or standard (e.g., of or from a section, clause, paragraph, sentence, phrase or combination). See also, *specification*.
Specification Sheet, Spec Sheet	Not a preferred phrase – see *cut sheet*.
Specifications	*Specifications* are a detailed description details regarding a product's appearance, construction, materials, operation, performance, or standard of workmanship required for production. See also, *cut sheet* and *specification*.

Phrase or Term	Definition

Spillover Effect	A *spillover effect* is:
	• In business, when an event influences beliefs regarding attributes that are not directly associated with the event itself. For example, in April 2017 a United States federal judge ordered German vehicle manufacturer Volkswagen (VW) to pay a criminal fine of USD$ 2.8 billion for *rigging diesel-powered vehicles to cheat government emissions tests*. This scandal, also known as Dieselgate, caused great financial loss to VW, damaged the *made in Germany* image, and harmed VW's reputation – all of which are spillover effects of cheating the tests;
	• In economics, when seemingly unrelated economic events have an economic impact elsewhere. For example, most nations experience negative economic significant spillover effects during a downturn in one of the world's two largest economies: China or the United States;
	• In the knowledge economy, when information facilitates the exchange of ideas to promote creativity and innovation. For example, one company's innovation may stimulate related inventions and technical improvements by other companies. For example, the advent of Apple's Netflix streaming service in 2007, and the subsequent spillover effect that was the creation of Amazon Prime Video, Apple TV, CuriosityStream, Disney+, Hulu Plus, Stan, Vudu, and other direct competitors;
	• In technology, when peers of innovators can acquire technology at a lower cost than the cost of inventing it themselves. For example, a company can enhance their innovative capability and productivity by using:
	  o Involuntarily diffusion. For example, through imitation and learning, labour force mobility, or downstream or upstream linkages in the supply chain or value chain; and,
	  o Voluntarily diffusion. For example, research and development alliances or technical licensing.
	• In psychology, when one's emotions affect how others around them feel. For example, if one experiences workplace bullying, it may trigger negative emotions – which may cause the spillover effect of domestic incivility (or worse).
	There are positive spillover effects; however, this phrase is frequently used to describe negative spillover effects. See also, *bully*, *Cobra Effect*, *entrepreneur*, *externality*, *knock-on effect*, *knowledge economy*, *second-order effect*, *supply chain*, *trail blazer*, *unanticipated consequences*, and *value chain*.

Phrase or Term	Definition
Spin	*Spin* is information presented with an emphasis, interpretation, or point of view to influence public opinion, understanding, or both. Celebrities, companies, governments, and political parties use spin. Traditional advertising and public relations may manage the presentation of facts, but spin often implies the use of disingenuous, deceptive, emotional, and manipulative tactics that are biased and therefore a form of propaganda. It is a polite word for deception. For example, a political party will always *try to spin* everything they do in a positive way. Many other examples may be found in Roman literature, which is filled with propaganda that portrays Rome as civil, peace loving, and righteous while portraying adversaries as dangerous, unruly, and savage. To be Roman was to be in a *commonwealth of the civilized* while others were *barbarians* – this is Roman spin. In modern times, individuals and groups spin ideals of opponents as substandard when compared to their own. See also, *euphemism*, *false narrative*, *propaganda*, and *spin doctor*.
Spin Doctor	A *spin doctor* is a spokesperson who is skilled in public relations and spin. They give a favorable interpretation of actions and events to the media and public, especially on behalf of a political party (e.g., the United States White House press secretary). These personnel may also be referred to as a *spin merchants* or *spinmeisters*. See also, *spin*.
SPIN Selling	*SPIN Selling* was the brainchild of Neil Rackham, who authored a book of the same name, *SPIN-selling*. It is based on extensive research by Rackham and his company Huthwaite, in which they examine large, complicated sales scenarios. After analyzing more than 35,000 sales calls, they debunked a variety of myths about closing sales. SPIN Selling proposes there are four types of questions; thus, SPIN stands for: • Situation Questions Deal with the facts about the buyers existing situation. • Problem Questions Ask about the buyer's pain and focus the buyer on this pain while clarifying the problem, before asking implication questions (i.e., identify implied needs). • Implication Questions Discuss the effects of the problem, before talking about solutions, and identify the seriousness of the problem to increase the buyer's motivation to change. • Need-Payoff Questions Get the buyer to probe for and explain their explicit needs and the benefits that the solutions offer, rather than forcing the seller to explain the benefits to the buyer. Getting the buyer to state the benefits of a product, instead of the salesperson, has greater impact because it sounds *less pushy,* and reveals the buyer's need for a product as described in the buyer's terms. See also, *close*.

Phrase or Term	Definition
Spirit of the Law	To obey the *spirit of the law* is to follow the intent of a rule, sometimes by considering those who established it (i.e., the general purpose or policy) but not necessarily the literal interpretation of the words. This phrase may be used in a legal or non-legal context. It is the antithesis of the letter of the law. For example, the rules for a marathon prohibit certain types of running shoes but do not prohibit barefoot running. Thus, by following the spirit of the law, a sprinter is permitted to run without running shoes. Another example. A parent instructs their child Sarah not to watch cartoons on television until her homework is completed. Later, the parent finds Sarah watching cartoons on a mobile device with their homework unfinished. In this case, Sarah did not follow the spirit of the law because the homework was unfinished. However, Sarah did follow the letter of the law by not watching cartoons on television and instead choosing to watch them on a smart phone. See also, *antithesis*, *letter of the law*, and *technicality*.
Splaining	*Splaining* is sarcastically calling out someone for explaining something without taking the speaker's comments into consideration or in an extensive and sometimes condescending way. The informal *splain* is derived from the root word *explain* and has been around for about 200 years. Many variations exist, including *leftsplaining*, *straight-splaining*, and *whitesplaining*. These terms refer to either the person who does the splaining or to the topic being splained. *See also, discrimination*, *mansplaining*, *racism*, and *sexism*.
Splice	A *splice* joins two shorter lengths of material together (e.g., beam or column). See also, *beam* and *column*.
Sponsored Content	*Sponsored content* is material: • In a publication which resembles the publication's editorial content but is paid for by an advertiser to promote products or services; • Paid for by an advertiser, but created and shared by another brand, influencer, or publisher (e.g., a paid partnership on social media); or, • That serves as entertaining or useful information to favourably influence the perception of the sponsor's brand (i.e., advertising a brand rather than products or services to increase sales). See also, *brand*, *commercial*, *influencer*, *infomercial*, *native advertising*, and *sponsored content*.
Spot Check	A *spot check* is a random inspection.
Spot Rate, Cash Value	The *spot rate* or *cash value* is the current market value (e.g., for a currency value or an interest rate applicable to an immediate financial transaction). See also, *forward rate* and *theories of exchange rates*.
Spotted Trailer	A *spotted trailer* is a stationary trailer that is ready for loading or unloading. For example, the spotted trailer is parked at the loading dock.

Phrase or Term	Definition
Spray and Pray	*Spray and pray* is a phrase used to describe an approach to communication, where a message is broadcasted or emails are dispersed to a mass audience (millions). The Disney Corporation refers to organizations that use a spray and pray approach towards training, as *spraying the training on and praying it sticks*. It is also a derisive military term for firing an automatic firearm towards an enemy in long bursts, without trying to line up each shot or burst of shots. This method of firing is prevalent for those who have not had the benefit of training.
Spray Foam Insulation	*Spray foam insulation* is a polyurethane product that can be applied to equipment and used as a barrier to: • Air; • Fire; • Heat; • Moisture; • Smoke; or, • Sound. Common industrial applications include: • External surfaces of a shop-fabricated tank for thermal insulation; • Under a pipeline – pre-formed or sprayed in-situ as a protective pillow; and, • Under a skidded buildings as thermal insulation.
Spread	A *spread* is a manageable section of a large pipeline construction project. The approach to constructing a spread is assembly-line like. Each spread requires highly specialized crews, each with their own responsibilities, to construct the spread in an organized sequence.
Squad Check	A *squad check* is a review of project (e.g., client or engineering and procurement company) or supplier documentation by engineering or other disciplines to ensure that project, purchase order, and other requirements are met. See also, *cold eyes review*, *discipline review*, and *peer review*.
Square Peg in a Round Hole	*Square peg in round hole* is a phrase that: • Can be a metaphor for something that has a poor fit, such as inserting a square wooden stick into a round hole or adding a clown to a business meeting; • Describes someone who does not fit into, or is uncomfortable with, a group, position, place, or situation; and, • Describes someone who is the only incompetent team member. The metaphor was originated by Sydney Smith's lecture on moral philosophy entitled *On the Conduct of the Understanding*, that he delivered at the Royal Institution between 1804 and 1806.

Phrase or Term	Definition
Squaring the Circle	*Squaring the circle* is: • Is a metaphor for attempting to accomplish the impossible; or, • The challenge proposed by ancient geometers, of constructing a square with the same area as a given circle by using only a compass and straightedge with a finite number of steps. The task was proven to be impossible by the Lindemann–Weierstrass theorem in 1882, because pi (π) is a transcendental, rather than an algebraic irrational number. This means it is not the root of any polynomial with rational coefficients. Approximate squaring with non-perfect accuracy is possible in a finite number of steps because there are rational numbers *arbitrarily* close to pi.
Stage Gate	*Stage gate* is an engineering methodology that features the design and engineering of an asset in a progressive step-like fashion, with periodic gates to approve or suspend work during development.
Stainless Steel	*Stainless steel* is an alloy of iron, chromium, and nickel. Additional elements may also be used. There are over 150 grades of stainless steel, of which 15 are the most commonly used. These alloys are most notable for corrosion resistance. There are five main groups, which are primarily classified by their microstructure. 1. Austenitic stainless steels have an austenitic microstructure, are classified as SAE 200 and 300 Series, and are weldable; 2. Ferritic stainless steels have a ferrite microstructure, are classified as SAE 400 Series, and are not weldable; 3. Martensitic stainless steels have a very hard microstructure, are classified as SAE 400 Series, and are not weldable; 4. Duplex stainless steels have a mixed microstructure of austenite and ferrite, and are weldable; and, 5. Precipitation hardening stainless steels can be hardened by heat treatment. Stainless steel is also known as inox steel or inox, from the French *inoxydable* (inoxidizable). See also, *alloy*, *microstructure*, and *steel*.
Stakeholder	A *stakeholder* is a person, group, or organization with a financial or other interest in a project. A stakeholder can be affected by another organization's actions, such as: • A non-governmental organization or community; • An individual or organization, including a government, that claims some part of a project's value; • An investor or owner; or, • The project teams and team members (e.g., requisition team, discipline engineering, subject matter experts, supplier quality representatives, completions, operations and maintenance, or other personnel). The claim on value does not need to take the form of monetary payment. For example, an environmental organization may take payment in the form of project reconfiguration that is, or is perceived to be, less damaging to the environment. See also, *discipline*, *non-governmental organization*, *owner*, *project team*, and *requisition team*.

Phrase or Term	Definition
Stand Down, Stand-down	*Stand down* or *stand-down* means: • A meeting between an employer or management and personnel that addresses safety (e.g., the team stood down to honor colleagues injured in the tank rupture last year); • To relax or cause to relax after a state of readiness (e.g., the troops stood down after the threat of a missile strike was averted); or, • To resign or withdraw from a position or office (e.g., after intense public pressure, the vice president finally stood down). See also, *sit down*, *stand up*, and *where you stand determines where you sit*.
Stand Up	To *stand up* means to: • Defend against attack or criticism, face boldly, or meet fairly and fully (e.g., the professor *stands up* for his beliefs in physics); • Have a high degree of personal integrity or loyalty (e.g., he is a *stand up* professional); or, • Remain intact, sound, or valid after analysis, attack, scrutiny, or stress (e.g., the building stood up to the earthquake but sustained minor damage). Another example: the engineer's argument *stood up* against the technical authority's rebuttal. See also, *sit down*, *stand down*, and *where you stand determines where you sit*.
Standalone	*Standalone* means to be separate. In the context of industry and projects, standalone can mean: • A business that can operate on its own without needing other assistance; • Computer hardware or software that can operate independently of other hardware or software, or a device or program; or, • Not connected to anything or depending on anything.
Standard	A *standard* is a set of industry-specific criteria relating to function, operation, production, and other generally-accepted requirements. A standard may be recommended or adopted for use by regulations, owner specifications, or both. See also, *specification*.
Standard of Workmanship	A *standard of workmanship* is a benchmark established by a code, specification, or standard that requires a certain level of qualifications from the fabricator, installer, manufacturer, and trades persons, and acceptable tolerances of the finished work. See also, *benchmark*, *specification*, *standard*, *trades person*, and *workmanship*.
Standard Operating Procedure	A *standard operating procedure* is a step-by-step process that must be followed to complete complex or routine operations. A standard operating procedure standardizes aims to achieve efficiency, quality output, safety, and uniformity of performance, while simultaneously reducing miscommunication and failure to comply with industry regulations. For comprehension, procedures must be written to commensurate with appropriate literacy and readability levels of personnel. See also, *safe work procedure*.

Phrase or Term	Definition
Standard Temperature and Pressure	*Standard temperature and pressure* is commonly used to define standard conditions for temperature and pressure, and measure chemical and physical processes. The International Union of Pure and Applied Chemistry defines this as air at 0°C (32°F) and 100 kPa (14.5 psi). The Imperial and USA System of Units defines it as air at 15.6°C (60°F) and 101 kPa (14.7 psi).
Standard Welding Procedure Specification	A *standard welding procedure specification* is a welding procedure specification qualified according to the requirements of AWS B2.1/B2.1M and made available for production welding by companies or individuals other than those performing the qualified test (i.e., pre-qualified).
Standards of Journalism, Journalistic Ethics and Standards	*Standards of journalism* or *journalistic ethics and standards* comprise the principles of good morals and practices for journalists. This is known as journalism's professional *code of ethics* and *the canons of journalism*. These standards may vary between organizations or places, but most share common elements. The principles include the common elements of accountability, accuracy, diversity, fairness, humanity or limitation of harm, impartiality, independence, objectivity, privacy, transparency, and truthfulness for the acquisition and dissemination of information or news. The standards also respect the presumption of innocence in a case or matter that is proceeding in court (e.g., on trial). Some standards prohibit discriminatory references based on mental and physical disability, race, religion, and sexual orientation. See also, *ethics* and *ethical*.
STANDATA	*STANDATA* are standardized technical bulletins that provide advice relevant to various regulatory disciplines (e.g., building, electrical, fire, and boilers and pressure vessels for pressure equipment). These bulletins are created jointly by *Alberta Safety Services* and the *Safety Codes Council*. Some bulletins are issued under the authority of a code, standard, or the *Safety Codes Act*. These bulletins contain interpretations, clarifications, recommended practices, and province-wide variances on codes and standards matters related to the *Safety Codes Act*. Find STANDATA on the *Alberta Safety Services* website: https://www.municipalaffairs.alberta.ca/cp_electrical_standata.cfm See also, *Safety Codes Act* and *Safety Codes Council*.
Standby	*Standby* means: • Available for use or to be relied upon as an alternate or in an emergency (e.g., a genset); or, • Present or held in reserve and ready for deployment (e.g., personnel).
Starting Pipe	The *starting pipe* is the original raw pipe used to manufacture another finished product (e.g., a bend, fitting, or smaller pipe or tube). See also, *mother pipe*.

Phrase or Term	Definition
Start-up	*Start-up* is: • A transitional phase that encompasses activities between facility construction completion and commercial operations. Start-up occurs after commissioning, and includes systems turnover, introduction of feedstock, and performance testing; or, • An early stage in the life cycle of an enterprise where an entrepreneur moves from the idea stage to securing financing, defining the basic structure of the business to develop a unique product or service, bring a product or service to market and, ultimately, initiate operations or trading. See also, *commissioning*, *evolution*, and *feedstock*.
State-of-the-Art, Cutting Edge, Leading Edge	*State-of-the-art*, *cutting edge*, or *leading edge* is the highest or most recent level of development (e.g., of a device, procedure, process, science, or technique) reached using modern methods. It incorporates the newest features, ideas, and technologies. For example, Moderna's COVID-19 vaccine *mRNA-1273* what considered state of the art in 2020 because it was 94.5% effective when released. See also, *best-in-class*.
Static Head Test	A *static head test* is used to ensure that an item (e.g., drain piping or a tank) is leak-free. The item is filled with water and held for a set period (i.e., without pressure being applied).
Static Mixer, In-line Mixer, Motionless Mixer	A *static mixer*, *in-line mixer*, or *motionless mixer* is device that continuously blends, disperses, and mixes fluid materials (e.g., at an injection point) without moving parts. The fluid moves through the mixer in an alternating or continuous motion (e.g., clockwise/counterclockwise flow or turbulent flow) to produce product homogeneity. The energy for mixing is obtained from the loss of pressure as fluids flow through the device and a pump provides fluid flow. These devices are: • A fin-, plate- or wafer-type fitting; or, • Elements contained (e.g., welded) in a cylindrical (e.g., pipe) or square housing. See also, *fitting*.

Phrase or Term	Definition

Station	A *station* is:
	• A complete assembly of telecommunications equipment for receiving or transmitting;
	• A facility such as a building, place, or site, where a specified activity or service is based, or equipment is located (e.g., a pump station);
	• A place:
	  o Established to provide a public service (e.g., a fire station); or,
	  o For specialized observation and study of scientific phenomena (e.g., seismology);
	• A regular stopping place on a public transportation route (e.g., along a bus route or railroad line);
	• Any of the places in a facility at which one part of the work is done (e.g., a painting or welding station);
	• The equipment used for performing a specific work (e.g., painting or welding equipment); or,
	• The place or position in which something or someone stands or is assigned to stand or remain (e.g., to operate a facility).
	See also, *facility* and *local control station*.

Status Quo, Business as Usual	*Status quo* or *business as usual* refers to the current or existing situation, especially regarding business, political, or social issues. For example: • In business, things continue working normally, despite difficulties or disturbances; • In politics, how conditions invite a bad or good or analysis; and, • In sociology, maintaining an existing social structure, values, or both. See also, *game changer* and *unlearning*.
Status Report, Tracker Report	A *status report* or *tracker report* summaries progress recurrently (e.g., weekly, bi-weekly, or monthly). A status report is a critical part of an effective project communications and management strategy used to update the project management or team, stakeholders, or both. It may contain general, specific, or summative information (or a combination) and is often issued by a specific discipline (e.g., expediting, procurement, or supplier quality surveillance). See also, *progress report*.
Steam-assisted Gravity Drainage	*Steam-assisted gravity drainage* is an enhanced oil recovery technology that uses steam to produce heavy crude oil and bitumen. See also, *enhanced oil recovery* and *underground burning*.
Steel	*Steel* is a hard, gray or bluish-gray, and comprises strong alloys of iron with carbon and other elements. It is used extensively as a fabricating and structural material. See also, *alloy*, *carbon steel*, and *stainless steel*.

Phrase or Term	Definition
Steel Man Argument, Steel-manning	A *steel man argument* or *steel-manning* is to find the best form of an opponent's argument or improve their argument and test opposing opinions to determine the truth (rather than simply to win the argument). It is an antidote for the tendency to caricaturize an opponent. When an opponent feels understood, strengthened, or both, they naturally become less defensive. This means that proponents accept or concede a point to their opponent instead of attempting to attack the weakest point or present a distorted, exaggerated, or misrepresented version of their argument (i.e., straw manning). For example, in a debate about personal choice vs. vehicle safety: • The proponent for personal choice concedes to their opponent that the cost of treating injuries and trauma from accidents are greatly reduced when airbag installation is mandatory; and, • The proponent for vehicle safety concedes to their opponent that air bags should not be mandatory because people may drive less safely and vehicle cost is increased. See also, *hollow man argument*, *ideological Turing test*, *straw man argument*, and *weak man argument*.
Steep Learning Curve	A *steep learning curve* is an expression that indicates that a task or activity is difficult to learn or knowledge is difficult to gain due to the amount, complexity, or both. By contrast, *a learning curve with a steep start* indicates rapid initial progress. See also, *learning curve*.
Steganography	*Steganography* is the practice of concealing a file, image, message, or video within another file, image, message, or video. This hides the fact that a secret message is being sent because the message appears to be part of something else. The message may be concealed by many methods (e.g., code, colour, invisible ink, microdots, non-printing Unicode characters, and text converted into a sound file). See also, *cryptography*, *encryption*, and *microdots*.
Stencil	A *stencil* is spray-painted identification applied to pipe by the manufacturer. It usually identifies the heat number, hydrotest pressure, grade, method of manufacture, normal mill characters and identification, size, specification, and wall thickness. Stencil details vary by purchase order and specification requirement.
Stick File	A *stick file* is drawing clamp or vertical drawing file that holds and organizes the current revisions of issued drawings. Projects use a stick file to record drawing mark-ups that typically comprise interim changes and corrections (before the next issue of the drawing). A stick file is also known as a drawing stick, hanging clamp, plan clamp, or plan stick. With the advent of electronic editing tools, drawing markups are now mostly completed and issued electronically; stick files are less commonly used.
Stick Welding	Not a preferred phrase – this is slang for *shielded metal arc welding*.
Stick-built Construction	*Stick-built construction* involves building a structure or piping system entirely or largely on-site. See also, *prefabricated construction* and *modular construction*.

Phrase or Term	Definition
Stickhandle	In ice hockey, *stickhandle* means to skillfully control and maintain possession of a hockey puck using a hockey stick. For example, *his stickhandling ability only enhanced his ability to score*; and, In management, *stickhandle* means to capably and swiftly manage or skillfully resolve a challenge. For example, *he stickhandled the problem to the client's satisfaction*. See also, *hockey stick*.
Sticktoitiveness, *Stick-to-it-iveness*, *Stick-to-itiveness*	*Sticktoitiveness*, *stick-to-it-iveness*, or *stick-to-itiveness* is an informal expression derived from the expression *stick to it*. *Sticktoitiveness* describes these human characteristics: determination, dedication, perseverance, persistence, and stamina. The ability to stick to it enables a task to be completed despite all challenges. For example, the project team's sticktoitiveness ensured that the budget was prepared on time.
Stock Item	A *stock item* is something that is approved for issue, procurement, or storage, and usually kept available in a seller's inventory for immediate sale or use.
Stock Keeping Unit Number	A *stock keeping unit number* is a product and service identification code that is often displayed as a machine-readable barcode to track the item for inventory. A stock keeping unit number does not need to be assigned to physical products in inventory. These numbers are also applied to intangible, but billable, products (e.g., units of repair time or a warranty).
Stockpile	A *stockpile* is an accumulated inventory of materials or product meant for future use (e.g., held in reserve as a contingency for emergencies, or for anticipated or unanticipated shortages).
Stop Doing List	A *stop doing list* is a list of things to stop doing. This means that resources previously allocated to items on the stop doing list are channelled into fewer business activities. For example, Walgreens stopped (exited) the profitable food-service business to focus its resources on one idea: to be the best and most convenient drugstore. Their stop doing list was *stop selling food*. Consequently, between 1975 and 2000, Walgreens' multiplied their cumulative stock value over 15 times. The concept of a *stop doing list* is discussed in the book *Good to Great: Why Some Companies Make the Leap ... and Others Don't* by Jim Collins. See also, *budgeting*, *future-back process*, and *to-do list*.
Stopgap, Stop-gap	Something that is *stopgap* or *stop-gap* is a makeshift or temporary solution to a problem or need. For example, the act of repeatedly refilling a tire with air instead of fixing or replacing it is a stopgap. See also, *jerry-build*, *jerry-rig*, and *workaround*.

Phrase or Term	Definition

Story	A *story* is:
	• A horizontal division of a building's interior (e.g., a floor or level);
	• A falsehood, legend, lie, or rumor (e.g., the story he told was obviously fabricated);
	• A statement regarding the facts about a matter or situation (e.g., if you want the real story, read the book instead of listening to the advertisers); or,
	• An account or report of imaginary or real characters and events or incidents that is told to entertain, educate, inform, instill moral values, preserve culture, or a combination. A story may be:
	  o Fiction that describes imaginary events and people (e.g., a fable, myth, or parable); or,
	  o Non-fiction that describes facts, real events, and real people (e.g., an autobiography, biography, or historical narrative).
	See also, *anecdote*, *backstory*, *founder story*, *story form*, *storytelling*, and *writing style*.

Story form	*Story form* is the structural skeleton of a story, that is stripped of all its details, and includes elements of the:
	1. Plot structure. This is the same, regardless of story length (e.g., for a 2-page fable and a 400-page novel). Plot comprises:
	  A beginning with the development of the main character(s) and the basic situation;
	  A middle with the development of the challenge and its resolution; and,
	  An end with the conclusion.
	2. Main characters such as animals, humans, imaginary beings, or a combination;
	3. Setting with the place and timeline; and,
	4. Theme and point of the story.
	See also, *story* and *storytelling*.

Storytelling	*Storytelling* is the activity of telling or writing stories. At its core, it is the art of using gestures, language, physical movement, and vocalization, or a combination to reveal the elements and images of a story to an audience. A central and unique aspect of storytelling is its reliance on the audience to develop specific details and visual imagery to co-create and complete the story. See also, *story*, *story form*, and *wordplay*.
Straddle	To *straddle* is to be on both sides (and typically on centre), or to extend over or across (e.g., a flange that straddles the centreline or to straddle a horse). See also, *centreline* and *two hole*.
Straight Thread	A *straight thread* is used on a bolt that will pull tight, but will not make a fluid-tight seal. See also, *national pipe thread*.
Strainer	A *strainer* is a device with holes punched in it or made of wire mesh that is used to separate solid matter from a liquid (e.g., to strain debris from oil or water before it enters the suction nozzle of a pump). It may be inserted (e.g., into a piping system) or an appurtenance or fitting (e.g., pressure equipment). See also, *appurtenance*, *fitting*, and *witches' hat*.

Phrase or Term	Definition
Strategic Lawsuit Against Public Participation, SLAPP	A *strategic lawsuit against public participation* or *SLAPP* is a lawsuit or lawsuits used by a plaintiff to censor, intimidate, and silence opponents by forcing them into a legal defense. The punishment is in the process, not the verdict. Rather than winning a lawsuit, the plaintiff's goals are accomplished if the defendant abandons their criticism or opposition when they succumb to exhaustion, fear, intimidation, legal costs, or a combination. This action may also intimidate others from participating in a debate. See also, *anti-strategic lawsuit against public participation law* and *lawfare*.
Straw Man Argument, Straw Man Fallacy, Straw-manning, Wicker Man Argument	A *straw man argument*, *straw man fallacy*, *straw-manning*, or *wicker man argument* is a logical fallacy that is a distorted, exaggerated, or misrepresented version of an argument or position that is substituted for an actual argument or position. The typical straw man argument creates the illusion of having completely refuted or defeated an opponent's proposition through the covert replacement of it with a different proposition (i.e., to stand up a straw man) and to then refute that false argument (i.e., to knock down a straw man) instead of the opponent's proposition. Straw man arguments are used in polemical debate, particularly with highly charged emotional subjects. An example is United States President Richard Nixon's 1952 Checkers speech. When campaigning for vice president, he was accused of illegally appropriating $18,000 in campaign funds for personal use. He spoke, however, about another gift, a dog he had been given by a supporter. This was a straw man response because his critics had never criticized the dog as a gift or suggested he return it. This argument was successful at distracting many people from the issue and portraying his critics as nitpicking and heartless. See also, *bait and switch*, *circular reasoning*, *either/or*, *hollow man argument*, *false equivalence*, *logical*, *logical fallacy*, *paper tiger*, *red herring*, *righteousness fallacy*, *slippery slope*, *steel man argument*, and *weak man argument*.
Street Smart, Street Smarts	To be *street smart* or to have *street smarts* is to be intelligent and have knowledge from experience with practical methods (especially in an urban environment), with little or no understanding of related academic and theoretical topics. For example, someone with streets smarts could likely survive a walking all night in a dubious urban environment. See also, *book smart*, *easy way*, *hands on*, *on-the-job learning*, and *hard way*, and *school of hard knocks*.
Strength Weld	A *strength weld* is any weld designed primarily to provide pressure containment or structural support. The strength is determined by either the strength of the base or weld metal. This type of weld may or may not provide a specific degree of leak-tightness by sealing the joint. See also, *base metal*, *joint*, *leak-tight*, *seal weld*, and *weld metal*.
Strengths, Weaknesses, Opportunities, and Threats Analysis	*Strengths, weaknesses, opportunities, and threats analysis* is a strategic planning technique used to help an individual or organization identify strengths, weaknesses, opportunities, and threats related to business competition or project planning. It is intended to specify the objectives of the business venture or project and identify the internal and external factors that are favorable and unfavorable to achieving those objectives.

Phrase or Term	Definition
Stress	*Stress* is a constraining force or influence that is: • A chemical, emotional, or physical factor, or a combination, that causes bodily or mental tension that may contribute to disease; • A mutual force exerted when one body or part presses on, pulls on, pushes against, another body or part; • The deformation caused in a body or part by the intensity of a mutual force; or, • The intensity of a mutual force exerted by a body or part (e.g., expressed in pounds per square inch or psi). In humans, stress is a normal biological reaction to a potentially dangerous situation. It is a natural feeling of not being able to cope with specific demands or events. It is a feeling of emotional or physical tension and how the brain and body respond to a demand or challenge (e.g., caused by extreme exercise, financial pressures, overwork, public speaking, or significant life-changing or traumatic events). Stress can affect one's physical and mental health, so it is important to address stressors and seek help if needed. There are many well-known quotations that describe stress, including: • *If the problem can be solved, why worry? If the problem cannot be solved, worrying will do you no good.* Buddha • *Rule number one is don't sweat the small stuff. Rule number two is it's all small stuff.* Robert Eliot • *Stress is caused by being here but wanting to be there.* Eckhart Tolle • *You cannot always control what goes on outside. But you can always control what goes on inside.* Wayne Dyer • *Working hard for something we don't care about is called stress. Working hard for something we love is called passion.* Simon Sinek See also, *fight-or-flight mechanism*, *public speaking,* and *stress test*.

Phrase or Term	Definition
Stress Test	A *stress test* is a form of deliberately intense or thorough testing used to determine the stability of an entity or system (i.e., hardware or software). It involves testing beyond normal operational capacity, often to a breaking point, to observe the results. A stress test is performed to: • Confirm that a mathematical model or software is accurate enough when predicting breaking points or safe usage limits; • Confirm that intended specifications are met; • Determine the breaking points or safe usage limits; • Determine the modes of failure (i.e., exactly how a system fails); • Identify hidden vulnerabilities in institutions and investment portfolios by using computer simulations to evaluate how well they might withstand adverse events or market conditions (i.e., assess and manage risk); or, • Test stable operation of a part or system outside standard use. The term *stress* may have a more specific meaning in certain industries, such as material science or testing – where it may have a technical meaning such as fatigue testing. See also, *destructive testing* and *stress*.
Stretch Target Paradox, Stretch Goal Paradox	The *stretch target paradox* or *stretch goal paradox* is that organizations which: 1. Would most benefit from this strategy seldom use it; and, 2. Are not suited for this strategy often use it in a desperate attempt to generate breakthroughs. Neither approach is likely to be successful. See also, *breakthrough*, *paradox*, and *stretch target*.
Stretch Target, Stretch Goal	A *stretch target* or *stretch goal* differ from ordinary challenging targets or goals in two important respects • Extreme difficulty with radical expectations that go beyond current capabilities and performance. For example, Southwest Airlines' early stretch goal of achieving a 10-minute turnaround at airport gates. A familiar task was involved, but the target was a drastic departure from the industry standard at the time, which was close to one hour; and • Extreme novelty so that brand-new approaches and paths must be found to bring a stretch goal within reach. In other words, working differently, not simply working harder, is required. For example, to get gate turnarounds down to 10 minutes, Southwest Airlines had to completely overhaul its work practices and reimagine customer behaviour. The airline did, however, famously determine how to reach this goal. Stretch goals are often viewed as truly important sources of individual and organizational achievement and motivation. See also, *stretch target paradox*.
Stretch the Truth	To *stretch the truth* means to: • Exaggerate, often to the point where the truth is obscured or lost; or, • Say something that is not exactly truthful. This may be done to deceive or to embellish. For example, the shoulder of the horse I rode was taller than me. See also, *bullshit*, *bullshit detector*, *disinformation*, *ethical*, *ethics*, *fake news*, *false narrative*, *misinformation*, *smells fishy*, *truthiness*, *white lie*, and *untruth*.

Phrase or Term	Definition
String Test	Not a preferred phrase – see *complete unit test*.
Strike	Not a preferred phrase – see *labour disruption*.
Strong Artificial Intelligence, True Artificial Intelligence	*Strong artificial intelligence or true artificial Intelligence* is: • A hypothetical machine that: o Exhibits behaviour at least as flexible and skillful as humans; or, o Possesses awareness of external objects, ideas and/or self-awareness. • The philosophical position that human minds are computer programs, as named by John Searle (i.e., strong artificial intelligence). See also, *artificial intelligence* and *weak artificial intelligence*.
Strong Presence	Having a *strong presence* in business is to dominate or be noticed by competitors in the market (e.g., the investment banking industry has a strong presence in the city). Having a strong presence in social or professional circles means to be impactful and noticeable (e.g., most public speakers have a strong presence). However, verbal expression is not a prerequisite to having a strong presence. In his article *How to communicate a strong presence without saying a word*, Jake Daghe developed the concept of using body language to communicate a strong presence and make positive impressions without speaking by being: • Engaged (i.e., make eye contact); • Intrigued (i.e., ignore distractions like your telephone); • Invested (i.e., lean forward on the balls of your feet); • Open to opportunity (i.e., uncross your arms); and, • Solid (i.e., stand straight, with the correct posture). See also, *body language* and *Personality Type Theory*.
Structural Steel	*Structural steel* is a category of steel used for making construction materials in a variety of shapes, with a profile of a specific cross-section (e.g., angle, beam, pipe, plate, or tube). Structural steel shapes are usually very stiff with respect to their cross-sectional areas and thus can support a high load without excessive sagging.
Sub-assembly	A *sub-assembly* is a portion of an assembly, package, or system comprising one or more components or parts. See also, *assembly*.
Sub-assignment	A *sub-assignment* is an inspection assignment issued for another inspection discipline or sub-supplier's scope of supply. The primary assignment is usually issued for the prime supplier's scope of supply. Some purchase orders have only one inspection assignment (e.g., mechanical), while others have more than one (e.g., mechanical and electrical). See also, *assignment instruction*.
Sub-contract	Not a preferred phrase – see *outsourcing*.
Sub-contractor	A *sub-contractor* works for, and has a contract with, a prime contractor. See, also *contractor* and *prime contractor*.
Subject Matter Expert	A *subject matter expert* is a person knowledgeable and experienced with a specific body of knowledge or discipline (e.g., electrical and motors, mechanical and pumps, or materials and welding). This person may be a consultant or an employee.
Submittal	A *submittal* is a document or drawing that is submitted for approval, consideration, or review.

Phrase or Term	Definition
Substrate	A *substrate* is any surface to which a coating is applied (e.g., concrete or steel). See also, *coating* and *surface preparation*.
Sub-supplier	A *sub-supplier* works for a prime supplier and has a purchase order from a prime supplier. See, also *prime supplier* and *supplier*.
Sub-vendor	Not a preferred term – see *supplier*.
Success	*Success* is: • A lack of failure; • A person or thing that achieves desired aims or attains prosperity; • The accomplishment of an aim or purpose (e.g., a desired or favourable outcome); or, • The attainment of popularity or profit. See also, *failure*.
Suitcase Words	*Suitcase words* are words into which multiple meanings are attributed. Misunderstanding can result when listeners assume that everyone attributes the same meaning to them. For example, to some, the phrase *social justice* implies a need for more social welfare programs and a more progressive tax code. However, for others, it has the opposite meaning. Marvin Minsky, in *The Emotional Machine*, first coined this phrase. See also, *social justice*.
Summative	Something that is *summative* is brief, short and contains the main points of a larger story or body of work. For example, a table of contents in a book.
Sunk Cost Fallacy, Concorde Fallacy	*The sunk cost fallacy or Concorde fallacy* is the false assertion that investments (i.e., sunk costs) justify further expenditures. The effect is that people demonstrate a greater tendency to continue an endeavour after making an investment in money, effort, time, or combination. Those who follow this fallacy are often described as colloquially as *throwing good money after bad* (money) when they make expenditures. Refusing to succumb to this fallacy and not making additional expenditures is described colloquially as *cutting one's losses*. See also, *sunk cost*.
Sunk Cost, Retrospective Cost	The *sunk cost* or *retrospective cost* is the amount that has already been incurred or paid and cannot be recovered. It is a sum paid in the past that is no longer relevant to decisions about the future. See also, *prospective cost*.
Sunshine Clause, Sunset Provision	A *sunset clause* or *sunset provision* is a requirement within a law, regulation, or statute that it shall cease to have effect after a specific date, unless further legislative action is taken to extend or reauthorize its authority. This compels the government to periodically re-examine its delegations of authority and legislation. See also, *sunshine law* and *sunshine list*.
Sunshine Law	A *sunshine law* is a regulation requiring disclosure and transparency of official business or government actions. These laws make information such as deliberations, meetings, records, votes, and other official actions available for public examination, observation, participation, or a combination. See also, *freedom of the press*, *sunset clause*, and *sunshine list*.

Phrase or Term	Definition
Sunshine List	A *sunshine list* is a record of government or other personnel (e.g., non-profit employees) whose benefits, salary, and severance payments exceed the specified annual threshold (e.g., CAD$ 100,000). The purpose of the list is to provide accountability and transparency for organizations that receive public funding. See also, *non-profit* and *sunshine law*.
Super Majors, Big Oil	*Super majors* and *big oil* are a group of international oil companies, that have notable age, market position, or size. These are BP, Chevron Corporation, Eni, ExxonMobil, Royal Dutch Shell, and Total. ConocoPhillips has also been included in this group. These are typically integrated companies, with divisions in exploration, production, marketing, refining, transportation, and distribution (i.e., upstream, midstream and downstream). The use of this phrase excludes national oil companies and the Organization of Petroleum Exporting Countries (OPEC) oil companies, who have a much greater role in setting prices than the super majors. Non-super-major Sinopec Group, for example, is a state-owned Chinese oil company with greater revenues in 2018 than any super major. In the maritime industry, the six or seven large oil companies (e.g., including BP, Chevron Corporation, ExxonMobil, Royal Dutch Shell, and Total) that control the chartering of most of the world's crude oil tankers are also called oil majors. See also, *downstream*, *international oil company*, *midstream*, *national oil company*, and *upstream*.
Supercritical Carbon Dioxide, Supercritical CO_2	*Supercritical carbon dioxide* or *supercritical* CO_2 is the fluid state of carbon dioxide where it is held at or above its critical temperature of 31°C (88°F) and critical pressure of 7,377 kPa (1,070 psi). Carbon dioxide behaves as a gas in air at standard temperature and pressure or as a solid (i.e., dry ice) when sufficiently cooled, pressurized, or a combination. If the temperature and pressure of carbon dioxide are both increased from standard temperature and pressure to be at or above the critical point, it can adopt properties midway between a gas and a liquid. Specifically, above its critical temperature and critical pressure, it behaves as a supercritical fluid to expand to fill its container like a gas but with a density like that of a liquid. See also, *standard temperature and pressure*.
Supercritical Steam	*Supercritical steam* is a supercritical water because at supercritical pressures fluid it is considered as a single-phase substance. It is water in a state that exceeds its critical point of 22,118 kPa, 374°C (3,208 psi, 705°F). At the critical point, the latent heat of steam is zero, and its specific volume is the same whether considered gaseous or liquid. In other words, water that is at a higher pressure and temperature than the critical point is in an indistinguishable state that is neither liquid nor gas. Because supercritical steam is hotter and under more pressure than regular sub-critical steam, it can release more energy, which converts more heat into usable electricity. There is no definitive standard for ultra-supercritical steam, which typically means a steam temperature of greater than 600°C (1,112°F) or 593°C (1,100°F), according to the *Organisation for Economic Co-operation and Development*, and steam pressure in excess of 24,000 kPa (3,480 psi). See also, *standard temperature and pressure*.

Phrase or Term	Definition
Superforecasting	*Superforecasting* is the art and science of prediction, which is a skill that can be cultivated. It is the product of a way of thinking that is careful, curious, open-minded, and self-critical. Superforecasters can measurably predict events in advance. This concept was developed in the book *Superforecasting: The Art and Science of Prediction* by Philip E. Tetlock and Dan Gardner. They reported on data from *The Good Judgment Project*, which demonstrated that a group of selected *amateur forecasters* were often more accurate than individual *analysts*, *insiders*, and *experts*. For example, the result of the 2014 vote on the referendum for Scotland to separate from the United Kingdom – which was successfully predicted by the Good Judgment Project. By contrast, a forecaster plans a budget for personal finances or a capital project. See also, *creative estimating*, *crystal ball gazing*, *extrapolation*, *forecasting*, *Good Judgment Project*, *Monte Carlo simulation*, *second-order effect*, and *top-down estimating*.
Supervisory Control and Data Acquisition	*Supervisory control and data acquisition* is a control system architecture that uses computer networked data communications and graphical user interfaces for high-level process supervisory management. It also uses other peripheral devices such as a programmable logic controller and a discrete proportional–integral–derivative controller to interface with the process machinery or plant. It collects data about equipment performance such as energy consumption, levels, temperatures, and throughput. See also, *programmable logic controller*.
Supermajority	A *supermajority* is a voting requirement to obtain a specified level of support that is greater than an absolute majority or simple majority (e.g., >60% or 80%). See also, *first-past-the-post*, *absolute majority*, and *simple majority*.
Supplier	A *supplier* is a company that has been awarded a purchase order from an engineering and procurement company or their client, to provide products, services, or both, for a project. This also includes all sub-suppliers, from either another division of the same company, or a separate company. A procurement supply chain may also have sub-sub-suppliers. See also, *contractor* and *supply chain management*.

Phrase or Term	Definition
Supplier Change Request	A *supplier change request* is an offer (e.g., with the bid) or inquiry (e.g., after purchase order award) by the supplier for the project to accept an alternative to, or to change or override a requirement in the procurement documents (e.g., description or specification). This may be to: • Offer a more advantageous solution (e.g., economical, technical, or time-related); • Overcome mismatched requirements (e.g., the specified design, material, or product is not applicable or available); • Request information for clarification before proceeding with the work (i.e., request for information); or, • Request to accept a non-conforming product with a specified disposition (i.e., non-conformance report). Industry does not typically use the phrase *supplier change request*. In this document, *design change notice* is used to describe the project's authorization and record of a change to a design or specification. This procedure is usually identified in corporate or project instructions or specifications, or both. The terms deviation and waiver are therefore identified as not preferred. This procedure, however, may not address how the supplier will receive authorization to accept an alternative to, or to change or override a requirement or specification in the procurement documents. The terms used for a supplier change request may vary significantly between the owner, engineering and procurement company, and suppliers. These terms are often used interchangeably, even though others may use the same terms differently. Often, the purchase order will have boilerplate text that describes these scenarios and offers a defined process and terminology to be used. Because the purchase order is the commercial document between the project and the supplier, it is recommended that the terminology therein be used by the project, for clear understanding and consistency. Furthermore, this is an important subject that should be discussed at bid clarification and kick-off meetings. The terms that are often used to describe a supplier change request include a: • Concession request; • Deviation request; • Exception request; • Non-conformance report; • Request for information; and, • Waiver request. See also, *bid clarification meeting*, *design change notice*, *kick-off meeting*, *non-conformance report*, and *request for information*.
Supplier Document Index	The *supplier document index* is a list of supplier documents that are submitted to fulfill the supplier document requirements in a purchase order. This list correlates the supplier's document codes with the project document codes and includes metadata. A supplier's document codes may be categorized per a pre-determined list using alphabetic and numeric characters to facilitate document processing by type (e.g., in descending order and by sequence). The document codes may be assigned by a document control or engineering system, or another resource. Regardless, the intent is to effectively communicate document metadata. See also, *document code*, *metadata*, and *supplier document requirements*.

Phrase or Term	Definition
Supplier Document Requirements	*Supplier document requirements* specify the documentation deliverables in the purchase order, including drawings and data to be submitted for review and record, and with shipment. These usually appear in Section 3 of the purchase order. See also, *review code*, *supplier document code*, and *supplier document index*.
Supplier Document Requirements Code	A *supplier document requirements code* is a list containing a series of unique alphabetic and numeric characters that are assigned to each specific type of document submitted by a supplier (e.g., the *D01* in D01 mechanical calculation, the *M08* in M08 welding procedure specification, and the *Q02* in Q02 inspection and test plan). A company or project includes this list in purchase orders to systematically identify all document types. The code is assigned by an engineering and procurement company or project. See also, *supplier document requirements*.
Supplier of Choice Agreement, Supplier Frame Agreement, Master Service Agreement, Master Service and Supply Agreement. Multiple Use Contract	A *supplier of choice agreement*, *supplier frame agreement*, *master service agreement*, *master service and supply agreement*, or *multiple use contract* is a legally-binding contract between two parties in which the parties agree that the terms documented therein will govern future transactions or future agreements (i.e., to establish the terms and conditions or pricing of future purchase orders). This is not the same as a purchase order. See also, *purchase order*.
Supplier Quality Representative	The *supplier quality representative* or SQR receives the inspection assignment and performs inspection activities at the supplier's facility. The SQR may be a contractor assigned through an inspection agency or work directly for the project. The SQR may be a client or engineering and procurement company employee, including someone assigned or designated by the client, or the supplier quality surveillance coordinator. See also, *supplier quality surveillance*.
Supplier Quality Surveillance	*Supplier quality surveillance* is the selective audit, evaluation, inspection, observation, and review, of engineering and procurement processes, fabrication and manufacturing operations, and quality management systems to determine a supplier's compliance with purchase order requirements. These activities may be conducted before and after the award of a purchase order. See also, *quality management*, *supplier quality surveillance coordinator*, and *risk management*.
Supplier Quality Surveillance Administrator	A *supplier quality surveillance administrator* provides data management and administrative support to the SQS coordination lead and team. Some projects are large enough to have a team of SQS coordinators; hence, there is enough work to warrant a dedicated person for this role.
Supplier Quality Surveillance Coordinator (SQS Coordinator)	A *supplier quality surveillance coordinator* or *SQS coordinator* is a subject matter expert responsible for planning and supervising quality surveillance activities. The SQS coordinator provides technical and quality support to the project team and may also perform inspection activities. See also, *coordinate* and *supplier quality surveillance*.
Supplier Quality Surveillance Folders	*Supplier quality surveillance folders* are electronic folders that reside on a project's shared network. They are used to store electronic versions of project information, including documents, drawings, and other information.

Phrase or Term	Definition
Supplier Quality Surveillance Log (SQS Log)	A *supplier quality surveillance log* or *SQS log* is a spreadsheet or other resource (e.g., Microsoft Access, Excel, or SharePoint file or site) that is used to record and analyze information about inspection assignments. A log may have few or many data entry points. For example, purchase order and assignment numbers, supplier name and location, equipment description, report number, hours budgeted and used, quality surveillance representative and agency names, and status as pending, open, or closed. See also, *supplier quality surveillance*.
Supplier Team	A *supplier team* comprises personnel responsible for preparing, issuing, and managing requisitions and purchase orders. This team typically employs an account manager, engineer or technical specialist, expediter, production manager, quality manager, salesperson, subject matter experts, and other personnel (depending on project size, composition, and complexity). See also, *project team* and *requisition team*.
Supply Chain	*A supply chain* is a system of organizations, people, activities, information, and resources involved in producing and distributing or moving a product or service from the supplier to a customer. Supply chain activities involve the transformation of natural resources, raw materials, and components into a finished product. See also, *customer*, *supply chain management*, and *value chain*.
Supply Chain Management	*Supply chain management* is the process of strategically managing business relationships that involve the flow of goods, services, finance, and knowledge to realize economic value. The core activities for a project are procurement, expediting, and logistics. Supplier quality surveillance activities may also be included. See also, *goods*, *management*, and *supply chain*.
Supply Chain Risk	*Supply chain risk* is the risk of disruption in the quality or to the quantity of products or services sourced from suppliers. See also, *supplier* and *supply chain*.
Supply-side Economics	*Supply-side economics* is an economic theory that proposes greater tax cuts for investors and entrepreneurs provide incentives to save and invest, which produce economic benefits that trickle down into the overall economy. This means that the production of goods or services, or supply, is of primary importance in economic growth. The factors of production are capital, entrepreneurship, labour, and land. Supply-side fiscal policy focuses on creating a better climate for businesses. This is also known as Reaganomics or trickle-down economics. See also, *classical economic theory*, *demand-side economics*, fiscal policy, and *goods*.
Support Group	A *support group* is a group of peers with common concerns or experiences who provide mutual support for one another (e.g., coping strategies, emotional and moral support, relevant information, and a sense of community). A support group may or may not be led by a professional. For example, a career search group that helps members find employment. See also, *self-help*.
Surface Preparation	*Surface preparation* is the process of treating (e.g., cleaning or preparing) a substrate or surface for coating or other use. Surface preparation can be performed mechanically, chemically, or both. See also, *coating* and *substrate*.
Surveillance Point	Not a preferred phrase – see *verification point*.

Phrase or Term	Definition
Sustainability	*Sustainability* is: • A focus on meeting the needs of the present generation without compromising the ability of future generations to meet their needs. This concept comprises three pillars: economic, environmental, and social, which are also known informally as profits, planet, and people; • Avoiding depleting natural resources and instead maintaining an ecological balance; • The ability of something to be maintained at a certain level (e.g., a waterline) or rate (e.g., the number of trees harvested compared vs. the number of new trees planted to replace harvested trees); or, • The ability to co-exist constantly. In the 21st century, this refers generally to the *co-existence capacity* of the biosphere and human civilization. Sustainability is a major component of business, government, environmentalism, productivity and many other areas. See also, *environmentalism, natural resource, self-sustaining, sustainable development, Sustainable Development Goals*, and *three Ps of sustainable development*.
Sustainable Development	*Sustainable development* is the holistic approach and temporal processes that leads to the end point of sustainability. See also, *sustainability, Sustainable Development Goals*, and *three Ps of sustainable development*.
Sustainable Development Goals	The *Sustainable Development Goals* are an urgent call for action by all countries - developed and developing - into a global partnership. It is recognized that ending poverty and other deprivations must go hand-in-hand with strategies to improve health and education, reduce inequality, and spur economic growth – all while tackling climate change and working to preserve forests and oceans. The *2030 Agenda for Sustainable Development*, which includes 17 sustainable development goals adopted by all United Nations' Member States in 2015, provides a shared blueprint for peace and prosperity for people and the planet, now and into the future. See more information at the United Nations Sustainable Development website: https://sustainabledevelopment.un.org/post2015/transformingourworld. See also, *Millennium Development Goals, sustainability, sustainable development*, and *three Ps of sustainable development*.
Swage	A *swage* is: • A shaped tool or die for giving a desired form to metal by hammering or pressure; or, • A groove, ridge, or other molding on an object. See also, *Swagelok fitting*.
Swagelok® Fitting	A *Swagelok® fitting* is part of a proprietary leak-tight mechanical joining system that uses Swagelok technology for tubing systems. Swagelok manufactures many other types of components, fittings, and instruments. See more information at the *Swagelok®* website: https://www.swagelok.com. See also, *leak-tight, proprietary* and *swage*.
Sweet Service	*Sweet service* describes equipment (e.g., pressure vessel or tank) or piping system that processes or transports crude oil or natural gas that is not sour. See also, *sour service*.

Phrase or Term	Definition
Swiss Cheese Model, Cumulative Act Effect	The *Swiss cheese model* or *cumulative act effect* is a representation of accident causation used in risk analysis and risk management. It acknowledges that, although many layers of defense or security may lie between hazards and accidents, there are flaws in each layer that, if aligned, will allow an accident to occur. This model correlates defense systems with multiple slices of swiss cheese (which contain holes of various sizes) that are layered behind each other, so that the risk of a threat becoming a reality is prevented by the different layers and types of defenses (where the cheese holes are not aligned). The theory proposes that lapses and weaknesses in one defense do not permit a risk to materialize because other defenses exist to prevent a single point of failure. However, sometimes the holes do line up. This model was originally proposed by Dante Orlandella and James T. Reason. See also, *risk*, *risk assessment*, and *risk management*.
Synchronized	To be *synchronized* is to behave in complementary, tandem, or a simultaneous rate or time. See also, *unsynchronized*.
Synergy	*Synergy* is the interaction or cooperation of two or more organizations to produce a combined effect greater than the sum of their individual efforts. For example, the simultaneous procurement, supply, and delivery of equipment create synergy with cost-savings for a project and supplier, including the synergies of staggering production and delivery over time. See also, *win-win*.
Syngas, Synthesis Gas, Substitute Natural Gas	*Syngas* (an abbreviation for synthetic natural gas), *synthesis gas* or *substitute natural gas* is a fuel-gas mixture comprising primarily hydrogen and carbon monoxide. It may also contain carbon dioxide, nitrogen, and small particles of ash, char, oil, and tar. It can be produced from any hydrocarbon or organic material (e.g., biomass, coal, liquid petroleum gas, naphtha, refuse, and wood). It is produced using these methods: • Auto-thermal reforming (a combination of gasification and steam-methane reforming); • Gasification; or, • Steam-methane reforming. See also, *biomass*, *gasification*, *hydrocarbon*, and *hydrogen*.
Synthesis	*Synthesis* is the combining of two or more things to create something new. Synthesis is about understanding the parts and the whole at the same time, along with the connections and relationships that make up the dynamics of the whole. Essentially, it is the ability to see interconnectedness. See also, *analysis*, *causality*, *emergence*, *feedback loop*, *interconnectedness*, *systems mapping*, *and systems thinking*.
Synthetic Crude Oil	*Synthetic crude oil* is a manufactured crude oil comprised of naphtha, distillate, and gas-oil-boiling range material. It can range from high quality and light sweet bottomless crude to heavy and sour blends.
Synthetic Methane	*Synthetic methane* is methane produced from a biomethanation process using hydrogen produced by electrolysis and carbon dioxide from a source such as anaerobic digestion, industrial processes, or water treatment. See also, *methane*, *natural gas*, and *renewable natural gas*.
System Integration and Functionality Test	A *system integration* or *system integration and functionality test* is carried out in an integrated hardware and software environment to verify the behaviour of the complete system.

Phrase or Term	Definition
System Test	A *System test* analyzes control safety, trip or shutdown, electrical or telecommunications systems. This test may include integration of all interfaces and connectivity to demonstrate compliance with purchase order or project requirements (i.e., overall test with all systems fully assembled and interconnected in the supplier's facility, including central equipment with a representative number of field equipment items).
System Thinking, Systems Thinking	*System thinking* or *systems thinking* is a method of enabling people to view systems from a broad perspective that includes seeing overall cycles, patterns, and structures in systems, rather than only specific events. This broad overview allows for the quick identification of the real causes of an organization's challenges and where work is needed to address them and improve performance. Personnel are surrounded by systems that manage processes, products, and projects, which are increasingly complex and interrelated. For example, Google began with two people and an algorithm. As the business grew, so too did its tangible footprint. Today, Google owns enormous data centres that use significant amounts of energy, which has driven them to become the world's largest investor in renewable energy. The most important themes in this method are: 1. Causality to decipher the way things influence each other in a system; 2. Emergence: seeing larger things emerge from smaller connecting parts; 3. Feedback loops to observe, understand, and intervene; 4. Interconnectedness to shift mindsets from linear to circular; 5. Synthesis to combine two or more things and create something new; and, 6. Systems mapping to identify and map the elements of *things* within a system, and understand how they interact, interconnect, and interrelate. Systems thinking is referred to as *the fifth discipline* because it integrates the four other disciplines of a learning organization. See also, *causality*, *emergence*, *feedback loop*, *interconnectedness*, *learning organization*, *synthesis*, *systems mapping*, and *the fifth discipline*.
Système international d'unités	*Système international d'unités* is the modern form of the metric system. See also, *International System of Units*.
Systems Completion	The *systems completion* process is the sequential activities within a project to verify and prove that the selection, design, construction, installation, integration, testing, and preparation of systems have been completed correctly. It confirms the facility is ready for start-up and operations. It is the structured process through which the technical integrity of a newly-built or modified facility is verified. Assets are ultimately turned over from the (commissioning) project team to base business owners (facilities engineering, maintenance, and operations) for the life of the assets. Industry-wide application of this process is guided by API RP 1FSC, *Facilities Systems Completion Planning and Execution*.

Phrase or Term	Definition
Systems Mapping	*Systems mapping* identifies and maps the elements of things within a system to understand how they act, interconnect, and relate in a complex system. A system map shows the components and boundary of a system and the components of the environment at a point in time. Unique insights and discoveries can then be used to develop interventions, shifts, or policy decisions that will dramatically change the system in the most effective way. Methods include analog cluster mapping and complex digital feedback analysis. See also, *causality*, *emergence*, *feedback loop*, *interconnectedness*, *synthesis*, and *systems thinking*.

Phrase or Term	Definition
T	1 A B C D E F G H I J K L M N O P Q R S **T** U V W X Y Z
Tack Weld	A *tack weld* is a weld made to hold metal parts (e.g., pipe and fittings) in alignment until the final weld is made. A tack weld may be removed or incorporated into the final weld.
Tag	A *tag* is a small piece of stainless steel or sheet metal that contains a tag number (e.g., written, marked or engraved), which is attached to equipment (e.g., with adhesive, rivets, or wire). See also, *tag number* and *name plate*.
Tag Line	A *tag line* is a catchphrase or slogan, especially as used in advertising. See also, *motto* and *slogan*.
Tag List	A *tag list* is a record or register of tag numbers, which may also include other information about the equipment. See also, *tag number*.
Tag Number	A *tag number* is a unique identifier (number) designated by the project for each piece of equipment and is used throughout an asset's entire life cycle. Typically, a tag number is etched or stamped on stainless steel or engraved on a lamacoid name plate. The tag number may also be marked on the equipment name plate. The tag number will also be identified in the purchase order (e.g., scope of supply and data sheet) and in other supplier and owner documents. Instead of a tag number, a commodity number may be used for bulk valves. See also, *asset*, *life cycle*, *name plate*, *serial number*, *tag*, and *tagged equipment*.
Tag Out	*Tag out* is a labelling process that is always used when lockout is required. The process of tagging out a system involves attaching or using an information tag or indicator (usually a standardized label) that includes the following information: • The name of the authorized person who attached the tag and lock to the system; • Time of application of the lock and tag; and, • Why the lockout/tag out is required (e.g., maintenance, repair, or testing). Only the authorized individual who placed the lock and tag onto the system is permitted to remove it, to ensure that the system cannot be energized without the authorized individual's knowledge. See also, *car seal* and *lockout*.
Tagged Equipment	*Tagged equipment* is identified with a tag number. Some tagged equipment have a data sheet (e.g., gauges and pumps), while others do not (e.g., ladders and platforms). See also, *data sheet*, *tag number*, and *untagged equipment*.
Tail Wagging the Dog	*Tail wagging the dog* is an old expression used to describe a backwards situation in which someone or something important or powerful such as a dog (e.g., an organization or person), is being controlled by someone or something that is much less important or powerful such as the dog's tail. For example, when a small vocal minority acquires what it demands even though in is against the wishes of the silent majority. The phrase originated in Tom Taylor's play *Our American Cousin*. See also, *to put the cart in front of the horse* and *wag the dog*.
Tailings	*Tailings* are a combination of water, sand, silt, and fine clay particles that are a by-product of extracting bitumen from oil sands. See also, *fine tailings*, *mature fine tailings*, *slurry*, and *tailings*.

Phrase or Term	Definition
Tailings Dam, Tailings Pond	A *tailing dam* or *tailings pond* is an area where slurry or tailings are drained or pumped, to allow sedimentation (separation) of the solids from the water. See also, *slurry* and *tailings*.
Take It to The Grave, Carry It To Your Grave	*Take it to the grave* means to keep something a secret for the rest of your life and to never reveal it. For example, the escape artist Harry Houdini never published or revealed his methods; thus, he took his secrets to the grave. A synonymous version of this expression is *carry it to your grave*.
Talent	*Talent* is: • A person or people engage in a common goal, activity, or discipline (e.g., a project team); • A person's natural aptitude, endowments, or skills (e.g., Jill excels at math); or, • Intelligence or mental power (i.e., an artistic or creative aptitude). See also, *resource* and *talent stack*.
Talent Stack	A *talent stack* is a set of ordinary skills that, combined, make someone more valuable than others in the marketplace. For example, Scott Adams is a famous syndicated cartoonist who has little artistic talent and never took a college-level writing class. But few people are good at *both* drawing and writing. These skills, together with his business skills, sense of humor, risk tolerance, and strong work ethic, make him unique. That uniqueness has commercial value. This concept was developed by Adams, who is the author and creator of the comic strip Dilbert. See also, *Dilbert* and *talent*.
Talking Head	A *talking head* is: • A television personality who appears in a headshot; • Someone who appears in a discussion program or interview to give their opinion; or, • The televised headshot of a person who is talking. This can be a derogatory phrase that describes someone who is boring or never stops talking. See also, *headshot*, *thinkfluencer*, and *thought leader*.
Talking Your Book	*Talking your book* is an expression that means someone: • Has a vested interest in an outcome, and speaks in an authoritative way to present something as factual but only to support their position; • Is arguing in favour of something (e.g., an ideology or investment) that would benefit them (e.g., arguing that physical fitness is a noble goal when selling physical fitness equipment); or, • Is literally talking about a book they published and recommending that others buy it. A true authoritative expert will objectively and scientifically discuss the facts and not *talk your book*. See also, *subject matter expert*.
Tank	A *tank* is a container designed to hold gases or liquids at a pressure substantially the same as the ambient pressure (e.g., internal or external). See also, *aboveground storage tank*, *atmospheric pressure*, *field-erected tank*, *shop-fabricated tank*, *pressure vessel*, and *underground storage tank*.

<table>
<tr><th>Phrase or Term</th><th>Definition</th></tr>
<tr><td>Tank Car</td><td>A tank car is railroad car used for transporting bulk materials (e.g., gaseous, granular, liquid, or slurry).</td></tr>
<tr><td>Tar Sands</td><td>Not a preferred phrase – see oil sands.</td></tr>
<tr><td>Task</td><td>A task is an activity or unit of work that needs to be completed to progress project goals. Typically, a task must be completed by a set deadline. Tasks may be further broken down into assignments or subtasks. See also, task analysis and work.</td></tr>
<tr><td>Task Action Results Formula, TAR Formula</td><td>The task action results formula or TAR formula is a method of writing the experience section in a resume. It uses the following structure to provide information relating to the capabilities and competencies demonstrated in previous positions or roles:<ul><li>Task (or action which is often the same thing); and,</li><li>Results.</li></ul>For example:<blockquote>Using proprietary datasets and software, I conceived, designed, and implemented a full suite of estimating guidelines, presentations, processes, procedures, templates, and workflows to ensure standardized estimating practices across all Canadian and US offices.
In the previous paragraph, the unitalicized text is the task; italicized text is are the results.</blockquote>See also, resume, situation objective action results aftermath method, and situation task action result method.</td></tr>
<tr><td>Task Analysis</td><td>A task analysis details the actions or resources required to complete a task.
See also, task.</td></tr>
<tr><td>Team</td><td>A team is a group of people with complementary skills required to complete a task, job, or project. Team members:<ol><li>Operate with a high degree of interdependence;</li><li>Share authority and responsibility for self-management;</li><li>Are accountable for the collective performance; and,</li><li>Work toward common goals and shared rewards.</li></ol>A team becomes more than just a collection of people when a strong sense of mutual commitment creates synergy; thus, generating greater performance than the sum of the performance of its individual members. Teams are sometimes referred to as A, B and C teams:<ul><li>An A Team comprises elite members of an organization; and,</li><li>B Team and C Team comprise non-elite team members.</li></ul>It is unlikely that a project is staffed with only elite personnel. The reality is that personnel will always bring a variety of abilities, experience, knowledge, and skills. Smart projects assign the right team – a team that possesses adequate and complementary skills, and ensures it remains in place for the duration of the project. According to Simon Sinek: a team is not a group of people who work together. A team is a group of people who trust each other. See also, Dunbar's number and tribe.</td></tr>
</table>

Phrase or Term	Definition
Technical Bid Evaluation	A *technical bid evaluation* is the comparison or review of a bid per technical related items or conditions. It is performed based on criteria that include: • Applied, practical, or scientific, capabilities; • Compliance with regulations and specifications; and, • Operating cost, quality, and support (service during or after sales). See also, *bid* and *commercial bid evaluation*.
Technical Design Change	Not a preferred phrase, see design change.
Technical Deviation	Not a preferred phrase – see *design change*.
Technical Deviation Notice	Not a preferred phrase – see *design change notice*.
Technical Document	A *technical document* is a digital or printed record that is referenced in a purchase order. These may be industry documents (e.g., code, standard, and regulation) or owner documents (e.g., data sheet, drawing, and specification). Typically, industry documents are not attached (i.e., obtained separately by the supplier) to the purchase order. Owner documents are attached to the purchase order. See also, *hard copy*, and *technical document list*.
Technical Document Control	Not a preferred phrase – see *document control*.
Technical Document List	A *technical document list* details all the referenced documents that are attached to or referenced in a purchase order. Technical notes may also be listed or attached. This usually appears in Section 2 of the purchase order. See also, *technical document* and *technical note*.
Technical Note	A *technical note* is a statement that provides additional technical details or information for an application, product, or technology in the scope of supply in a purchase order. It may amend, correct, explain, supersede, or supplement requirements stated in the technical documents. A technical note may be listed in the *technical document list*, provided as a separate attached document or listed in a data sheet. See also, *data sheet*, *technical document*, and *technical document list*.
Technicality	A *technicality* is: • A point of law or a small detail in a set of rules or specification (i.e., based on a strict interpretation). For example, if a sprinter crossed the finish line first but was disqualified because the running shoes used were prohibited; • Something technical, especially a detail meaningful only to a specialist. For example, there are 43 species of New World sparrows commonly found in Canada and the United States (Order: Passeriformes, Family: Passerellidae). Technicality does not have a legal definition or an exact meaning. A technicality can impeded or clarify strict adherence to the letter of the law and often prevents the spirit of the law from being enforced. It also denotes any point that is advantageous for, or interferes with, the user's desired outcome. See also, *academic*, *letter of the law* and *spirit of the law*.
Technician, Engineering Technician	A *technician* or *engineering technician* is a professional with a practical understanding of technology. In Canada, a certified engineering technician may have a Certified Technician designation. See also, *engineer* and *technology*.

Phrase or Term	Definition
Technologist, Engineering Technologist	A *technologist* or *engineering technologist* is a professional who follows as a profession or is trained in a branch or discipline of applied engineering, or technology. In Canada, a certified engineering technologist may have the following designations: • Applied Science Technologist; • Certified Engineering Technologist; • Professional Technologist; • Registered Engineering Technologist; or, • Technologue Professionnel. See also, *applied engineer*, *applied engineering*, *engineer*, and *technology*.
Technology	*Technology* is: • A method to accomplish an objective or task, or produce a product or service, using knowledge, processes, techniques, or skills; • The practical application of knowledge, especially in a particular area; or, • The specialized aspects of a particular discipline or field of endeavour. See also, *applied engineering*, *discipline*, *engineering technology*, and *engineering*.
Technology Hub, Tech Hub	A *technology hub* or *tech hub* is a group of buildings near amenities that encourage collaboration and networking, resulting in numerous small technological start-ups growing into midsized and larger organizations. For example, the downtown core of Calgary, Alberta, Canada has been a technology hub for the creation and growth of oil companies since the 1970s. See also, *cluster*.
Technology Quotient	*Technology quotient* is a person's ability to adapt, integrate, and manage technology based on the needs of the current organization or project. See also, *technology*.
Technology Transfer	*Technology transfer* is the process of moving inventions into commercial production. See also, *technology*.
Technophobe	*A technophobe* is a person who a person who avoids, dislikes, or fears complex devices or new technology (e.g., computers or smart phones). Technophobes may also have concerns or fear about: • Corporate and government tracking of individuals; • Data collection, misinformation, and security; or, • Freedom or privacy. See also, *Luddite*.
Teflon® or Polytetrafluoroethylene	*Teflon®* or *polytetrafluoroethylene* is a tough synthetic resin that repels liquid. It is made by polymerizing tetrafluoroethylene and used to make bearings, gaskets, and seals. See also, *gasket*.
Teflon® Tape	*Teflon® tape* is a proprietary tape that used to seal pipe threads. See also, *proprietary*, *Teflon®*, and *thread*.
Teleology	*Teleology* is the study of design or purpose as an explanation of a natural phenomenon or processes (i.e., being directed toward an end or shaped by a purpose). The Greek philosopher Aristotle's classic example is that the oak tree is the natural purpose of the acorn. See also, *etiology*.
Telltale Hole	Not a preferred phrase – see *vent hole*.

Phrase or Term	Definition
Temperature Stick, Temp Stick, Tempilstik®	A *temperature stick* or *temp stick* is a temperature-indicating crayon that leaves a mark on a substrate (e.g., pipe) that is used to measure temperature quickly. When the substrate is heated, it melts the mark at a specified minimum temperature. Temp sticks provide a broad range of temperature measurements: from 38°C (100°F) to about 1,370°C (2,500°F). They are used during welding to check preheat, inter-pass, or post-weld temperatures, and for other heat treatments. Several marks from several crayons identify ranges of temperatures. For example, when the one temp stick mark reaches the desired temperature, it melts. If the second temp stick mark, which melts at a higher temperature, does not melt the welder knows the approximate temperature range of the substrate. *Tempilstik®* is the tradename for the original temperature-indicating stick.
Template	A *template* is: • A form or a sample document that has already some details in place, with placeholders for information to be added, changed, completed, or removed by hand or with software. After completion, the form can be edited, managed, saved and printed as hard copy, and emailed as an electronic file. This standard documentation approach allows initial setup and configuration time to be minimized when creating standardized documents and forms; or, • A shaped piece of metal, paper, plastic, wood, or other material used as a gauge, mold, or pattern to guide cutting, drilling, measuring, painting, and shaping processes. See also, *form*.
Tender	Not a preferred phrase – see *bid*.
Tender Loving Care, TLC	*Tender loving care* or *TLC* is considerate and compassionate care or extra attention to make someone or something feel or look better. For example, a sad friend or sick child may need TLC. Another example: an old house or old vehicle may need TLC (e.g., a fresh coat of paint).
Tenth Man, Inner Contrarian, Devil's Advocate	*Tenth Man*, *inner contrarian*, and *devil's advocate* describes a situation where someone who, not unlike a lawyer, argues a position they do not necessarily agree with or argues an unpopular position (e.g., a project procurement plan cannot be improved because it contains adequate procedures with no erroneous assumptions or incomplete sections, and is missing no critical information). The objective of this exercise is to debate and examine a point of view using valid reasoning. A contentious opinion may be expressed to provoke debate or test the strength of opposing arguments. If all members in a group (e.g., ten members) agree that a plan or strategy is correct, then the tenth person should take a contrarian view to facilitate thorough consideration of a plan or strategy. This is effective at preventing groupthink, where flaws may be willingly overlooked to acquire group consensus. See also, *diffusion of responsibility* and *groupthink*.
Termination for Cause, Termination With Cause	*Termination for cause* or *termination with cause* is when an employee's employment or service is ended due to their egregious actions in the workplace. This may be from an assault, dishonesty, fraud, insubordination, misconduct, negligence, poor performance, theft, or violation of a law, regulation, or rule. Being terminated and being fired are synonymous and indicate the end of employment. See also, *termination without cause*.

Phrase or Term	Definition
Termination Without Cause	*Termination without cause* is when an employee's employment or service is terminated due to reasons other than their own actions. For example, due to a business closure, the completion of a project, or the lack of work. Employment agreements are not perpetual, so employment is not guaranteed. Without cause termination is a method that allows an employer to manage their labour force. This is also referred to as being *laid off*. See also, *termination for cause*.
Terms and Conditions	*Terms and conditions* are the general and special arrangements, provisions, requirements, rules, specifications, and standards that form an integral part of an agreement, contract, or purchase order. See also, *contract* and *purchase order*.
Test Report	A *test report* is a document that records data obtained from an experiment or evaluation in an organized manner (date of test and identification of the item tested), describes the environmental or operating conditions, and includes a comparison of test results. It may also identify the test criteria, objectives or purpose, names of personnel who performed the test, and referenced procedures.
That is Right, You are Right	The phrases *that is right* and *you are right* sound very similar but during negotiations have very different meanings. • That is Right Indicates that one has assessed what someone else has said and pronounced it correct by their own free will. It means that rapport and trust has been established, which leads to the best outcomes. • You are Right Used to make someone leave or stop talking. It means that there is no agreement with what someone else has said and nothing will change. *That is right* may also be expressed in a way with statements of synonymous meaning, such as: *that is it exactly* or *you hit the nail in the head*. *That is correct* is a more powerful statement than *yes* during a negotiation, because *yes* is meaningless without an agreement about *how* (e.g., something will be done). This concept was developed by Chris Voss in the book *Never Split The Difference: Negotiating As If Your Life Depended On It* (with Tahl Raz). See also, *win-win*.
The Chair Test	*The Chair Test* is a simple but powerful decision-making exercise that uses two chairs when one is at a crossroads, to remain true to one's purpose. One chair is the *should* chair and the other chair is the *must* chair. When an important decision shall be made, one sits in the should chair and writes down the reasons why it should be done. Then, one sits in the must chair, and writes down the reasons it must be done. The reasons why something must be done are those that align with one's passion or purpose. This concept was developed by Elle Luna in her book *The Crossroads of Should and Must: Find and Follow Your Passion*. See also, *purpose* and *why*.

<table>
<tr><th>Phrase or Term</th><th>Definition</th></tr>
<tr><td>The Dirty Words</td><td>The dirty words are seven vulgar English words that cannot be spoken on television or in polite conversation. These words were first listed in 1972 by American comedian George Carlin, in his monologue Seven Words You Can Never Say on Television in the recording Class Clown. These words are listed at https://en.wikipedia.org/wiki/Seven_dirty_words. There are also many other words that are considered inappropriate and too vulgar for television or polite conversation.
In his book The Dirty Words: Change Your Language, Change Your Life, Randy Lennon describes dirty words as those that are ambiguous, damaging, disempowering, and destructive to, and undermining of oneself. These are common words that may be used every day without an awareness of their impact including: but, cannot, have to, hope, should, try, and wish.</td></tr>
<tr><td>The 11 Laws of the Fifth Discipline</td><td>The 11 Laws of the Fifth Discipline are advantageous habits or mindsets developed by Peter Senge in his book The Fifth Discipline: The Art and Practice of the Learning Organization. These habits or mindsets support a learning organization:
1. Today's problems come from yesterday's solutions;
2. The harder you push, the harder the system pushes back;
3. Behaviour grows better before it grows worse;
4. The easy way out usually leads back in;
5. The cure can be worse than the disease;
6. Faster is slower;
7. Cause and effect are not closely related in time and space;
8. Small changes can produce big results ... but the areas of highest leverage are often the least obvious;
9. You can have your cake and eat it too ... but not all at once;
10. Dividing an elephant in half does not produce two small elephants; and,
11. There is no blame.
See also, learning organization, the fifth discipline, and the learning disabilities.</td></tr>
</table>

<table>
<tr><th>Phrase or Term</th><th>Definition</th></tr>
<tr><td>The Customer is Always Right</td><td>The customer is always right is a phrase that encourages service personnel to give customer satisfaction the highest priority. This phrase was coined in 1905 by Marshall Field, the founder of the Marshall Field and Company, a Chicago-based department store. This phrase acknowledges that:
• A dissatisfied customer can be bad for business (e.g., by giving a poor testimonial);
• Customers are needed, so they must feel appreciated and welcomed; and,
• It is ethical to offer excellent customer service, loyalty, and respect.
The customer is always right advocates good intentions, however, this guideline may outdated because:
• Bad customers receive an unfair advantage because, by complaining, the merchant will likely acquiesce;
• Service personnel may feel their expertise and opinions are not valued;
• Some customers are abusive or frustrating, and bad for business (e.g., by disrupting personnel and processes that may reduce profits); and,
• Sometimes customers are wrong.
American business magnate Steve Jobs did not believe in this saying. He said I think Henry Ford once said: If I asked customers what they wanted, they would have told me 'a faster horse'. People do not know what they want until you show it to them. See also, ethical and let the buyer beware.</td></tr>
<tr><td>The Fifth Discipline</td><td>The Fifth Discipline integrates the four disciplines of a learning organization, which are:
1. Personal mastery
Continually clarifying and deepening one's vision, focusing energies, developing patience, and perceiving reality objectively.
2. Mental models
Deeply ingrained assumptions or generalizations that influence how one understands the world and acts.
3. Building shared vision
Unearthing a shared vision of the future that fosters genuine commitment and enrollment rather than compliance.
4. Team learning
The capacity of team members to suspend assumptions and genuinely think collaboratively.
Systems thinking is the discipline that integrates the other four disciplines. This concept was developed by Peter Senge in his book The Fifth Discipline: The Art and Practice of the Learning Organization. See also, learning organization, the 11 Laws of the Fifth Discipline, learning organization, the learning disabilities, and system thinking.</td></tr>
<tr><td>The Golden Rule</td><td>The Golden Rule, which is a biblical reference, is: do unto others as you would have them do unto you. Put another way: treat other people as you would like to be treated. The Golden Rule is closely associated with Christian ethics, and prominent in many ethical traditions and religions. It was endorsed in 1993 by 143 leaders of the world's major faiths as part of the Towards a Global Ethic: An Initial Declaration, which was written by Hans Küng. See also, ethics, ethical and honest.</td></tr>
</table>

Phrase or Term	Definition
The Good Judgment Project	*The Good Judgment Project* began in 2011 to harness crowd wisdom to forecast world events. It was co-created by Barbara Mellers, Don Moore, and Philip E. Tetlock, who were professors at the University of Pennsylvania. They started a commercial spin-off called Good Judgment Inc. in 2015. For example, the Good Judgment Project's superforecasters correctly forecasted (as a group) *no* as the result of the 2014 vote on the referendum for Scotland to separate from the United Kingdom. The no side won with 55.3%; the yes side had 44.7%. The project even beat the odds of the British betting markets that made bets with real money. See also, *superforecasting*.
The Learning Disabilities	*The learning disabilities* are deleterious habits or mindsets developed by Peter Senge in his book *The Fifth Discipline: The Art and Practice of the Learning Organization*. These habits or mindsets do not support a learning organization: 1. I am my position; 2. The enemy is out there; 3. The Illusion of taking charge; 4. The fixation on events; 5. The parable of the boiling frog; 6. The delusion of learning from experience; and, 7. The myth of the management team. See also, *boiling frog syndrome*, *learning organization*, *the 11 Laws of the fifth discipline*, and *the fifth discipline*.
The Left Hand Does Not Know What the Right Hand is Doing	*The left hand does not know what the right hand is doing* is a criticism that describes uncoordinated and especially contrary actions, especially in a large organization with different business units or departments that communicate poorly. For example, an invoice cancelled by customer services that still charges the customer for accounting interest charges. This phrase originated from the *Bible* (New Testament Matthew 6:3) and is part of the Sermon on the Mount. In this context, it had a different meaning – to deliberately not let the left hand know what the right hand is doing (i.e., to *not* publicize one's good deeds).
The Press	*The press* refers to the personnel (e.g., editors, photographers, and reporters) who work for media companies (e.g., magazines, newspapers, radio, television, website, or another outlet). See also, *alternative media*, *freedom of the press*, and *mainstream media*.

Phrase or Term	Definition
Theories of Exchange Rates	The top four *theories of exchange rates* for currencies are: • Interest rate parity or covered interest parity This theory applies to financial securities, assumes an investor can invest in the currency of the country of residence or a foreign currency, money can freely flow between both the countries, and transaction costs are zero. It states that there is a link between the nominal interest rates in two countries and the exchange rate between their currencies. • International fisher effect or uncovered interest parity This theory states that the forward rate and the spot rate will be identical because, even without covering exchange rate risk in the forward (i.e., future) market, actions of market participants will make them equal. • Purchasing power parity This theory applies to commodities, assumes conversion and transportation costs are zero, and there are no restrictions on the movement of commodities between countries. There are two variant theories: o Absolute theory An identical product should have the same price in two countries. o Relative theory When the inflation rate is higher in country X than in country Y, the price of goods in country X will increase more than the price of goods in country Y. • Unbiased forward rate This theory states that the forward rate is an unbiased predictor of the expected spot rate because the actions of market participants make the forward rate equal to the expected future spot rate. See also, *Big Mac Index*, *exchange rate*, *forward rate*, *random walk theory*, and *spot rate*.
Theory of Constraints	The *theory of constraints* explains that any process is limited from optimum performance by its weakest link or links. For example, this can be a: • Market force Sales are slow but more goods are continually produced. • Paradigm A culture or mindset that can be difficult to overcome. • Resource Lack of equipment, personnel, or space. • Policy Cumbersome contracts, procedures, or regulations. This theory is a methodology used for process or quality improvement. The main goal is to remove barriers, bottlenecks, and constraints. See also, *bottleneck*, *goods*, and *theory of constraints methodology*.
Theory of Constraints Methodology	The *theory of constraints methodology* involves identifying the weak links in a process using a strategy called focusing, and then improving them until they no longer limit performance. See also, *theory of constraints*.

Phrase or Term	Definition
Thick Wall Pressure Vessel, Thick-walled Pressure Vessel	A *thick wall pressure vessel* or *thick-walled pressure vessel* has a ratio between its radius and wall thickness that is less than 10, using this formula: radius / wall thickness < 10 During operation (e.g., while under internal pressure), greater circumferential stress is developed on the inside surface of a thick wall, with reduced stress towards the outside surface. See also, *thin wall pressure vessel.*
Thin Wall Pressure Vessel, Thin-walled Pressure Vessel	A *thin wall pressure vessel* or *thin-walled pressure vessel* has a ratio between its radius and wall thickness that is greater than 10, using the formula: radius / wall thickness > 10 During operation (e.g., while under internal pressure), the circumferential stresses developed in a thin wall are assumed to be uniform. See also, *thick wall pressure vessel*.
Think Tank, Policy Institute, Research institute	A *think tank*, *policy institute*, or *research institute* performs advocacy and research on topics including economics, politics, and technology. Most are non-governmental organizations, but some are associated with particular businesses or political parties, or are semi-autonomous within government. Funding may include government grants, individual contributions and donations, or a combination. Think tanks publish articles, draft legislation, and studies on policy matters of interest. Think tanks yield information that is used by businesses, governments, interest groups, media, and social movements to advance goals. See also, *non-governmental organization*.
Thinkfluencer	A *thinkfluencer* is someone with an online influence massively greater than the thoughtfulness of their ideas. Their rhetorical flourishes and style grab attention but do not stand up to scrutiny. For example, the fake Twitter account Professor Jeff Jarvis https://twitter.com/ProfJeffJarvis was a created by Rurik Bradbury to satirize the current affairs of the media and technology sectors one tweet at a time. The term was coined by Rurik Bradbury in 2012. Thinkfluencer is a portmanteau of *think* and *influence*. See also, *influencer*, *jargon*, *talking head*, and *thought leader*.

<table>
<tr><th>Phrase or Term</th><th>Definition</th></tr>
<tr><td>Thinking Mode, Thinking Method</td><td>A thinking mode or thinking method is the way someone tends to think based on their beliefs, experience, learning, and reason. Thinking may be described in various ways, including these thinking methods and modes:
<ul>
<li>Two thinking methods:
<ol>
<li>System 1 – instantaneous thinking that is driven by instinct and prior learning; or,</li>
<li>System 2 – slower thinking that is driven by deliberation and logic or rational consideration.</li>
</ol></li>
<li>Two thinking modes:
<ol>
<li>Focused – when concentrating intently on something that is being learned, solved or understood; or,</li>
<li>Diffused – when not consciously thinking of a problem that is being solved, with the mind in a relaxed state.</li>
</ol></li>
<li>Three thinking modes:
<ol>
<li>Convergent – using logic or analytical, critical linear, and vertical thinking;</li>
<li>Divergent – using imagination, or creative or horizontal thinking; or,</li>
<li>Lateral – using both convergent and divergent thinking, or thinking outside the box.</li>
</ol></li>
<li>Four thinking modes:
<ol>
<li>Defeatist – feeling frustration or worry and the fear of what might go wrong;</li>
<li>Sustainer – going through the motions or sustaining the status quo, multitasking or living on autopilot;</li>
<li>Dreamer – producing many ideas with little effort; or,</li>
<li>Opportunist – operating at peak performance and building on the dreamer mode, with the added element of action-taking.</li>
</ol></li>
</ul>
See also, abundance mindset, bias, black and white, cognitive dissonance, cognitive flexibility, convergent thinking, creative thinking, creativity, critical thinking, dichotomy, diffusion of responsibility, divergent thinking, grey area, groupthink, false dichotomy, flexibility, forward-thinking, inside-out thinking, lateral thinking, lucky fool syndrome, mental prison, metacognative thinking, loserthink, one-variable thinking, outside the box, outside-in thinking, Pollyanna principle, scarcity mindset, vertical thinking, and what you see is all there is.</td></tr>
<tr><td>Third-hand</td><td>Third-hand is something that is received from or through two intermediaries. For example, my brother gave me a book that he bought at a second-hand store; thus, the book became third-hand. See also, first-hand and second-hand.</td></tr>
<tr><td>Third-party</td><td>A third-party is any supplier of packaged equipment to a project that can communicate with a DeltaV™ control system.</td></tr>
<tr><td>Third-Party Inspector</td><td>Not a preferred phrase – see quality surveillance representative. The phrases quality surveillance representative and third-party inspector may be used interchangeably in industry; however, inspection may also be completed by other project personnel (e.g., engineering and procurement or client employees).</td></tr>
</table>

Phrase or Term	Definition
Thought Leader	A *thought leader* is an individual or organization that: 1. Clients, colleagues, competitors, intermediaries, prospects, and referral sources recognize as authoritative, brilliant, and influential, and a foremost authority in an area of specialized expertise, so that they are the go-to for that expertise; and, 2. Significantly profits from being recognized as a foremost authority. This term may be used to describe someone with more emphasis on the 1st part of the definition and less emphasis om the 2nd part, especially within an organization. For example, *our business manager is a though leader for innovation and start-ups*. The term's earliest use was in *The Theistic Annual* of 1876, when it was applied to Ralph Waldo Emerson, who was said to *manifest the wizard power of a thought-leader*. See also, *competitor*, *prospect*, *talking head*, and *thinkfluencer*.
Thought Crime, Crimethink, Wrongthink	*Thought crime*, *crimethink*, and *wrongthink* describe a person's unspoken beliefs, doubts, opinions, or thoughts that are contrary to accepted norms of society or official ideologies and policies. For example, to hold the opinion that new pipelines are a benefit to society and good for the economy when many others believe that new pipelines are no longer needed and will threaten efforts to fight climate change The use of unacceptable words imply that freedom of opinion is being suppressed. The term thoughtcrime was coined by George Orwell in the book *Nineteen-Eighty Four*. See also, *brainwashing*, *disinformation*, *misinformation*, *propaganda*, *thought control*, and *thought crime*.
Thought Control, Mind Control	*Thought control* or *mind control* is the imposition of opinions (e.g., state-run television), restriction of ideas through censorship (e.g., control of the press), and use of authoritarian techniques (e.g., propaganda) to prevent subversive and undesirable ideas that would challenge official ideologies and policies from spreading. See also, *brainwashing*, *disinformation*, *doublethink*, *misinformation*, *propaganda*, and *thought crime*.
Thread	A screw *thread* is used to fasten components and pipe for hydraulic or pneumatic systems. There are many thread types and designations, and two types of threads: 1. Jointing threads are for joints made leak-proof or pressure-tight by sealing on the threads. These are either taper external and parallel or taper internal threads. The sealing effect is improved by using a jointing compound; and, 2. Fastening threads do not make a pressure-tight joints on the threads. Both threads are parallel and sealed by compression of a soft material onto the external thread or a flat gasket. See also, *fastener*, *gasket*, *leak-proof*, *national pipe thread*, *straight pipe thread* and *threaded end*.
Threaded End	*Threaded end* is a pipe end preparation with a threads cut into the pipe. See also, *bevelled end*, *plain end*, and *thread*.
Threat	A *threat* is a negative risk that could adversely affect project objectives. See also, *risk*.

<table>
<tr><th>Phrase or Term</th><th>Definition</th></tr>
<tr><td>Three Bids and a Buy</td><td>Three bids and a buy is a procurement method for a small purchase (e.g., less than $250,000 or as defined by the project) of products or services that is required to:<ul><li>Be free of anti-competitive practices;</li><li>Use competitive bidding (i.e., a minimum of three price quotes); and,</li><li>Use procurement documents.</li></ul>This method is considered a less formal or less sophisticated form of competitive bidding.
See also, bid, competitive bid, and procurement documents.</td></tr>
<tr><td>Three Bs Rule</td><td>The three Bs rule guides the determination of what attire is and is not considered appropriate to wear in an office setting. The three Bs rule teaches that it is unacceptable to wear at the office what one would wear to:<ul><li>A bar;</li><li>A beach; or,</li><li>Bed.</li></ul>See also, business casual.</td></tr>
<tr><td>Three Ps of Sustainable Development</td><td>The three Ps of sustainable development are people, planet, and profit. See also, sustainability, sustainable development, and Sustainable Development Goals.</td></tr>
<tr><td>Three-point Estimate</td><td>A three-point estimate is a superset of estimating techniques that use averages, or weighted averages, of the most likely, optimistic, and pessimistic costs, with the duration estimates, to form the final estimates. See also, definitive estimate.</td></tr>
<tr><td>Three Types of Leverage</td><td>When negotiating, the three types of leverage are:<ul><li>Negative
The ability to make a counterpart suffer or use a threat (i.e., with or without a direct attack). For example, to make a report to the credit bureau if rent is unpaid or register a lien if a debt is unpaid.</li><li>Normative
The ability to use the counterpart's norms and standards to advance one's own position because nobody wants to look like a hypocrite. For example, to suggest that a customers overdue account should be paid within 30 days because they demand the same of their customers.</li><li>Positive
The ability to provide or withhold what the counterpart wants. For example, to have a vehicle for sale and an interested buyer.</li></ul>This concept was developed by Chris Voss in the book Never Split The Difference: Negotiating As If Your Life Depended On It (with Tahl Raz). See also, leverage.</td></tr>
<tr><td>Through the Trees Syndrome, Can't See the Forest for the Trees</td><td>The through the trees syndrome or the expression can't see the forest for the trees describe how it is difficult to see the overall goal (i.e., the forest) when focusing on details (i.e., individual trees). This expression teaches that it is more important to focus on the overall goal than on the necessary minor steps to accomplish it. Another variation of this expression is: can't see the wood for the trees, which has a synonymous meaning.</td></tr>
</table>

Phrase or Term	Definition
Tick-the-box	*Tick-the-box* refers to simply checking a box on a check list for a completed task, without considering that the task is in fact complete and correct. For example, checking that a vehicle was fueled but not verifying that the tank is full. This occurs when prescribed practices are followed in form rather than intent and substance, creating the illusion that good practices are being used when they are not. For example, completing the work and then checking all the boxes on the form to *prove* work was completed (instead of checking the boxes as each step of the work is completed). In some organizations, the process is seen as a substitute for competence and hard work. Following the process seemingly carries a cloak of immunity from accountability, even when decisions are made that common sense cannot support. This trap is virtually guaranteed when internal or external groups are developed to provide assurance that a process is followed rather than assessing the quality of what is produced.
Tie-in Crew	A *tie-in crew* welds the special short sections of a pipeline that are welded by the poor-boy crew into the main sections that were welded by the main gang crew to complete the pipeline. See also, *crew*.
Tie-in Weld	A *tie-in weld* is a connection weld: • Between pressure tested sections, pre-tested piping and other piping, new facilities and existing piping systems; or, • Two lengths of piping that are fixed at opposite ends or are long enough to act as if they are fixed. It is a weld that cannot be hydrostatically tested (e.g., due to the restriction where test fluid cannot be introduced into the system because it could contaminate the product, system, or both). Also, the weight of the test fluid could exceed the weight capacity of the equipment or piping system that supports it. The code of construction or design specification typically requires that a tie-in weld undergoes additional inspection and non-destructive examination in lieu of hydrotesting. A highly-skilled and qualified welder is required to complete tie-in welds.

Phrase or Term	Definition
Tier Supplier	A *tier supplier* is a member of the supply chain in a hierarchy of suppliers and sub-suppliers. Categories of tier suppliers include: 1. Tier 1 A Tier 1, Tier One, or Top Tier supplier is the most important member of the supply chain because they supply critical components and sub-assemblies directly to the original equipment manufacturer. A Tier 1 supplier may: o Complete the manufacturing process with the original equipment manufacturer acting as the marketing organization; and, o Work closely with the original equipment manufacturer to develop joint strategies and plans to improve supply chain efficiencies. 2. Tier 2 A Tier 2 or Tier Two supplier is often an expert supplier that: o Supplies products to customers in multiple industries; and, o Does not have the ability or desire to supply industry-specific or industry-standard products. 3. Tier 3 A Tier 3 or Tier Three supplier supplies raw or close-to-raw materials, such as metal or plastic. All higher-tier suppliers require raw materials; hence, a Tier 3 supplier feeds product to all higher tier levels. See also, *manufacturing*, *original equipment manufacturer*, and *supply chain*.
Time is Money	*Time is money* is an expression that means time is a valuable commodity or resource, so it is better to do things as quickly and not to waste time. Hence, someone's time *is as valuable* as money.
Time to Recover	*Time to recover* is the period required to resume current operations after an asset, facility, or organization experiences a major disruption. For example, a facility experiences an earthquake and shuts down all operations as a cautionary action.
Time-impact Analysis	A *time-impact analysis* is a scheduling technique used to assess and quantify the effects of an unplanned event – namely a change that increases work scope. It is forward-looking and can also be used in other ways. For example, to evaluate potential impacts to the schedule for an acceleration or delay. This process can be a preferred method to promote negotiation and later agreement on delay claims. Most construction contracts require this analysis when requesting a time extension.
Tip of the Iceberg Syndrome	The *tip of the iceberg syndrome* is when a project or task appears easy but consumes many more hours, days, weeks, or months than were originally planned. Because up to 90% of most icebergs are below the surface, this is a metaphor for having a limited understanding of what is hidden from view. For example, the leaking windows in the house are the tip of the iceberg; renovation scope will likely expand and the costs will likely increase greatly.
TKY Joint	A *TKY joint* is the connection formed when a hollow structural steel member intersects with a similar member or the planar surface of another member (e.g., an I-beam or plate) to form a T shape, K shape, or Y shape. These are some of the most difficult welded joint designs to produce and non-destructively examine. See also, *hollow structural steel* and *member*.

Phrase or Term	Definition
To-do List	A *to-do list* is a list of things to do (e.g., action items or tasks). See also, *future-back process* and *stop doing list*.
To Put the Cart in Front of the Horse	*To put the cart in front of the horse*: • Is an old expression that indicates doing things in the wrong order (e.g., contrary to a conventional or culturally expected sequence or relationship); • An idiom that confuses cause and effect; or, • Describes doing the last steps first, making a strategy ineffective. It is common sense that a cart must be pulled by a horse; attempting to pull the horse with the cart is as useless as it is futile. For example, before marketing a brand (the cart) the brand must be created (the horse). Putting the horse before the cart in this case means marketing a brand that does not exist (yet) – which makes no sense and guarantees failure. See also, *common sense*.
Tolerance	*Tolerance* refers to the total allowable error or permissible variation within an item. It is a specified allowance +/- of a given dimension or value of an item, to allow for acceptable inaccuracies during production or use. It is an acceptable amount of variation that will still allow an item to function correctly. Tolerances are usually quite small (measured in thousands of an inch or in very small percentages) and stated by the applicable reference standard (e.g., ASME, ASTM, API or CSA). Tolerances can be applied to any calibration or dimension. Tolerances can compound to create an *out of tolerance condition* (e.g., two pipes are both 12 mm [0.5 in.] too long, or *cancel each other out* (e.g., one pipe is 12 mm [0.5 in.] too long and the other is 12 mm [0.5 in.] too short). See also, *calibration* and *out of tolerance*.
Tone-deaf	*Tone-deaf* describes a person who: • Cannot accurately perceive differences in musical pitch; • Communicates using the correct words but in the wrong way; • Has a leadership style that treats everyone the same regardless of differences; • Is emotionally aloof or unengaged and disregards other people; or, • Shows a lack of perception or sensitivity, particularly in matters of public opinion, sentiment, or taste. For example, it is tone deaf to ask for permission to leave work early when an assignment is overdue and incomplete. See also, *active listening, dialogue of the deaf*, and *paraverbal communication*.
Tool Time, Wrench Time	*Tool time* or *wrench time* is a measure of labour force productivity. This involves performing time studies (e.g., for trades personnel and technicians) to determine what proportion of their time is spent with *hands on tools* and what proportion is spent on other tasks (e.g., administrative paperwork or unproductive tasks such as waiting for materials). See also, *hands on*, *labour force*, *on-the-job training*, *trades person*, and *wrench*.
Toolbox Talk	A *toolbox talk* is an informal group discussion about health and safety topics that promotes safety culture on-the-job site, in the workplace or both, immediately before the work is performed. It is intended to focus the working group on safety. See also, *safety moment*.

Phrase or Term	Definition
Top Down Analysis	*Top down analysis* is a method of market analysis that calculates market share or a total sales figure as a percentage of the total market. This method takes less effort than a bottom up analysis, but the result is usually less accurate. See also, *bottom up analysis* and *marketing*.
Top of the Market	The *top of the market* refers to the high-end products that are the costliest or most advanced in a company's product range or the market. For example, a vehicle manufacturer introduces an expensive and luxurious model that only a few customers can afford. See also, *bottom of the market*, *market trend*, and *niche market*.
Topcoat, Finish Coat	The *topcoat* or *finish coat* is the final coat (e.g., paint) of a coating system. See also, *coating system* and *primer coat*.
Top-down Estimating	*Top-down estimating* uses historical data from similar projects to compute the approximate size (i.e., duration, effort, and cost) and risk of a new or proposed project. The comparison to previous projects may be made directly using an algorithm, lessons learned reports, by way of personnel recollection, or a combination. See also, *creative estimating*, *definitive estimate*, and *lessons learned*.
Topping Out, Topping Off	*Topping out* or *topping off* is a builders' rite or ceremony, traditionally held when the last beam (or its equivalent as the highest structural feature) is placed atop a building or structure to symbolize – figuratively and literally –construction completion. Today, this ceremony is often parlayed into a media event for public relations and legacy purposes.
Topside	*Topside* refers to the above-water portions of an offshore oil rig or ship (i.e., any parts that are above the splash zone or waterline). The structures that comprise topsides are typically modular and installed on a fixed or floating underwater structure. See also, *modular*.
Torque Wrench	A *torque wrench* is a tool used to tighten bolts to a specific, measurable tightness. See also, *wrench*.
Total Cost of Ownership	The *total cost of ownership* estimates the sum of direct and indirect costs incurred in the purchase, operation, and maintenance throughout an asset's life. See also, *asset*.

Phrase or Term	Definition
Total Quality Management, TQM	*Total quality management* or *TQM* is a philosophy and a system of organizational management. Consequently, is has a cultural dimension (i.e., organizational behaviour) and a technical dimension (i.e., the processes by which the product or service is produced and delivered). It is the continual process to: • Detect and eliminate or reduce errors; • Ensure personnel are trained; • Focus on and improve the customer experience; • Involve all personnel in continuous improvement; and, • Streamline supply chain management. It uses data, communication, and strategy to integrate the quality discipline into the activities and culture of the organization. The eight principles of total quality management are: 1. Communication; 2. Continual improvement; 3. Customer-focus; 4. Fact-based decision-making; 5. Integrated system; 6. Personnel involvement; 7. Process-centric; and, 8. Strategic and systematic approach. The concepts were developed by quality leaders including Philip B. Crosby, W. Edwards Deming, Armand V. Feigenbaum, Kaoru Ishikawa, and Joseph M. Juran. See also, *continuous improvement*, *customer centric*, *doing the right things right the first and every time*, *quality*, *quality management system*, *Six Sigma*, and *supply chain management*.
Touch Up	A *touch up* is a repair of small blemishes to a coating system (e.g., repair of minor defects and damage such as chips and scuff marks). See also, *coating system*.
Touchpoint	A *touchpoint* is any method with which a consumer can interact with a business (e.g., an application, face-to-face, Internet call, telephone, video conference, website, or any form of communication).
Town Hall, Town Hall Forum, Town Hall Meeting	A *town hall, town hall forum, or town hall meeting* is: • A political event at which a politician or public official meets their constituents or members of the public. The purpose is to answer questions without decision-making (e.g., to legislate budget or policy). In contrast to a town meeting, attendees do not vote on issues; or, • An informal corporate event held to share information with personnel such as business challenges, changes, or results. Despite the name, town hall meetings need not take place in a town hall. A political town hall may be held in many venues such as a church, library, municipal building, or school. A corporate town hall may be held in company workspace or at a hotel meeting room. A digital format may also be used. Historically, no specific rules or guidelines have defined the town hall meeting (i.e., a specific agenda or format).

Phrase or Term	Definition
Toxic Behaviour	*Toxic behaviour* is destructive actions or attitudes that create unnecessary distraction and drama that erodes the ability to perform. Personnel with toxic behaviour are often actors, manipulators, and skilled liars that poison personal and professional relationships, and impair the ability of others to contribute, grow, or both. Toxic behaviour may: • Be boastful or shamelessly self-promotional; • Be inappropriate verbally (e.g., shouting) and nonverbally (e.g., body language or email); • Constantly blame, complain, and criticize; • Express indirect passive-aggressive tendencies; • Focus on organizational barriers and impediments or negativity, and why things fail; • Have a complete disregard for decorum or the feelings of others; or, • Twist benign comments and ideas into personal attacks. Allowing toxic behaviour to persist breeds cynicism, especially when consistent standards of behaviour and values are not enforced. See also, *body language*, *bully*, and *office jerk*.
Traceability	*Traceability* is the ability in: • Accounting, to track a specific piece of financial information by means of recorded data or audit trail; • Cost accounting, to assign a cost directly to an activity or cost object on the basis of a cause-and-effect (causal) relationship; • Quality control, to trace the application, location, and/or history of an activity or item by means of recorded data; and, • Systems management, to track system requirements from a system function to all those elements that individually or collectively perform a function. See also, *bill of material* and *traceable record*.
Traceable records	A *traceable record* is documentation about an equipment item or facility that can be clearly linked to original information. See also, *traceability* and *verifiable record*.
Tracking Capability	*Tracking capability* is a function of an electronic data management system that provides a record of document's history including status, users, and versions. See also, *electronic data management system* and *record*.
Trade	*Trade* is: • The act of buying, exchanging, and selling products and services; or, • An occupation or skilled profession, especially one that requires manual or mechanical skills and special training. See also, *apprentice*, *journeyman*, and *trades person*.
Trade Secret	A *trade secret* is a confidential formula, technique, or process known and used to benefit only one manufacturer, which often provides an advantage over competitors. For example, significant business intelligence, original product design, or a newly-developed technology, There is no formal intellectual property process for protecting a trade secret. See also, intellectual property.

<table>
<tr><th>Phrase or Term</th><th>Definition</th></tr>
<tr><td>Trademark</td><td>A trademark is a symbol, word, or words legally established or registered for use by a company that identifies a company or product. See also, tradename.</td></tr>
<tr><td>Tradename</td><td>A tradename is:<ul><li>A name by which something is known in a particular profession or trade (e.g., Esso);</li><li>A name or designations used by a company to identify and differentiate business and business products from competitors; or,</li><li>A name that has the status of a trademark (e.g., Caterpillar).</li></ul>See also, trademark.</td></tr>
<tr><td>Trade-off</td><td>A trade-off is the balance achieved between two desirable but incompatible features (i.e., a compromise or giving up of one thing in exchange for another). For example, the trade-off of shortening the delivery schedule is faster delivery – but with increased costs.</td></tr>
<tr><td>Trades Person</td><td>A trades person is a worker in a skilled trade. See also, skilled trade, trade, and unskilled trade.</td></tr>
<tr><td>Traditional Economy</td><td>A traditional economy is an economic system that is rural, second- or third-world, and closely tied to the land, usually through agriculture. Each person in this system has a pronounced and specific role. These societies tend to be very close-knit and may be socially satisfied; however, they typically lack access to advanced medicine and technology. Bhutan and Haiti are countries with traditional economies. See also, centrally planned economy, economic system, market economy, mixed economy, and world model.</td></tr>
<tr><td>Traditional Non-destructive Examination</td><td>Not a preferred phrase – see conventional non-destructive examination.</td></tr>
<tr><td>Traffic Lights,
Traffic Signals</td><td>Traffic lights or traffic signals are a best practice that leverages the well-understood traffic light system for automobiles to quickly determine and report on the health or status of a project. Colours used may have these or similar meanings:<ol><li>Green – All good. Status quo. Performance or progress is on track.</li><li>Yellow – Potential issues. Proactive attention and possible corrective action are required.</li><li>Red – Serious issues. Immediate attention and corrective action is required. The project is at risk of failure.</li></ol>See also, dashboard and scorecard.</td></tr>
<tr><td>Trail Blazer</td><td>A trail blazer:<ul><li>Is an innovator, inventor, or a pioneer. For example, trailblazer Thomas Edison invented the electric light bulb, phonograph, and many other innovations; or,</li><li>Makes a new track through wild country. For example, David Thomson is a famous trailblazer who paddled or walked at least 80,000 km (49,710 miles), mapping most of Western Canada, and parts of Eastern Canada and the Northwestern United States.</li></ul>See also, early adopter and entrepreneur.</td></tr>
</table>

Phrase or Term	Definition
Transactional Customer Service	*Transactional customer service* involves building sales by offering customer service and short-term solutions (e.g., a salesperson's service when buying a pair of shoes). While efficient, transactional customer service does not emphasize customer experience, other customer needs, or foster customer loyalty. For example, while you may shop at the same shoe store for shoes, it is unlikely that you buy your shoes exclusively from the same salesperson. Thus, for transactional customer service, the service provider or supplier is differentiated from competitors only by price and quality. See also, *customer centric*, *outside-in*, *positive order entry*, and *relational customer service*.
Transmittal	A *transmittal* is a document that records information about the distribution of document(s) sent from one organization to another. The transmittal contains details about documents being distributed, including company or project-related requirements (such as document processing instructions for the recipient).
Transparent	*Transparent* describes: • A document or scenario that is free of conditions, deceit, hidden agendas, prejudice, and pretense; • A situation when people are open and honest with each other. Transparency in business dealings is ethical; • A situation when people are provided with information, even though it might be inappropriate or irrelevant. For example, if a company's management allows information about the organization's business to become public knowledge (e.g., a specific challenge or opportunity). As a result, personnel might become distracted by things they should not, instead of focusing on the organization's core deliverables; • Accessibility or visibility of information, especially for business practices; • Compliance with regulatory disclosure requirements by intentions and words; and, • Content that may be readily understood. See also, *ethics*, *ethical*, and *honest*.
Trenchless Technology	*Trenchless technology* is a type of subsurface construction work that requires few trenches or no continuous trenches. It uses equipment, materials, and methods, capable of installing new or rehabilitating or replacing existing underground infrastructure with minimal disruption to surface business, traffic, and other activities. These methods are used for the installation of cables and pipelines below the ground, including: • Horizontal directional drilling, also known as directional boring, horizontal auger boring, moling, pipe jacking, or pipe ramming; • Micro tunnelling; and, • Tunneling. Large diameter tunnels, such as those constructed by drilling and blasting techniques or tunnel boring machines, are larger versions of subsurface construction. The difference between trenchless and other subsurface construction techniques depends upon the size of the passage under construction.

<table>
<tr><th>Phrase or Term</th><th>Definition</th></tr>
<tr><td>Trend</td><td>A trend is a:<ul><li>Formal, early warning method used to identify and track possible deviations from the project baseline. For example, increased pricing, a known future project change, or a projection of increased costs based on experience. Typically, any project stakeholder may identify and document a trend. The trend goes through a review and approval process to determine its validity and impact on the project, if any. A trend may ultimately result in a baseline change; or,</li><li>Pattern of gradual change in a condition, output, or process, or an average or general tendency of a series of data points to move in a certain direction over time, represented by a line or curve on a graph.</li></ul>Se also, trend analysis.</td></tr>
<tr><td>Trend Analysis</td><td>Trend analysis is the practice of collecting information and attempting to spot a pattern. In some fields of study, the term has more formally defined meanings. Although trend analysis is often used to predict future events, it could be used to estimate uncertain events in the past. For example, historical cost data (monthly actuals) are used to predict future spending trends. While this may not be 100% accurate, it will be more indicative of future spending than a plan that was created many months or years previously. See also, trend.</td></tr>
<tr><td>Trespass</td><td>To trespass is to illegally enter private or public property. Trespass and trespassing have many legal meanings. See also, right of entry.</td></tr>
<tr><td>Triage</td><td>A Triage can be described as:<ul><li>Assigning an order of priority to projects based on where resources can be best used, are most needed, or are most likely to achieve success; or,</li><li>Prioritizing the sequence of treating patients by urgency of needed care (e.g., in an hospital emergency room).</li></ul></td></tr>
<tr><td>Trial Balloon</td><td>A trial balloon is information (e.g., about a new policy or product) issued publicly or privately (e.g., to the media by a leak or press release) to gauge an audience's reaction. It can be used by companies to determine customer's reaction or by politicians to determine the public's reaction. The phrase refers to sending up a small balloon before a manned ascent in a larger balloon, to determine wind direction, wind speed, and other environmental factors that may affect or impede a manned flight. If the reaction is positive, the idea may be more easily accepted or implemented. However, if the reaction is negative, the idea may not be accepted or implemented. In the United Kingdom this is known as kite-flying. See also, run it up the flagpole.</td></tr>
<tr><td>Trial Fit-up</td><td>Not a preferred phrase – see fit-up test.</td></tr>
<tr><td>Triangle Method</td><td>The triangle method is a document markup convention that identifies recent or important revisions by outlining the revisions with a triangle-shaped object. This method allows readers to disregard portions of the document and focus on recent or important revisions. The triangle shape may be drawn by hand or using software, and is removable to accommodate future revisions. This method may be used to standardize mark-ups, improve quality control, increase the predictability of how markups are delivered, and for many other review purposes (e.g., to flag or question a section, to identify a section as on hold, and to draw attention to detail). See also, on hold and cloud method.</td></tr>
</table>

Phrase or Term	Definition
Tribal Knowledge	*Tribal knowledge* is unwritten information that may or may not be commonly known by others within a company or organization. Tribal knowledge is required to efficiently produce a quality product or service. This knowledge may or may not be a best practice. Because tribal knowledge is undocumented, it is prone to change and variation over time. See also, *codification, knowledge transfer*, *personalization*, *tribalism*, and *tribe*.
Tribalism	*Tribalism* is: • The state of being organized by, or advocating for, tribes or tribal lifestyles; • A way of behaving or thinking in which people are loyal to their social group above all else; or, • A type of animosity or discrimination based upon group differences. Humans naturally maintain a small social network, so human evolution has primarily occurred in groups or tribes rather than mass societies. See also, *Dunbar's number*, *diffusion of responsibility*, *groupthink*, *tribal knowledge*, and *tribe*.

Tribe	A *tribe* is a:
	• Group of:
	  o Business or workplace personnel united by a shared culture, goal, leader, organizational boundary, or purpose;
	  o Friends that become like family;
	  o People connected to an idea, a leader, and one another; or,
	  o Personnel that have a common character, interest, or occupation.
	• Social:
	  o Division in a traditional society consisting of clans, families, or generations with their adoptees, dependents, and friends, united by economy, heredity, and religion with a common culture and dialect, typically with a recognized leader; or,
	  o Organization based on a set of smaller groups with a temporary or permanent political integration that is defined by common traditions (e.g., culture, history, ideology, and language).
	See also, *dialect*, *Dunbar's number*, *team*, *tribalism*, and *tribal knowledge*.

Phrase or Term	Definition
Tricks of the Trade	*Tricks of the trade* are special ingenious skills or techniques used in a profession or trade, especially those that are not or little known by others. Tricks of the trade are acquired through experience, knowledge transfer, the making of mistakes, and the motivation to perform a task using better, faster or more efficient methods than conventional methods. These may be informal and based on a belief, compromise, judgement, logic, or rule. Example 1 A project may receive hundreds of manufacturing record books from suppliers, each of which comprising hundreds and thousands of pages. Receiving these documents for turnover as incomplete and incorrect can cause a massive amount of rework that becomes difficult or impossible to finish. It is also becomes a liability when necessary information (e.g., a manufacturing data report) is missing. These two tricks of the trade effectively minimize and mitigate this risk: • First, the manufacturing record book indexes, which are comprised of only a few pages, can be reviewed by the responsible engineer before the documentation is compiled; and, • Second, the quality surveillance representative can review the compiled documentation at the supplier's facility before it is submitted for turnover. Example 2 Box breathing is helpful before public speaking, or a stressful business meeting or personal situation (e.g., a challenging meeting or a visit to the doctor). See also, *box breathing*, *manufacturing record book*, and *turnover*.
Triple bottom line	Coined by John Elkington in 1994, the *triple bottom line* is an accounting framework that uses three criteria (profits, the improvement of peoples' lives, and the improvement of the planet) instead of one (profits) to measure corporate performance. This concept has been described in these two ways: • Economic, and environmental (or ecological), and social; or, • People, planet, and profit. See also, *bottom line*.
Triple Random	Not a preferred term – see *random length*.
Trolling	*Trolling* is the act of causing discord to intentionally upset others, or start flame wars by posting or sending inflammatory and digressive messages over the Internet. A troll intentionally provokes emotional responses from readers and viewers for the troll's amusement or a specific purpose (e.g., social engineering). See also, *bully*, *cyberbullying*, *flame war*, *flaming*, *office jerk*, *social engineering* and *trolling team*.
Trolling Team	A *trolling team* is coordinated group of trolls, used to maximize efficiency and effect. A trolling team is often used to create the illusion of mass hatred toward another group or person. See also, *trolling*.
Troubleshoot	To *troubleshoot* is to: • Solve problems; or, • Trace and correct faults in a mechanical, electrical, electronic, or other systems.

Phrase or Term	Definition
True North	*True north* is located in the earth's rotational axis at the geographic north pole where the lines of longitude (meridians) converge in the north – 90° north of the equator. See also, *plant north*.
Trust	*Trust* is a firm belief in the ability, reliability, strength, or truth, of someone or something. If your conduct is exemplary, people will trust you. People tend to be loyal to the people they trust. A client or customer's trust is gained with honesty. Customers with distrust are difficult to retain. See also, *honest* and *vulnerable*.
Truth	*Truth* is: • An idea, judgment, or proposition, that is true or accepted based real events, facts, and things; or, • Something that is accurate or faithful to an original or standard. See also, *brainwashing, bullshit, directionally accurate, disinformation, factfulness, fake news, false narrative, ground truth, media bias, misinformation, propaganda, smells fishy, stretch the truth, truth, truth filter, truthiness, and untruth.*
Truth Filter	A *truth filter* comprises six filters used to evaluate information and help determine if something is true. Generally, one source cannot produce absolute truth, so information from any one of those single filters is not enough to be considered true. When seeking truth, the best method is to confirm the existence of at least two of these six dimensions: 1. Personal experience, but human perceptions are doubtful; 2. Experience of people you know, but human experience is unreliable; 3. Experts, but they work for money, not truth; 4. Scientific studies, but correlation is not causation; 5. Common sense, but a good way to be mistaken is with complete confidence; and, 6. Pattern recognition, but coincidence, patterns, and personal bias all look alike. The more filters information can pass through and remain consistent, the more likely the information is true: consistency is the key. For example, if a study indicates that eating ice cream is an excellent method to lose weight. But when you try the diet, you gain weight. Thus, two dimensions are out of agreement (or three if you count common sense!). This concept was developed by Scott Adams. See *also, brainwashing, bullshit, bullshit detector, iffy, truth, truthiness*, and *untruth*.
Truthiness	*Truthiness* is the assertion or belief that a statement is true based on the intuition or perceptions of an individual or individuals, without regard to evidence, facts, intellectual examination, or logic. Truthiness can range from ignorant assertions of falsehoods to deliberate duplicity or propaganda intended to persuade. Results produced by (unbiased) software have more truthiness than results produced by (biased) humans. See also, *brainwashing, bullshit, disinformation, fake news, false narrative, media bias, misinformation, propaganda, truth, and untruth.*
Tube	A *tube* is a round, square, or rectangular hollow structure that is used in architectural, pressure, and structural applications. A tube is measured by outside diameter and wall thickness. Pipes and tubes may look similar, yet they have many differences and are rarely interchangeable. See also, *hollow structural steel*, *pipe*, and *structural steel*.
Tube Bundle	A *tube bundle* transfers heat using a set of tubes in a shell and tube heat exchanger. See also, *heat exchanger*, *heat transfer* and *tube*.

Phrase or Term	Definition
Turbomachinery	*Turbomachinery* refers to machines that transfer energy between a rotor and fluid. Fans, compressors, and pumps absorb energy to increase the fluid pressure; turbines produce energy by expanding fluid to lower pressures. The former transfers energy from a rotor to a fluid; the latter transfers energy from a fluid to a rotor. Propellers, unshrouded fans and windmills are open machines that act on an infinite quantity of fluid; whereas, compressors, pumps, and turbines are closed machines that operate on a finite quantity of fluid as it passes through a housing or casing. See also, *driver*.
Turing Test	In a *Turing test*, a computer is presented as a human. A human judge engages in a natural language conversation with one human and one machine, each emulating human responses. All participants are separated from one another. If the judge cannot reliably tell the machine from the human, the machine is said to have passed the test. See also, *ideological Turing test*.
Turnaround	Not a preferred term – see *shutdown*.
Turnkey	*Turnkey* is something (e.g., packaged equipment) that is supplied or installed complete and ready to operate or use (e.g., a new vehicle at a dealership). No assembly is required. A project may be considered turnkey if commissioning and start-up are included in the contract (i.e., the contractor turns over an operating facility). See also, *commissioning*, *module*, *out of the box*, *packaged equipment*, and *plug and play*.
Turn-of-Nut Bolting Method	The *turn-of-nut bolting method* is used to apply the correct tension when tightening structural bolts. It is performed by rotating the nut or bolt of a fastener assembly a specific turn angle based on the fastener's length and diameter. The unturned element is restrained from rotating. See also, *direct tension indicating washer*, *fastener*, and *twist-off bolt*.
Turnover	*Turnover* is a project phase where final supplier documentation is given to the engineering and procurement company. Turnover can also mean that final project documentation is permanently given to the client. See also, handover.
Turnover Data Book	Not a preferred phrase – see *manufacturing record book*.
Twenty-twenty	*Twenty-twenty* means: • Marked by facilely accurate assessment, discernment, or judgment; • Meeting a standard of perfect visual acuity. For example, testing vision by reading letters at 6.1 m (20 ft); or, • Perfect. See also, *twenty-twenty hindsight*.
Twenty-twenty Hindsight	*Twenty-twenty hindsight* means having the complete understanding and full knowledge that one has about an event, only after it has happened. For example, with twenty-twenty hindsight, it can be seen where the strategy went wrong. See also, *twenty-twenty*.
Twist-off Bolt, Tension-Control Bolt	A *twist-off bolt* or *tension-control bolt* is used to apply the correct tension when tightening structural bolts. The torque needed to twist off a splined extension manufactured into the bolt shank is calculated so that the *twist-off* occurs at a tension above the minimum required. See also, *direct tension indicating washer* and *turn-of-nut bolting method*.

Phrase or Term	Definition
Two Hole	*Two hole* refers to a bolt hole configuration that conforms to the industry standard for all flange bolt holes to straddle the centreline. This means that for a vertical flange face, with the flange face in the vertical and the line in the horizontal, bolt holes are orientated to straddle the vertical and horizontal centreline. For a horizontal flange face, the flange face is horizontal and the line is vertical above or down, and the bolt holes are orientated to straddle the north centreline. According to ASME B16.5 and similar standards, the number of bolt holes divided over the diameter of the bolt circle is always an even number (i.e., 4, 8, and 12). This allows a flange to be installed level, using a spirit level across the tops of the bolt holes, flange bolts, or pipefitting pins. See also, *centreline*, *one hole*, and *straddle*.
Two-faced	Someone who is *two-faced* is deceitful, hypocritical, untrustworthy, insincere, or double-dealing. This means that they: • Say they believe or do one thing when their behaviour or words reveal the opposite or something different; or, • Say something to one group or person and say something different (often conflicting) to another group or person. The roots of this phrase are from the Roman God Janus, who had two faces: one at the front and one at the back of his head. See also, *double-dealing*.
Two-way Street	A *two-way street* is a metaphor for a relationship or situation involving mutual or reciprocal action or obligation (i.e., give-and-take). For example, *my boss asked me to work over the weekend but let me take Monday and Tuesday off because our professional relationship is a two-way street*.
Type	The *type* of steel is determined by steel standards (e.g., ASME, ASTM, or CSA). The type of steel guides end users, metallurgical laboratories, fabricators, and manufacturers in their proper application, methods of processing, and use. The basic steel types include: • Alloy; • Carbon; • Stainless; and, • Tool. See also, *grade*, *material*, and *material classification*.

Phrase or Term	Definition
Type A/B/C	*Type A/B/C* is a classification system for deficiencies or hazards used to assign a level of seriousness or severity and establish priorities for corrective action. For example: • Type A A Type A classification is any condition or practice with the potential for causing bodily injury or loss of assets or life, so that immediate action is required. • Type B A Type B classification is any condition or practice with the potential for causing a serious illness or injury, or damage to assets so that attention is urgently required. • Type C A Type C classification is any condition or practice with the potential for causing a non-disabling injury or non-disruptive damage to assets, but is not an emergency, so that action is required as soon as possible. Deficiency or hazard classification may use an alternate system (e.g., ascending or numerical) and identify appropriate actions or timelines as defined by the user. For example, if a certain category of deficiency or hazard exists then specific activities may or may not occur. For instance, personnel may not operate an asset until they have been trained. Or, if an asset is missing a component that makes it inoperable then it may not be released for shipment. See also, *deficiency*.

Phrase or Term	Definition
Ulterior Motive	An *ulterior motive* is an alternative or extrinsic reason for doing something, especially when it differs from the apparent reason or stated reason or is a secret. For example, to be especially agreeable and kind to a manager to increase the chances of getting a better assignment, pay raise, or promotion. See also, *virtue signal*.
Ultrasonic Test	An *ultrasonic test* is a non-destructive method for examining the internal structure and integrity of steel and other metals using very short or ultrasonic sound waves. This test can also be used on composites and concrete, albeit with less resolution. This test is based upon the propagation of ultrasonic waves in the test specimen. The test instrument converts electrical pulses into mechanical vibrations or pulses that travel through the test specimen. Discontinuities reflect these signals because of their different acoustic nature. The returning reflected pulses are re-converted to electric energy and displayed as signals on a cathode ray tube. The position and size of signals correspond to the position and size of the flaws. This test method can also measure material thickness, with returning reflected pulses displayed as a measurement (e.g., inches or millimeters). See also, *discontinuity* and *non-destructive examination*.
Unanticipated Consequences, Unforeseen Consequences, Unintended Consequences	*Unanticipated consequences*, *unforeseen consequences*, and *unintended consequences* describe results that were not intended or foreseen. Though unintended consequences may be anticipated or unanticipated, they are the product of specific actions. Unintended consequences can be grouped into these three types: 1. Unexpected benefit A positive outcome that was not anticipated. Also referred to as luck, serendipity or a windfall; 2. Unexpected drawback A detrimental outcome in addition to the desired effect; 3. Perverse result An outcome contrary to what was originally intended. Also referred to informally as *a backfire* (i.e., the intended solution makes a problem worse). See also, *Cobra Effect*, *knock-on effect*, *second-order effect*, and *spillover effect*.
Unbelief	*Unbelief* is, during negotiation, the active resistance and complete rejection of what a counterpart says. This is the usually the point at which negotiations start.
Unconscious Bias Training	*Unconscious bias training* exposes personnel to their unconscious biases, and provides tools to adjust automatic patterns of thinking and eliminate discriminatory behaviour. Understanding unconscious preferences allows organizations to create innovative teams, harness the potential of diversity, and make sound business decisions. See also, *bias* and *unconscious bias*.

Phrase or Term	Definition
Unconscious Bias, Implicit Bias, Implicit Stereotype, Explicit Stereotype	An *unconscious bias*, *implicit bias*, or *implicit stereotype* is the unconscious attribution of qualities to a member or members of a certain social group (e.g., based on gender or race). This bias is a learned through the social experience and is automatic, deeply engrained, unintentional, universal, often subconscious, and influences behaviour. These attributes are based on learned associations between qualities and social categories. By contrast, an *explicit stereotype* is the result of conscious, controllable, and intentional, belief or thought. This stereotype is usually directed toward a group (e.g., people or things). See also, *bias* and *unconscious bias training*.
Unconventional Gas	*Unconventional gas* is natural gas obtained from sources of production considered non-conventional (e.g., different, difficult, or new). For example, coalbed or coal seam methane and shale gas. See also, *easy oil and gas*.
Unconventional Oil	*Unconventional oil* is oil obtained from sources of production considered non-conventional (e.g., different, difficult, or new). For example, frontier (e.g., the Arctic), oil sands mining, tight or shale oil, and ultra-deep off-shore wells (non-conventional) vs. drilling a conventional well. See also, *easy oil and gas*.
Under-commit and Over-deliver	To *under-commit and over-deliver* is a strategy in which a service provider strives for excellent customer service and satisfaction by doing more than they say that they will or exceeding their customer's expectations. See also, *over-commit and under-deliver*.
Underdog	An *underdog* is a: • Competitor thought to have little chance of winning a contest, fight, or struggle (i.e., a predicted loser); or, • Victim of injustice or persecution. See also, *victory snatched from the jaws of defeat*.
Underground Burning, Fire Flooding	*Underground burning* or *fire flooding* is an enhanced oil recovery technology that injects compressed air or oxygen into a petroleum reservoir and ignites it. The flame burns some of the oil and produces gases that: • Drive the remaining underground oil into producing wells; and, • Heat and soften bitumen (or heavy oil), making it less viscous and easier to pump; or, • Include hydrogen, which can be extracted continuously and in massive quantities. See also, *enhanced oil recovery* and *steam-assisted gravity drainage*.
Underground Storage Tank	An *underground storage tank* stores products (e.g., chemicals, fertilizer, fuel, oil, or water) below grade (i.e., buried). See also, *aboveground storage tank* and *tank*.
Underground, Below Grade	To be *underground* or *below grade* is to be situated below the ground. See also, *aboveground*.

Phrase or Term	Definition
Unfocus Group	An *unfocus group* is a diverse group of people who develop ideas about a product or service they are passionate about, even though they do not have expertise related to the topic of discussion. For example, artisan, client, and marketing personnel could be included in a group to develop new ideas for improving project execution strategies. Participants are people who are outsiders to the topic but who can offer a different perspectives, with fresh insights and innovative suggestions, than those immersed and learned about the topic. See also, *brainstorm*, *creative thinking*, *focus group*, and *whiteboard session*. See also, *brainstorm*, *creative destruction*, *creative thinking*, *focus group*, and *whiteboard session*.
Unicorn, Purple Squirrel	A *unicorn* or *purple squirrel* is an informal expression used to describe a job candidate with a very desirable skill set. It is a metaphor that describes the rare candidate who has exactly the right experience, and hard and soft skills. A *unicorn* is also a digital start-up with a valuation of a billion dollars or more. See also, *start-up*.
Unified Numbering System	The *unified numbering system* is an alloy designation system widely accepted in North America. It consists of a prefix letter and five digits designating a material composition. For example, the prefix *S* indicates a stainless steel alloy. Stainless steel Type 310 in the original 3-digit system became S31000 in the UNS System. Often, the suffix digit is chosen to represent a material property. UNS S31008 is stainless steel Type 310 that is assigned the number *-08* because the maximum allowed carbon content is 0.08%. A UNS number alone does not constitute a full material specification because it establishes no requirements for material properties, heat treatment, form, or quality. The system is managed jointly by ASTM International and SAE International.
Unincorporated	An *unincorporated* company or other organization is not a legal corporation. See also, *corporation* and *sole proprietorship*.
Uninterruptable Power Supply, Uninterruptible Power Source	An *uninterruptable power supply* or *uninterruptible power source* is electrical equipment that provides temporary emergency power when a facility power source or main power fails. This differs from an auxiliary or emergency power system or standby generator because it provides near-instantaneous protection from input power interruptions. Energy is stored in batteries with a run-time that is relatively short (e.g., only a few minutes); however, this is enough power to accommodate energizing a standby power source or to accommodate safe and correct equipment shutdown. This power source is typically used to protect hardware such as computers, data centres, telecommunication equipment, and other sensitive electrical or mechanical equipment (e.g., valves or motors) that could experience an unexpected power disruption and would result in data loss, injuries, fatalities, or serious disruptions.
Uniqueness Bias	*Uniqueness bias* describes: • The tendency to view personal attributes, qualities, and traits and as unique, when they are not; and, • How project personnel view their project as singular because new projects often use non-standard designs and technologies. This leads them to think that their project is more different from other projects than it is. It impedes learning because personnel think that they have nothing to learn from other projects because theirs is unique. Very few projects are truly unique. See also, *bias*.

Phrase or Term	Definition
United Nations Declaration on the Rights of Indigenous Peoples, UNDRIP	The *United Nations Declaration on the Rights of Indigenous Peoples* or *UNDRIP* was adopted by the United Nations General Assembly in 2007. It is the most comprehensive international instrument on the rights of indigenous peoples. It establishes a universal framework of minimum standards for the survival, dignity, and well-being of the indigenous peoples of the world and it elaborates on existing human rights standards and fundamental freedoms as they apply to the specific situation of indigenous peoples. See also, *Indigenous*.
Universal Product Code, UPC	A *universal product code* or *UPC* is a barcode used for tracking products. It comprises a strip of black bars and white spaces above a sequence of 12 numerical digits that are uniquely assigned to each product. There is a one-to-one correspondence between the strip of black bars and white spaces and the 12-digit number (i.e., there is only one way to represent each 12-digit number visually and each strip of black bars and white spaces numerically). A 6-digit barcode may be used for small products. This barcode was designed and implemented for commercial use in the early 1970s by the grocery industry. Similar automated checkout systems were developed beginning in the early 1930s. See also, *barcode* and *quick response code*.
Unlearning	*Unlearning* is: • Learning (i.e., by an individual or organization) to behave, perceive, or think differently, when there are conflicting assumptions, behaviours, or beliefs already in place; • The process of breaking down the origins of one's attitudes, behaviours, biases, feelings, and thoughts; • To discard a habit of, or undo the effect of, something; or, • To put something out of one's knowledge or memory. Unlearning may become necessary if something has a bad influence, is false or incorrect, or when what is known prevents gaining additional knowledge or learning. For example, when driving a left-handed vehicle, manually gear shifting requires the right hand. To drive a right-handed vehicle, shifting gears with the left hand and unlearning the instinctive use of the right hand for this purpose is necessary. Alvin Toffler said that *the illiterate of the 21st century will not be those who cannot read and write, but those who cannot learn, unlearn, and relearn* (undated). For example, unlearning is: • Fearing failure, yet knowing it often results in better ideas or opportunities; • Recruiting talented personnel and then instructing them about how to do something; • Understanding that a sales problem could be a culture problem; or, • Understanding that solutions are often on the inside (i.e., internal) rather than on the outside (i.e., external). Another example of unlearning is learning to drive a right-handed drive vehicle after only driving left-handed vehicles, or vice versa. Confirmation bias can make it harder to unlearn something that contradicts one's current understanding. See also, *active learning*, *Bloom's Taxonomy*, *cone of learning*, *confirmation bias*, *knowledge transfer*, *metacognative thinking*, *passive learning*, *status quo*, and *unsee*.

Phrase or Term	Definition
Unobtainium, Unobtanium	*Unobtainium* or *unobtanium* is a material that is costly, extremely rare, fictional, hypothetical, scientifically impossible, or a combination. It may also refer to a device with desirable engineering properties, which is impossible to use in an application. For example, in the science fiction movie Avatar it is the fictitious mineral mined on the planet Pandora that is used for energy generation. Unobtainium is a portmanteau of unobtainable and the suffix -ium (i.e., the conventional suffix for a chemical element). See also, *fool's gold*, *hopium*, *mineral*, *moonshot*, *panacea*, *pipe dream*, and *when pigs fly*.
Unorthodox	Something *unorthodox* is unconventional; that is, something contrary to what is accepted, normal, or traditional. Unorthodox behaviour is to *not* follow an authorized or generally-accepted doctrine, interpretation, practice, or theory. For example, Andy Kaufman was known for his unorthodox approach to humour. See also, *dogma*, *orthodox*, and *outside the box*.
Unpriced Purchase Order	An *unpriced purchase order*: • Does not include pricing information. This information is hidden or redacted from the scope of supply and revision notes, and may include lot and unit rates, tax amounts, subtotals, totals and grand totals; and, • Includes necessary product, schedule, service, and technical information. Most projects make pricing information confidential to prevent unauthorized disclosure or use. See also, *scope of supply* and *purchase order*.
Unsee	To *unsee* is to: • Erase the image or memory of something; or, • Forget having seen something distasteful, disturbing, or unpleasant. For example, after seeing a mistake in a PowerPoint presentation, it is difficult not to remember (unsee) it. See also, *unlearn*.
Unskilled Trade	An *unskilled trade* is an unskilled labour occupation that may require work experience or on-the-job training, but not an apprenticeship or formal vocational education. For example, a hydrotest technician, machine operator, shipper/receiver, swamper, or trades person's helper. See also, *apprenticeship*, *trades person*, and *skilled labour*.
Unsynchronized	To be *unsynchronized* is to behave disharmoniously, incongruously, or at a different rate or time. See also, *synchronized*.
Untagged Equipment	*Untagged equipment* is not identified with a tag number. See also, *bulk items*, *tag number*, and *tagged equipment*.
Untruth	An *untruth* is: • A false statement or lie; • An ethical, factual, or logical falsehood; or, • Something that is not real but instead imagined or invented. See also, *brainwashing*, *bullshit*, *bullshit detector*, *disinformation*, *ethical*, *ethics*, *fake news*, *false narrative*, *media bias*, *misinformation*, *propaganda*, *smells fishy*, *stretch the truth*, *truth*, *truth filter*, *truthiness*, and *white lie*.

Phrase or Term	Definition
Unzip	*Unzip* is a type of progressive failure that occurs when a stress concentration at one end of a weld is high enough to rupture the weld without first distributing additional load to the remaining weld. While the overall weld may be sized to provide required strength, if there is not enough ductility to distribute the load across the entire weld and after the weld fractures in one area – it will continue to break (or *unzip*) along the entire length of the weld. For example: • Aluminium alloy welds subjected to dynamic loading can unzip. This failure is a mechanism of unstable crack growth along the heat affected zone-weld metal interface; and, • Piping with a welded seam (i.e., not seamless) has unique reliability and safety risks. Upon failure, the tendency is for the entire seam to unzip.
Upcycle, Creative Re-use	*Upcycle* or *creative re-use* is the transformation of by-products (e.g., discarded objects or waste materials) into new products of perceived higher quality or value than the original. The goal is to prevent wasting potentially useful materials by making use of existing materials. For example, the use of brewer's spent grain, a waste by-product of the brewing processes, as a feedstock for renewable natural gas. Refurbishment or re-manufacturing of second hand products are upcycling. See also, *downcycle*, *feedstock*, *renewable natural gas*, and *second-hand*.
Upgrading	*Upgrading* is the process of converting heavy oil or bitumen into synthetic crude oil through the removal of carbon (coking) or the addition of hydrogen (hydro-conversion). See also, *synthetic crude oil*.
Upside	*Upside* means: • An upward trend (e.g., prices or stock value); • The favorable or positive aspect of something; or, • Potential or promise. See also, *down-side*.
Upside Down	*Upside down* means with the top side facing down. For example, do not place the box upside down. See also, *right-side up*.
Upstairs, On the Carpet, Mahogany Row	*Upstairs*, *on the carpet*, and *mahogany row* refer to the offices of senior management where these: • Are at a higher level in the building, which is accessed by stairs (or an elevator); • Have carpet floor covering which is more expensive, instead of linoleum or bare concrete, which is less expensive; or, • Have furniture made from mahogany, which is more expensive, instead of metal, oak, or particle board, which is less expensive. See also, *senior management*.
Upstream	The *upstream* or exploration and production sector of the oil and gas industry searches for crude oil and natural gas fields, and drills and operates wells to bring crude oil or raw natural gas to the surface. See also, *downstream* and *midstream*.

Phrase or Term	Definition
Upturn	*Upturn* is: • An improvement or upward trend, especially in business, economic conditions, or someone's fortunes; or, • To turn something upward or upside down. See also, *downturn* and *upside*.
Use As-is	The phrase *use as-is*, is used to disclaim an implied warranty for an item being sold. This means that the buyer accepts the current condition of an item, including all faults and potential faults – whether immediately apparent – as a term of the sale of the item. The seller may or should warrantee items sold *use as is*. For example, a piping spool length exceeds the maximum tolerance by a small amount (e.g., 12.7 mm or ½ in), but repair will delay the schedule and the over tolerance can be accommodated during construction. This is also a possible disposition for a non-conformance. See also, *non-conformance*, *tolerance*, and *use as-is*.
User Experience Design, User Experience	*User experience design* or *user experience* are phrases that describe a website developer's efforts to design with the user in mind. It includes content management, information architecture, interaction design, interface design, usability, and visual design. The role of user experience design is to take the user's needs into account at every stage of the product life cycle – from advertising through purchasing to delivery and more. See also, *user-centred design*.
User-centred Design, Usability	*User-centred design and usability* were terms used in the 1990s to describe a website developer's efforts to design with the user in mind. It focused on designing the right product and ensuring it was useable. This meant making sure that: • The content is organized in a way that allows personnel to find what they need (i.e., information architecture); and, • Things were designed in a way that enabled personnel to use them successfully. See also, *user experience design*.

Phrase or Term	Definition
V	1 A B C D E F G H I J K L M N O P Q R S T U **V** W X Y Z
Vacuum	A *vacuum* is negative (i.e., external) pressure applied to pressure equipment. See also, *pressure equipment* and *vacuum test*.
Vacuum Test	A *vacuum test* uses negative pressure to test for leaks. See also, *vacuum*.
Value Chain	A *value chain* is a set of activities that an organization performs to deliver a product, service, or both, to the market. A well-run value chain encourages useful activities and attempts to acquire completive advantage. It also seeks to eliminate wasteful activities that hamper business for each step of the product development process. A value chain is based on the idea that if a value is added during each process step, then greater overall profit margins are achieved. For example, Starbucks has an impressive value chain. It: 1. Buys coffee beans from farmers worldwide that are transported for storage; 2. Roasts and packages beans that are transported for distribution; 3. Sells beans and other products exclusively in Starbucks stores or licensed stores; 4. Provides franchisees with resources for design, finance, management, and legal requirements; 5. Motivates employees with benefits and incentives; and, 6. Uses technology for coffee-related processes to ensure consistent quality and taste, and reduce costs and connect with customers. The concept comes from business management and was first described in *Competitive Advantage: Creating and Sustaining Superior Performance* by Michael Porter. See also, *supply chain*.
Value for Money Ratio	The *value for money ratio* is the comparison of financial and other benefits to the resources expended for a project.

Phrase or Term	Definition
Value Improving Practice	A *value improving practice* is an out-of-the-ordinary practice used during front end loading to improve capital project performance above that attained through good engineering and project management practices. It addresses the value process during the concept, definition, and implementation phases of a project. Some well-recognized value improving practices include: • Classes of facility quality; • Constructability; • Design to capacity; • Energy optimization; • Optimization of standards and specifications; • Predictive maintenance; • Process reliability simulation; • Process simplification; • Review of 3D CAD drawings; • Technology selection; • Value engineering; and, • Waste minimization. See also, *front end loading* and *value engineering*.
Value Innovation	*Value innovation* is the alignment of innovation with cost, price, and utility positions to create uncontested market space and make the competition irrelevant. See also, *blue ocean strategy*, *disruptive innovation*, *red ocean strategy*, and *value innovation*.
Value Ladder	*A value ladder* is a business structure with a tiered range of products and services that ascend in price, from the lowest value offer to the highest-priced and premium offerings. A value ladder makes a customer's first purchase easy, helps build trust, and provides entry-level products or services for free or at a low-cost. Additionally, a value ladder increases customer awareness and interest, total number of products and services offered, and proves value. When a customer is satisfied with a purchase, they will likely return to buy more. The business then guides the customer through the value ladder to maximize the customer's value to the business. See also, *value ladder-less* and *value proposition*.
Value Ladder-less	*Value ladder-less* is when a business structure does not boost profits by offering various product or service options at different tiers that ascend in price. See also, *value ladder* and *value proposition*.
Value Proposition	*Value proposition* refers to a company business or marketing statement used to summarize why a consumer should buy a product or procure a service. It is an easy-to-understand reason or reasons why a customer should purchase a product or service from a specific business. It is also a promise by a company to a customer or market segment. See also, *Business Model Canvas*, *value ladder*, and *value ladder-less*.
Value Stream	*Value stream* is the end-to-end sequence of activities – value added and non-value added – required to design, produce, and provide a specific product or service. It includes the flow of information that creates an overall deliverable or result for a customer, stakeholder, or end-user that can be sold.

<table>
<tr><th>Phrase or Term</th><th>Definition</th></tr>
<tr><td>Value-added</td><td>Value-added has many meanings. For a project it is:<ul><li>A cost that is included or hidden in the base price such that customers will (feel as though they) get more than they paid for;</li><li>Having features added to a basic model for which the buyer is prepared to pay extra;</li><li>The amount by which the value of a product is increased at each stage of its production, exclusive of initial costs; or,</li><li>To connect an additional attribute or service to a product, at no cost to the buyer.</li></ul>See also, value-added engineering.</td></tr>
<tr><td>Value-added Engineering</td><td>Value-added engineering is a systematic method of improving the value of products and services by using examining function. Value in this context is the ratio of function to cost to provide the most efficient and cost-effective results. See also, holistic engineering, independent project review, value-added, value-based decision-making, and value improving practice.</td></tr>
<tr><td>Value-based Decision-making</td><td>Value-based decision-making is a method of making critical organizational decisions in an informed and timely manner. These decisions often directly affect the value generated or lost by an organization. It deals with making the organization better in a meaningful way by delivering values to the marketplace.
In entrepreneurship, there is a risk to decide in favour of something that does not generate business values or goes against something that does. Decisions are often made based on personal beliefs and experience. However, the problem with this approach is that these beliefs are based on past information that is projected into the future. Values are concepts that transcend contexts. Beliefs, on the other hand, are context-dependent. The reasons why value-based decision-making is important are:<ul><li>Individuation and self-actualization, to transcend the belief structures and cultural conditioning;</li><li>Institutionalization and development of democracy to transcend cultural and ethnic belief structures by uniting people around basic shared human principles; and,</li><li>Reduce reliance on bureaucratic procedures that set out what people should or should not do in a specific situation so personnel become accountable and responsible for their behaviours.</li></ul>See also, bureaucracy, decision-making, entrepreneur, decision-based evidence making, evidence-based decision-making, evidence-based management, opinion-based decision-making, personal values, value-based decision-making, and values.</td></tr>
<tr><td>Value-driven</td><td>A value-driven business or project focuses on a premium value proposition, often with a high degree of customization or personalization (e.g., asset management and designer or luxury items). See also, cost-driven, schedule-driven, and value proposition.</td></tr>
</table>

Phrase or Term	Definition
Values Statement	A *values statement* is a declaration about how an organization values internal and external stakeholders. A value statement explicitly defines what an organization's personnel believe and how they will behave. For example, Ericsson's value statement is: *At Ericsson we share three core values: respect, professionalism, perseverance. These values matter because they have become established within our business over 140 years of history.* See also, *values*.
Values, Core Values	*Values* or *core values* describe an organization's behavioural compass or desired culture. See also, *mission*, *personal values*, *purpose*, *values statement*, and *vision.*
Van Stone, Conrac	A *Van Stone* or *Conrac* is a mechanical piping connection made without using a conventional screwed or welded flange. The pipe is formed (e.g., flanged, flared or lapped) to provide a sealing face that is connected using a rotating bolted flange. This system is used with lined, metallic, and non-metallic piping systems.
Variable Costs	*Variable costs* are expenses that vary in proportion to the volume of products or services (e.g., materials, labour, and salaries). See also, *cost plus* and *fixed costs*.
Variable Renewable Energy	*Variable renewable energy* is renewable energy that fluctuates depending on resource availability, such as solar, tidal or wave, and wind power. See also, *renewable energy*.
Variance Analysis	*Variance analysis* is the practice of investigating deviations between planned and actual performance. See also, *negative variance* and *positive variance*.
Vehicle-to-Grid	*Vehicle-to-grid* is a technology that enables energy to be pushed back to the power grid from the battery of an electric vehicle. This allows a vehicle battery to be charged and discharged based on different signals – to balance variations in nearby energy consumption or production. See also, *smart charging*.
Vendor	Not a preferred term. *Sub-vendor* is also not preferred. See *supplier*.
Vendor Data Book	Not a preferred phrase – see *manufacturing record book*.
Vendor Document Review Code	Not a preferred phrase – see *supplier document review code*.
Vendor Drawing and Data Requirements	Not a preferred phrase – see *supplier document requirements*.
Vendor document index	Not a preferred phrase – see *supplier document index*.
Vent Hole	A *vent hole* is a hole in reinforcing plates or pads and other structural attachments that, when attached by welding, create an interstitial or void space. A vent hole prevents the build-up of pressure in the space, if a leak occurs through the pressure-retaining component of storage or pressure equipment (e.g., a tank or pressure vessel). A vent hole may or may not be tapped with threads, and may be open or plugged when the equipment is in service. If the hole is plugged, the plugging material used shall not be capable of sustaining pressure between the reinforcing plate and the equipment. See also, *weep hole* and *weld access hole*.
Verifiable Record	A *verifiable record* is information about equipment, materials or a facility that is confirmed by complementary, but separate, documentation. A verifiable record must have at least two different documented means of arriving at the same conclusion. See also, *traceable record* and *traceability*.

Phrase or Term	Definition
Verification	*Verification* is the determination if products and services conform to specified requirements. See also, *verification document*.
Verification Document	A *verification document* provides proof and a record of conformance to specified requirements. See also, *verification*.
Verification Point	A *verification point* is a key step in a work process in which quality surveillance representative verification is necessary. This activity is completed in-process during an inspection visit, so advance notification is not required. Verification points may be identified on the inspection and test plan or inspection assignment. See also, *non-witnessed point*.
Vermin	*Vermin* are unwanted pests and wild animals (e.g., animals, birds, insects, and rodents) that are harmful to animals, buildings, coatings, crops, electrical wiring, facilities, humans, and insulation because they: • Carry disease or cause injury; and/or, • Destroy materials and property (e.g., through infestation, waste discharge, nesting and other deleterious behaviour). An insurance policy may define vermin in specific terms that can exclude or include certain damages and damages from certain vermin species.
Vernacular	*Vernacular* is: • An expression, or mode of expression that occurs in ordinary speech rather than formal writing. For example, *y'all* is a contraction of you and all. It is the main second-person plural pronoun used in Southern American English. *All y'all* is used to indicate a larger group; or, • A local dialect or language that is distinct from the standard or mainstream language (i.e., some aboriginals in Taiwan speak a local dialect called *Haka*, which differs dramatically from Mandarin). See also, *colloquial*, *dialect*, *Ebonics*, *jargon*, and *Plain English*.
Vertical Integration	*Vertical integration* is a strategy whereby a company owns or controls its suppliers, distributors, or retail locations to control its value or supply chain. Vertical integration benefits companies by allowing them to control processes, improve efficiencies, and reduce costs. However, vertical integration has its disadvantages, including the significant amounts of required capital investment. Netflix is a prime example of vertical integration whereby the company started as a digital video disc or DVD (i.e., compact disc) rental company supplying movie and television content. The company's executive management realized they could generate more revenue by shifting to original content creation. Today, Netflix uses its distribution model to promote their original content alongside third party movie and television content.
Vertical Slice	*Vertical slice* is a performance indicator that demonstrates progress across all project components or performance areas on a certain date or within a date range.

Phrase or Term	Definition
Vertical Thinking	*Vertical thinking* involves using analytical, selective, sequential, and external data, facts, and knowledge to resolve a challenge. For example, solving this math equation requires vertical thinking: (21 + 3) x 4 A vertical thinker is required to solve the challenge in the correct sequence; doing otherwise will yield an incorrect answer. See also, *cognitive dissonance*, *cognitive flexibility*, *convergent thinking*, *divergent thinking*, *lateral thinking*, and *thinking mode*.
Victaulic® Coupling	A *Victaulic® coupling* is a proprietary mechanical joining system that uses gasket and grooved technology for piping systems up to 9.58 kPa (200 psi). See more information at the Victaulic® website https://www.victaulic.com. See also, *proprietary* and *Victaulic® groove*.
Victaulic® Groove	A *Victaulic® groove* is rolled or cut grooves in the end of a pipe designed to accommodate a Victaulic coupling. See also, *Victaulic® coupling*.
Vicious Circle, Vicious Cycle	A *vicious circle* or *vicious cycle* is: • A chain of events or a self-sustaining feedback loop with undesirable effects. It creates a continuing unpleasant situation, wherein one problem causes another problem – which that makes the first problem worse. For example, a weakening economy leads to lower home prices, which leads to less new home construction, which further weakens the economy, and so on; or, • When the conclusion of an argument is appealed to as one of the truths or principles upon which the argument itself rests. For example, Picasso was a great artist because everything he created was great art. See also, *Catch-22*, *feedback loop,* and *virtuous circle*.
Virtual Assistant	A *virtual assistant* is an employee or independent contractor who provides various services from a remote location to entrepreneurs, small and medium-sized businesses, and solopreneurs. Their focus is on administrative tasks (e.g., administration, answer or make telephone calls, bookkeeping, business development, data entry, digital marketing, event management, run personal errands, research, respond to emails, and schedule appointments). This provides the benefits to: • Acquire skills and resources as-needed; • Allow other personnel to focus on critical or high-value tasks; • Improve customer experiences and service; • Increase efficiency, flexibility, productivity, and quality; and, • Provide professional and timely communications; and, • Reduce labour costs. See also, *consultant*, *entrepreneur*, *full-time equivalent*, *gig*, *independent contractor*, *labour force*, *liquid labour force*, *side hustle*, and *solopreneur*.

Phrase or Term	Definition
Virtual Design and Construction	*Virtual design and construction* is the management of integrated multi-disciplinary performance models of design-construction projects, including the product (i.e., facility), work processes, and organization of the design, construction, and operation team to support explicit and public business objectives. It uses building information modelling tools that focus on designable and manageable aspects of projects to create integrated models that predict project performance. See also, *building information modelling* and *virtual reality*.
Virtual Private Network, VPN	A *virtual private network* or *VPN* is an encrypted connection over the Internet from a device to a network to ensure that sensitive data is safely transmitted. It allows users to conduct work remotely and prevents unauthorized users from eavesdropping.
Virtual Reality	*Virtual reality* is a computer-generated simulation of a 3D environment or image. See also, *augmented reality*, *digital reality*, *mixed reality*, and *virtual design and construction*.
Virtual Team	A *virtual team* comprises people from different organizations, locations, or hierarchies. It is not necessarily the same as a remote team. See also, *remote team*.
Virtue Signalling	*Virtue signalling* is: • Attempting to show others that one is a good person by expressing opinions that are acceptable to them. For example, telling others that you donate to charities often when you never donate to charities; or, • Taking a conspicuous, but essentially useless actions, ostensibly to support a good cause but with the intent of showboating one's perceived moral superiority over others. For example, posing with a bicycle for pictures to be shared on social media to showboat personal commitment to energy conservation; yet, not owning or riding a bicycle and taking vacations using a private jet. See also, *ulterior motive*.
Virtuous Circle, Virtuous Cycle	A *virtuous circle* or *virtuous cycle* is a chain of events or a self-sustaining feedback loop with favourable effects. It creates a continuing desirable situation, wherein one occurrence causes another occurrence – which makes the first occurrence better. For example, a strengthening economy leads to higher home prices, which leads to more new home construction, which further strengthens the economy, and so on. See also, *feedback loop* and *vicious circle*.
Viscosity	*Viscosity* is the ability of a liquid to flow. The lower the viscosity, the more easily liquid will flow.
Vision	*Vision* is: • A concept, object, or thought formed by imagination; or, • The ability to plan or think about the future with foresight, imagination, or wisdom. See also, *mission*, *purpose*, *values*, and *vision statement*.

Phrase or Term	Definition
Vision Statement	A *vision statement* is a declaration of an organization's mid-term and long-term goals, stating what they want to become in the future. It is an inspirational statement of an organization's emotional or idealistic future. For example, Coca-Cola's vision statement is: *Our vision serves as the framework for our roadmap and guides every aspect of our business by describing what we need to accomplish in order to continue achieving sustainable, quality growth.* See also, *mission statement* and *vision*.
Visual Workplace	A *visual workplace* employs devices designed to visually share information about organizational operations. It seeks to make human-machine performance as exact, reliable, repeatable, and safe as possible.
Visualization	*Visualization* is the techniques and technologies used to present information in ways that enable analysis and understanding.
Voice of the Customer	*Voice of the customer* is the in-depth process of capturing customer aversions, expectations, and preferences. Specifically, it is a qualitative and quantitative market research technique that produces a detailed set of customer wants and needs, organized into a hierarchical structure, and then prioritized in terms of relative importance and satisfaction with current alternatives. Common sources include sales and technical reports, social media, user support forums or help lines, and warranty claims.
Voluntary Sector	The *voluntary sector* is the part of the economy that is operated or owned by charities and non-profit organizations. See also, *philanthropy*, *private sector* and *public sector*.
Volunteering	*Volunteering* is the act giving labour or time for free to benefit a cause. For example, helping cook in a kitchen built to feed homeless people. Volunteers may be unskilled or skilled (e.g., construction, education, emergency rescue, or medicine). See also, *donation* and *pro bono*.
Vulnerable	*Vulnerable* is: • A state of being that may be used with other people who are trusted. A person's decision to become vulnerable is generally a risky one that needs courage and the comfortable feeling of a safe space. Being honest with other people allows them to feel a safe connection and secure enough to be genuine in their interactions; • An asset or system that is susceptible to attack; a cybercriminal could exploit a pipeline control system to disrupt or damage a pipeline. For example, in April 2018, cyberattacks caused the shutdown of communication systems at four of the largest natural gas pipeline companies in the United States; • Capable of being emotionally or physically wounded; and, • The degree to which personnel, property, resources, systems, and cultural, economic, environmental, and social activity is susceptible to damage, degradation, destruction, or harm by being exposed or open to an attack by a hostile agent or factor. See also, *honest* and *trust*.

Phrase or Term	Definition
Wag the Dog	*Wag the dog* is an expression used in politics that refers to the creation of a diversion to distract attention from a damaging issue or scandal. For example, the prime minister wags the dog with threats of sanctions to divert media attention away from their personal scandal. This phrase gained notoriety with Larry Beinhart's novel *Wag the Dog* and movie by the same name. See also, *tail wagging the dog* and *smoke and mirrors*.
Waiver	Not a preferred term – see *design change notice*.
Walk Down	A *walk down* is an inspection activity performed by a client, project, stakeholder, team, or combination for large packaged equipment and modules or systems. Walk downs typically occur during the final project phase or at completion of fabrication or construction, or before commissioning.
Walk the Line, Walking the Line	To *walk the line* or *walking the line* is to: • Behave in an authorized or socially accepted manner, or exercise self-control, especially as prescribed by law or morality (e.g., this soldier walks the line during war); or, • Maintain an intermediate position between contrasting choices or opinions (e.g., deeply religious, the scientologist walks the line between theism and secularism). See also, *sit on the fence*.
Walkaway	A *walkaway* is: • A deliverable that a supplier cannot or refuses to supply. For example, due to the failure of the business (i.e., going into receivership) or a purchase order that was unprofitable, the manufacturing record book is not compiled or submitted after completion of the work. • A failed project wherein the project asset cannot be salvaged. Thus, the asset is abandoned and the stakeholder suffers a complete loss. For example, roads are constructed and the power supply is upgraded for construction at site, but the project is later cancelled; thus, the assets become useless. See also, *manufacturing record book* and *receivership*.
War Room	A *war room* is a room (i.e., a headquarters) where a: • Business or political campaign or strategy is planned; or, • Military operation is directed (i.e., a war). See also, *business wargame* and *wargame*.

Phrase or Term	Definition
Wargame	A *wargame* is a strategic game that simulates warfare realistically (i.e., operationally, strategically, or tactically), as opposed to an abstract strategic game such as chess. A wargame may be played: • For recreation (e.g., paintball or the boardgame Risk); • To study the nature of potential conflict (e.g., a simulated wargame on a computer); • By gaming (e.g., using wargame software on an arcade-style machine or laptop gaming computer); • To train military personnel in the art of strategic thinking (e.g., training combat helicopter pilots in advanced flight simulators); and, • To simulate combat by air, land, or naval forces, or a combination (e.g., live wargames performed by actual United States' troops in the Nevada desert – often referred to as field exercises). Modern wargaming was invented in Prussia during the turn of the 19th century. See also, *business wargame*, *game*, *games against nature*, *gamesmanship*, *gamification*, *lose-lose*, *war room*, *win-win*, and *zero-sum game.*
Warm Call	A *warm call* is contact made with someone (e.g., by email or telephone, or in-person) with whom there has been previous interaction, interest, or referral – to offer them a product or service. See also*, referral*, *sales lead*, and *cold call*.
Warranty	A *warranty* describes the conditions under which, and the duration of, compensation that will be supplied by the producer (e.g., fabricator, manufacturer, or supplier) to repair or replace a defective item or product without cost or with a discounted cost to the consumer. A warranty may also specify rights and obligations of both parties as legal guidance for claims or disputes between parties under warranty. See also, *extended warranty* and *guarantee*.
Wastewater	*Wastewater* is a by-product of agricultural, domestic, commercial, or industrial activities that contains biological, chemical, or physical pollutants, or a combination. It may be described as either: • Black water Black water is generated by businesses or domiciles, and includes fecal contamination from toilets; or, • Grey water Grey water is generated by businesses or domiciles and does not include fecal contamination from toilets. See also, *potable water* and *potable water*.

<table>
<tr><th>Phrase or Term</th><th>Definition</th></tr>
<tr><td>Waterfall Model</td><td>The waterfall model is a breakdown of project activities into linear sequential phases, where each phase depends on the deliverables of the previous one and corresponds to a specialization of tasks. Each development process may begin only if the previous phase is completed. This progress is seen as flowing steadily downwards, like a waterfall, through these phases:
1. Conception;
2. Initiation;
3. Analysis;
4. Design;
5. Construction;
6. Testing;
7. Production/implementation; and,
8. Maintenance.
This development model originated in the manufacturing and construction industries, where the highly structured physical environments meant that design changes became prohibitively expensive much sooner in the development process. When first adopted for software development, there were no recognized alternatives for knowledge-based creative work. As this process is illustrated in a linear sequential flow, it is also referred to as a linear-sequential life cycle model.</td></tr>
<tr><td>Watershed Event</td><td>A watershed event is a historic moment, life-changing experience, or turning point. For example, being hired for your first job as a teenager or in your field of study after completing an education. Watershed also describes:
• The area that drains into a single river; and,
• A chain of mountains where water drains into a different river on either side.</td></tr>
<tr><td>Weak Artificial Intelligence, Narrow Artificial Intelligence</td><td>Weak artificial intelligence or strong artificial intelligence is artificial intelligence that is focused on one narrow task. Many currently existing systems that claim to use artificial intelligence are likely operating weak artificial intelligence that is focused on a narrowly defined and specific problem. For example, Siri is narrow intelligence because it operates within a limited pre-defined range of functions. See also, artificial intelligence and strong artificial intelligence.</td></tr>
<tr><td>Weak Man Argument</td><td>A weak man argument selects only the weakest and worst parts of an opponent's argument for attack, and then implies that these are the best or the only arguments they have. See also, straw man argument and steel man argument.</td></tr>
</table>

Phrase or Term	Definition
Wealth	*Wealth* is an abundance of valuable possessions (e.g., a collection of antique or exotic vehicles, or real estate) or resources (e.g., business assets or money), which can be converted into a form that can be used for transactions (e.g., cash). Wealth may be obtained by being self-made (e.g., as an entrepreneur or investor), by inheritance (e.g., from family), or a combination. To accumulate wealth over time, one needs to do these things: 1. Make money; 2. Save money; and, 3. Invest money. The meaning of wealth is context dependent. For example, a wealthy person in one country may not be considered wealthy in another. See also, *net worth* and *wealth pyramid*.
Wealth Pyramid	The *wealth pyramid* is a graphical depiction of the inverse relationship between two financial variables where one increases as the other decreases (i.e., the percentage of world wealth and the percentage of world population controlling it). This concept was developed by the professor Coimbatore Krishnarao (CK) Prahalad in his book *The Fortune at the Bottom of the Pyramid*: • The base or bottom of the wealth pyramid is the largest and comprises the poorest socioeconomic group, who live on less than USD$ 2.50 a day (USD$ 1,000 per year); and, • The top of the wealth pyramid is the smallest and comprises the richest socioeconomic group (e.g., millionaires and billionaires). The gap between the top and bottom of the wealth pyramid has been widening over time. Today, 1% of the world's population controls 50% of the wealth. The remaining 99% of the world's population has access to the remaining 50% of wealth. Credit Suisse publishes an annual *Global Wealth Report* with a wealth pyramid infographic that explains the world distribution of wealth. See also, *factfulness*, *Gapminder*, *wealth*, and *world model*.
Weaponization	*Weaponization* is the act or process of: • Adapting or making something ready to use as a weapon (e.g., adding missiles to a drone); or • Using something as a weapon (e.g., using a political message to insult political adversaries). Weaponize originated as technical jargon in the United States military. Today, it is regularly used outside of military contexts. For example, *the report identified both pros and cons, but the information was weaponized to attack the project*. See also, *jargon*, *pros and cons*, *political*, *politicization*, and *radicalization*.
Wear and Tear	*Wear and tear* is the normal depreciation or deterioration that occurs as something ages over time during ordinary use. This damage or degradation occurs even when something is properly cared for and well maintained. For example, normal wear and tear is the abrasion (from dust) and fading (from sunlight) of a vehicle's paint that occurs every year. Damage from a vehicle collision, for example, is not normal wear and tear. A manufacturer's warranty usually excludes covering damage from normal wear and tear because wear and tear is impossible to prevent.

Phrase or Term	Definition
Weasel Out	To *weasel out* means to avoid doing something or taking responsibility by being clever, dishonest, or persuasive. Informal. See also, *sea-lawyering* and *weasel words*.
Weasel Words	*Weasel words* are words that are: • Intentionally deceptive, disingenuous, ambiguous, or misleading; • Supposed to sound authoritative and meaningful but lack comprehensive meaning and substance; or, • Used to avoid a commitment, clear position, or statement. This phrase originates from a weasel's ability to suck the contents out of an egg while leaving the shell superficially intact. See also, *sea-lawyering* and *weasel out*.
Web-based Business	A *web-based business* (e.g., Amazon) offers products and services to customers over the Internet. Its advantage over the traditional brick-and-mortar businesses model is lower overhead, lower operating costs, and, often, lower taxes. See also, *brick-and-mortar business*.
Webinar	A *webinar* is an online seminar with a real-time conversation or presentation. Webinar is a portmanteau of *website* and *seminar*.
Weep Hole	A *weep hole* is a drainage opening in a hollow structural member (e.g., a ¼" hole at the lowest point of a structural member). See also, *vent hole* and *weld access hole*.
Weight	To *weight* is to: • Attach importance or value to something (e.g., factor a calculation or identify more or less important items in a list). For example, the budget carries weight; or, • Hold (something) down by placing a heavy object on top of it (e.g., precast concrete pipeline weights). See also, *precast concrete*.
Weld Access Hole	A *weld access hole* is a design technique in which a part of the web of an I-beam or T-beam is cut out at the end or ends of the beam. The hole in the web allows a welder to weld the flange to another part of the structure with a continuous weld extending across the full width on the top and bottom sides of the flange. Without a weld access hole, the middle of the flange would be blocked by the web and inaccessible for welding. See also, *vent hole* and *weep hole*.
Weld Metal	*Weld metal* is the *weldment*. See also, *weldment*.
Weld Overlay	*Weld overlay* is a welding process where a consumable electrode with specific characteristics (e.g., corrosion or erosion resistance) that is welded to a base metal to improve desirable properties or restore original dimensions of the part.
Weld Reinforcement	*Weld reinforcement* is weld material in excess of the specified weld size. See also, *weldment*.

Phrase or Term	Definition
Weldability, Joinability	*Weldability* or *joinability* refers to a material's ability to be welded. Many metals and thermoplastics can be welded, but some are easier to weld than others. Some materials cannot be welded at all (i.e., non-weldable reinforcing bar or martensitic stainless steels SAE 400 series) or welded to each other (e.g., aluminum and steel). A material's weldability: • Determines the welding process that will be employed; and, • Provides a comparison of the final weld quality in relation to other materials. Quantitatively, weldability is often difficult to define; thus, most standards define it qualitatively. For steel there are three major failure modes by which weldability can be measured: 1. Hydrogen-induced cold cracking; 2. Lamellar tearing; and, 3. Spot-weld peeling. See also, *carbon equivalent*, *carbon steel*, *hydrogen embrittlement*, *low hydrogen*, *low hydrogen electrode*, *reinforcing bar*, and *stainless steel*.
Welder	A *welder* performs a manual or semi-automated welding operation. See also, *pressure welder* and *welding*.
Welding	*Welding* is process that joins materials such as metals or thermoplastics, typically using high heat, pressure, or both, to melt the parts together. When the parts cool, the parts bind through fusion. A filler metal or electrode may or may not be used. See also, *bonding*, *brazing*, *electrode*, *filler metal*, *fusion*, *soldering*, and *weldment*.
Welding Operator	A *welding operator* operates automatic welding equipment or machinery. See also, *welding*.
Welding Procedure Data Sheet	A *welding procedure data sheet* is the formal written document describing welding procedures, that provides direction to the welder or welding operators for making sound and quality production welds as per code requirements.
Welding Procedure Specification	A *welding procedure specification* provides guidelines for the control of production welding according to the code of construction by addressing all essential and non-essential (and if required, supplementary essential) variables that shall be used.
Welding Rod	Not a preferred phrase – this is slang for an *electrode*.
Weldment	A *weldment* is a part created from an assembly of smaller components (e.g., base metals) that are joined by welding. See also, *base metal*, *heat affected zone*, and *welding*.
Well to Wall	Not a preferred term – see *life cycle*.
Well to Wheels	Not a preferred term – see *life cycle*.
Wet Film Thickness	*Wet film thickness* is the thickness of a coating measured immediately after application and before any appreciable solvent has evaporated or drying has occurred. See also, *film*.
Wet Signature	A *wet signature* is created when someone physically marks a document with a pen (i.e., using *wet ink*) or seal to make it legally binding. A wet signature can be mandated by law or demanded from a party. See also, *authentication*, *digital signature,* and *electronic signature*.

Phrase or Term	Definition
Wetland	Wetland is land consisting of marsh or swamp that is saturated with water. See also, *dryland*.
Wetwear	*Wetwear* is human brain power, which is distinct from hardware (e.g., chips) and software (e.g., code).
Whataboutism, Whataboutery, Tu Quoque	*Whataboutism*, *whataboutery*, or tu quoque (Latin for *you to*) is a technique of responding to an accusation or difficult question by making a counteraccusation or raising a different issue. It often begins with the words *what about* and suggests that something wrong is not wrong because of the existence some other wrong. This means that nothing may be deemed to be wrong if there is another example that is worse. This method of arguing or debating attempts to establish an equivalence between two or more disparate actions, thereby defaming the accuser with the insinuation that their priorities are backwards. It is like asking *who are you to call the kettle black?* For example, to argue that environmental regulations are pointless because other countries are polluting the environment and lack similar regulations. This practice was first used by the Soviet Union as a propaganda tactic during the cold war, but it is also used by individuals and organizations (e.g., governments). See also, *double standard*, *forced empathy*, and *propaganda*.
What Is In It For Me?	*What is in it for me?* Is a question about the advantages of a new idea or method where the benefits are not obvious. It asks how one will personally benefit or gain from an action, activity, or situation.
What you see is all there is	*What you see is all there is*, is a cognitive bias or an egocentric worldview whereby a person only sees the world as it is visible from their perspective. Judgements and impressions are normally made according to the information that is available or presented. In general, time is not spent thinking about what is unknown; thus, whatever is known is asserted. For example, when meeting someone for the first time, it takes only seconds to gain an impression of them. Immediately, people decide if a person is kind and pleasant or dominant and hostile, and if they are likeable. See also, *bias*, *cognitive bias*, and *thinking mode*.
Whatchamacallit, Thingamabob, Thingamajig, Gizamatron, Gizmo, Doohickie	*Whatchamacallit*, *Thingamabob*, *Thingamajog*, *Gizamatron*, *Gizmo*, and *Doohickie* are informal terms used to refer to a thing whose name cannot be recalled, is unknown, or is unspecified. Whatchamacallit derives from the informal phrase *what-you-may-call-it*.
What-If Scenario Analysis	A *what-if scenario analysis* is a simulation technique that allows a project manager to determine and compare specific conditions' effects on project schedules and objectives.
When Pigs Fly	*When pigs fly* is an expression that means something is impossible or will never happen. For example, *when pigs fly I will be able to wake up at 5:00 am*. This sarcastic phrase has been used in various forms since the middle ages. Its origin is unknown. There are similar phrases in many other languages such as *kun lehmät lentävät* (Finnish, when cows fly) or *quand les poules auront des dents* (French, when hens will have teeth). The meaning of the expression *when it snows in hell* is synonymous. See also, *fool's gold*, *hopium*, *pipe dream*, and *unobtainium*.

Phrase or Term	Definition
Where You Stand Determines Where You Sit, Miles Law	*Where you stand determines where you sit* or *Miles Law* is the principle in politics that position in a bureaucracy determines position on an issue. This means that perspective is shaped by environment. This phrase was coined by Rufus Edward Miles, Jr. in the late 1940s. See also, *sit down*, *stand down*, and *stand up*.
Whistleblower, Whistle Blower, Whistle-blower	A *whistleblower*, *whistle blower*, or *whistle-blower* is someone who reports activity or information deemed abusive, fraudulent, illegal, illicit, unsafe, wasteful, or unethical. For example, to report workplace theft to an employer or the police. A whistleblower is often legally protected against retaliatory action (e.g., a whistleblower typically may not be terminated for reporting facts). United States consumer advocate Ralph Nader used this term in the early 1970s to avoid negative connotations associated with words such as *informer* and *snitch*. Use of this word dates to the 19th century. It is linked to blowing a whistle to alert others of a crime in progress, danger, or rule-breaking in sports. See also, *termination without cause*.
Whistle Stop Tour	A *whistle stop tour* is the practice of making political campaign speeches in many towns in a short time, often during a single day. When politicians traveled by train, small towns were called whistle-stops. Politicians would use each stop to deliver a quick speech, often from the back of the train, before travelling to the next stop.
White Collar	*White collar* constitutes or relates to the class of personnel whose duties, generally: • Are performed inside an office setting; • Do not perform manual labour; • Earn more than blue collar wage earners; • Require a post-secondary education (e.g., diploma or degree); and, • Require business attire (e.g., a white shirt with a suit and tie). Writer Upton Sinclair contributed to the modern understanding of the phrase white collar by using this phrase to describe administrative work. See also, *blue collar*, *pink collar*, and *white-collar crime*.
White-collar Crime, Corporate Crime	A *white-collar crime* or *corporate crime* is a financially motivated and nonviolent crime committed by a business or government professional. It was first defined by the sociologist Edwin Sutherland in 1939 as a crime committed by one of respectable and high social status during their employment. White-collar crime has no official legal classification, but it is used to describe a group of crimes such as bribery, copyright infringement, cybercrime, embezzlement, forgery, fraud, identity theft, insider trading, labor racketeering, money laundering, Ponzi schemes, and theft. See also, *blue-collar crime*, *bribe*, *kickback*, *money laundering*, and *white collar*.

Phrase or Term	Definition
White Elephant	A *white elephant* is something that is too expensive or difficult to maintain but is difficult or impossible to abandon. Originating from ancient Thailand, the King of Siam would gift an actual white elephant to those who displeased him. A white elephant was known as rare and a revered symbol in Buddhism. Whoever was gifted a white elephant from the King was obliged to care for it, which would be far too expensive for most. Some modern-day examples of a white elephant: • A historic building that cannot be updated without great expense; • A business that is so expensive to operate and maintain that its expenses consume all profit; • An enterprise or venture on which large sums of money are spent, but is somehow abandoned or does not produce a profit. For example, the Ryungyong Hotel in Pyongyang North Korea (1987). Its 3,000 rooms have never been used; or, • Possessing a motorhome that is very expensive to maintain, store, and operate. See also, *green elephant*.
White Flag	A *white flag* is an internationally recognized protective sign of a truce or request for negotiation. It is also a sign of surrender or yielding by a weaker party (e.g., waving a white flag signals surrender to an offensive military force). The white flag and its uses are described in the Hague Conventions of 1899 and 1907. See also, *false flag* and *red flag*.
White Hat Marketing	*White hat marketing* is the use of ethical and legal search engine optimization techniques to promote online business. White hats write web websites effectively so that the search engines can accomplish the intended task and ethically give users the results they want. This uses keywords to identify the actual information on the website and writes content that focuses on those keywords. Keywords, in this case, are used the way the search engine expects them to be used, as index words, or as titles in a table of contents. See also, *black hat marketing*, *ethics*, *ethical*, and *search engine optimization*.
White Label Branding, White Labelling	*White label branding* or *white labelling* is a manufacturing-marketing practice wherein a product or service that is produced by one company (Company A) is rebranded, repackaged or to make the product appear to be produced by another company (Company B).

<table>
<tr><th>Phrase or Term</th><th>Definition</th></tr>
<tr><td>White Lie</td><td>A white lie is an untruth about a trivial or unimportant matter that is told for a benign (i.e., not an evil or malicious) purpose A white lie tends to benefit the person being lied to or the relationship, while a real lie tends to benefit the liar. A white lie may be told when the truth is complicated, tedious, or uncomfortable to:<ul><li>Be agreeable with a professional superior (e.g., avoid the appearance of insubordination). For example, to agree to perform work over the weekend work even though it means cancelling personal plans;</li><li>Be courteous, not cause trouble, or avoid hurting someone's feelings. For example, to tell a spouse their breakfast soup is delicious when it is not;</li><li>Maintain relational harmony and prevent conflict. For example, to tell a spouse that it would be fun to jog at 6:00 am when it is raining or -20° C (-4° F); or,</li><li>Protect an ego or image. For example, telling others that you quit you job when you are were dismissed (e.g., fired).</li></ul>See also, bullshit, bullshit detector, ethical, ethics, false narrative, smells fishy, stretch the truth, truthiness, and untruth.</td></tr>
<tr><td>White Paper, Whitepaper</td><td>A white paper or whitepaper is an informational document issued to highlight or promote the features of a methodology, product, service, solution, or technology. It is an authoritative guide or report that concisely informs the reader about a complex challenge or problem and presents the author's opinion or philosophy about it.
A white paper is a marketing tool that may be flashy and include an obvious sales pitch to promote a product or viewpoint. It is intended to:<ul><li>Help the reader make a purchasing decision, find a solution, or understand something; and,</li><li>Provide factual and persuasive evidence that an offering provides a superior solution – but is not a user manual or technical document that provides post-purchase support.</li></ul>White papers are also used by governments as a method to gauge public reaction to proposed legislation or policies.</td></tr>
<tr><td>Whiteboard</td><td>A whiteboard is a:<ul><li>An erasable board with a white surface that is used for presentations and teaching; or,</li><li>An interactive area on a display screen common to several users, on which users can draw and write.</li></ul>See also, whiteboard session.</td></tr>
<tr><td>Whiteboard Session</td><td>A whiteboard session is a collaborative meeting that facilitates a common understanding of challenges that need resolution and milestones to be reached. This meeting supports discovery and learning processes because the focus is on action, energy, and visualization. It can also be a brainstorming session. This session is called whiteboard because of the placement of files on a shared digital screen or whiteboard. Data and video conferencing software allows the host to annotate files, like a physical whiteboard. Using software, numerous personnel can contribute content to the whiteboard simultaneously. The whiteboard reveals revisions to all participants in real time or near-real time. See also, brainstorm, creative destruction, focus group, whiteboard, and unfocus group.</td></tr>
</table>

<table>
<tr><th>Phrase or Term</th><th>Definition</th></tr>
<tr><td>Whiteout</td><td>Whiteout is a:<ul><li>Blizzard (i.e., snowstorm) that reduces visibility to near zero;</li><li>Material used to blot out or obliterate something (e.g., to apply white correction fluid to cover printing, typing, or writing mistakes); or,</li><li>Surface weather condition in a snow-covered area where only dark objects are discernible, the horizon cannot be seen, and there are no shadows.</li></ul>See also, blackout and brownout.</td></tr>
<tr><td>Whitewashing</td><td>Whitewashing is to:<ul><li>Alter something in a way that caters to, favours, or features white people;</li><li>Cover up or gloss over something (e.g., a crime, scandal, or vice);</li><li>Exonerate someone by means of a perfunctory investigation or using biased information; or,</li><li>Paint (e.g., a building, fence, or room) with white paint or whitewash (i.e., chalked lime).</li></ul>See also, greenwashing and woke washing.</td></tr>
<tr><td>Why Movement</td><td>The Why Movement is a person or organization's path to long-term fulfillment and success — their purpose. The Why Movement was started by Simon Sinek in 2009 to help people become more inspired at work, and in turn to inspire their colleagues and customers. He started with these fundamental questions about organizations and people:<ol><li>Why are some more influential, innovative, and profitable than others?</li><li>Why do some command greater loyalty from customers and employees alike?</li><li>Why are so few able to repeat their success over and over?</li></ol>This concept was developed by Simon Sinek, Peter Docker, and David Mead in their book Find Your Why a Practical Guide for Discovering Purpose for You and Your Team. See also, purpose.</td></tr>
<tr><td>Wideband Delphi</td><td>Wideband delphi is an estimation technique based on expert consensus. Each member of an estimation team uses a work breakdown structure to create anonymous estimates of the effort required to complete each project element or work package. The estimates are then reviewed as a group before the experts create new estimates, and the process is repeated for several rounds until a consensus is reached. See also, delphi technique and work breakdown structure.</td></tr>
<tr><td>Wifi, Wi-Fi</td><td>Wifi is a family of wireless network protocols that allow devices (e.g., computers, drones, printers, smartphones, and tablets) to communicate with one another or connect to the Internet wirelessly within a particular geographic area. Wi-Fi is a trademark of the non-profit organization Wi-Fi Alliance, which restricts the use of the term Wi-Fi Certified to products that successfully complete interoperability certification testing. See also, Internet.</td></tr>
</table>

Phrase or Term	Definition
Wiki	A *wiki* is a website that can be used to assemble content from numerous sources, capture and collect organizational knowledge, and share ideas and plans. For example, a company-wide wiki can be used to share information about corporate or project activities, benefits, and services with personnel. It can also be used to build a glossary of terms. After a wiki page is created, other personnel can add more content or supporting links, or edit the content. The community of authors helps ensure accuracy and content relevance. Special editing tools are not required. Personnel can easily create links to pages for others to finish creating later, or links to existing pages without using long, onerous web addresses. See also, *SharePoint*.
Wildcat Strike	A *wildcat strike* is a labour disruption (e.g., a walk out) undertaken by unionized workers without union leadership's approval, authorization, or support by quitting work and leaving the workplace. A wildcat strike could be an illegal strike, which may result in significant penalties (e.g., contempt of court fines, fines against the union or individual members, suspension of union dues, or a combination). A wildcat strike may or many not later be supported by union leadership. For example, on October 26, 2020 hundreds of Alberta Union of Provincial Employee members went on a wildcat strike at healthcare centres across the province. This action protested an announcement made by the Alberta government two weeks prior about their intent to outsource services to private companies as a cost-saving measure (i.e., privatization of services). The union members returned to work the next day after the Alberta Labour Relations Board issued a cease and desist order – which made the strike illegal. See also, *labour disruption*, *labour union*, and *work to rule*.
Wildcat Well	A *wildcat well* is an exploratory test bore to verify the commercial quantities of hydrocarbon in an underground reservoir. Typically, little is known about the subsurface geology until it is drilled (e.g., the pressure regime). Thus, the drilling crew must be experienced and skilled (e.g., employ careful planning and risk assessments) and be aware of what well parameters indicate about the formation while it is being drilled. They must also use suitable equipment and materials (e.g., blowout preventers and drilling fluids). This phrase originates from Texas where, in the early 1920s, drilling crews encountered wildcats as they cleared locations for exploratory wells. The wildcats were killed and hung on the oil derricks; hence, the wells became known as wildcat wells.
Will	In this document, the word *will* is used where a provision is mandatory. See also, *shall* and *must*.
Wimp	A *wimp* is a cowardly, fearful, ineffectual, unadventurous, or weak person. See also, *wimp out* and *wimp-win*.
Wimp out	To *wimp out* means to fail to do or not do something because of fear or a lack of confidence. For example, I wimped out and failed to ask my manager for a pay raise. See also, *wimp* and *wimp-win*.

Phrase or Term	Definition
Wimp-win	*Wimp-win* is a negotiating result where the negotiator focuses on their best alternative to a negotiated agreement or bottom line, which is a lower goal and less than the best possible agreement. A wimp-win is accepting a less-than-ideal result because it is easier to acquire (e.g., requires less negotiation effort or time) than the desired result. This is a halfway house between total success and complete failure. Wimp-win yields only modest results, The saving grace of a wimp-win is that it generates some results, which are better than no results. People who articulate and expect more from a negotiation generally get more. See also, *best alternative to a negotiated agreement*, *bottom line*, *halfway house*, *wimp*, *wimp out*, and *win-win*.
Wind Farm, Wind Park, Wind Power Plant, Wind Power Station	A *wind farm*, *wind park*, *wind power plant*, or *wind power station* is a group of wind turbines used to produce electricity. A wind farm can be offshore or onshore, and vary in size from a few turbines to several hundred. How closely the turbines are spaced together is a major factor in wind farm design. Typically, wind turbines require a minimum distance of 3.5 times the rotor diameter between each turbine. Spacing turbines closer together reduces the available wind power for the turbines in front of other turbines. Conversely, spacing turbines farther apart increases costs by increasing the amount of infrastructure and land required.
Window of Opportunity, Critical Window	A *window of opportunity* or *critical window* is short period during which a desirable or rare action can be taken achieve a specific outcome. After the (metaphorical) window closes, the opportunity may never come again. It is sometimes possible to anticipate and plan for a window of opportunity and then act accordingly when the window opens. Or, an unforeseen opportunity may arise, so individuals must identify the opportunity and then to act on it. For example, a window of opportunity occurs before real estate is formally listed for sale on the Internet or a publicly-available real estate system (e.g., the Multiple Listing Service in Canada). At this point, only the seller and realtor (and few – if any) others know that it will be listed. Another window of opportunity occurs when the real estate is formally listed, and it becomes public knowledge. Once it has been sold, the real estate may not be available for purchase again for many years or ever. See also, *window of vulnerability*.
Window of Vulnerability	A *window of vulnerability* is a time frame within which defensive measures are diminished, compromised, or lacking. See also, *vulnerable* and *window of opportunity*.
Winterization	*Winterization* is to adapt or prepare something for use in cold or winter weather. This is typically accomplished with desiccant, heat tracing, insulation, dry-out, the use of materials rated for use at low temperature, or a combination. See also, *heat tracing*, *insulation*, and *preservation*.

Phrase or Term	Definition
Win-win, Positive Sum Game	*Win-win* or *positive sum game* is a situation where all participants win, or the sum of winnings (positives) and losses (negatives) is positive. It is a situation in which each party realizes a benefit or gain (e.g., of money, property, rights, ownership, etc.). For example, a discounted price is acquired for a large volume purchase. This benefits the seller (because the seller sells a large volume of product) and the buyer (because the buyer acquires product at a lower price allowing higher profit margins). Win-win implies that two parties benefit or gain from a situation. The phrase win-win-win has the same meaning, but with more than two benefitting parties. See also, *best alternative to a negotiated agreement*, *business wargame*, *game*, *finite mindset*, *forced empathy*, *infinite mindset*, *lose-lose*, *non-zero-sum*, *synergy*, *that is right*, *wargame*, *wimp-win*, and *zero-sum game*.
Witch Hunt	A *witch hunt* is: • An attempt to find and punish a group of people who are blamed for something because of their opinions; not because they have done something wrong. It is a public and vindictive investigation that harasses others with unpopular views (e.g., political opponents) and preys on public fears; or, • The investigation, trial, and punishment of persons accused of witchcraft. The source of this phrase are the Salem Witch Hunts in 17th century Salem, Massachusetts where innocent women accused of witchcraft were persecuted and executed. See also, *fishing expedition*, *fool's errand*, and *goose chase.*
Witches' Hat	A *witches' hat* is a conical strainer. See also, *strainer*.
Witness Point	A *witness point* is an important quality verification point in which quality surveillance representative attendance is necessary. Work may proceed past a witness point after notification has been provided. The supplier shall provide advance notification for attendance to be scheduled. Also, a *witnessed test* generally identifies any activity at a supplier's facility that is attended (witnessed) by project personnel (e.g., factory acceptance test, hold point, observation point, and walk down). See also, *hold point* and *witnessed test*.
Witnessed Test	A *witnessed test* is API's phrase for a hold point. See also, *factory acceptance test*, *hold point*, *observation point*, *walk down*, and *witness point*.
Woke	To be *woke* is to have an awareness of, and pay attention to, facts and issues in society, especially those of racial and social justice. See also, *woke washing*.
Woke Washing	*Woke washing* is when a corporation, individual, or institution does or says something to signal advocacy for a marginal cause or vulnerable group but continues with the status quo. It is the appropriation of ethical and progressive values as a form of advertising to make more profit in the guise of ethical business practices. For example, for the 2017 Super Bowl, vehicle manufacturer Audi spent millions of dollars on a feminist ad, which focused on *equal pay*. Yet, Audi has no women on its executive team. See also, *brand activism*, *conscious capitalism*, *crony capitalism*, *ethics*, *ethical*, *greenwashing*, *status quo*, *virtue signalling*, *whitewashing*, values, and *woke*.
Womb to Tomb	Not a preferred term – see *life cycle*.

Phrase or Term	Definition
Word of Mouth Advertising, Word of Mouth Marketing	*Word of mouth advertising* or *word of mouth marketing* is an organic marketing method that promotes a product using casual conversations (e.g., referrals or testimonials) and social interactions (e.g., likes and shares on social media) of satisfied customers. It is commonly believed to be one of the most effective forms of marketing, which is made even more powerful with social media. For example, American comedian Conan O'Brien sold out all the tickets in each of the 42 comedy venues for his *The Legally Prohibited from Being Funny on Television Tour*, with one tweet on Twitter. This tweet was shared on other social networks and re-tweeted innumerable times by his fanbase.
Word Salad	A *word salad* is a confused or unintelligible mixture of seemingly random phrases and words. The words may or may not be grammatically correct but are semantically confused to the point that the listener cannot extract meaning or that the meaning is ambiguous. The phrase is used in: • Logic To describe utterances perceived by others as nonsense—especially when the speaker is trying to make a serious impression or speak about a serious topic. Since political debates are frequently about the differences between nonsense and sense, as perceived by one side or another, the phrase has entered the political arena. • Psychiatry and theoretical linguistics To describe a person's mental state, a grammatical acceptability judgement by native speakers, and textual randomization for computer programming. An example of a word salad is: *going forward, we must bring a sense of urgency to the table and, quite frankly, unpack the issues as we step up to the plate and meet the challenges head-on*. See also, *gibberish*, *jargon*, *logical*, *non sequitur*, and *plain English*.

Phrase or Term	Definition
Wordplay, Play on words	*Wordplay* or *play on words* is the clever and humorous use of ambiguities and meanings of words for amusement. For example: • Double entendre or innuendo A phrase or word with a double interpretation. The first is straightforward while the second is humorous and often risqué or rude. For example, *I shot an elephant in my pajamas. How he got into my pajamas, I will never know* (Groucho Marx). • Homograph Words that are spelled alike but differ in derivation, meaning, or pronunciation. For example, the lead hand led the lead workers to the lead lead mine. • Homonym Words that are spelled and pronounced alike but differ in meaning; these are both a homograph and a homophone. For example, the bank of a river accrues no interest and a savings bank holds no water. • Homophone Words that are pronounced alike but differ in derivation, meaning, or spelling. For example, what is more effective, putting a carat, caret, or carrot before the horse? • Hyperbole Extreme exaggeration for effect. For example, the house is as big as a mountain. • Pun An expression that exploits the different possible meanings of a word or the fact that there are words which sound alike but have different meaning. For example, *are you working hard or hardly working?* • Spoonerism An expression where the initial sounds or letters of two or more words are transposed. For example, a bottle in front of me is better than a frontal lobotomy. See also, storytelling.
Wordsmith	To *wordsmith* is to skillfully use words (e.g., to edit or write).
Word-thinking	*Word-thinking* is trying to understand the world or win a debate by focusing on the technical definition of words instead of using facts or logic (e.g., using highly subjective and vague words such as normalize or problematic). See also, *hollow man argument*, *loserthink*, *mental prison*, *one-variable thinking*, and *professional jealousy*.
Work	*Work* is the amount of effort needed to complete a task. See also, *task*.

Phrase or Term	Definition
Work at Home, Work from Home	*Work at home* or *work from home* means performing job activities at home instead of at an employer's office or work location – a phenomenon that became prevalent with COVID-19 in 2020. With the advent of authenticator applications, cloud computing, technologies like Facetime, Google Hangouts, Skype, Slack, and Zoom, and the use of group chat, email, and text communications – full-time office attendance is no longer necessary. Pros to working at home include: • Increased flexibility; • Increased independence; • Increased productivity (e.g., fewer interruptions); and, • No commuting, parking cost savings, fuel savings. Cons to working from home include: • Feeling disconnected; • Lack of access to technology platforms and physical offices; • Lack of formality or routine; and, • Reduced productivity (e.g., taking too many breaks to perform personal activities). Workplace expenses paid by the employee (e.g., a portion of costs for cleaning supplies, electricity, heating fuel, insurance, light bulbs, mortgage interest, minor repairs, office suppliers, taxes, and other expenses) may be claimed as a tax deduction, reimbursed by the employer, or a combination. Terminology with similar meanings include flexible work, mobile work, remote work, smart work, telecommute, and telework. See also, *COVID-19*.
Work Breakdown Structure	A *work breakdown structure* is a key project deliverable that organizes the work into manageable sections. It is a deliverable-oriented hierarchical decomposition of the work to be executed by the project team. It visually defines the project scope so a project team can understand it. Each level of the work breakdown structure provides further definition and detail. See also, *scope* and *work package*.
Work Package	A *work package* is a group of related tasks that can be managed by a project. Smaller sets of tasks are delegated to departments, disciplines, geographical locations, or technology. For example, the insulation and painting of piping at a pump station may comprise two separate work packages because the work will be performed by two different trades and two different service providers. A work package is project's lowest-level deliverable and the smallest element of the work breakdown structure. See also, *work breakdown structure*.
Work Permit	A *work permit* is a document that grants formal permission to proceed with work. See also, *cold work* and *hot work*.
Work to Rule, Rule-book Slowdown, Slowdown, Virtual Work Stoppage, Blue Flu	*Work to rule*, *rule-book* slowdown, *slowdown*, and *virtual work stoppage* are expressions that mean to follow official working hours and rules exactly, and only produce or service the required minimum. The purpose of this activity is to reduce efficiency and output to collectively voice disagreement as a form of labour disruption. For example, refusing to perform duties (e.g., replying to email and answering telephone calls) during unpaid time (e.g., before or after a shift, during lunch, breaks, or weekends). *Blue flu* (i.e., influenza) is a labour disruption that occurs when many police officers simultaneously use sick leave. See also, *letter of the law* and *wildcat strike*.

Phrase or Term	Definition
Workaholic	A *workaholic* is someone who compulsively or obsessively works long hours to the detriment of their health and relationships. Workaholics tend to be less effective than others because they have difficulty with delegation, teamwork, and trust, or have organizational problems because they undertake too many responsibilities. Technology (e.g., laptops, tablets, and smartphones) provides opportunities to work from anywhere and at anytime. For some people, an interconnected and fast-paced world with demanding deadlines means working all the time. There is no generally-accepted medical definition for workaholic. This term is a portmanteau of work and alcoholic. See also, *Personality Type Theory*.
Workaround	A *workaround* is a method for overcoming a limitation or problem using an effective or non-standard method that may or may not be temporary (e.g., a welding procedure specification does not meet the specification's low temperature requirement, so the lifting lugs are bolted instead of welded). It is a response to an unidentified risk that frequently involves outside the box thinking, which can be more cost effective, creative, and efficient than a conventional solution. See also, *contingency plan*, *lifting lug*, *outside the box*, *shortcut*, and *stopgap*.
Workface	*Workface* is a: • Customer-initiated engagement technology that helps companies' market, sell and support their products and services on the Internet. It combines profile technology with real-time, two-way communications (e.g., audio/ text/video chat) so that companies can engage with customers and prospects in real time; • Description of the immediate work objectives of personnel or a labour force. It is focused on productivity, and the difference between productivity measures made by planning and direct tool time at the workface. The complete application of workface planning provides a definitive view of what should be done, what can be done and what is being done; or, • Surface (e.g., a sheet of steel or a quarry excavation) to be worked on or measured from. See also, *labour force* and *tool time*.
Workflow	Not a preferred term – see *squad check*.
Workforce	Not a preferred term – see *labour force*.
Work-life Balance	*Work–life* balance describes the balance that working personnel need between time allocated for work and other aspects of life (e.g., family, leisure, social activities, and personal interests). A *good* work-life balance comprises flexible work arrangements that allow personnel time to enjoy personal life. That phrase *work-life balance* was first used in the 1970s in the United Kingdom and, later, in the 1980s in the United States.

Workmanship	*Workmanship* is the:
	• Art, quality, or skill:
	  ○ Of an artisan, craftsperson, trades person, or working person; or,
	  ○ With which something is made or a job is completed.
	• Degree of art, quality, or skill exhibited in the finished product.
	See also, *standard of workmanship*.

Phrase or Term	Definition
Workshop, Shop	A *workshop* or *shop* is a: • Building or room in which products are manufactured or repaired; or, • Meeting at which a group engages in an educational or training activity, or collaborative discussion regarding a particular project or subject – with a focus on practical work (e.g., techniques and skills). See also, *boot camp*, *continuing education*, *cross-train*, *facility*, *lifelong learning*, *lunch and learn*, *on-the-job training*, *professional*, and *professional development unit*.
World Class	*World class* means processes, products, and services that are ranked by customers and industry-experts to be among the best of the best. This designation denotes standard-setting excellence in terms of customer satisfaction and value, and for design, quality, and performance when compared to similar items anywhere in the world. See also, *best in class* and *best practice*.

Phrase or Term	Definition
World Model	The *world model* is a classification of the geopolitical world from the time of the mid-1900s. It is a rough and somewhat outdated model and the origin of this concept is unclear. It has been attributed to: • Alfred Sauvy, a French demographer; and, • Charles de Gaulle, a French army officer, president, and statesman, however, he may have only quoted Sauvy. After World War 2, the world split into two major geopolitical blocs and spheres of influence, with conflicting political views about government and society. This produced a model with different countries classified as the: • First World This is the Western bloc of capitalist, democratic, developed, and industrial countries, generally aligned with the North Atlantic Treaty Organization and United States. Some neutral European countries such as Austria, Finland, Ireland, Switzerland, and Sweden are also classified as First World. Some African countries are or were classified as First World because of their links with Western countries. • Second World This is the Eastern bloc of communist-socialist and less industrialized countries, generally aligned with the Soviet Union. The Asian communist states in China's sphere of influence are also included; and, • Third World These are countries that do not belong to First World and Second World classifications. These include agricultural and underdeveloped countries of Africa, Asia, and Latin America. Today this term is often replaced by the phrase *least developed countries* (United Nations) or *low-income countries* (World Bank). • Fourth world These are the cultural entities living across or within national state boundaries (e.g., the Kurdish people who currently reside in Iraq, Iran, Syria, and Turkey, but do not have an established and internationally recognized state based on their national identity) or indigenous peoples. This term was coined by Shuswap Chief George Manuel in the early 1970s. See also, *economic system, factfulness, Group of 77, Group of Eight, Group of Eight + Five, Group of Seven, Group of Six, Group of Twenty, indigenous peoples, and wealth pyramid.*
Worst Alternative to a Negotiated Agreement	A *worst alternative to a negotiated agreement* is the least advantageous alternative agreement or worst possible outcome if a negotiation fails and an initial agreement is not reached. For example, the buyer and seller of a business cannot agree on price and, consequently, the sale cannot be completed. This concept was developed in the book *Getting to Yes* by Roger Fisher and William Ury. See also, *best alternative to a negotiated agreement*.
Worst-case Scenario	The *worst-case scenario* principle applies to a plan that considers the most severe possible outcomes that could reasonably occur. It assumes that an activity or project will encounter significant risks that need supplementary actions to mitigate risk and ensure success. See also, *everything goes according to plan* and *most likely development*.

<table>
<tr><th>Phrase or Term</th><th>Definition</th></tr>
<tr><td>Wrench</td><td>A wrench is a tool used to turn, loosen or fasten objects such as rotary fasteners (e.g., nuts and bolts). The most common wrenches are open-end and box-end. See also, fastener and torque wrench.</td></tr>
<tr><td>Writing</td><td>Writing is a means of rendering a language into a form that can be reconstructed and understood by other humans separated by time, space, or both. It is:
<ul>
<li>Something that is a printed or a written composition;</li>
<li>The act or process of one who writes;</li>
<li>The activity of creating pieces of written work (e.g., articles, books, or reports);</li>
<li>The profession of authorship; or,</li>
<li>The written work of one or more authors.</li>
</ul>
See also, penmanship and writing style.</td></tr>
<tr><td>Writing Style, All Caps, All Capitals</td><td>Writing style is the method a writer employs to write, including:
<ul>
<li>Writing Syntax
Writing syntax includes casing:
<ul>
<li>Uppercasing
Using capital, all caps, all capitals means using uppercase letters (e.g., A, B, and C). Using uppercase letters indicates to the reader that this text is:
<ul>
<li>Emphasized (e.g., do NOT enter the building);</li>
<li>An abbreviation (e.g., P&ID or piping and instrumentation diagram); and,</li>
<li>Being yelled or shouted at a large volume (in most contexts, this is considered rude).</li>
</ul></li>
<li>Lowercasing
Using lowercase letters (e.g., a, b, and c).</li>
</ul>
Casing usage may be prescribed by conventions of a discipline (e.g., engineering or typography), be dictated by personal preference, or guided by a style book (e.g., The Chicago Manual of Style).
Cursive or script writing is individual letters joined together in a flowing manner to speed writing. For example, writing with cursive looks like this.</li>
<li>Writing expression
These are the writer's chosen words, chosen paragraph and sentence structure, and chosen tense, tone, structure, and logic used to effectively convey meaning. The rules (e.g., grammar, punctuation, and spelling) are about what a writer does, while the style is about how the writer does it. The point of a good writing style is to express the message to the reader clearly, convincingly, and simply, and to keep the reader attentive, engaged, and interested.</li>
</ul>
See also, acronym, copywriting, penmanship, plain English, story, and writing.</td></tr>
<tr><td>Wrought</td><td>A wrought product is made from pipe, plate, or tube by bending, extruding, forging, forming, fusion welding, hammering, machining, pressing, piercing, rolling, shaping, or upsetting, or a combination of two or more of these operations (e.g., a fitting or pipe). See also, forging and seamless.</td></tr>
</table>

Phrase or Term	Definition
X	1 A B C D E F G H I J K L M N O P Q R S T U V W **X** Y Z
X-ray, X-radiation	An *X-ray* or *X-radiation* is: • A radiograph produced by an X-ray, or an informal name for a radiograph produced by a gamma rays; or, • A penetrating electromagnetic wave of high energy with a very short wavelength, produced by electricity, which can pass through many materials that are opaque to light. In some languages, X-rays are referred to as Röntgen radiation after the German scientist Wilhelm Röntgen who discovered this energy in 1895. He named it X-radiation to signify it as an unknown type of radiation. See also, *gamma ray*, *radiograph*, and *radiographic test*.

Phrase or Term	Definition
Y	1 A B C D E F G H I J K L M N O P Q R S T U V W X **Y** Z
Yes	*Yes* means to: • Provide assent or agree; • Contradict or correct a negative assertion or direction; • Indicate attentiveness, polite interest, or uncertainty; or, • Introduce a more emphatic or explicit phrase. The word yes can have one of these three meanings: • Commitment yes An affirmation offered as a sign of agreement or respect (e.g., yes, we agree and can continue to negotiate). • Confirmation yes A true agreement that leads to action (e.g., yes, we can sign the contract). • Counterfeit yes Means no because of a deflection, lie, or wishful thinking (e.g., yes, our negotiation has been fair). This concept was developed by Chris Voss in the book *Never Split The Difference: Negotiating As If Your Life Depended On It* (with Tahl Raz). See also, *no*, *rule of three*, and *win-win*.
Yes-oriented Question, Yes-seeking Question	A *yes-oriented question* or *yes-seeking question* is asked to obtain a positive answer (i.e., *yes*). Most questions asked are yes-oriented questions, so people are *yes-battered*. For example, a yes-oriented question could be: • Do you want me to succeed? • Is my idea good? • Is this a good time to talk? See also, *no-oriented question™* and *yes*.
Y-lateral, Lateral	A *Y-lateral* or *lateral* is a branch connection fitting where the branch direction is not 90° to the main pipe (e.g., 45°). See also, *branch connection* and *fitting*.
You Can Lead a Horse to Water but You Cannot Make It Drink	*You can lead a horse to water, but you cannot make it drink* means you can give someone an opportunity to do something or to make it easy for them, but you cannot make them to do it. For example, you can encourage an unemployed colleague to pursue a job opportunity, but cannot make them apply for it. This is an ancient English proverb that originated in the Middle Ages.

Phrase or Term	Definition
Z	1 A B C D E F G H I J K L M N O P Q R S T U V W X Y **Z**
Zenith	The *zenith* is the highest point. For example, the zenith of Neil Armstrong's career was being the first person in history to walk on the moon. See also, *nadir*.
Zero Defects	*Zero Defects* is a method used to prevent and reduce defects, which was began development in the early 1960s. It motivates personnel to develop a conscious and constant desire to do the job right the first time. This method challenges the attitude that the number of mistakes personnel make does not matter because quality control (i.e., others) will eliminate all defects before they reach the customer. It requires the mindsets of personnel to be conditioned to take a personal interest in everything they do. They need to be convinced that their job is as important as any other. Philip Bayard "Phil" Crosby, a quality control department manager at the Martin Company, developed the Zero Defects concept. See also, *do it right the first time* and *double-check*.
Zero-sum Bias, Zero-sum Thinking	*Zero-sum bias* or *zero-sum thinking* is the perception that a situation is as zero-sum game, where one's gain is another's loss. See also, *zero-sum*.
Zero-sum Game, Constant Sum Game	*Zero-sum game* or *constant sum game* is a situation where one or more participants' gain or loss equals the loss or gain of other participants. Thus, a gain or loss for one results in a loss or gain to one or more participants. Adding the total gains and subtracting the total losses of all participants always equals zero. For example, when eating cake: taking more than one piece of cake reduces the total amount of cake available to others. Conversely, not eating a piece of cake makes the uneaten piece of cake available to others. Eating the cake is zero-sum if each participant places equal value on each piece of cake. See also, *business wargame*, *game*, *lose-lose*, *non-zero-sum*, *wargame*, and *win-win*.
Zone of Possible Agreement	The *zone of possible agreement* is where a buyer and seller have an overlapping area of interest during a negotiation. For example, if the seller of a vehicle will not sell for less than CAD$ 9,000 and the buyer will not buy for more than CAD$ 10,000. In this context, the zone of possible agreement is CAD$ 1,000. A negotiated deal may or may not have a zone of possible agreement. For example, the owner of a business will sell it if the buyer pays 100% of the sale price. A skilled negotiator may use an extreme anchor rather than focus on a zone of possible agreement and win-win scenario. See also, *extreme anchor* and *win-win*.

Appendix A Abbreviations and Glossary

Table A-1 lists and provides the meanings for abbreviations used in this document (excluding sections 3 and 4, and all appendices).

See Section 3 for common industry and project abbreviations. See Section 4 for definitions of document and project glossary terms.

Table A-1: Abbreviations

Abbreviation	Meaning
AB	Alberta
API	*American Petroleum Institute*
ASME	*American Society of Mechanical Engineers*
CQI	Knowledge Transfer (KT) Project
CSA	*Canadian Standards Association*
EP	Engineering and procurement
ESL	English as a second language
IEEE	*Institute of Electrical and Electronics Engineers*
IMS	Integrity management system
ISO	*International Organization for Standardization*
ITP	Inspection and test plan
KT	Knowledge transfer
KTP	Knowledge Transfer Project
NASA	National Aeronautics and Space Administration
PO	Purchase order
QA	Quality assurance
QC	Quality control
QMS	Quality management system
QS	Quality surveillance
QVP	Quality verification point
SI	International System of Units (Système international d'unités)
SQR	Supplier quality representative
USD	United States Dollars

Appendix B References

This appendix lists references used in this document.

1. API RP 1FSC, *Facilities Systems Completion Planning and Execution*
2. API RP 2218, *Fireproofing Practices in Petroleum and Petrochemical Processing Plants*
3. API RP 572, *Inspection Practices for Pressure Vessels*
4. API RP 581, *Risk-Based Inspection Methodology*
5. API SPEC 5L, *Line Pipe*
6. API SPEC 6D, *Specification for Pipeline and Piping Valves*
7. API SPEC 6FA, *Standard for Fire Test for Valves*
8. API STD 607, *Fire Test for Quarter-turn Valves and Valves Equipped with Nonmetallic Seats*
9. API STD 650, *Welded Tanks for Oil Storage*
10. ASME B16.5, *Pipe Flanges and Flanged Fittings: NPS ½ through NPS 24 Metric/Inch Standard*
11. ASME B30.2-3.2.1.1, *Planned Engineered Lifts*
12. ASME B31, *Code for Pressure Piping* (Sections 1 through 12 and G)
13. ASME SA350, *Standard Specification for Carbon and Low-Alloy Steel Forgings, Requiring Notch Toughness Testing for Piping Components*
14. AWS A3.0, *Standard Welding Terms and Definitions*
15. AWS A4.3, *Standard Methods for Determination of the Diffusible Hydrogen Content of Martensitic, Bainitic, and Ferritic Steel Weld Metal Produced by Arc Welding*
16. AWS A5.1, *Specification for Carbon Steel Electrodes for Shielded Metal Arc Welding*
17. CSA W48, *Filler metals and allied materials for metal arc welding*
18. CSA W59, *Welded Steel Construction (Metal Arc Welding)*
19. CSA Z245.12, *Steel Flanges*
20. CSA Z662, *Oil and Gas Pipeline Systems*
21. ISO 10005, *Quality Management Systems — Guidelines for Quality Plans*
22. ISO 14001, *Environmental Management Systems – Requirements with Guidance for Use*
23. ISO 3690, *Welding and allied processes - Determination of hydrogen content in arc weld metal*
24. ISO 45001, *Occupational Health and Safety Management Systems – Requirements with Guidance for Use*
25. ISO 9000, *Quality Management Systems – Fundamentals and Vocabulary*
26. ISO 9001, *Quality Management Systems – Requirements*
27. KTP-GEN-001, *Introduction to the Knowledge Transfer Project*

28. KTP-GEN-003, *Project-Supplier Meetings*
29. KTP-GEN-005, *Knowledge Transfer Project Key Knowledge*
30. NACE MR0175/ISO 15156, *Petroleum and natural gas industries—Materials for use in H_2S-containing environments in oil and gas production*
31. NB-383, *Rules for Certified Individuals*
32. UL 1709, *Standard for Rapid Rise Fire Tests of Protection Materials for Structural Steel*

Appendix C Referenced Works

This appendix lists the works referenced in this document.

1. Adams, Scott. *Loserthink: How Untrained Brains Are Ruining America*. New York, New York: Portfolio/Penguin, 2019.
2. Anderson, Lorin W. and Krathwohl, David R. *A taxonomy for learning, teaching, and assessing: a revision of Bloom's taxonomy of educational objectives*. New York, New York: Longman, 2001.
3. Badaracco, Joseph. *Managing in the Gray: Five Timeless Questions for Resolving Your Toughest Problems at Work*. Boston, Massachusetts: Harvard Business Review Press, 2016.
4. Bolton, Patrick., Despres, Morgan., Da Silva, Pereira., Awazu,Luiz., Svartzman, Romain., and Samama, Frederic. *The Green Swan: Central Banking and Financial Stability in the Age of Climate Change*. Basel, Switzerland: Bank for International Settlements, 2020.
5. Barber, Michael. Kihn, Paul. and Moffit, Andy. *Deliverology: From Idea to Implementation*. Public & Social Sector Insights (blog), February 1, 2011. https://www.mckinsey.com/industries/public-and-social-sector/our-insights/deliverology-from-idea-to-implementation.
6. Barber, Michael. Kihn, Paul. and Moffit, Andy. *Deliverology 101: A Field Guide for Educational Leaders*. Thousand Oaks, California: Corwin Press, 2011.
7. Beinhart, Larry. *Wag the Dog: a Novel* (Original title, *American Hero*). New York, New York: Nation Books, 2004.
8. Blanchard, Kenneth H., Johnson, Spencer. *The One Minute Manager*. New York, New York: William Morrow and Company, 1984.
9. Bloom, Benjamin. Engelhart, Max. Furst, Edward. Hill, Walter. and Krathwohl, David. *Taxonomy of educational objectives: The classification of educational goals*. Handbook I: Cognitive domain. New York, New York: David McKay Company, 1956.
10. Braungart, Michael. and McDonough, William. *Cradle to Cradle: Remaking the Way We Make Things*. London, United Kingdom: Vintage, 2019.
11. Carlin, George, perf. *Class Clown*. Rec. 27 May 1972. Monte Kay, Jack Lewis, 1972. Vinyl recording.
12. Carroll, Lewis. *Through the Looking Glass, and What Alice Found There*. London, United Kingdom: Macmillan, 1872.
13. Christensen, Clayton M. *The Innovators Dilemma: When New Technologies Cause Great Firms to Fail*. Boston, Massachusetts: Harvard Business Review Press, 2016.
14. Cipolla, Carlo. *Basic Laws of Human Stupidity*. Place of Publication Not Identified: Doubleday, 2020.
15. Collins, James C. *Good to Great: Why Some Companies Make the Leap ... and Others Don't*. New York, New York: Collins, 2009.
16. Collodi, Carlo. *The Adventures of Pinocchio: The Story of a Marionette* (Italian, *Le Avventure di Pinocchio: La storia di un burattino*). Place of Publication and Publisher Not Identified, 1883.
17. Crisis Prevention Institute. *Seven Principles for Effective Verbal Intervention*. https://educate.crisisprevention.com/Refresh--VerbalInterventionv2-GEN.html. 2021.

18. Daghe, Jake. *How to Communicate a Strong Presence without Saying a Word*. Career Advice, Ladders, 15 November 2020, www.theladders.com/career-advice/how-to-communicate-a-strong-presence-without-saying-a-word.

19. Dale, Edgar. *Audio-Visual Methods in Teaching*. New York, New York: Dryden Press, 1954.

20. Darwin, Charles. *The Expression of the Emotions in Man and Animals*. London, United Kingdom: John Murray, 1872.

21. Darwin, Charles. *The Origin of Species*. London, United Kingdom: John Murray, 1859.

22. Dass, Ram. *Remember, Be Here Now*. San Cristobal, New Mexico: Lama Foundation, 1973.

23. Davitz, Joel R., and Beldoch, Michael. *The Communication of Emotional Meaning*. Westport, Connecticut: Greenwood Press, 1976.

24. Drucker, Peter F. *The Practice of Management*. Place of Publication Not Identified: Routledge, 2017.

25. Eichele, Tom, et. Al. *Prediction of Human Errors by Maladaptive Changes in Event-related Brain Networks*. Research Article. 16th ed. Vol. 105. Washington, District of Columbia: Proceedings of the National Academy of Sciences, 2008. 6173-178.

26. Etam, Terry. *From Bitcoin to Canada's New Grantcoin – the Earth-shaking, Energy-sucking Impact of The New Money*. Columns (blog), December 16, 2020. https://boereport.com/2020/12/16/column-from-bitcoin-to-canadas-new-grantcoin-the-earth-shaking-energy-sucking-impact-of-the-new-money/.

27. Fisher, Roger. and Ury, William. *Getting to Yes*. London, United Kingdom: Penguin Group, 1991.

28. Frankfurt, Harry G. *On Bullshit*. Princeton, New Jersey: Princeton University Press, 2005.

29. Galai, Dan, and Orly Sade. *The 'Ostrich Effect' and the Relationship between the Liquidity and the Yields of Financial Assets*. Journal of Business, Sept. 2006, pp. 2741–2759.

30. Gapminder. *Factfulness*. Factfulness (blog) February 27, 2018. www.gapminder.org/factfulness/.

31. Gerber, Michael E. *The E Myth: Why Most Small Businesses Don't Work and What To Do About It*. New York, New York: HarperBusiness, 2001.

32. Goleman, Daniel. *Emotional Intelligence: Why It Can Matter More than IQ*. London, United Kingdom: Bloomsbury, 2010.

33. Grotius, Hugo. *Mare Liberum, Sive De Jure Quod Batavis Competit Ad Indicana Commercia Dissertatio*. Lugduni Batavorum: Ex officina L. Elzevirii, 1609.

34. Hamel, Gary. and Prahalad, Coimbatore K. *Competing for the Future*. Boston, Massachusetts: Harvard Business School Press, 2010.

35. Hamilton, W. D. *Geometry for the Selfish Herd*. Publication. Department of Zoology, Imperial College. 2nd ed. Vol. 31. United States: Journal of Theoretical Biology/Academic Press, 1971. 295-311.

36. Harnish, Verne. *Scaling Up: How a Few Companies Make It ... and Why the Rest Don't*. Ashburn: Gazelles, 2014.

37. Henkin, Hilary. Mamet, David. Levinson, Barry. Hoffman, Dustin. De Niro, Robert, Heche, Anne. Leary, Denis. et al. 1998. *Wag the Dog*. California: New Line Home Video.

38. Herman, Dan. *Introducing short-term brands: A new branding tool for a new consumer reality*. Journal of Brand Management. J Brand Manag 7, 330–340 (2000).

39. Hill, Napoleon, Hartley, Ann. and Hartley, Bill. *The Law of Success: The 21st Century Edition*. Los Angeles, California: Highroads Media, 2004.

40. Hill, Napoleon. *Think and Grow Rich: The Secret to Wealth, Updated for the 21st Century*. United States: SoHo Books, 2014.

41. Hull, Raymond. And Laurence, Peter J. *Peter Principle Why Things Always Go Wrong*. London, United Kingdom: Profile Books, 2020.

42. Jevons, William Stanley. *The Coal Question; an Inquiry concerning the Progress of the Nation, and the Probable Exhaustion of Our CoalMines*. London, United Kingdom: Macmillan & Co., 1865.

43. Kerr, Michael. *Are You or a Loved One a Victim of Secondhand Stress?* Humor at Work (blog), October 7, 2020. https://mikekerr.com/e-zine-archive/

44. Kim, W. Chan. and Mauborgne, Renée. *Blue Ocean Strategy: How to Create Uncontested Market Space and Make The Competition Irrelevant*. Boston, Massachusetts: Harvard Business Review Press, 2016.

45. King John of England. Langton, Stephen. *Magna Carta Libertatum*. Runnymedem, United Kingdom 1215.

46. KPMG International, *Resourcing the Energy Transition: Making the World Go Round*. Insights. https://assets.kpmg/content/dam/kpmg/xx/pdf/2021/03/resourcing-the-energy-transition.pdf

47. Krug, Steve. *Don't Make Me Think!: A Common Sense Approach to Web Usability*. Berkley, California: New Riders, 2017.

48. Kucharska, Justyna. *Two superpowers for 2021: psychological flexibility and sense of direction*. Blog (blog). Accessed May 5, 2021. https://www.jk-elate.com/blog/psychological-flexibility

49. Küng, Hans. *Towards a Global Ethic: An Initial Declaration*. Chicago, Illinois: Parliament of the World's Religions, 1992.

50. Langley, Ann. Between *"Paralysis by Analysis" and "Extinction by Instinct" (management decision-making approaches)*. Sloan Management Review (pp 63+), vol. 36, no. 3, 1995.

51. Lasswell, Harold. *The Structure and Function of Communication in Society*. In L. Bryson (Ed.), The Communication of Ideas (pp. 37-51). New York, New York: Harper and Row, 1948.

52. Lennon, Randy. *The Dirty Words: Change Your Language, Change Your Life*. Kindle: Randy Lennon, 2020.

53. Luna, Elle. *The Crossroads of Should and Must: Find and Follow Your Passion*. New York, New York: Workman Publishing, 2015.

54. Lyman, Robert. *The Clean Fuel Standard – The Latest in Climate Pancaking*. Friends of Science Calgary (blog). July 20, 2020. https://blog.friendsofscience.org/2020/07/20/the-clean-fuel-standard-the-latest-in-climate-pancaking/.

55. M., P. C. *The Theistic Annual. Published on the Occasion of the Forty-sixth Anniversary of the Brahmo Somaj*. Calcutta, Pakistan: Calcutta Central Press, 1876.

56. Mackey, John. and Sisodia, Raj. *Conscious Capitalism: Liberating the Heroic Spirit of Business*. Boston, Massachusetts: Harvard Business Review Press, 2014.

57. Maeda, John. *The Laws of Simplicity*. Place of Publication Not Identified: The MIT Press.

58. Maxwell, John C. *The 5 Levels of Leadership*. United States: BookBaby, 2014.

59. Melville, Herman. *Moby-Dick*; or, *The Whale*, London, England: Richard Bentley, 1851.

60. Misner, Ivan. *'Givers Gain' Is a Standard, Not a Sword*. Dr. Ivan Misner® (blog). http://ivanmisner.com/givers-gain-is-a-standard-not-a-sword/.

61. Moore, Don A. *Perfectly Confident: How to Calibrate Your Decisions Wisely*. New York, New York: Harper Business, an Imprint HarperCollins Publishers, 2020.

62. Nakamoto, Satoshi. *Bitcoin: A Peer-to-Peer Electronic Cash System*. Report. White paper, 2009.

63. Neumann, John Von, and Morgenstern, Oskar. *Theory of Games and Economic Behavior.* Princeton, New Jersey: Princeton University Press, 1944.

64. Niemiec, Ryan. *How to Meet Your Future Self.* All Articles (blog). https://www.viacharacter.org/topics/articles/how-to-meet-your-future-self.

65. Northcutt, Wendy. *The Darwin Awards VI: Countdown to Extinction*. Camberwell, Victoria: Penguin Group, 2010.

66. O'Brien, Conan. *The Legally Prohibited from Being Funny on Television Tour.* Canada and United States, 42 Venues, 2010. Also known as the *Prohibited Tour.*

67. Orban, Bill. *The 5BX Plan: For Physical Fitness*. Ottawa, Ontario: Queens Printer, 1958.

68. Orban, Bill. *The XBX Plan: For Physical Fitness*. Ottawa, Ontario: Queens Printer, 1962.

69. Orwell, George. (pen name Eric Arthur Blair) *Nineteen Eighty-Four.* London, United Kingdom: Secker & Warburg, 1949.

70. Osterwalder, Alexander, and Pigneur, Yves. *Business Model Generation: A Handbook for Visionaries, Game Changers, and Challengers*. Hoboken, New Jersey: Wiley, 2010.

71. Pakula, Alan J., Redford, Robert., Hoffman, Dustin., Warden, Jack., Shire, David., Bernstein, Carl., and Woodward, Bob. 1976. *All the President's Men*. Burbank, California: Warner Brothers.

72. Parkinson, Cyril Northcote. and Lancaster, Osbert. *Parkinsons Law: The Pursuit of Progress*. London, United Kingdom: Readers Union John Murray, 1959.

73. Porter, Eleanor H. *Pollyanna*. Boston, Massachusetts: L.C. Page & Company, 1913.

74. Prahalad, Coimbatore K. *The Fortune at the Bottom of the Pyramid*. Upper Saddle River, New Jersey: Wharton School Publishing, 2004.

75. Prasad, Vinay. *Op-Ed: What Does 'Follow the Science' Mean, Anyway?*. MedPage Today (blog). https://www.medpagetoday.com/blogs/vinay-prasad/89856

76. Quinn, Daniel. *The Story of B*. New York, New York: Bantam, 1997.

77. Rackham, Neil. *SPIN-selling*. London, United Kingdom: Routledge, 2017.

78. Rath, Tom. *Strengthsfinder 2.0*. New York, New York: Gallup Press, 2017.

79. Ries, Al. and Trout, Jack. *The 22 Immutable Laws of Marketing: Violate Them at Your Own Risk*. New York, New York: HarperBusiness, 1994.

80. Robbins, Tony. *How Should I Start Each Day? What is Priming?*. Ask Tony (blog). https://www.tonyrobbins.com/ask-tony/priming/

81. Roosevelt, Franklin D., and Moly, Raymond. *Inaugural Address of the President*. Address, The First Inauguration of Franklin D. Roosevelt as the 32nd President of the United States, East Portico of the United States Capitol, Washington, D.C., March 4, 1933.

82. Rosling, Hans, et al. *Factfulness: Ten Reasons Were Wrong about the World - and Why Things Are Better than You Think*. Flatiron Books, 2020.

83. Rymer, Thomas. *The Tragedies of the Last Age Considered ...: In a Letter to Fleetwood Shepheard, Esq*. London, England: Printed for Richard Tonson, 1678.

84. Schwartz, Barry. *The Paradox of Choice: Why More Is Less*. New York, New York: Ecco, an Imprint of HarperCollins Publishers, 2016.

85. Senge, Peter M. *The Fifth Discipline: The Art and Practice of the Learning Organization: First Edition*. Cornerstone Digital, 2010.

86. Shorrock, Anthony. Davies, James. Lluberas, Rodrico. *Global Wealth Report 2020*. Report. October 2020. https://www.credit-suisse.com/about-us/en/reports-research/global-wealth-report.html.

87. Sinek, Simon. Docker, Peter. and Mead. David. *Find Your Why a Practical Guide for Discovering Purpose for You and Your Team*. London, United Kingdom: Portfolio Penguin, 2017.

88. Sinek, Simon. *The Infinite Game*. London, United Kingdom: Penguin Business, 2020.

89. Slotosch, Andreas. *5 Communication Disasters in History Beekeeper Could've Helped*. Beekeeper Blog (blog). Accessed August 8, 2020. https://www.beekeeper.io/blog/4-historical-communication-failures-prevented-beekeeper/.

90. Solnit, Rebecca and Fernández, Ana Teresa. *Men Explain Things to Me*. Chicago, Illinois: Haymarket Books, 2015.

91. Solnit, Rebecca. *River of Shadows: Eadweard Muybridge and the Technological Wild West*. New York, New York: Penguin Books, 2010.

92. Stern, William. *Die Psychologischen Methoden Der Intelligenzprüfung Und Deren Anwendung An Schulkindern*. Leipzig, Germany: Johann Ambrosius Barth, 1912.

93. Sutton, Robert I. *The No Asshole Rule: Building a Civilized Workplace and Surviving One That Isn't*. New York, New York: Business Plus, 2010.

94. Taleb, Nassim Nicholas. *Fooled by Randomness*. New York, New York: Random House, 2016.

95. Taupin, Bernie. and John, Elton. 1973. *Daniel From The Album Don't Shoot Me I'm Only The Piano Player*. [Record] MCA (United States) DJM (United Kingdom).

96. Taylor, Tom. *Our American Cousin*. Laura Keene's Theatre, New York, New York, 1858.

97. Tetlock, Philip. and Gardner, Dan. *Superforecasting the Art and Science of Prediction*. London, United Kingdom: Random House Books, 2016.

98. Thalheimer, Will. *Mythical Retention Data & The Corrupted Cone*. Will's Blog, Work-Learning Research, 5 January 2015, www.worklearning.com/2015/01/05/mythical-retention-data-the-corrupted-cone/.

99. Thomas, Northcote Whitridge. *Crystal Gazing: Its History and Practice, with a Discussion of the Evidence for Telepathic Scrying*. New York, New York: Dodge Publishing Company, 1905.

100. Turchin, Peter. *Political Instability May Be a Contributor in the Coming Decade*. Nature, February 3, 2010, 26-32.

101. United Nations. *Transforming Our World the 2030 Agenda for Sustainable Development*. United Nations, 2015.

102. Van Swol, Lyn M., Braun, Michael T., and Malhotra, Deepak (2012) *Evidence for the Pinocchio Effect: Linguistic Differences Between Lies, Deception by Omissions, and Truths*. Discourse Processes 49, No. 2: 79–106. 2012.

103. VIA Institute on Character. *The 24 Character Strengths*. Character Strengths (blog). https://www.viacharacter.org/character-strengths.

104. Voss, Chris. Raz, Tahl. *Never Split The Difference: Negotiating As If Your Life Depended On It*. London, United Kingdom: Random House Business Books, 2016.

105. Whipple, Guy Montrose. Trans. *The Psychological Methods of Testing Intelligence*. Baltimore, Maryland: Warwick & York, 1914.

106. Wiefling, Kimberly. *Scrappy Project Management: The 12 Predictable and Avoidable Pitfalls Every Project Faces*. Cupertino, California: Scrappy About, 2008.

107. Wilson, James Q., and Kelling, George L. *Broken Windows - The Police and Neighborhood Safety*. The Atlantic, March 1982. https://www.theatlantic.com/magazine/archive/1982/03/broken-windows/304465/

108. Wojick, David. *It Takes Big Energy to Back Up Little Wind and Solar*. Townhall Columnists (blog), March 30, 2021. https://townhall-com.cdn.ampproject.org/c/s/townhall.com/columnists/davidwojick/2021/03/30/it-takes-big-energy-to-back-up-little-wind-and-solar-n2587070

109. Woolsey, Barbara. *The German 'Quereinsteiger' Ending Career-changing Stigma*. Worklife (blog), January 6, 2021. https://www.bbc.com/worklife/article/20201222-the-german-quereinsteiger-ending-career-changing-stigma

110. Young, Michael Dunlop. *The Rise of the Meritocracy*. London, United Kingdom: Routledge, 2017.

Appendix D Industry Resources that Define Project Terminology

Table D-1 lists resources that provide meanings of abbreviations and definitions of glossary terms for various related industries.

Table D-1: List of Industry Resources that Define Project Terminology

Organization	Publication	Subject	Resource
ABB	*Glossary of technical terms commonly used by ABB*	Electrical	https://new.abb.com/media/glossary
AISI	*Steel Glossary*	Steel Production	https://www.steel.org/steel-technology/steel-production/glossary
AIST	*Steel Glossary*	Steel	http://apps.aist.org/asp/glossary/
AGA	*Glossary*	Natural Gas	https://www.aga.org/natural-gas/glossary/
ArchDaily	*45 Construction Terms & Concepts All Architects Should Know*	Construction	https://www.archdaily.com/898221/45-construction-terms-and-concepts-all-architects-should-know
Arena	*Manufacturing Glossary*	Manufacturing	https://www.arenasolutions.com/resources/articles/manufacturing-glossary
ASNT	*NDT Handbook Glossaries Volume 1 to 10 and Aerospace NDT*	Non-destructive Testing	https://asnt.org/MajorSiteSections/Learn/NDT_Handbook_Glossaries.aspx
ASQ	*Quality Glossary*, 3rd Edition	Quality	https://asq.org/quality-resources/quality-glossary
Australian Government	Guidance Note 1 Language in Alliance Contracting: A Short Analysis of Common Terminology	Alliance Contracting	https://www.infrastructure.gov.au/infrastructure/ngpd/files/NACG_GN1.pdf
AWS	AWS A3.0M/A3.0, *Standard Welding Terms and Definitions, Including terms for Adhesive Bonding, Brazing, Soldering, Thermal Cutting, and Thermal Spraying*	Welding, Joining, and Cutting	https://pubs.aws.org/p/874/a30ma302010-standard-welding-terms-and-definitions-including-terms-for-adhesive-bonding-brazing-soldering-thermal-cutting-and-thermal-spraying

Organization	Publication	Subject	Resource
BDC	*Glossary*	Accounting, Banking, and Business	https://www.bdc.ca/en/articles-tools/entrepreneur-toolkit/templates-business-guides/glossary/pages/default.aspx
CAPP	*Glossary of Industry Terms*	Upstream Oil and Natural Gas	https://www.capp.ca/publications-and-statistics/glossary
CII	*Knowledge Base Glossary*	Construction	https://www.construction-institute.org/resources/knowledgebase/about-the-knowledge-base/glossary
Commissioning & Startup	*Commissioning Terminology*	Commissioning and Startup	https://commissioningandstartup.com/commissioning-dictionary-acronyms/
Control and Instrumentation	*Control and Instrumentation Glossary*	Control and Instrumentation	http://www.controlandinstrumentation.com/glossary/
Control and Instrumentation	*Glossary of Relief Valve Terms*	Relief Valves	http://www.controlandinstrumentation.com/glossary/relief.html
CRSI	*Glossary of Reinforced Concrete*	Reinforced Concrete	http://www.crsi.org/index.cfm/glossary (link deleted, ask the KT Project for an archived copy)
CSCMP	*Supply Chain Management Definitions and Glossary*	Supply Chain	https://cscmp.org/CSCMP/Educate/SCM_Definitions_and_Glossary_of_Terms/CSCMP/Educate/SCM_Definitions_and_Glossary_of_Terms
DTCS	*Blog*	Business English and Writing	http://dalicetrost.com/blog
ESAB	*Glossary of Technical Terms* 07SEP05	Welding	http://www.esabna.com/eu/literature/esab_glossary_of_technical_terms.pdf
Fastenal	*Engineering Glossary*	Engineering	https://www.fastenal.com/en/81/engineering-glossary
FDRsafety	*Safety Terms Glossary*	Safety	https://www.fdrsafety.com/about/safety-terms-glossary/
Federal Court	*Glossary of Terms*	Law (Canada)	http://www.fct-cf.gc.ca/fc_cf_en/Litigants/Glossary.html
Hobart IWT	*Pocket Welding Guide*, 32nd Edition	Welding	https://www.welding.org/product/pocket-welding-guide-31st-edition/ (USD$ 8.00)
Hydrogen Tools Portal	*Hydrogen Glossary and Acronyms*	Hydrogen	https://h2tools.org/hyarc/hydrogen-glossary-and-acronyms
ICC	*Incoterms® Rules 2010*	International Trade	https://iccwbo.org/publication/incoterms-rules-2010
ICC	*Incoterms® Rules 2020*	International Trade	https://iccwbo.org/media-wall/news-speeches/icc-releases-incoterms-2020/
INCOSE	*Systems Engineering Glossary*	Systems Engineering	https://www.incose.org/about-systems-engineering/system-and-se-definition/se-glossary

Organization	Publication	Subject	Resource
Inspectioneering	*Integripedia*	Asset Integrity	https://inspectioneering.com/topics
JWN a division of Glacier Media Inc.	*Glossary of Oil Sands Terms*	Oil Sands	http://www.oilsandsreview.com (link deleted, ask the KT Project for an archived copy)
LafargeHolcim	*Glossary of Terms 2017*	Concrete	https://www.lafargeholcim.com/sites/lafargeholcim.com/files/atoms/files/glossary_of_terms_v5.pdf
Lexicool	*Mechanical engineering dictionaries*	Multi-lingual Dictionary	https://www.lexicool.com/online-dictionary.asp?FSP=C15&FKW=mechanical-engineering#
Lincoln Welding	*Basic Welding Terms*	Welding	https://www.lincolnelectric.com/en-ca/support/process-and-theory/Pages/basic-welding-terms-detail.aspx
Logistics Bureau	*Supply Chain Glossary of Terms*	Supply Chain	https://www.logisticsbureau.com/supply-chain-glossary/
MarketingTerms .com	*Glossary of Digital Marketing Terms*	Digital Marketing	https://www.marketingterms.com/dictionary/
Miller Electric Mfg. LLC	*Introduction to Welding*	Welding	https://www.millerwelds.com/equipment/training-solutions/training-materials/introduction-to-welding-247358
NACE	*Glossary of Corrosion-Related Terms 2002*	Coatings	https://www.nace.org/resources/general-resources/corrosion-glossary (registration and login required)
NDT.net	*NDT Encyclopedia*	NDE	https://www.ndt.net/ndtaz/ndtaz.php
Onsite Safety Management	*Glossary of Essential Safety Terminology*	Safety	http://www.onsitesafetymanagement.com/blog/2018/7/31/a-glossary-of-essential-safety-terminology
Ohio DOT	*Bridge Inspectors Reference Manual*	Bridges	http://www.dot.state.oh.us/Divisions/Engineering/Structures/bridge%20operations%20and%20maintenance/Bridge%20Inspectors%20Reference%20Manual/Glossary.pdf
Piping Designers	*Piping Terms Glossary*	Piping	http://pipingdesigners.com/contents/tips-tools-training/section-1-glossary-terms-definitions/101-section-1a-piping-terms-glossary
Plastics Pipe Institute	*Plastic Pipe Glossary*	Plastic Pipe	https://plasticpipe.org/pdf/glossary.pdf
Project Management Institute	*A Guide to the Project Management Body of Knowledge*	Project Management	https://pmi.bookstore.ipgbook.com/-products-9781628251845.php
Praxair	*Welding Terms Glossary*	Glossary	https://www.praxairdirect.com/Industrial-Gas-and-Welding-Information-Center/Technical-Data/technical-terms-glossary.html

Organization	Publication	Subject	Resource
Pricewaterhouse Coopers	*Glossary of Terms*	Oil and gas, utilities and mining commodities	https://www.pwc.com/gx/en/energy-utilities-mining/pdf/eumcommoditiestradingriskmanagementglossary.pdf
Project Manager	*Project Management Terms*	Project Management	https://www.projectmanager.com/blog/glossary-of-project-management-terms
Reliable Plant	*Glossary of Terms*	Equipment and Machinery	https://www.reliableplant.com/Glossary
Research Council on Structural Connections	*Specification for Structural Joints Using High-Strength Bolts*	Bolted Fasteners	http://www.boltcouncil.org/files/2009RCSCSpecification.pdf (See Glossary, page x)
S&P Platts	*Steel Glossary*	Steel	https://www.steelbb.com/steelglossary/
Safeopedia	*Safety Dictionary*	Safety	https://www.safeopedia.com/dictionary
Schlumberger	*Oilfield Glossary*	The Oil Field Meets the Dictionary	https://www.glossary.oilfield.slb.com/
Schneider Electric	*Glossary of Power Terms*	Electricity	https://www.schneider-electric.ca/en/faqs/FA229166/
Smartsheet	*Project Management Terminology*	Project Management	https://www.smartsheet.com/complete-glossary-project-management-terminology
SSPC Protective Coatings	*Glossary*	Coatings	Download the app, "SSPC Victory Enterprises" then select Glossary in the Menu (members only)
The Office Life	*The Ridiculous Business Jargon Dictionary*	Business Jargon	http://www.theofficelife.com/business-jargon-dictionary-A.html; https://gumroad.com/l/jargon
The Law Dictionary	*Black's Law Dictionary*	Law (USA)	https://thelawdictionary.org/
The Project Definition	*Project Definitions*	Definition of Words	http://www.theprojectdefinition.com
TechTarget	*Acronyms Used in Manufacturing*	Manufacturing	https://searcherp.techtarget.com/definition/Fast-Guide-to-acronyms-used-in-manufacturing
Topcorrect	*50 Common Business Idioms*	Business Jargon	https://www.topcorrect.com/blog/50-common-business-idioms/
UNFCCC	*Glossary of Climate Change Acronyms and Terms*	Climate Change	https://unfccc.int/process-and-meetings/the-convention/glossary-of-climate-change-acronyms-and-terms
US DOT	*Glossary of Terms*	Shipping	https://origin-www.marad.dot.gov/wp-content/uploads/pdf/Glossary_final.pdf

Organization	Publication	Subject	Resource
Wikipedia	*Glossary of Electrical and Electronics Engineering*	Electricity and Electronics	https://en.wikipedia.org/wiki/Glossary_of_electrical_and_electronics_engineering
Wikipedia	*Glossary of Mechanical Engineering*	Mechanical Engineering	https://en.wikipedia.org/wiki/Glossary_of_mechanical_engineering
Wikipedia	*Glossary of Project Management*	Project Management	https://en.wikipedia.org/wiki/Glossary_of_project_management
Workforce Compliance Safety Ltd.	*Glossary Of Workplace Safety Terms*	Safety	https://workforcecompliancesafety.ca/resources/workplace-safety-glossary-of-terms/

Notes:

1. See codes of construction (e.g., API, ASME, CSA, and IEEE) for more terminology.
2. For a general description of NDE (NDT) and related terminology, visit the CINDE website at: http://www.cinde.ca/ndt.php

Appendix E Pipeline Welding Crews

Table E-1 lists the sequence of work and defines the typical welding crews that construct a pipeline.

Table E-1: Pipeline Welding Crew Descriptions

Sequence	Term	Definition
First	Main Gang Crew	A *main gang crew* completes the bulk of the mainline welds along a right-of-way. Welds that would result in the inefficient use of their time are welded by other crews.
Second	Poor-boy Crew	A *poor-boy crew* or mini gang crew completes the special short sections of a pipeline at steep hills and rail or road crossings, and prefabricated fittings, and any other welds that the main gang did not complete.
Third	Tie-in Crew	A *tie-in crew* welds the special short sections of a pipeline that are welded by the poor-boy crew into the main sections that were welded by the main gang crew to complete the pipeline.
Fourth	Repair Crew	A *repair crew* performs repair welding whenever non-destructive examination (e.g., visual, ultrasonic, or radiographic testing) identifies welds with defects that do not meet the acceptance criteria of the code(s) of construction (e.g., CSA Z662, *Oil and Gas Pipeline Systems*) or owner specifications.

Appendix F Document Review Codes

An engineering and procurement (EP) company's Document Control department applies a document review stamp to all supplier documents it receives. The stamp may be ink on hard copy or electronic and included with appropriate electronic files.

The stamp comprises the EP company name and a disclaimer. A typical disclaimer states that *the EP company's review of supplier documents does not in any way waive the supplier's responsibility for design or compliance with all procurement documents.*

The document is then issued to the responsible engineer for review or a squad check. The responsible engineer reviews and consolidates all comments, if any, before assigning a review code. The responsible engineer also signs and dates the review stamp and returns the document to the supplier through the Document Control department.

Table F-1 lists document review codes used by a responsible engineer to summarize the results of a supplier document review process. These may be alphabetic or numeric characters and may be used in ascending or descending order. Table F-1 uses an alphabetic-descending order.

An EP company procedure may define the codes differently than what is shown in Table F-1. Regardless, the intent is to effectively communicate the same information.

Table F-1: Document Review Codes

Code	Description
1	May proceed without changes
2	May proceed subject to making changes and re-submitting
3	May not proceed
4	Review not required, for as-built, turnover, or record only
5	Superseded
6	Void

Appendix G Procdeure to Prepare, Implement, and Maintain an Effective Glossary

90% of all management problems are caused by miscommunication.
Dale Carnegie, 1892 – 1955

Ineffective communication is a huge challenge for capital projects in every industry. This challenge has negative, and sometimes dire, impacts on budgets and schedules. With project success at risk, the importance of clear communication cannot be understated. Ambiguous terminology frustrates project success and is a time thief.

Using a glossary as the single authority for all company, industry, and project terminology greatly reduces the risk of ineffective communication and unnecessary costs. The procedure in Appendix G.1 is used to prepare, implement, and maintain an effective glossary for your organization. Appendix G.2 contains additional, related resources.

G.1 Procedure

Follow these steps to prepare, implement, and maintain an effective glossary for your organization.

1. Gather resources and organize glossary components.

   Acquire and assign experienced, accountable, and responsible personnel (assigned resources) to prepare, implement, and maintain the glossary. This is typically a technical writer or technical editor retained as an employee or contractor. Work with assigned resources and:

   a. Determine if the glossary shall be prepared as:

      i. A controlled document (i.e., managed only by the assigned resources); or,

      ii. A wiki (i.e., other personnel can edit, or add content or supporting links).

   b. Establish and make available existing company-project internal glossary resources (e.g., company, historical, or project) to assigned resources;

   c. Work with assigned resources to establish all relevant external glossary resources for reference (e.g., client specifications, industry organizations or standards, other stakeholders, professional associations, and regulatory documents); and,

   d. Purchase or subscribe to a glossary with standardized industry and project terminology (if needed) and make these available to assigned resources.

2. Assigned resources prepare these glossary items:
    a. Abbreviations and acronyms;
    b. Brand names, proprietary names, and trademarks;
    c. Company-, discipline-, industry-, and project-specific terminology;
    d. Expressions, historical use of terms, etymology, and social influencers of jargon and slang;
    e. Phrases and terms used by third parties; and,
    f. Preferred and non-preferred terminology.
3. Assigned resources refine glossary to include:
    a. Key terminology for communicating typical and a-typical concepts; and,
    b. Terminology that is frequently mis-used or misunderstood (e.g., the same terminology used interchangeably to describe different things, or different terminology used to mean the same or a similar thing).
4. Assigned resources publish suitable external glossaries as additional reference documents (e.g., frequently used or relevant) – to help reduce the size of the glossary by avoiding repetition. The glossary does not need to regurgitate terminology that is already defined in other glossaries that are widely used. For example, the Canadian Association of Petroleum Producers has a comprehensive glossary of oil and gas industry terminology: https://www.capp.ca/resources/glossary/. See a list of examples of industry resources that define project terminology in Appendix D.
5. Assigned resources:
    a. Edit and format the glossary using the approved corporate template and styles;
    b. Acquire approval from stakeholders;
    c. Submit glossary work to stakeholders who reject or approve their work;
    d. Publish the glossary (on the Internet, intranet, SharePoint, or other location) after receiving approvals;
    e. Distribute the glossary;
    f. Provide personnel access to the published glossary (usually by way of hyperlink), informing them to reference it for correct and preferred terminology, and encouraging them to use it;
    g. Provide training to inform personnel about the proper use of the glossary as the single authority for company, industry, and project terminology; and,
    h. Update the glossary on a regular basis, or as needed, to ensure it remains current and relevant. This may include preparing a list of old terminology to be excluded, new terminology to be included, and existing terminology to be revised or updated.

6. Update business documents, including:
    a. Review and edit business, project, and strategic documents (business documents and templates) using the glossary as the authority;
    b. In business documents, refer to the glossary when discussing terminology and include hyperlinks to it; and,
    c. To avoid rework (e.g., changes due to updated terminology), do not include verbatim glossary excerpts in business documents; instead, simply refer to the glossary.
7. Assigned resources establish a method of receiving and implementing glossary feedback.

The KT Project can prepare a customized glossary for client-specific use. Contact us for details.

G.2 Additional Related Resources

Refer to these additional, related resources for information about the importance of an effective glossary:

1. Articles by Roy O. Christensen https://ktproject.ca/articles/
2. KT Project Downloads https://ktproject.ca/downloads/

Appendix H Contact

H.1 For Assistance

For assistance with this document, or implementing, editing, and preparing an effective glossary for your organization, contact the KT Project at:

Email: hello@ktproject.ca

Phone: 403.703.2686

Web: https://www.ktproject.ca

Appendix I How to Submit Feedback about this Glossary

Your feedback is extremely important for improving this glossary. Please contact the KT Project (see Appendix G) to submit suggestions for improvements and new glossary entries.

The KT Project thanks you in advance for your contributions. The KT Project will credit all significant revisions contributed to this glossary.

Appendix J Glossary Contribution Credits

Table J-1 lists contributors that helped improve this glossary. Your contribution are extremely important. Please keep them coming by sending a suggestion or contribution to hello@ktproject.ca! See Appendix H for contact information.

Table J-1: Glossary Contribution Credits

No.	Date	Name	Description
1.	January 01, 2019	Roy Christensen	Issue first draft. Added new terms-phrases (70 pp).
2.	January 01, 2020	Roy Christensen	Issue second draft. Added new terms-phrases (220 pp).
3.	January 01, 2021	Roy Christensen	Issue third draft. Added new terms-phrases (270 pp).
4.	February 23, 2021	Robert Blaskowski	Project cancers, MODER
5.	January 09, 2022	Roy Christensen	Issue fourth draft. Pending.
6.	October 21, 2024	Roy O. Christensen	Issue for publication

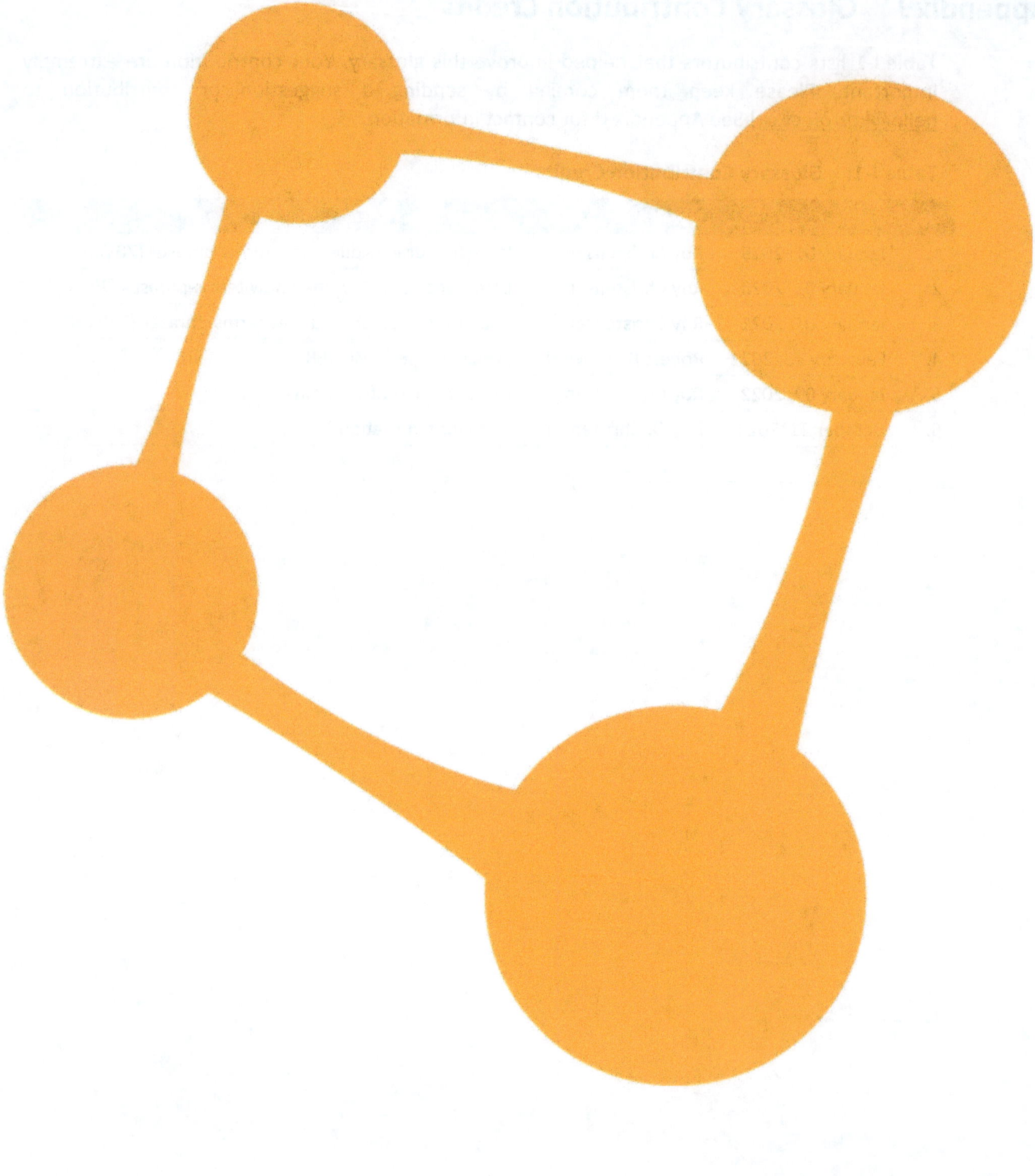

www.ingramcontent.com/pod-product-compliance
Lightning Source LLC
LaVergne TN
LVHW060011210926
848481LV00013B/3687
9781068976612